GLOBAL CLIMATE CHANGE: IMPLICATIONS, CHALLENGES AND MITIGATION MEASURES

The Pennsylvania Academy of Science Publications
Books and Journal

Editor: Shyamal K. Majumdar
Professor of Biology, Lafayette College
Easton, Pennsylvania 18042

1. *Energy, Environment, and the Economy,* 1981. ISBN: 0-9606670-0-8. Editor: Shyamal K. Majumdar.
2. *Pennsylvania Coal: Resources, Technology and Utilization,* 1983. ISBN: 0-9606670-1-6. Editors: Shyamal K. Majumdar and E. Willard Miller.
3. *Hazardous and Toxic Wastes: Technology, Management and Health Effects,* 1984. ISBN: 0-9606670-2-4. Editors: Shyamal K. Majumdar and E. Willard Miller.
4. *Solid and Liquid Wastes: Management, Methods and Socioeconomic Considerations,* 1984. ISBN: 0-9606670-3-2. Editors: Shyamal K. Majumdar and E. Willard Miller.
5. *Management of Radioactive Materials and Wastes: Issues and Progress,* 1985. ISBN: 0-9606670-4-0. Editors: Shyamal K. Majumdar and E. Willard Miller.
6. *Endangered and Threatened Species Programs in Pennsylvania and Other States: Causes, Issues and Management,* 1986. ISBN: 0-9606670-5-9. Editors: Shyamal K. Majumdar, Fred J. Brenner, and Ann F. Rhoads.
7. *Environmental Consequences of Energy Production: Problems and Prospects,* 1987. ISBN: 0-9606670-6-7. Editors: Shyamal K. Majumdar, Fred J. Brenner and E. Willard Miller.
8. *Contaminant Problems and Management of Living Chesapeake Bay Resources,* 1987. ISBN: 0-9606670-7-5. Editors: Shyamal K. Majumdar, Lenwood W. Hall, Jr. and Herbert M. Austin.
9. *Ecology and Restoration of The Delaware River Basin,* 1988. ISBN: 0-9606670-8-3. Editors: Shyamal K. Majumdar, E. Willard Miller and Louis E. Sage.
10. *Management of Hazardous Materials and Wastes: Treatment, Minimization and Environmental Impacts,* 1989. ISBN: 0-9606670-9-1. Editors: Shyamal K. Majumdar, E. Willard Miller and Robert F. Schmalz.
11. *Wetlands Ecology and Conservation: Emphasis in Pennsylvania,* 1989. ISBN: 0-945809-01-8. Editors: Shyamal K. Majumdar, Robert P. Brooks, Fred J. Brenner and Ralph W. Tiner, Jr.
12. *Water Resources in Pennsylvania: Availability, Quality and Management,* 1990. ISBN: 0-945809-02-6. Editors: Shyamal K. Majumdar, E. Willard Miller and Richard R. Parizek.
13. *Environmental Radon: Occurrence, Control and Health Hazards*, 1990. ISBN: 0-945809-03-4. Editors: Shyamal K. Majumdar, Robert F. Schmalz and E. Willard Miller.
14. *Science Education in the United States: Issues, Crises, and Priorities,* 1991. ISBN: 0-945809-04-2. Editors: Shyamal K. Majumdar, Leonard M. Rosenfeld, Peter A. Rubba, E. Willard Miller and Robert F. Schmalz.
15. *Air Pollution: Environmental Issues and Health Effects,* 1991; ISBN: 0-945809-05-0. Editors: Shyamal K. Majumdar, E. Willard Miller, and John J. Cahir.
16. *Natural and Technological Disasters: Causes, Effects and Preventive Measures*, 1992; ISBN: 0-945809-06-9. Editors: Shyamal K. Majumdar, Gregory S. Forbes, E. Willard Miller, and Robert F. Schmalz.
17. *Global Climate Change: Implications, Challenges and Mitigation Measures,* 1992; ISBN: 0-945809-07-7. Editors: Shyamal K. Majumdar, Laurence S. Kalkstein, Brent M. Yarnal, E. Willard Miller, and Leonard M. Rosenfeld.

GLOBAL CLIMATE CHANGE: IMPLICATIONS, CHALLENGES AND MITIGATION MEASURES

EDITED BY

S.K. MAJUMDAR, Professor
　Department of Biology
　Lafayette College
　Easton, PA 18042

L.S. KALKSTEIN, Professor
　Center for Climatic Research
　University of Delaware
　Newark, DE 19716

B.M. YARNAL, Associate Professor
　Earth Science System Center
　The Pennsylvania State University
　University Park, PA 16802

E.W. MILLER, Professor and Associate Dean (Emeritus),
　The Pennsylvania State University
　University Park, PA 16802

L.M. ROSENFELD, Assistant Professor
　Department of Physiology
　Jefferson Medical College
　Philadelphia, PA 19107

Founded on April 18, 1924

A Publication of
The Pennsylvania Academy of Science

Library of Congress Cataloging in Publication Data

Bibliography
Index
Majumdar, Shyamal K. 1938-, ed.

Library of Congress Catalog Card No.: 92-85374

ISBN-0-945809-07-7
Copyright © 1992 By The Pennsylvania Academy of Science, Easton, PA 18042

All rights reserved. No part of this book may be reproduced in any form without written consent from the publisher, The Pennsylvania Academy of Science. For information write to The Pennsylvania Academy of Science, Attention: Dr. S.K. Majumdar, Editor, Department of Biology, Lafayette College, Easton, Pennsylvania 18042. Any opinions, findings, conclusions, or recommendations expressed are those of the authors and do not necessarily reflect the views of the Editors or The Pennsylvania Academy of Science.

Printed in the United States of America by

Typehouse of Easton
Phillipsburg, New Jersey 08865

PREFACE

It was once thought that the atmosphere was so vast that human activity could not alter it. Since the beginning of the Industrial Revolution about two centuries ago, it is now recognized that there has been a substantial increase in the global atmospheric content of carbon dioxide (CO_2) and other trace gases such as methane. The concentration of CO_2, for example, has been observed to be increasing at locations as far apart as Mauna Loa in Hawaii and Alert in the Canadian Arctic Archipelago, prompting many climatologists to deduce that global warming will occur. The potential global impacts of a climate change have enormous implications on how we may live in a warmer world. Mass migrations of plants and animals, changes in crop yields, coastal flooding and habitat loss represent only some of the long-term possibilities.

Since the Environmental Protection Agency sponsored its initial sea level rise conference in 1983, global climatic warming and its possible greenhouse effect has moved from the scientific realm to a subject of concern to the world's political leaders. Global climatic change was a major policy issue of the world's political leaders who met at the environmental conference of 1992 in Rio de Janeiro. Although over 100 nations produced a framework agreement on climate change at the recent United Nations Conference on Environment and Development, many developing nations who participated in the negotiations have little information on how they could be affected by climate change. This underscores the necessity of this volume, where a range of possible scenarios are provided, and a plethora of mitigating actions are suggested.

Although there is no consensus among scientists at the present time concerning climate change for it is difficult to determine long range alteration, the problem cannot be ignored. If the world's energy system is altered, vast economic, political and cultural changes will occur. There is a need to develop energy systems that will not pollute the atmosphere, including the use of solar energy, wind power, energy from the oceans and the development of fusion power.

There has been much recent attention to a short-term global cooling which has occurred in June 1991, and some scientists attribute this cooling to the eruption of Mount Pinatubo. In certain cases, climate skeptics have utilized this cooling to assuage public fears about global warming. It is imperative that we recognize that the climate impacts of a volcanic eruption are short-term, and these must be separated from the much greater threat of a trace gas-induced global warming. We must not permit short-term climate changes to deter further research on the potential long-term impacts of an anthropogenic global warming.

This book presents a broad perspective of the potential problem of global climate change. Consequently, it will be of value to a wide audience who are awakening to the perils of a worldwide change in climate. It also provides information to specialists who want to broaden their understanding of global climatic change.

Global climatic change will affect every human being on earth, thus this volume has special interest to the general public. It should be in the library of every college and university so that the students and public can become better informed about this essential of environmental issue.

We express our deep appreciation for the cooperation and dedication of contributors, who recognize the importance of investigating this vital environmental problem. In addition to the authors, many other individuals in the Environmental Protection Agency and the Pennsylvania Academy of Science made contributions. Gratitude is extended to Lafayette College, The Pennsylvania State University, University of Delaware and Jefferson Medical College for providing facilities for the editors.

Shyamal K. Majumdar
Laurence S. Kalkstein
Brent M. Yarnal
E. Willard Miller
Leonard M. Rosenfeld
Editors
November, 1992

ACKNOWLEDGMENTS

The publication of this book was aided by contributions from The Pennsylvania Power and Light Company, Allentown, Pennsylvania and E. Willard and Ruby S. Miller book publication fund.

Global Climate Change: Implications, Challenges and Mitigation Measures

Table of Contents

Preface V
Acknowledgments VI
Foreword — Dennis Tirpak, Director, Climate Change Division, U.S. EPA, Washington, D.C. XII
Introduction — R.A. Reinstein, Deputy Assistant Secretary, Environment, Health and Natural Resources, U.S. Department of State, Washington, D.C. XIII
Message — Governor Robert P. Casey, Commonwealth of Pennsylvania, Harrisburg, PA XV

Part One: Natural Climatic Fluctuations

Chapter 1: THE CLIMATE SYSTEM: AN OVERVIEW
Peter J. Robinson, Department of Geography, University of North Carolina, Chapel Hill, NC 1

Chapter 2: THE PAST AS A GUIDE TO THE FUTURE
John E. Oliver, Department of Geography and Geology, Indiana State University, Terre Haute, IN 14

Chapter 3: SHORT-TERM CLIMATIC VARIABILITY
Brent Yarnal, Earth System Science Center, The Pennsylvania State University, University Park, PA 27

Chapter 4: CONTEMPORARY GLOBAL WARMING: ARE WE SURE?
Thomas R. Karl, National Climatic Data Center, Global Climate Laboratory, Federal Building, Asheville, NC 37

Chapter 5: HYDROLOGICAL IMPACTS OF GLOBAL WARMING OVER THE UNITED STATES
Johannes J. Feddema, Department of Geography, University of California, Los Angeles, CA and John R. Mather, Department of Geography, University of Delaware, Newark, Delaware 50

Part Two: The Greenhouse Effect

Chapter 6: THE PHYSICS OF THE GREENHOUSE EFFECT
Thomas P. Ackerman, Department of Meteorology and Earth System Science Center, The Pennsylvania State University, University Park, PA 63

Chapter 7: GLOBAL WARMING: A CASE FOR CONCERN
Stephen H. Schneider, National Center for Atmospheric Research, P.O. Box 3000, Boulder, CO 81

Chapter 8: GLOBAL WARMING: BEYOND THE POPULAR VISION
Patrick J. Michaels, Department of Environmental Sciences, University of Virginia, Charlottesville, VA 100

Chapter 9: THE CARBON CYCLE AND THE CARBON DIOXIDE PROBLEM
Berrien Moore III, Director, Institute for the Study of Earth, Oceans and Space, University of New Hampshire, Durham, New Hampshire and Robert Kaufman, Center for Energy and Environmental Studies, Boston University, Boston 117

Part Three: Monitoring and Modeling

Chapter 10: CLIMATE MODEL DESCRIPTION AND IMPACT ON TERRESTRIAL CLIMATE
Roy L. Jenne, National Center for Atmospheric Research, Box 3000, Boulder, CO 145

Chapter 11: MONITORING GLOBAL TEMPERATURE CHANGES FROM SATELLITES
John R. Christy, University of Alabama, Atmospheric and Remote Sensing Lab, Johnson Research Center, Huntsville, AL 165

Chapter 12: THE URBAN HEAT ISLAND: CONTAMINANT TO THE GLOBAL TEMPERATURE RECORD?
Robert C. Balling, Jr., Office of Climatology, Department of Geography, Arizona State University, Tempe, Arizona 179

Chapter 13: GENERAL CIRCULATION MODEL STUDIES OF GLOBAL WARMING
Robert G. Crane, Earth System Science Center and Dept. of Geography, The Pennsylvania State University, University Park, PA 189

Chapter 14: CHANGES IN CLIMATE VARIABILITY WITH CLIMATE CHANGE
Linda O. Mearns, National Center for Atmospheric Research, Box 3000, Boulder, CO 209

Chapter 15: ZONAL COMPARISONS OF GLOBAL CIRCULATION MODEL: TEMPERATURE AND PRECIPITATION DATA WITH HISTORICAL CLIMATE FOR NORTH AMERICA AND EURASIA
Anthony J. Brazel, Office of Climatology, Arizona State University, Tempe, Arizona and Robert A. Muller, Department of Geography and Anthropology, Louisiana State University, Baton Rouge, Louisiana 227

Chapter 16: SATELLITE MONITORING OF KUWAIT'S EFFLUENT: SMOKE FROM THE OIL FIRES
Sanjay S. Limaye, Space Science and Engineering Center, University of Wisconsin - Madison, 1225 West Dayton Street, Madison, WI 242

Part Four: Biophysical Impacts

Chapter 17: SEA LEVEL RISE: IMPLICATIONS AND RESPONSES
Stephen P. Leatherman, Department of Geography, LeFrak Hall, University of Maryland, College Park, MD 256

Chapter 18: USING A GIS TO ANALYZE THE EFFECTS OF SEA LEVEL RISE
Gary Ostroff, Coastal Environmental Services, Inc., 2 Research Way, Princeton, NJ and Susan Tucker, Hunter College, Department of Geography, New York, NY 264

Chapter 19: FORESTS AND GLOBAL CLIMATE CHANGE
Daniel B. Botkin, Robert A. Nisbet, and Lloyd G. Simpson, Department of Biological Sciences and Environmental Studies Program, University of California at Santa Barbara, Santa Barbara, CA 274

Chapter 20: ECOLOGICAL EFFECTS OF RAPID CLIMATE CHANGE
Dexter Hinckley, Climate Change Division, U.S. EPA, 401 M Street, SW, Washington, DC and Geraldine Tierney, Bruce Co. 291

Chapter 21: GENERAL CIRCULATION MODEL ESTIMATES OF REGIONAL PRECIPITATION
David R. Legates, Department of Geography, University of Oklahoma, Norman, OK and Gregory J. McCabe, Jr. US Geological Survey, Denver Federal Center, Mailstop 412, Denver, CO 302

Part Five: Socioeconomic Impacts

Chapter 22: FEDERALISM: A REGIONAL APPROACH TO GLOBAL ENVIRONMENTAL PROBLEMS
James L. Huffman, Northwestern School of Law, Lewis & Clark College 10015 SW Terwilliger Boulevard, Portland, OR 315

Chapter 23: THE NEED FOR HIGHER RESOLUTION IN THE ECONOMIC ANALYSIS OF GLOBAL WARMING: AN ASSESSMENT AND EXPANDED RESEARCH AGENDA
Robert Ayres, Department of Engineering and Public Policy, Carnegie-Mellon University, Pittsburgh, PA and Adam Rose, Department of Mineral Economics, The Pennsylvania State University, University Park, PA 329

Chapter 24: PREDICTED EFFECTS OF CLIMATIC CHANGE ON AGRICULTURE: A COMPARISON OF TEMPERATE AND TROPICAL REGIONS
Cynthia Rosenzweig, GISS, New York, NY and Diana Liverman, Dept. of Geography, The Pennsylvania State University, University Park, PA 346

Chapter 25: IMPACTS OF GLOBAL CLIMATE CHANGE ON HUMAN HEALTH: SPREAD OF INFECTIOUS DISEASE
Robert Shope, Yale Arbovirus Research Unit, Box 3333, New Haven, CT 363

Chapter 26: IMPACTS OF GLOBAL WARMING ON HUMAN HEALTH: HEAT STRESS-RELATED MORTALITY
Laurence S. Kalkstein, Center for Climatic Research, University of Delaware, Newark, DE 371

Part Six: Impact Mitigation, Adaptation, and Emissions Limitation Strategies

Chapter 27: THE COSTS OF CLIMATE CHANGE TO THE UNITED STATES
James G. Titus, Office of Policy Analysis, US EPA, PM 221, 401 M Street, SW, Washington, DC 384

Chapter 28: ENERGY SUPPLY TECHNOLOGIES FOR REDUCING
GREENHOUSE GAS EMISSIONS
*Barry D. Solomon, US EPA, ANR 445, 401 M Street, S.W.,
Washington, DC* 410

Chapter 29: IMPLEMENTATION OF MITIGATION AT THE
LOCAL LEVEL: THE ROLE OF MUNICIPALITIES
*L.D. Danny Harvey, University of Toronto, 100 St. George Street,
Toronto, Ontario Canada M5S 1A1* 423

Chapter 30: FORESTRY RESPONSES TO MITIGATE THE EMISSION
OF GREENHOUSE GASES
*Steven M. Winnett, Office of Policy Analysis, US EPA, PM 221, 401 M
Street, SW, Washington, DC* 439

Part Seven: Policy, Regulation, and Legal Issues: Domestic and International Perspective

Chapter 31: ENERGY POLICY RESPONSES: CONCERNS ABOUT
GLOBAL CLIMATE CHANGE
*Thomas J. Wilbanks, Energy Division, Oak Ridge National Laboratory,
P.O. Box 2008, MS184, Oak Ridge, TN* 452

Chapter 32: ENERGY PRODUCTION, ECONOMY AND
GREENHOUSE GAS EMISSIONS IN HUNGARY:
POSSIBLE REGIONAL CONSEQUENCES OF GLOBAL
CLIMATE CHANGE
*E. Mészáros and Á. Molnár, Institute for Atmospheric Physics, H-1675,
Budapest, P.O.B. 39, Hungary* 471

Chapter 33: GLOBAL CLIMATE CHANGE AND HUMAN
POPULATIONS IN THE CENTRAL AND WEST
AFRICAN REGION: AN EXAMINATION OF THE
VULNERABILITIES, IMPACTS AND POLICY
IMPLICATIONS
*Stella C. Ogbuagu, Department of Sociology, University of Calabar,
Calabar, Nigeria* 481

Chapter 34: REGIONAL EFFECTS OF GLOBAL WARMING:
ISRAEL AND EASTERN MEDITERRANEAN BASIN
*Ariel Cohen, Department of Physics and Atmospheric Science, Drexel
University, Philadelphia, PA* 493

Chapter 35: GLOBAL CLIMATE CHANGE: THE CALIFORNIA
PERSPECTIVE
*James M. Strock, Secretary for Environmental Protection, California
Environmental Protection Agency, Air Resource Board, 2020 L. Street,
P.O. Box 2815, Sacramento, CA* 498

Chapter 36: COMBATTING THE CLIMATE CHANGE: A CHALLENGE
FOR INTERNATIONAL COOPERATION IN THE '90s
*Aira Kalela, Special Advisor, Ministry of the Environment, Ratakatu 3,
00120 Helsinki, Finland* 522

Chapter 37: A CANADIAN PERSPECTIVE ON CLIMATE CHANGE
*Ian Burton and Deborah Herbert, AES - La Salle Academy, 373 Sussex
Drive, 3rd Floor Block E, Ottawa, Ontario, Canada K1AOH3* 529

Chapter 38: UK POLICY ON CLIMATE CHANGE
Simon Oliver, Department of the Environment, Global Atmospheric Division, Room A103, Romney House, 43 Marsham Street, London, SW1P 3PY, England 539

Chapter 39: JAPAN'S ACTION TO COMBAT CLIMATE CHANGE
Shinichi Isashiki, Director and Kazuto Suda, Global Environmental Affairs Division, United Nations Bureau, Ministry of Foreign Affairs, 2-2-1, Kasumagaseki, Chiyoda-KU, Tokyo, Japan 543

Chapter 40: UNITED NATIONS SPONSORED 1992 EARTH SUMMIT IN RIO DE JANEIRO: WHERE DO WE GO FROM HERE?
E. Willard Miller[1], Michelle A. Baker[2] and Shyamal K. Majumdar[3]
[1]Department of Geography, The Pennsylvania State University, University Park, PA,
[2]Department of Biology, University of New Mexico, Albuquerque, NM,
[3]Department of Biology, Lafayette College, Easton, PA 551

Subject Index 555

FOREWORD

Throughout its history, the earth has withstood changes in climate resulting from a large variety of natural processes. However, the threat of a human-induced climate change is precedent setting, not only because it implies alteration of natural processes, but also because of the unusually *rapid* rate of change which is forecast by many climatologists. The issue of global climate change has contributed to a plethora of ideas on a course of action to be taken which might mitigate this problem.

Possibly the greatest disagreement among scientists involving the climate change issue revolves around the prudence of developing mitigating policy since there is uncertainty about the rate and magnitude of climate change. There are also significant long range economic and developmental issues which will affect the developed countries, fossil fuel producers, and developing countries. A small minority of scientists believe that both policy development and implementation should proceed only after the science of the issue has been resolved. Improving our knowledge of climatic processes and *simultaneously* developing and implementing a series of cost-effective policy options ought to be proposed. Many studies suggest that insurance against climatic change would carry significant benefits as efforts toward energy efficiency, water conservation, and other environmental technologies are desirable even if global warming does not occur. This year's (1992) United Nations Conference on Environment and Development (UNCED) in Rio de Janeiro was attended by thousands of scientists, politicians, and environmentalists who share in the belief that an international coordinated effort is necessary to diminish the risk of a global warming.

The editors of this book have taken great pains to produce an extensive volume which discusses many facets of the climate change issue. The chapters and introductory materials were written by experts who represent the breadth of the climate change issues: economists and meteorologists, skeptics and advocates, domestic and international specialists are all represented. An unusual aspect of this work is the attempt to emphasize natural climatic fluctuations against the backdrop of potential human-induced climatic alterations. In addition, the text contains over 20 chapters on impacts and 10 chapters on mitigation and policy development. The Pennsylvania Academy of Science is to be commended for developing possibly the most comprehensive volume on climate change to date.

Possibly the most difficult challenge of the climate change issue is to foster the unprecedented international cooperation necessary to link new scientific information with the international policy process. The authors and editors of this book understand the challenge. The UNCED Conference has provided the momentum. Implementing the Rio recommendations should be a top priority for the coming decade.

Dennis Tirpak
Director, Climate Change Division
U.S. Environmental Protection Agency
Washington, DC

UNITED STATES DEPARTMENT OF STATE
BUREAU OF OCEANS AND INTERNATIONAL
ENVIRONMENTAL AND SCIENTIFIC AFFAIRS
WASHINGTON, D.C. 20520

R.A. Reinstein
Deputy Assistant Secretary
U.S. Department of State

INTRODUCTION

It is a pleasure to introduce this volume on the subject of global climate change prepared by the Pennsylvania Academy of Science. The subject is of great interest to environmentalists, scientists of several kinds, economists, energy analysts, land-use planners, specialists in agriculture, forestry, transportation and numerous other fields, and the general public. It is often described as an "environmental" issue, but is in fact a much more fundamental issue, bringing into play virtually all aspects of society and our economies, with quite significant implications for the shape of our world in the next century and beyond.

The science of global climate change is a dynamically changing discipline. While the basic fact of the greenhouse effect and the build-up of greenhouse gases in the atmosphere is accepted by all, the outlook for the future and the implications for public policy have been debated, one may say, heatedly. It is generally acknowledged that continued increases in greenhouse gas concentrations are likely to disrupt the climate system, if not mitigated in some manner. The timing and magnitude of the effects is still a matter of debate, not likely to be resolved for several years. Moreover, while the effects would clearly be negative for some countries and regions, they may be positive for others. Countries also differ in their ability to adapt to climate change, and some adaptation may be required in any case. All this makes it difficult for policymakers to identify the best responses to the situation.

The United Nations established the Intergovernmental Panel on Climate Change (IPCC) in 1988 to try to assist governments in understanding all the aspects of the issue. Working under the auspices of the World Meteorological Organization (WMO) and the United Nations Environmental Program (UNEP), the IPCC completed its first assessment report in mid 1990, based on efforts by three working groups dealing with science, impacts, and response strategies. A supplemental report updating the first assessment was completed in February 1992. As chairman of the IPCC working group on response strategies, I am confident that these efforts over the longer term will provide important information to help guide public policy on this issue and its many facets.

Drawing on the IPCC's first assessment and other developments, the UN General Assembly in December 1990 established the Intergovernmental Negotiating Committee (INC) for a Framework Convention on Climate Change and authorized it to work toward completion of an international treaty (convention) which would create the longer-term formal context for addressing the issue. The goal was to have the initial framework convention completed in time for signing at the UN Conference on Environment and Development (UNCED) — the so-called Earth Summit — at Rio de Janeiro, Brazil, in June 1992.

The negotiation process has revealed clearly the broad range of issues and interests involved in the question of global climate change. Nations are being asked to consider fundamental changes in the way they produce and use energy, raise livestock and grow food, manage their forests, transport people and goods, and use the land. These questions have impacts not only on both the local and the global environment but also on basic economic health and security, on the quality of our lives. Governments will need to consider all of these issues very carefully as the process established by the framework convention on climate change plays out over the longer term, and will need to review continuing developments in science, technology and economics as an integral part of this process.

R.A. Reinstein

Deputy Assistant Secretary
Environment, Health and Natural Resources
U.S. Department of State

Chief U.S. Negotiator
INC Climate Negotiations

Chairman
IPCC Response Strategies Working Group

Commonwealth of Pennsylvania
Office of the Governor
Harrisburg

Robert P. Casey
Governor
Commonwealth of Pennsylvania

MESSAGE FROM THE GOVERNOR

Three centuries ago our Commonwealth's founder, William Penn, wrote of a "good fruitful land" where "fowl, fish and wild deer" were plentiful and "the air is sweet and clear, the heavens serene."

The environment of Pennsylvania has been severely tested since the first European settlers arrived and began using the Commonwealth's natural resources to fuel an industrial revolution. It is now becoming increasingly clear that the unfettered burning of fossil fuels may cause long-term changes in the global climate.

Global warming would significantly alter the environment of Pennsylvania as we now know it. New weather patterns could significantly affect agriculture, river systems, forests, wetlands, energy use and human health if current levels of consumption continue and the warming predictions are correct.

We've been told this is the price of progress, but it's a price we cannot afford. What is at stake now is the health of the environment and the economy of our Commonwealth. Our ability to create jobs and to attract and expand business and industry is now linked to the environmental quality of life in our communities. That's why I've made the environment a major part of my agenda.

Over the past four years, Pennsylvania has led the nation in recycling, toxic waste cleanup programs and the improvement of water and sewer systems. We are now also taking steps to become a national leader and model for clean air protection.

Automobiles are a major source of "heat-trapping gases" such as carbon dioxide and nitrous oxide which contribute to urban smog, acid rain, and toxic air and water pollution. It's clear that tough, new measures are needed to make automobile emissions cleaner.

In the fall of 1991, I directed the Departments of Transportation and Environmental Resources to develop and implement some of the toughest new car emissions standards in the country in line with the new requirements of the federal Clean Air Act. Beginning in the fall of 1995, we will see new cars on Pennsylvania's highways with enhanced air pollution control systems, and many running on alternative fuels. By that time, we also hope to reduce the number of inefficient older cars on the road which can pollute ten times as much as the average new car.

Cleaning vehicle emissions isn't enough, however, if we are to make a measurable difference in the quantity of heat-trapping gases and other air pollutants being released in Pennsylvania. We must increase the use of mass transit in our urban areas to reduce the number of emission-producing vehicles. A dedicated funding source for mass transit is now in our state budget to help promote this effort.

We're also increasing our efforts to protect our trees which naturally reduce atmospheric carbon dioxide and cleanse the air. Pennsylvania already has one of the nation's largest forestry programs, and we've begun developing a strategy that will protect and manage our two million acres of state forestland well into the next century. In addition, we are placing greater emphasis on urban forestry to increase tree planting in our communities.

We are also encouraging Pennsylvania's farming community to be responsive to new techniques and practices as research uncovers ways to reduce the release of agricultural methane — another recognized heat-trapping gas.

All of these measures will not be easy and will require sacrifice and commitment. But, in Pennsylvania, our mission is certain: we are crafting our own environmental legacy for the future. Our state constitution promises our people that the natural environment we inherited from our parents and pass on to our children will be unspoiled.

I commend the Pennsylvania Academy of Science on its timely review of this important issue which will serve to increase our understanding of the implications for the environment of our nation and planet. We must work to keep faith with our constitution and ensure a quality environment for the generations yet to come.

Robert P. Casey
Governor
Commonwealth of Pennsylvania
March 1992

Global Climate Change: Implications, Challenges and Mitigation Measures. Edited by S.K. Majumdar, L.S. Kalkstein, B. Yarnal, E.W. Miller, and L.M. Rosenfeld. © 1992, The Pennsylvania Academy of Science.

Chapter One

THE CLIMATE SYSTEM:
An Overview

PETER J. ROBINSON
Department of Geography
University of North Carolina
Chapel Hill, NC 27599

INTRODUCTION

Climate is the integrated expression of the day-to-day weather at any place. The weather itself includes elements such as wind speed and direction, cloud amount and type, visibility, precipitation type and intensity, humidity, and temperature. A comprehensive description of climate requires statements about the means, variances, and extremes of these weather elements, indications of their frequency distribution as individual elements and in various combinations, and consideration of sequences of weather events. Traditionally the description has been based on past weather information, assuming that the future will be similar to the recent past. Our increasing awareness that human activity might change the climate, however, is forcing us to investigate much more closely the processes acting to create and change the climate. Fundamental to this investigation is the definition and understanding of the climate system itself.

The climate system is composed of the interconnected aspects of the atmosphere, biosphere, cryosphere, hydrosphere, and lithosphere which create, maintain and modify the weather and its variability (Figure 1). All components of the system, acting on various space and time scales, influence climate. However, it is primarily through the energy and materials flows within and between the components that the system and its changes can be analyzed and understood. Hence this overview commences with consideration of global energy flows. Then the hydrologic cycle and other material flow systems are reviewed. These discussions lead to specification of the current climate and the nature and causes of changes in the system. The emphasis throughout is placed on the near-surface conditions, the area where most climatic impacts are felt and where the climate and its variability can be visualized most readily.

THE EARTH'S ENERGY BUDGET

On the most fundamental level, the climate of Earth is a response to energy coming from the sun. This energy is the incoming solar radiation, electromagnetic radiation containing that portion of the spectrum which we perceive as light. Part of this incoming radiation is absorbed by the earth and its atmosphere, leading to heating. This in turn leads to the emission of electromagnetic radiation from the earth, the terrestrial radiation which we cannot see but which acts in a way very similar to sunlight. The warmer the planet, the greater the emission. This emission leads to cooling. Consequently the planet is continually striving for an energy balance between the absorbed incoming solar radiation and the emitted outgoing terrestrial radiation. Changes in incoming solar radiation, such as the diurnal and seasonal cycles, together with changes in the components of the climate system, ensure that equilibrium is never fully achieved. Indeed, climate may be viewed as the atmospheric response to these continuous and various changes in the energy flows.

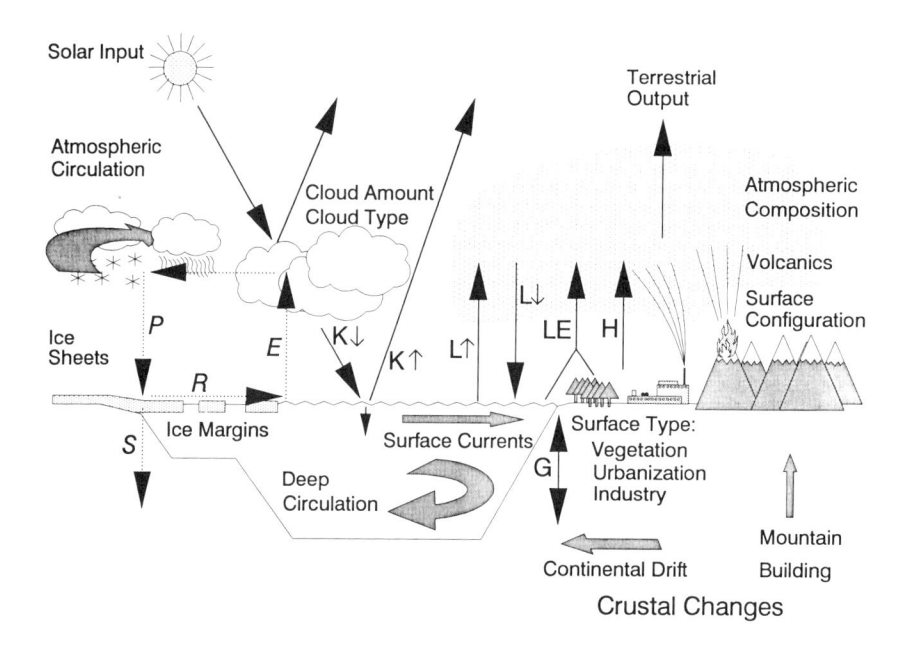

FIGURE 1. The components of the climate system, showing the main energy fluxes (solid arrows) and the hydrologic cycle (dotted arrows). Symbols are discussed in the text.

The energy flows are shown in general terms in Figure 1. The solar energy arriving from the sun passes into the atmosphere. Here some of the energy is reflected back to space, primarily from the tops of clouds, while a small amount is absorbed directly by the atmospheric gases. The solar radiation reaching the surface, K↓ in Figure 1, is, on a global annual average, about half of the energy arriving at the top of the atmosphere. Some reaches the surface as direct solar radiation, having passed unimpeded through the air, while the rest arrives after sideways and downward scattering, as the diffuse radiation we know as sky light. Once at the surface, some energy is reflected, K↑. The rest is absorbed. The proportion relected, the albedo α, varies with surface type.

Once absorbed by the surface layer, the radiant energy is transformed into heat energy. The resultant surface heating leads to thermal energy exchanges both with the underlying medium and the overlying air. For land surfaces the ground heat flux, G, has a distinct annual cycle. In spring and summer the warm surface leads to a downward flow into the cooler ground, while in fall and winter the warmer deep layers transfer energy towards the surface. Thus, the net annual flow is negligible. For water, with its horizontal currents and vertical mixing, the pattern is more complex but still the net annual exchange is small. This is not the case for the sensible heat flux into the air, H. The flow is predominantly from the warm surface into the cold air above, although downward flow can occur when a warm mass of air blows over a cold surface. The surface solar energy absorption also gives rise to evaporation and a latent heat flux, LE. This also is predominantly away from the surface with the latent energy being released aloft when condensation and cloud formation occurs. Occasional downward energy movement, and resultant dew formation, is possible. The final mode of energy loss from the surface is through the emission of terrestrial radiation, L↑. Much of this radiation is absorbed in the atmosphere, although some is transmitted directly to space. The portion transmitted depends on the state and composition of the atmosphere. Clouds enhance absorption. So do increasing concentrations of carbon dioxide. Indeed, this phenomenon is the basis for the greenhouse effect, details of which are given in later chapters. The energy which has been absorbed by the atmosphere, whether as the result of the sensible or latent heat flows or from radiation absorption, heats the atmosphere. This in turn emits radiation, some being sent to the surface as incoming terrestrial radiation, L↓, where it in turn is absorbed, helping to heat the surface and encouraging re-emission, while the rest is emitted to space. Once this occurs the energy is lost to the planet and the cycle is complete.

All components of the climate system can create temporal and spatial variations in energy flows. It is most useful, however, to consider surface energy exchanges and the associated surface temperatures and emphasize first those factors which have a major influence on the day-to-day weather and the seasonal cycle of climate. These can be approached through the generalized surface energy budget equation:

$$\Delta E = (1-\alpha)K\!\!\downarrow + L\!\!\downarrow - L\!\!\uparrow - H - LE - G, \qquad (1)$$

where ΔE is the change in surface energy storage. Although

$$\Delta T = \Delta E / \varrho C, \qquad (2)$$

where ΔT is the surface temperature change and ϱC is the heat capacity of the surface medium, Eq. (1) cannot be solved for ΔT directly since L↑, H, LE, and G are themselves functions of T.

The amount of incoming solar radiation at any location is a function of the time of day, the season of the year, and latitude. This, visualized as the path of the sun across the sky, leads to daily total amounts which are high throughout the year in equatorial regions, with an increasing seasonal effect towards higher latitudes (Figure 2). Annual total income decreases poleward. Largely as a result of this, more solar energy is absorbed in equatorial regions than in polar ones. The equatorial regions, being warmer, also emit more terrestrial radiation to space than do the poles. However, the long-term absorption and emission for each latitude band does not balance (Figure 3). There is a net radiant energy excess in the equatorial regions, a net deficit near the poles. Since no major long-term temperature changes occur in either region, other energetic factors must compensate for the imbalance. The necessary horizontal energy transports are performed by winds and ocean currents, through the General Circulation of the Atmosphere and of the Oceans, both of which will be considered later.

The other major factor to be considered as influencing day-to-day weather and seasonal climate is surface type. This affects the energy budget directly and, through the resultant horizontal temperature changes, plays a major role in defining the general circulation of the atmosphere. On a broad scale we can consider three surface types: land, water, and snow and ice. For the last the major characteristic is a high albedo and little energy absorption. It can be thought of as a self-perpetuating cold surface. Water, having a low albedo, a high heat capacity, and the ability to mix vertically, is capable of absorbing and storing a tremendous amount of energy without having a great temperature change. Such changes as do occur tend to affect a deep layer. Land, on the other hand, has a higher albedo but a lower heat capacity. In addition, solar radiation absorption is restricted to a very thin surface layer. So on balance at the surface land tends to heat and cool rapidly in response to energy changes. These contrasts between land and water give rise to the continentality effect, whereby the interiors of continents tend to warm more rapidly in spring and summer, and to reach a much higher summertime maximum temperature than does the ocean. The land then cools more rapidly to a much lower winter minimum temperature than does the ocean. One immediate consequence of this effect is that the cold continental interiors in mid-latitude winters tend to develop regions of high atmospheric pressure, while they generate low pressure regions in the heat of summer. This in turn influences the wind patterns. The most obvious result is the monsoon circulation, but the continentality effect has a profound influence on the whole of the general circulation.

Contrasts in surface type create similar phenomena on other temporal and spatial scales. Diurnal temperature ranges at interior stations tend to be higher than at coastal sites. Differential heating between land and water lead to the creation of sea or lake breezes. In short, many small scale phenomena, important for local climates, are influenced by land and sea contrasts. Thus the energy budget plays the major role in establishing temperatures and their spatial variations, and has a direct influence on wind patterns. However, it must not be inferred that all features of the climate can be explained by the energy budget alone, as indicated in the following section.

THE WATER BUDGET AND MATERIALS FLOW

Water, whether in the form of precipitation, soil moisture or atmospheric humidity, is a vital aspect of climate.[2] We can discuss water and its variability in an analogous way to the energy budget, first considering the general flows in the atmosphere and then the surface budget in rather more detail. The whole water flow of the planet is represented by the hydrological cycle (Figure 1). Discussion of this can start with evaporation, which is the practical expression of the latent heat flux mentioned earlier. Evaporation, E, can come directly from open water surfaces,

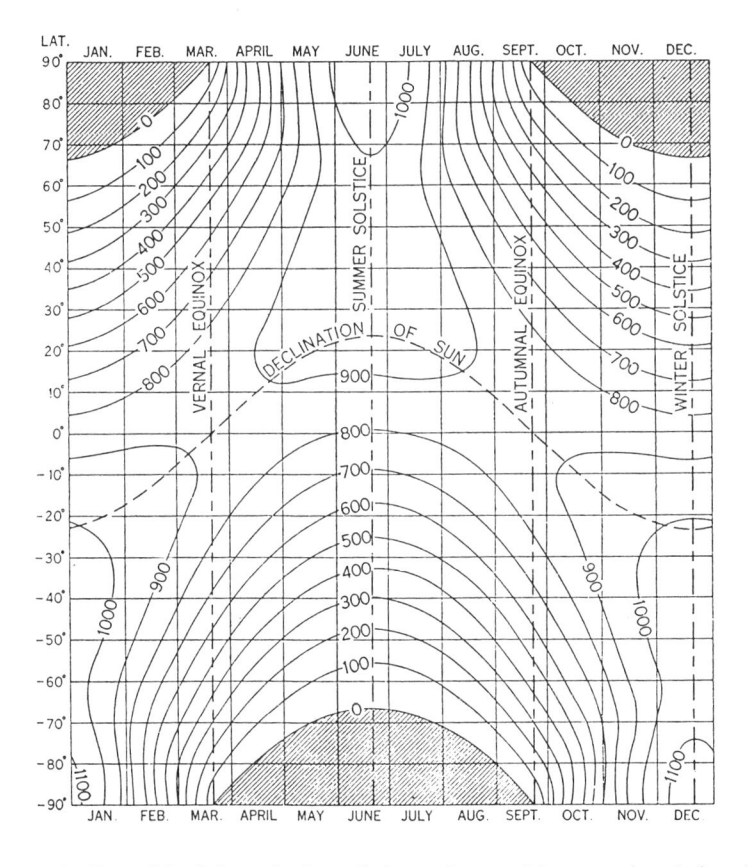

FIGURE 2. Chart of the daily total solar radiation at the top of the atmosphere (cal cm^{-2} day^{-1}), from *Smithsonian Meteorological Tables*, sixth revised edition, by Robert J. List. Table 143. Reprinted by permission of the Smithsonian Institution Press. © Smithsonian Institution 1984.

which can range in size from the pore spaces between soil particles, through puddles and lakes, to the world's major oceans. Or it can come through transpiration from vegetation. Both, usually treated together as evapotranspiration, place water vapor into the atmosphere, where it may be moved around by the general circulation. Eventually clouds are formed and any resulting precipitation, P, returns the water to the earth. Here it may sink into the soil or move across the land into the oceans, representing stored water which is available for evaporation again. The climatological and near surface aspects of this cycle can be expressed through a very general form of the water budget:

$$\Delta S = P - E - R \qquad (3)$$

where ΔS the change in storage at the surface of the earth and R is runoff. As formulated here it emphasizes the storage term, making it directly analogous to the energy budget. It is possible to refine this formulation for specific practical purposes. For example, treating ΔS as soil moisture in the crop root zone, and consequently adding an output term for deep drainage below the zone, Eq (3) can be used for analyses of drought conditions or as a means of assessing irrigation needs. Similarly, it is straightforward to rewrite it to emphasize runoff for flood studies or water supply planning.

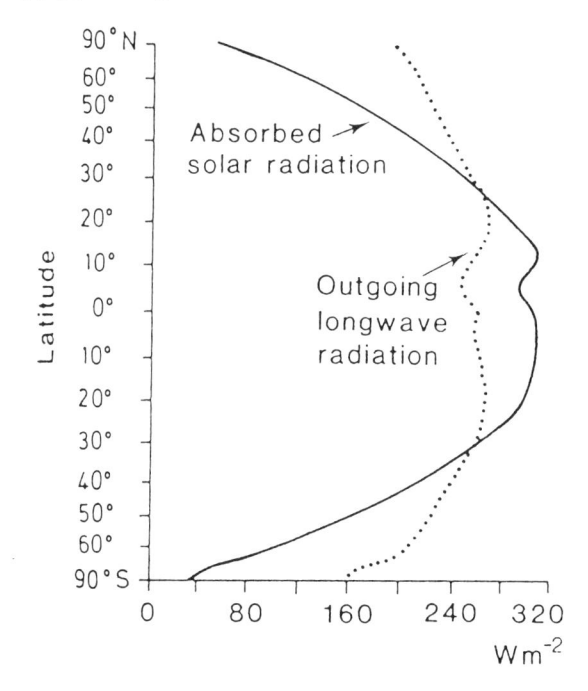

FIGURE 3. The latitudinal variation of absorbed solar and emitted terrestrial radiation. Data are from NOAA polar orbiting satellites for the June 1974-February 1978 period. From Ann Henderson-Sellers and Peter J. Robinson *Contemporary Climatology,* © 1986. Figure 2.24. Reprinted by permission of Longmans Group, U.K. Harlow, Essex.

As with the energy considerations, quantification of the components depends on a variety of factors. The surface type again, of course, plays a vital role. In the partitioning of storage and runoff there are obvious differences between rural and urban surfaces, between various soil types, and with topography. The amount, areal extent, timing, and intensity of precipitation may also affect partitioning. Since these factors depend on cloud formation, they are linked to the rest of the climate system in just the same way as for the energy budget.

The hydrologic cycle and the surface water budget are emphasized here since they are most directly connected with the precipitation, soil moisture and humidity we most readily associate with climate. The concepts, however, are directly applicable to the cycling of other materials between the earth and the overlying atmosphere. In particular, attention is increasingly being paid to the cycling of gaseous materials such as carbon dioxide and methane, the greenhouse gases, and the movement and fate of particulate and gaseous matter creating pollution.[3] Several of these are treated in more detail in later chapters.

The budget equations as written are perfectly general, and are not constrained in time or space. To be useful, either for understanding present and future climates or for the practical application of climate information, they must be considered within various time and space frames, and the following section looks at the global picture on the seasonal time scale.

THE GENERAL CIRCULATION OF THE ATMOSPHERE

The general circulation of the atmosphere represents the major wind systems of the world, along with the weather that they bring. The winds are a response to pressure forces created as a result of horizontal pressure differences arising from energy imbalances. The pressure force, from high to low pressure, controls windspeed, but because of the partial frictional uncoupling of the atmosphere from the rotating earth, wind direction is at right angles to the pressure gradient in mid- and high-latitudes and at an angle to the gradient in low latitudes.

The resultant interaction of pressure differences and rotation can be summarized using the 3-cell model of the general circulation of the atmosphere (Figure 4). In the tropics are the Hadley Cells. These extend from the Equator to about 30° latitude in each hemisphere, and are a thermally direct response to the heating caused by the solar radiation maximum near the Equator, enhanced by the release of latent heat in the upper level cloud formation in the near-equatorial band known as the Intertropical Convergence Zone, ITCZ. In this zone, the near-surface airflow from the two cells, the Northeast Trades and the Southeast Trades, converges, rises and creates the major area of tropical precipitation. The ITCZ location varies seasonally, being about 10-15°N in July and about 5°S in January. Away from the ITCZ descending air dominates, with little cloud formation or precipitation. Thus the types of tropical climates range from year round moist conditions in the equatorial zone with its continous ITCZ influence, through the tropical wet-dry climates slightly poleward of the equator to the desert conditions around 30°

latitude. In addition, the Hadley Cell is a region where disturbances such as the fronts and depressions of mid-latitudes, do not develop.

The Polar Cell of the three cell model is also thermally direct, with ascending air around 60° latitude and descending air at the poles. Here however, the earth's rotational effect is stronger, and the near-surface winds blow more nearly due west to east, giving the Polar Easterlies, with a west to east airflow aloft. At the surface friction slows the winds and gives them a north-south component, but they are commonly rather weak unless enhanced by local effects, while the whole cell covers a small geographical area. It can be regarded as the sink for the poleward energy transport.

The remaining cell, the mid-latitude cell, is much more complex. Indeed, it exists only as a concept indicating the need for energy transport across mid-latitudes. Rather than a vertically oriented circulation, there is an horizontally oriented one. The westerly airflow proceeds in a series of waves, the planetary waves. These vary in location from day to day because of flow instabilities, and from season to season because of changes in the continental-scale pressure distribution arising from the continentality effect. In addition, the horizontal distributions of pressure and

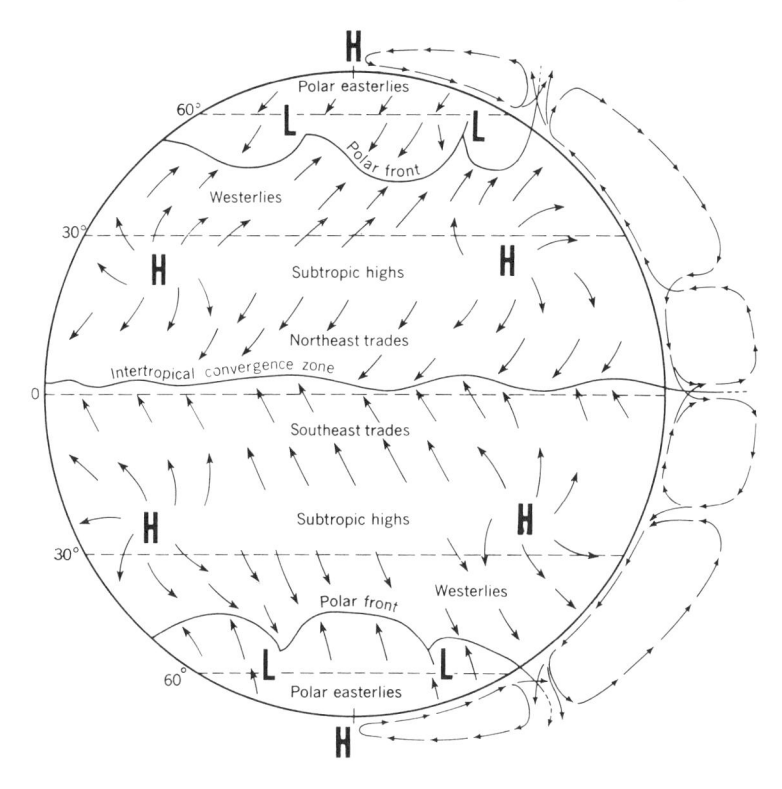

FIGURE 4. From: Howard J. Critchfield, *General Climatology*, 4e, © 1983, p. 95. Reprinted by permission of Prentice Hall, Englewood Cliffs, NJ.

temperature, and their variations with height, create strong airflow aloft at the northern and southern extremities of this pseudo-cell. These are the jet streams, ribbon-like regions of extemely fast moving air, frequently acting to steer lower level weather features. At the equatorward extremity, around 30°, is the subtropical jet. For the U.S. the influence of this is most often felt when it steers rain-bearing features into the Southwest. The poleward extremity has the Polar Front Jet, which is responsible for steering many weather features across much of the North American continent. Associated with this jet is the Polar Front itself. A front is a region of rapid horizontal temperature change, usually a few miles broad and hundreds, or even thousands, of miles long. The Polar Front separates cold polar air masses from the much warmer ones of tropical origin. This frontal region is commonly the breeding ground for smaller scale disturbances, the depressions, with their own embedded fronts, which are responsible for much of the day to day weather variability of the mid-latitudes. Certainly they are the primary source of precipitation for much of the area. However, the planetary waves, the jet stream, the polar front, the disturbances and the air masses all interact to produce the mid-latitude weather and climate.

Although the above paragraphs treat each cell in isolation, the interconnections between them are a vital part of the global climate system. For example, the Asiatic monsoon is driven by mid-latitude pressure contrasts but has a major impact on the climates of tropical Asia. Similarly, the El Niño-Southern Oscillation phenomenon is associated with the variations in oceanic and atmospheric circulation in the tropical South Pacific Ocean, but can influence weather in mid-latitudes.[4]

THE CHANGING CLIMATE

Climatic changes, on time scales ranging from minutes to millennia and influencing areas from individual fields to the whole globe, can occur as a result of changes in any and all components of the climate system, acting through the energy and water budgets and the general circulation of the atmosphere. While almost any change is likely to have important consequences for a specific activity, only selected examples can be considered here.

Changes in solar radiation, the ultimate climatic forcing function, will have major consequences. On the time scale of the evolution of the earth, incoming energy has increased by about 30%, but average surface temperatures have shown no distinct trend and have varied only between 275 K and 305 K. Reasons for this stability are unclear. The relationship between temperature changes, incoming radiation and variations in earth-sun geometry on times scales of tens to hundreds of thousands of years, however, are well-established.[5] Much shorter variations in solar output, covering tens of years, are associated with sunspots. It is unclear how the climate responds. There have been numerous attempts to find links using statistical methods, but no convincing physical connections have been demonstrated.

In contrast to these rather regular, cyclic changes, there are various irregular ones.

Over extremely long times, continental drift and mountain building periods change the configuration of the earth's surface, with profound implications for the operation of the general circulation of both the oceans and the atmosphere. On a much shorter time scale, volcanic activity may release gases which alter the composition of the atmosphere and modify the greenhouse effect for a short time. Additionally, volcanic eruptions also eject dust which collects at high levels and becomes distributed world-wide. This may raise the planetary albedo and lead to cooling. The dust may remain suspended for a few years. However, a few decades are needed for the whole system to come into equilibrium with changed conditions. Hence a single eruption is likely to produce only a transient change, and a great deal of continuous activity is needed to create permanent changes in the climate.

While volcanic effects are primarily transient, the gases and particulate matter placed in the atmosphere by anthropogenic activity are more permanent and much more likely to lead to climatic change. Many of the gases become distributed worldwide and enhance the greenhouse effect. Equally important, but currently more localized, are the effects of changes in atmospheric composition on the chemical climate.[3] Relatively little is known about this, but concerns with ozone depletion, acid rain and urban and regional air pollution are stimulating work incorporating the atmospheric composition and its chemical transformations into assessments of climatic variations on all time and space scales.

The other facet of the climate system where human activity is having a major impact is the land surface characteristic. Any change in surface character leads to change in albedo, in the partitioning of sensible and latent heat, and in the balance between infiltration and runoff. Until recently, the resultant climate changes have been on a local scale. The development of cities has led to increases in temperature over the cities but has had little impact elsewhere. Similarly, irrigated areas have increased humidities a short distance downwind, but have had no impact beyond that. Now, the increased scale of surface modification, whether through expanded city size or additional agricultural development, is becoming sufficient to have more than local consequences. Indeed, activities such as deforestation in Amazonia may have a global impact.[6] Climatologically, deforestation leads to increased runoff, decreasing the amount of water available for evaporation, which may lead to a decrease in cloud amount. This could increase the solar radiation receipt at the surface, enhancing evaporation of whatever water is available, and promoting drought. This change could in turn influence the Hadley Cell and the poleward transfer of energy, leading to climate changes outside, as well as within, the tropics.

Outside the tropics, one of the major concerns is snow and ice. This surface type has a high albedo, so that much of the solar radiation received at high latitudes is reflected back to space, and the polar regions remain cool. If for any reason the area of ice cover decreases, the uncovered surface will have a lower albedo and more energy absorption. This is likely to lead to further warming and further melting. This gives a positive "ice-albedo feedback", and indeed, it is easy to postulate that a small change in global temperature is rapidly amplified, the earth soon becoming deglaciated or, following a cooling, completely ice covered.

Missing from this simple "runaway greenhouse" suggestion is consideration of other feedbacks, especially those associated with clouds. The amount of evaporation on a warmer earth, for example, is likely to increase, leading to increased cloud formation. Clouds reflect incoming solar radiation, tending to cool the surface, but they also absorb and re-emit outgoing terrestrial radiation, serving to warm the surface. Which of these two dominates depends on the type and level of the clouds, and it is unclear whether heating or cooling results from increased evaporation.

One of the most important, but least understood, components of the climate system is the ocean. The ocean exchanges energy, water, and materials with the atmosphere in the same way as a land or ice surface, so that a major concern is with ocean surface temperatures. These are influenced by horizontal and vertical motions within the oceans themselves. The near-surface horizontal ocean currents are coupled to the atmospheric general circulation, and the two respond in tandem, but on somewhat different time scales, to climatic changes. The motions associated with the deep ocean circulation, however, are largely uncoupled from the atmospheric motions. Near-surface water bodies may sink and remain at great depth for thousands of years before returning to the surface a great distance from their original location. Since water very readily absorbs solar radiation, these water bodies store vast quantities of energy when they sink and release it when they rise.[7] Consequently, the energy exchange between the ocean and atmosphere may reflect the climate conditions many years ago and miles away superimposed upon the effect of the current climate.

UNDERSTANDING THE CLIMATE SYSTEM

Understanding of the climate system remains one of the major goals of climatology. Explanation of how the system operates provides not only insight into the nature and distribution of the present climate, but also the basis for prediction of future climates. The latter must now be seen not just as a scientific interest of climatologists, but as a vital concern for the future of the planet. The preceding discussion indicates that the system is understood in a qualitative way, but that a quantitative assessment, incorporating all of the potential processes and feedbacks, must at present be very tentative. Further, that tentative assessment is largely restricted to one climatic element, temperature. Increased quantitative understanding for all elements will come from the development of improved models of the system, from better observations of the current climate, and from refined analyses of past climates.

The climate models, called general circulation models, attempt to simulate the whole of the climate system.[8] The lack of complete physical understanding for many processes, combined with the lack of computer power needed to solve the governing equations at realistic temporal and spatial scales, prevents a complete system simulation, and many processes have to be simplified or approximated. The result is that the common outputs are monthly values averaged over grid-cells covering several degrees of latitude and longitude. Any system components and processes within the grid-cell, such as thunderstorms or hurricanes, cannot be explicitly simulated. However, the models allow explicit consideration of the influence of components covering more than one grid-cell. They provide, for example, the best information currently available on the climate which is likely to occur as a result of doubling atmospheric CO_2 concentrations.

Until relatively recently, our quantitative observation of the climate was confined to spot measurements at the earth's surface. Analysis of the resulting record led primarily to a description of the surface climate. The advent of satellite observations, and the resulting three dimensional data sets emphasizing energy flows, encouraged a more process-oriented viewpoint. The surface observational network will undoubtedly continue to be vital for specification of the present climate and its fluctuations and impacts, as well as for the validation of models run for current conditions. Continued observations from satellites will certainly provide increasing detail and insights concerning climate processes.

Past climates, revealed using proxy evidence involving sources as diverse as medieval monastic records, pollen grains in lake sediments, tree ring widths, or residual magnetism in ancient rocks, provide a means of specifying climate when conditions were very different from those of the present.[9] Establishment of the changes in the system which led to these past climate changes provide not only direct insight into the processes acting, but also assist in the validation of models when used to simulate the impact of system changes.[10]

The approaches available for enhancing our understanding of climate, therefore, are clear. Particular attention needs to be paid to the role of oceans in the climate system, the influence of cloud type and height on radiation exchanges, and the feedbacks within the system. Further, a major effort must be made to incorporate climatic elements in addition to temperature. Then we can use our knowledge of the climate system to understand and describe the full range of climate elements and their variability now and in the future.

REFERENCES

1. Sellers, W.D. 1965. *Physical climatology.* Univ. of Chicago Press, Chicago, IL, pp. 272.
2. Mather, J.R. 1978. *The climatic water budget in environmental analysis.* D.C. Health and Company, Lexington, MA, pp. 239.
3. Weubbles, D.L. and J. Edmunds. 1991. *Primer on greenhouse gases.* Lewis Publishers Inc., Boca Raton, FL, pp. 150.
4. Rasmusson, E.M. and J.M. Wallace. 1983. Meteorological aspects of the El Niño/Southern Oscillation. *Science* 222:1195-1202.
5. Kutzbach, J.E. and P.J. Guetter. 1986. The influence of changing orbital parameters and surface boundary conditions on climate simulations for the past 18,000 years. *J. Atmos. Sci.* 43:1726-1759.
6. Dickinson, R.E. (Ed.) 1987. *The geophysiology of Amazonia: Vegetation and Climate Interactions.* John Wiley & Sons, New York, NY, pp. 526.

7. Broecker, W.S. 1987. Unpleasant surprises in the greenhouse? *Nature* 328: 123-126.
8. Houghton, J.T., G.J. Jenkins and J.J. Ephraums. (Eds.) 1991. *Climate Change: The IPCC Scientific Assessment.* Cambridge University Press, New York, NY, pp. 403.
9. Hecht, A.D. (Ed.) 1985. *Paleoclimate analysis and modeling.* John Wiley, New York, NY, pp. 445.
10. COHMAP Members. 1988. Climatic changes of the last 18,000 years: Observations and model simulations. *Science* 241:1043-1052.
11. List, R. (Ed.) 1949. *Smithsonian Meteorological Tables.* Sixth Revised Edition. Smithsonian Institution, Washington, DC. pp. 527.
12. Henderson-Sellers, A. and P.J. Robinson. 1986. *Contemporary Climatology.* John Wiley, New York, pp. 439.
13. Critchfield, H.J. 1983. *General Climatology.* Prentice-Hall, Inc. Englewood Cliffs, NJ. pp. 453.

Global Climate Change: Implications, Challenges and Mitigation Measures. Edited by S.K. Majumdar, L.S. Kalkstein, B. Yarnal, E.W. Miller, and L.M. Rosenfeld. © 1992, The Pennsylvania Academy of Science.

Chapter Two

THE PAST AS A GUIDE TO THE FUTURE

JOHN E. OLIVER
Department of Geography and Geology
Indiana State University
Terre Haute, IN 47800

INTRODUCTION

The dictum "the past is a key to the present" is often and appropriately used to justify scientific inquiry into historic events. For the study of climate, which actually uses past events to establish the characteristics of the climate at a location, it is most apt. However, at the present there is both a need and a desire to look to future climates. It is thus interesting to speculate how climates past might serve as a key, or more realistically a guide, to what happens in the future. Clearly, if there are to be lessons learned from the past, then it is necessary to extend the view to the distant past, to look to the climates over the eons of geologic time.

To complete this task requires the assessment of what the climate of the earth was like, how has it changed over millions of years past and, significantly, how ancient climates may be deduced. Only by examining these is it possible to determine whether the past is, indeed, a guide to the future.

CLIMATE, FLUCTUATION, VARIABILITY, AND CHANGE

Like all areas of endeavor that encompass a number of disciplines, the terms that are used in the study of past climates have become somewhat confused and, in many instances, ambiguous. Before exploring the intricacies of climatic change, it is worth considering the meanings of some of the terms that are used.

Figure 1 shows the logical relationship between a number of terms used in dealing with climatic change. Below each term a graph (representing a variable that is continuous in time, e.g. temperature) indicates the nature of the change involved. The diagram shows that climatic change is an all-encompassing term that implies a number of changes regardless of the physical nature of the cause of the change. A *climatic trend* is a change characterized by smooth monotonic increase or decrease of values over a given time while a *climatic discontinuity* is a change that is abrupt or permanent during the period of record. *Climatic fluctuations* are treated either in terms of *climatic oscillation* or *climatic vacillation.* An oscillation comprises a fluctuation in which the variable tends to move gradually and smoothly between maxima and minima. This contrasts to a vacillation where the variable tends to dwell alternately around two (or more) mean values and to drift from one to the other at either regular or irregular intervals. A *climatic rhythm* can be applied to oscillations and vacillations as long as successive maxima and minima occur at about equal time periods. A *periodicity* describes the rhythm that has a constant or nearly constant time interval between maxima and minima.

Despite these accepted definitions, it is quite evident that the use of the general term climate change is time dependent. Changes are the result of forcing processes

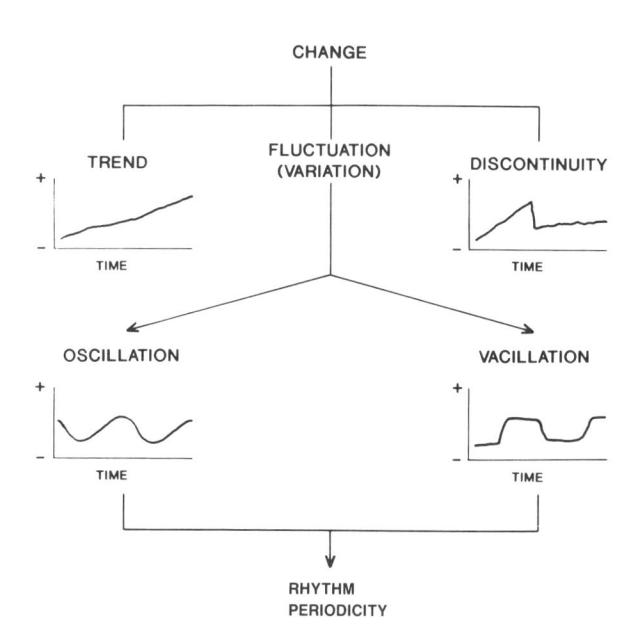

FIGURE 1. Relationships between some of the terms used to identify climatic changes and variations.

of the climate system, some are predictable and others not. For example, the annual variation of solar radiation gives rise to the seasons which, generally are predictable. Within the seasons, however, there are anomalies or small fluctuations from the seasonal conditions normally expected. In some cases these anomalies have extended disbenefits when droughts, floods and other hazards occur with undue frequency. The Asian, specifically the Indian Monsoon, is a recurring seasonal climatic phenomena which, when at variance with the "normal" monsoonal conditions, can cause widespread hardships. However, the failure of monsoon rains over a year, or even two years, is not described as a climatic change, although it may be part of a larger change that is obscured by the variability measures on the short-term basis.

Forcing mechanisms may also be episodic and influence climate for a given time then disappear, only to reappear at a later time. The well known effects of El Niño the oceanic component of the climatic southern oscillation is an apt example. El Niño is an irregular variation of an ocean current that, from January to March, flows off the west coast of South America, carrying warm, low-salinity, nutrient-poor water. Normally, it extends a few degrees south of the equator but, periodically, it penetrates to beyond 12°S to displace the cold Peru Current. If El Niño is sustained, the impacts range from a disastrous effect upon marine life to excessive rainfall and flooding in the normally dry coastal areas of South America.

The variability of climate over shorter time periods may or may not be associated with longer term climatic changes. To decipher whether the latter is taking place, the "noise" of the year to year variability must be masked. This is a very difficult task and one that makes forecasting the future very tenuous.

It is seen that the definition of climatic change is time dependent. Its identification need be separated from climatic noise created by anomalous weather conditions, short term fluctuations, and changes in variability. Climatic change is said to exist when a distinct signal can be easily separated from the noise of other variations in the existing climate. Many authorities reserve the term climatic change for those variations extending over centuries and millennia (Hare, 1985).

RECONSTRUCTING THE CLIMATIC RECORD

The reconstruction of climates over time is a fascinating puzzle. Instrumental records of meaningful spatial extent have become available only in very recent times and information about earlier climates require the use *proxy data*, observations of other variables that serve as a substitute or proxy for actual climatic records.

Proxies may be considered as paleoclimatological archives which provide indirect evidence of climatic trends over the eons of geologic time. The earliest scientific hints that the climates of the past differed from those of today was derived by geologists working with glaciers in Europe. They discovered that evidence of ice activity, such as moraines, till, and glacial erratics, could be explained only by assuming much more extensive ice cover at some time past. Evidence of glacial activity is still used to reconstruct climates, but it is now enhanced by many other interpretive methods.

Perhaps the most impressive record of past climates has been derived from drilling cores, some through ocean sediments, some through ice caps, and some through ancient lakes and bogs. Such cores sometimes provide a continuous and lengthy record of earth history and their contents can be analyzed in a variety of ways.

Evidence from oceanic cores may be obtained in a number of ways. Tiny primitive creatures called foraminifera, or forams, float at the surface of the world oceans. Some do well in warm water, others thrive in cold. When these forams die, their calcareous shells rain down to the ocean floor and settle layer upon layer. By taking a core through the seabed, which may have been undisturbed for millions of years, the types of forams that lived over a sequence of time can be identified. Some indication of the surface climate in which they lived may be derived by the preponderance of a given warm or cold water type. These clues are greatly aided by a process that compares the oxygen isotope content of the fossils. Forams extract oxygen from the ocean water and the ratio of oxygen 18, a heavy oxygen isotope, to the lighter oxygen 16, is a function of the surface temperature of the seawater. By comparing the ratios of one isotope to another in the ancient foram, a sequence of temperatures may be derived. Actually, the amount of a given isotope depends in part upon the amount of water tied in the major ice sheets. During glacial times, the lighter isotope is evaporated and tied up in ice, leaving more of the heavier in the oceans. This fact means that the oxygen 18/oxygen 16 ratios can also be studied in layers of glacial ice.

As the ocean becomes enriched in ocean waters, atmospheric precipitation will contain a higher proportion of the lighter oxygen 16. This is deposited as snow and eventually as ice in the ice formations. By examining cores taken through Greenland and Antarctic ice, the oxygen isotope ratios can be established and a chronology of temperature changes deciphered.

In 1916 Lennart von Post, a Swedish botanist, showed how paleoclimates could be interpreted by looking at pollen grains buried in lakes and bogs. Pollen is one of the most durable organic substances produced and does not easily decay even in acid swamps. Because pollen can be related to given plant species, it is possible to determine what type of plant was most prevalent when the pollen was produced and settled into the original body of water. Because different plants are identified with different climates, it becomes possible to identify the existing climate.

These examples of proxy sources are used with others (Table 1) not only to provide a sequence of events at the location where they are observed, but for more extensive use. By correlating the established chronology from, say, oxygen isotopes in Greenland, with the records derived from the extent of drift ice in the North Atlantic, the derived results can enhance the findings and be used to interpret what may have happened over large areas. Standard chronologies can thus be established for selected time periods (Fairbridge, 1987). However, to complete a chronology of climatic change over time requires not only reconstruction of that climate but a method whereby the reconstructed climate can be dated.

Relative dating, the age of one event to that of another in terms of before or after, has long been used. Relying upon the principle of superposition, geologists are able to determine relative ages of rock formations even in the most complex of geologic settings. Absolute dating is of more recent origin and provides actual, although sometimes approximate, dates and times when a given event took place. Of major importance is the use of isotopic dating. Some isotope elements are unstable and a small part of the nucleus may fly off, ultimately leading to the formation of a new element. The decay is measured in half-life. For example if we begin with 1 gram of Uranium 238, an element that decays to lead 206, after 1 half-life 0.5 grams remain. The decay represents the exponential decrease of the number of atoms of the parent material. Examples of some important isotopic age determination processes are shown in Table 2.

TABLE 1

Some Proxy Data Sources

Historical sources especially since 1000 A.D.
    Diaries, civil records, art, etc.

Faunal and floral information
    Terrestrial plants in fossil form
    Large reptiles, probably indicating warmth
    Corals, indicating warmth
    Giant shelled mollusca, indicating warmth
    Coal, indicating humidity
    Peat remains, indicating vegetation types
    Pollen analysis of sediments, indicating vegetation types
    Foraminifera in deep sea cores, temperature variation
    Tree rings, indicating warmth and moisture

Sediments and stratigraphic information
    Ironstones and bauxites, indicating tropical moisture
    Red beds, indicating tropical moisture and aridity
    Evaporites, indicating dry conditions
    Eolian sandstones, indicating desert conditions/loess
    Limestones, thick, pure sequences indicting warm seas
    Glacial material indicating cold conditions
    Iceberg rafted debris in deep sea cores, indicating cold conditions
    Oxygen isotope composition of fossils, indicating temperature
    Chemical composition of lake cores, indicating precipitation levels
    Oxygen isotopic and chemical composition of deep ice cores, indicating temperature
    Fossil soils

Geomorphological information
    Ice wedge polygons, etc., indicating permafrost
    Cirques, indicating snow line positions
    Closed lake basin shorelines, indicating precipitation
    Fossil dunes, indicating aridity and wind direction
    Caves, alternations of solution indicating moisture and precipitation indicating aridity
    Screes, indicating frost action
    Misfit meanders, indicating higher stream discharge levels
    Eolian erosion, indicating aridity, deflation, and wind direction

After Goudie (1987)

Apart from isotopic dating, other methods for absolute dating exist. Biological methods use growth of living things to decipher time with the study of tree rings (dendrochronology) probably the best known type. The analysis of tree rings itself provides important information about the environmental conditions under which the tree grew and is used to reconstruct past climates (Fritts, 1976). The study of lichens (lichenology) is becoming important. The method depends upon the success of lichen in growing onto fresh surfaces or boulders and rock surfaces. By establishing a growth curve for the slow growing lichen, it is possible to provide a date to the time that it first attached itself to, for example, a glacial boulder.

Paleomagnetic methods are based upon the known times of reversal of the earths magnetic field, the last reversal occurring some 730,000 years ago. By measuring the orientation of sediments influenced by magnetism, it is possible to identify sequences of deposition related to the reversal. Chemical methods require information concerning chemical change for either organic (e.g. amino-acid dating) or inorganic (e.g. tephra) samples. Proxy methods in themselves are intriguing areas of study and illustrate the great diversity of ways in which past climates may be deciphered (Bradley, 1985).

CLIMATE OVER GEOLOGIC TIME

Table 3 shows the geologic time divisions, naming the eras and periods, and identifies times at which the earth was gripped by ice ages. Perhaps the most significant idea expressed in the table is the fact that average global climates have been much warmer than they are today throughout much of geologic time. Periodically, the warmth has been interrupted by times of cooling, ice ages (Frakes, 1979).

Rocks of the Precambrian period, extending back to the origin of the earth, provide but few details of the climates that existed, and only the late Precambrian times

TABLE 2

Isotopic Age Determination[1]

Isotypes	Half Life/Dating Range[2]	Examples of Materials Dated
Uranium-238 /Lead-206	$4.50 \times 10^9 / 10^7$ to T_0	Zircon, Uraninite Pitchblende
Uranium-235 /Lead-207	$0.71 \times 10^9 / 10^7$ to T_0	Zircon, Uraninite Pitchblende
Potassium-40 /Argon-40	$1.30 \times 10^9 / 10^4$ to T_0	Muscovite, Biotite, Hornblende, Arkose, Siltstone
Rubidium-87 /Strontium-87	$4.7 \times 10^{10} / 10^7$ to T_0	Muscovite, Biotite, Microcline, Whole Metamorphic rock
Carbon-14	$5730 \pm 30/0$ to 50,000	Wood, Charcoal, peat, grain, tissue, bone, cloth, shells, water

[1]Thermoluminescence and fission-track dating are not listed.
[2]Time in years where T_0 = origin of earth.

can be reconstructed with any degree of confidence. It is known, however, that during late Precambrian much of the earth was glaciated, perhaps for the time span 950 million years before present (mbp) to 650 mbp. The Paleozoic (570-225 mbp) represents a time when fossil and sedimentary evidence became more widespread and more details of the Paleozoic environments may be derived. Organisms and rocks suggest that the Cambrian and early Ordovician were largely warm, although there may have been a glaciation toward the end of the Ordovician. The withdrawal of ice and deglaciation at the end of this event made the climates of the Silurian and Devonian similar to those of today. The Pennsylvanian/Mississippian periods were dominated by widespread humid climates with intermittent glaciation. Considerable evidence exists to indicate a major glaciation in the late Paleozoic.

The Mesozoic (225-65 mbp) was essentially a time of widespread warmth and aridity. Climates of the Triassic appear similar to those of the upper Paleozoic, cool and humid, but they gave way to a long period of warmth, especially marked during the Cretaceous. According to Fischer (1983), this time of earth history saw the world in its 'greenhouse mode' when climate was predominantly warm, polar ice caps nonexistent and sea level high. The change from this to an eventual "icehouse mode" may not have been smooth but rather episodic. The end of the Cretaceous, the so-called K/T boundary, is marked by mass-extinctions of selected life forms. Such episodes have become a center of scientific controversy. A debate over whether the extinctions were caused by earth changes, climatic change, or impacts on earth of extraterrestrial bodies continues.

The early Tertiary (65-22.5 mbp) is a time when the warm temperatures of the Mesozoic began to decline. Long episodes of relatively warm climates were punctuated by abrupt drops in temperature. During this time the first glaciers since the Paleozoic began to form in Antarctica. By the latter Tertiary (22.5-2 mbp) wide swings of temperatures occurred until in the Pliocene rapid alternations of warm and cold times, such as those associated with the Pleistocene are identified.

TABLE 3

ERA	PERIOD	BEGINNING (MBP)*	ICE AGES
Cenozoic	Quaternary	2-3	Pleistocene Ice Age
	Tertiary	65	
Mesozoic	Cretaceous	135	
	Jurassic	190	
	Triassic	225	
Paleozoic	Permian	280	
	Carboniferous	345	Ice Age at approx. 300 MBP
	Devonian	400	
	Silurian	440	
	Ordovician	500	Ice Age at approx. 450-430 MBP
	Cambrian	570	
Precambrian		>570	Ice Age at approx. 850-600 MBP

MBP = Millions of Years Before Present

The climates of the Pleistocene consisted of glacial and interglacial times, with polar ice advancing and retreating from numerous sources. The most recent full glacial, known as the Wisconsin in North America lasted from perhaps 30,000 to 12,000 years BP, with coldest temperature (4 to 6°C lower than present) occurring about 18,000 years ago. For such recent times much evidence is available and it is worth considering each separately. Perhaps, given that the objective is to gain information about climatic change in the future, it is meaningful to dwell upon selected time periods during this time. These slices through time provide views of the changing climates.

CLIMATE SINCE THE ICE RETREAT

The time 18,000 to 5,500 years ago corresponds to the deglaciation of the earth (See Figure 2). The northern hemisphere was deep in a full glacial time 18,000 years ago. Huge ice sheets over North America and Scandinavia extended as far equatorwards as 50°N in Europe and 40° in North America. Frigid polar water extended in the North Atlantic to 45°N. After this full glacial, there is clear indiction that deglaciation began and, by 12,000 years ago, only scattered remains of ice sheets were found in western North America, with the main ice sheet confined to eastern Canada.

About 10,200 years ago, a strange event occurred, especially in Scandinavia and Scotland. The margins of the remaining ice sheet expanded and some small ice sheets reappeared. This time, known as the Younger Dryas (named for a small flower found in cold climates), did not last long and shortly after, the northern hemisphere was approaching conditions of the present-day climate. But the rapidity of temperature decline of the Younger Dryas illustrates that climatic change need not be the long, slow non-impulsive change that it is generally thought to be. Such rapid changes in

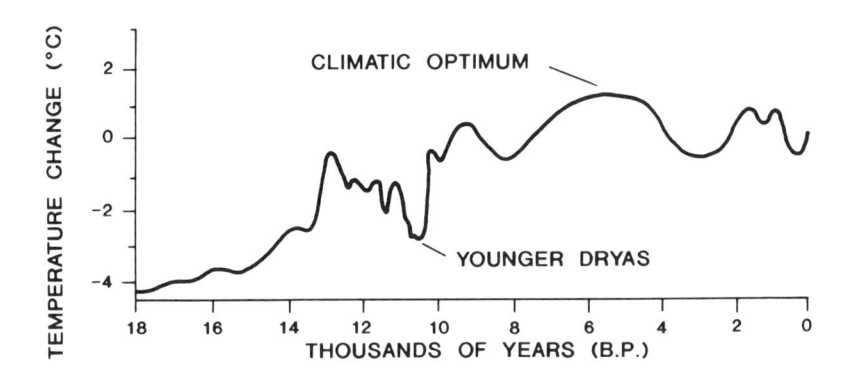

FIGURE 2. Patterns of surface temperatures of the last 18,000 years based largely upon Greenland oxygen isotope temperatures. Vertical axis provides temperature change from current average global temperature. (After various sources including Houghton, *et al.* 1990).

climates are obviously the center of research in climatic change for it illustrates that abrupt climatic change does occur.

By 7,000 years ago conditions had improved such that only remnants of ice remained and this, by 5,500 years ago had completely disappeared leaving only the Greenland Ice Sheet and the Arctic Ice that we have today. Temperatures were warmer than they are today.

After the cooling associated with the Younger Dryas, global climate warmed and reached a maximum at about 5,500 years BP. All evidence points to this being a time when mean atmospheric temperature of middle latitudes was about 2.5°C (4.5°F) above that of the present. This time has been described as the "Climatic Optimum," a term originally applied to Scandinavia when warmth would be highly favorable for varied flora and fauna. It might not be a good term for other world area whose climate may be far from optimum. This corresponds to early Neolithic times in Europe and the Early Kingdom cultures in Egypt. After this time, the temperature in middle latitudes varied from cool to warmer, but did not approach the levels of the Optimum until about 1,000 years ago.

THE LAST 1,000 YEARS

The time extending from about 950-1250 AD is known as the Little Climatic Optimum (Figure 3). Prevailing warm conditions in middle latitudes coincided with the remarkable tales of exploration and discovery by Vikings of Iceland and Greenland and in many facets of life in Europe. Compared to today, major differences are seen in agriculture with the best known being the widespread growth of vines, with wine production even in England (Lamb, 1982). But these conditions did not extend into the Middle Ages and between 1250 and 1450 AD climate deteriorated over wide areas. Iceland's population declined and grains, which had been grown in the 10th century, were no longer produced. The Viking settlements

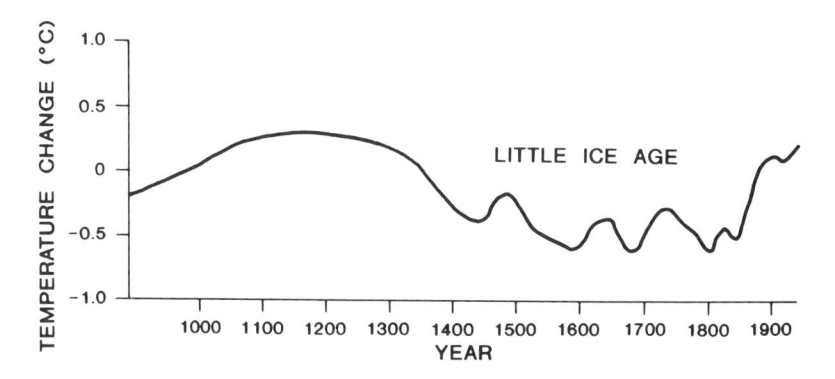

FIGURE 3. Variations from current average surface temperatures over the last 1,000 years. The Little Climatic Optimum was followed by a deterioration to the Little Ice Age. (Same source as Figure 2).

in Greenland were abandoned, while in Europe storminess resulted in the formation of the Zuider Zee and the excessively wet, damp conditions led to a high incidence of the horrifying disease St. Anthony's Fire (ergotism).

Conditions, however, got even worse and from about 1450 to 1880 AD the time known as the Little Ice Age occurred. The record of this has been well documented, and both physical and historical evidence are available (Grove, 1988; Wigley, 1985). Fortunately, by the end of the 19th century, the climate was again improving, a fact attested in the instrumental record.

It would seem given the availability of actual instrumental data that few problems would occur in reconstructing world climate over the last century. But reconstructing the average temperature of the earth's surface is not simply a matter of integrating the trends available from all existing temperature records. Some of the records are unreliable, for any number of reasons, while the distribution of recording stations is unevenly distributed throughout the world. In relation to unreliability, perhaps the most frequently cited problem is that many of the recorded data are perturbed by the urban heat island effect. In terms of distribution, industrialized nations with dense populations are more highly represented than remote land and great expanses of ocean. It is also necessary to note that within the records are seasonal and interannual variations that may obscure the presences of a trend in the data.

The reconstructed data are shown in Figure 4. As can be seen, in comparison to the 1951-80 mean (the period value on which the deviations are based), years prior showed a warming trend, those after, the 1980s, were much warmer. It is this latter trend that raises the specter of global warming.

While much of the speculation concerning future climates is based upon temperature, a consequence of the change is modification of circulation patterns and, eventually, precipitation amounts and distribution. Of course, the perception of rainfall has been greatly influenced by the widely reported droughts of the Sahel, Ethiopia and India. However, a number of studies of precipitation (e.g. Bradley

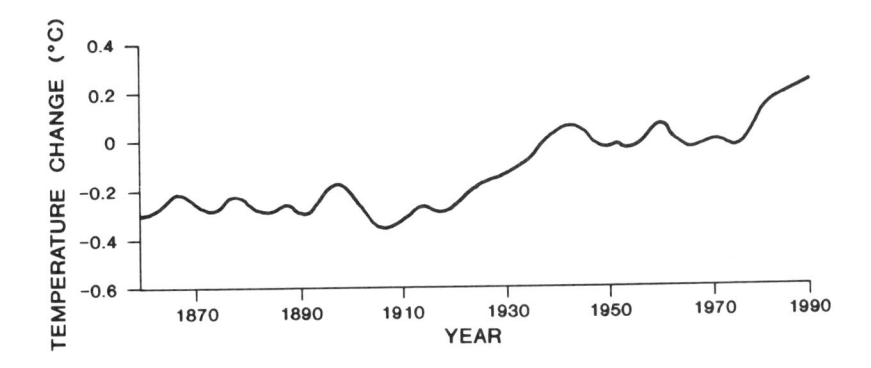

FIGURE 4. Globally averaged temperatures since about 1860 shown as a change from the 1951-80 average. (Same source as Figure 2).

et al., 1987) have shown that during the last few decades precipitation has (1) increased in mid-latitudes, (2) decreased in the northern hemisphere subtropics and (3) generally increased in the southern hemisphere. These findings, from land-based stations, must be viewed against the problems involved in analyzing precipitation data. Problems of the density of gauges, the measurement of snow, and the lack of instrument uniformity all contribute toward obtaining an accurate picture of precipitation distribution.

LEARNING FROM THE PAST

Given the reconstruction of climate over vast amounts of time, it becomes necessary to assess just what lessons can be learned from that history. Clearly, the most important point to emerge is that climate does change and that the rate at which the change occurs passes from millions of years, as seen in the Mesozoic, to very short time periods demonstrated in the evidence of the Younger Dryas.

Clearly though, many more lessons than this may be learned. Perhaps of importance is the identification of earlier times when the "greenhouse" climate prevailed. From an identified period, a guide to the environmental conditions that prevailed at those times could prove beneficial in assessing potential future "greenhouse" conditions.

Folland *et al.* (1990) suggest that there are three analogues of warmer climates. They identify 1) the Pliocene warm climate that occurred about 3.3 to 4.3 million years ago; this warm period was part of the wide swings of warm and cool climates that occurred during the Pliocene. 2) a Pleistocene interglacial, the Eemian, 125,000 to 130,000 BP, identified in Antarctic cores, experienced very warm conditions compared to other interglacials and 3) the climatic optimum that ocurred at 5,000 BP. Of these, the Holocene climatic optimum has been most closely examined.

This warm time saw summer temperatures in high latitudes some 3° to 4°C above modern values and increased precipitation in subtropical and high latitudes. COHMAP (1988) suggests that during this climatic optimum "In middle latitudes, summer temperatures were only 1-2°C higher and further south summer temperatures were often lower than today, for example in Soviet Central Asia, the Sahara, and Arabia. These areas also had increased annual precipitation." While there are uncertainties in this reconstruction, it does provide a relative guide to warmer conditions. The two older warm times, the Eemian and the Pliocene optimums have less evidence available and numerous uncertainties.

One important source of evidence of past warm and cold periods is that levels of atmospheric carbon dioxide can be obtained from ice cores. The analysis of air trapped in the layers of ice in the Antarctic Ice Cap allows such gases as carbon dioxide and methane to be measured. Recall that this ice can also be monitored for oxygen ratios that provide temperatures. A correlation between the two shows that ice core temperatures are closely related to the measured carbon dioxide present. Similarly, the relationship between temperature and carbon dioxide can be assessed using evidence of past vegetation over the globe. While the relationships are less exact, there is evidence that clearly relates the two.

One final lesson learned from the study of past climates is the impact that climatic change has upon the lives and activities of people. The very different accounts of the lives of people in Western Europe during the Medieval warm times and those of the Little Ice Age clearly illustrate this. Our ancestors experienced climate changes, and enough can be learned from the evidence they left to make us concerned about the future.

Whether or not this evidence is of value in looking at the potential warming forecast for the coming century is moot. Certainly, there is good agreement between atmospheric climatic models and inferred climatic change over the last 20,000 years (Climap, Cohmap), but according to Crowley (1989) there is no really good analogy to what is currently happening. The Pre-Pleistocene experiences, the Pliocene warming for example, have not been modelled in sufficient detail to ascertain their potential utility in assessing future climates. Clearly, a great deal of research remains to be completed.

REFERENCES CITED

Bradley, R.S. 1985. *Quaternary Paleoclimatology*, Allen and Unwin, Boston.

Bradley, R.S., Diaz, H.F., Eischied, J.K., Jones, P.D., Kelly, P.M. and C.M. Goodess. 1987. Precipitation fluctuations over the Northern Hemisphere land areas since the mid-19th century. *Science*, 237:171-175.

CLIMAP Project Members. 1976. The surface of the ice age Earth. *Science*, 191:1131-1136.

COHMAP Members. 1988. Climatic changes of the last 18,000 years: Observations and model simulation. *Science*, 241:1043-1052.

Crowley, T.J. 1989. Paleoclimate perspectives on greenhouse warming, pp. 179-210. In: A Berger, S. Schnieder and J.Cl. Duplessy (Ed.) *Climate and Geo-Sciences*. NATO ASI Series, Kluwer Academic Publishers, Dordrecht, pp. 724.

Fairbridge, R.W. 1987. Climatic variation: Geologic record. In: Oliver, J.E. and Fairbridge, R.W. (Ed.) *The Encyclopedia of Climatology*. Van Nostrand Reinhold, New York.

Folland, C.K. Karl, T. and Vinnikov, K.Y.A. 1990. Observed climate variations and change, pp. 195-238. In: J.T. Houghton, G.J. Jenkins and J.J. Ephraums (Ed.) *Climate Change: The IPCC Scientific Assessment*. Cambridge University Press, New York.

Frakes, L.A. 1979. *Climates Through Geologic Time*. Elsevier, New York, pp. 310.

Goudie, A.S. 1987. Paleoclimatology, pp. 660-668. In: J.E. Oliver and R.W. Fairbridge (Ed.) *The Encyclopedia of Climatology*. Van Nostrand Reinhold, New York.

Grove, J.M. 1988. *The Little Ice Age*, Methuen, London, pp. 498.

Hare, F.K. 1985. Climatic Variability and change, pp. 37-68. In: R.W. Kates, J.H. Ausubel and M. Berberian (Ed.) *Climate Impact Assessment*. SCOPE 27, John Wiley, New York, pp. 625.

Holcombe, T.L. 1989. Paleoclimate data for studies of global climate change and earth system science, pp. 173-178. In: A. Berger, S. Schneider and J.Cl. Duplessy (Ed.) *Climate and Geo-Sciences*. NATO ASI Series, Kluwer Academic Publishers, Dordrecht, pp. 724.

Houghton, J.T. Jenkins, G.F. and J.J. Ephraums. 1990. *Climatic Change: The IPCC Scientific Assessment*. Cambridge University Press, New York.

Quinn, W.H., Neal, V.T. and S.E. Antunez de Mayola. 1987. El Niño occurrences over the past four and a half centuries. *J. Geophys. Research*, 92:14,449-14,462.

Wigley, T.M.L., Huckstep, N.J., Ogilvie, A.E.J., Farmer, G. Mortimer, R. and Ingram, M.J. 1985. Historical Climate Impact Assessments. In: R.W. Kates, J.H. Ausubel and M. Berberian (Ed.) *Climate Impact Assessment*. SCOPE 27, John Wiley, New York, pp. 625.

Global Climate Change: Implications, Challenges and Mitigation Measures. Edited by S.K. Majumdar, L.S. Kalkstein, B. Yarnal, E.W. Miller, and L.M. Rosenfeld. © 1992, The Pennsylvania Academy of Science.

Chapter Three

SHORT-TERM CLIMATIC VARIABILITY

BRENT YARNAL
Department of Geography
and
Earth System Science Center
The Pennsylvania State University
University Park, PA 16802

INTRODUCTION

Global-scale climatic variations are never uniform over the earth's surface. For instance, when the average planetary temperature rises, some regions of the planet warm, others cool, and still others remain the same. Even the way in which individual regions warm can have quite different patterns of variation: one area might warm steadily over time; another might alternate equally between relatively warm and cool years, with a long-term trend toward warming; and a third might cool in all but the occasional devastating hot year, in which heating is so great that it swamps the accumulated effects of the cooler years.

The exact pattern of climatic variation over time and space is crucial to the biophysical environment and society. For example, each of the three hypothetical warming scenarios given above might have different economic impact on a particular agricultural region. If the main crop grown in that region thrives on warmth, but cannot stand excess heat, then (assuming moisture is always adequate): constant warming might steadily increase crop yields; alternating warm and cool conditions might cause alternating good and poor yields; and cooling with the occasional very hot year might produce a succession of poor yields punctuated by the isolated failed harvest. With these respective warming scenarios, therefore, the regional economy could be: 1) prosperous; 2) unstable, with economically-weaker enterprises failing to survive over the long haul; or 3) steadily declining, with the worst years eliminating all but the best or richest farmers.

The predicted sharp rise in surface temperature caused by increasing greenhouse-gas concentrations is an example of a global-scale climatic variation with regional implications. Most scientists agree that, for a doubling of atmospheric carbon dioxide concentrations, the surface warming will range from about 2 to 5°C. However, that single planetary temperature hides the spatial and temporal variation integral to climate. As shown above, the specific impacts of global warming on each region's biophysical and social systems will depend on how each region's climate responds to the global temperature rise. Although climatologists have devised several techniques to develop likely regional scenarios for the global warming (e.g., Lamb, 1987), nobody really knows the true climatic outcome for any single region. Accurately predicting the regional response to global warming is perhaps the most important challenge of climatology today.

Thus, to appreciate the biophysical and human implications of the global warming, it is necessary to understand climatic variation over time and space. Although future climate may be unlike anything we have experienced so far, it is possible that the climatic patterns of a warmer world will be similar to those of today. The aim of this chapter is to introduce the main ways in which present-day climate varies, providing a foundation on which to base an understanding of future climatic variation. Because atmospheric carbon dioxide concentrations are likely to double within the next century, it is appropriate to discuss only the shorter time scales of climatic variability; i.e., within-year, year-to-year and decade-to-decade variations. Climatic variations of centuries and longer may be important if the greenhouse-forced warming continues for hundreds to thousands of years, but the immediate concern is with the next few generations.

WITHIN-YEAR CLIMATIC VARIATIONS

Most regions can suffer considerable variation from month to month and season to season. Persistent climatic regimes, which can bring weeks to months of blistering heat, biting cold, drenching rains, or other weather-related miseries to a region, are one example of within-year variability (Namias, 1986). Curiously, persistent climate patterns often end abruptly; sometimes they can even be replaced by a strongly contrasting climatic regime (Namias, 1988a, 1988b, 1990). Predicting persistence, abrupt regime breaks, and transitions from one persistent regime to another present a formidable challenge to long-range weather forecasters.

Persistent climatic regimes can have huge impacts on society. In the eastern United States, for example, the persistent regime of December 1989 produced the coldest December on record. Amazingly, it was followed by an equally-persistent regime responsible for the warmest January-February ever. Energy stores dwindled rapidly in December and home heating costs for that month skyrocketed because of heavy demand. Heating bills in January and February were higher than they should have been because, despite the relatively low demand associated with the record warmth, the prices reflected the need to restock inventories depleted by the December cold wave.

Blocking in the extratropical westerlies is a special type of persistent weather regime. Blocking occurs when the normal westerly progression of long waves stalls and a stationary, exaggerated north-south flow develops over a large sector of the hemisphere. These events can last for periods of several days to months. Regional climatic anomalies are a notable part of blocking, with separate areas of extreme positive and negative temperature and precipitation departures occurring with each block (Rex, 1950 and 1951). Indeed, at monthly time scales, most record-setting climatic events for extratropical regions result from blocking activity. For example, two extended blocking events developed in May and June 1989, with an upper-level low-pressure area settling over the eastern United States. These blocks brought record-setting rainfall to this part of the country, saturating soils dried by the unprecedented drought of summer 1988, raising ground-water levels from all-time low levels to above normal, and nullifying other residual effects of the drought. Scientists dispute the causes of blocking (Held, 1983), but it appears that the presence of mountains and sea-surface-temperature anomalies greatly enhance the blocking process (Mullen, 1989a and 1989b).

In the tropics, the so-called 40-50 day oscillation often controls regional within-year climatic variability (Madden and Julian, 1971, 1972). This disturbance propagates from west to east and circumnavigates the globe every month or two. The 40-50 day oscillation enhances or suppresses convection and subsidence at the semi-permanent centers of equatorial climatic activity (i.e., the active convective centers over the Amazon, Congo and Indonesia, and the focus of strong subsidence over the eastern equatorial Pacific). The oscillation affects the onset, breaks and withdrawal of the monsoon systems of Asia, and is therefore critical to agriculture in these regions (Fein and Stephens, 1987). Krishnamurti and Gadgil (1985) have found evidence for 40-50 day periodicity in the extratropics, making this a global phenomenon, but one that is better expressed in tropical regions. The cause of the oscillation is poorly understood.

YEAR-TO-YEAR CLIMATIC VARIATIONS

The 2.2-year Quasi-Biennial Oscillation (QBO) is a reversal of winds that takes place in the stratosphere over the tropics and over the polar regions. Its signal is easily identified in many time-series analyses, such as those for Indian monsoon rainfall (Mooley and Parthasarathy, 1984). The phase of the QBO is a controlling factor of the number of tropical storms and hurricanes occurring in the tropical Atlantic (Gray, 1984). Recent work suggests that it has considerable impact on the global effects of the El Niño-Southern Oscillation (Barnston and Livezey, 1991; Barnston et al., 1991). The QBO's phase also appears to control the influence that solar cycles have on climate (see next section). Despite these important statistical relationships and the ubiquity of the QBO in the climate record, the reasons for the oscillation are poorly understood, as are the mechanisms that link the stratosphere with surface climate (Holton and Tan, 1980).

Teleconnections (i.e., correlations between climate variables separated by great distances) are another quasi-periodic form of climatic variation (Glantz et al., 1991). They occur because the earth's surface features, such as mountain chains, land-sea boundaries, and areas of strong surface heating, loosely anchor the general circulation of the atmosphere. Many teleconnections have been identified, but the three most important are the Southern Oscillation, the Pacific/North American pattern, and the North Atlantic Oscillation.

The Southern Oscillation (SO) is the surface-pressure component of the coupled ocean-atmosphere system of the tropical Indo-Pacific basin (Yarnal and Kiladis, 1985). It involves the inverse relationship of surface pressure over the *maritime continent* (i.e., the Indonesian archipelago, northern Australia and southeast Asia) and over the eastern tropical Pacific Ocean (centered near Tahiti). When the atmospheric circulation is strong, pressure is above average over Tahiti and below average over the maritime continent. This accompanies stronger surface easterlies over the equatorial Pacific, a stronger upper-tropospheric westerly jet stream over this area, enhanced convection over the maritime continent, and cooling in the eastern equatorial Pacific Ocean. Scientists call this oceanic cooling La Nina (Philander, 1990). Similarly, a weakening of the tropical circulation produces the following conditions: above-normal surface pressure over the maritime continent and below-normal pressure over the eastern tropical Pacific; a weakening of the Pacific trade winds and a decrease in the strength of the westerly jet stream; an eastward shift of the heaviest convection away from the maritime continent to the vicinity of the International Date Line; and a strong warming of the central and eastern equatorial Pacific Ocean. This oceanic warming is the infamous El Niño (Philander, 1990).

The coupled ocean-atmosphere system of the tropics, known as El Niño-Southern Oscillation (ENSO), fluctuates between El Niño and La Nina extremes every three to five years on average. Each extreme lasts about 18 months. Next to the seasonal cycle, the most important source of variation in the global climatic system is ENSO. The El Niño phase of ENSO coincides with below-normal rainfall (and often drought) in Hawaii, Australia, the maritime continent, India, southeastern Africa, and northeastern South America and adjacent areas of the Caribbean (Ropelewski and Halpert, 1987). At the same time, coastal Ecuador and Peru, Uruguay and northeastern Argentina experience above-normal rainfall. During the less-known La Nina phase of ENSO, areas that experience drought during El Niño have strong positive rainfall anomalies and flooding, while those areas that are wet during El Niño become abnormally dry during La Nina (Ropelewski and Halpert, 1989). Thus, tropical teleconnections have opposite regional climatic effects during the two extremes of ENSO.

The ENSO phenomenon is linked to extratropical climatic anomalies (Yarnal, 1985). Simple theory (Tribbia, 1991) and simulation modeling (e.g., Palmer and Mansfield, 1986) suggest that during El Niño winters, a Rossby wave-train should emanate from the equatorial Pacific and sweep in an arc over North America. However, observation demonstrates that this teleconnection is present in only about half of the El Niño winters (Yarnal and Diaz, 1986). Curiously, although neither theory nor simulation modeling predicts extratropical teleconnections with La Nina events, half of all La Nina winters have such teleconnections (Yarnal and Diaz, 1986). Recent theoretical and modeling work has not reconciled these inconsistencies. Because of scarce data, scientists know less about Southern Hemisphere teleconnections with ENSO (Karoly, 1989).

The Pacific/North American (PNA) pattern is an extratropical teleconnection covering much of the North Pacific Ocean and the North American continent (Wallace and Gutzler, 1981). The phase of the PNA pattern correlates strongly with temperature and precipitation fields over North America (Leathers et al., 1991). Furthermore, one phase of the PNA teleconnection is significantly associated with El Niño winters, while the other extreme phase is related to La Nina winters, suggesting forcing of the pattern by the tropics (Yarnal and Diaz, 1986). However, because this relationship breaks down in roughly one half of all ENSO winter months, long-range forecasters cannot use the tropical-extratropical association for predictive purposes. The PNA pattern appears to be the response of the atmospheric circulation to Northern Hemisphere geography, especially the shape and size of the North Pacific basin, North America, the Tibetan Plateau and the Rocky Mountains. Over North America, the two extreme phases of the PNA pattern represent the variations of the Polar Front jet stream around its average configuration. The extremes range from a large ridge over the western Cordillera and a deep trough over the eastern United States, to a dampening of this ridge-trough system (Yarnal and Leathers, 1988). The PNA pattern has a broad spectral peak of 38 to 40 months. This quasi-periodicity is associated with major elements of the global climate system, including ENSO variations, sea surface temperatures in the North Pacific, and climatic conditions over the Eurasian landmass (Leathers and Palecki, 1992).

The North Atlantic Oscillation (NAO) is a north-south dipole occurring in winter, centered over the North Atlantic Ocean, and extending over adjacent landmasses. One phase of the teleconnection represents coincident intensification of the Azores high and the Icelandic low, and the other denotes the simultaneous weakening of these two surface-pressure features (Rogers, 1984). The former phase of the NAO possesses intense westerly jet-stream flow and vigorous storm systems over the basin; the latter displays the weak southerly or northerly flow and has frequent blocking. As a result of these extremes, the intense-flow phase brings stormy, mild conditions to western Europe, while the weak-flow phase often generates prolonged periods of below-normal temperatures and above-normal snowfall. The NAO produces climate anomalies in other areas surrounding the North Atlantic basin, like those in northwestern Africa (Lamb and Peppler, 1987) and the eastern United States (Yarnal and Leathers, 1988). The cause of the NAO is not well understood.

Teleconnections are of special interest because they are global-scale climatic phenomena with regional impacts, and because they are products of the interaction between the dynamic atmosphere and global geography. The solid earth and large-scale oceanic features will not change appreciably with global warming. Therefore, it is possible that future teleconnections will resemble their present-day counterparts, although certain extreme phases may dominate over others.

DECADE-TO-DECADE CLIMATIC VARIATIONS

For nearly a century, scientists have recognized a significant statistical correlation between the periodic variation of many climate variables and either the 11-year sunspot cycle or the 22-year double-sunspot cycle. These correlations are intuitively appealing: if the sun is the engine that powers the climate system, then fluctuations in solar output (as denoted by sunspot activity) should cause variations in climate. However, until the mid-1980s few of the empirical relationships have maintained their significance over long data records. Worse, no mechanisms adequately explain the physical processes linking climate to solar variability (National Research Council, 1982). However, scientists have since uncovered strong empirical evidence for the sunspot-climate connection (Labitzke and van Loon, 1988, 1989; van Loon and Labitzke, 1988). Their work suggests that the solar cycle affects regional climates differently during each phase of the QBO. This statistical association is so strong, ubiquitous and consistent that the United States National Weather Service now uses it as an important factor in making its long-range climate forecasts (Livezey and Barnston, 1988; Barnston and Livezey, 1989). Unfortunately, physical links between solar cycles and climate are still missing.

Other climatic variations on decadal time scales are important. For instance, from the 1920s until mid-century, strong east-west (zonal) flow dominated the extratropical atmospheric circulation. Since then, the extratropical flow has been more north-south (meridional). Each of these decadal-scale flow regimes relates to distinct distributions of heating and moisture. During the zonal period the eastern United States was dramatically warmer than the long-term mean; since the onset of meriodional flow the East has been much cooler (Diaz and Quayle, 1980). This is easy to explain: with zonal flow the Polar Front jet stream is well to the north of its average position, placing the region in maritime-tropical air more often; with meridional circulation the region is usually under a trough of cooler continental-polar air (Yarnal and Leathers, 1988).

On decadal time scales, precipitation is more sensitive than temperature to atmospheric-circulation variations. The climate of the northeastern United States in the 1960s and 1970s illustrates this point. During the 1960s a deep trough, with its dry, cold polar air mass, frequently sat over the region. Because the average location of the storm track associated with this trough shifted southeast of its normal position, many storms dropped their moisture over the Atlantic Ocean instead of the Northeast. This combination of dry air and diverted storm track provided little precipitation to the region and resulted in the driest decade of the last 300 years (Cook and Jacoby, 1977). In contrast, during the 1970s the mean position of the trough migrated several hundred kilometers to the west of its 1960s location. The typical storm track now passed directly over this heavily populated region, producing record annual and decadal precipitation totals. In summary, relatively minor changes in average jet-stream positions over the eastern United States caused significant differences in the Northeast's water supply (Yarnal and Leathers, 1988).

Decadal-scale variability is an important climatic feature in other areas of the globe. For instance, the well-publicized drought in sub-Saharan Africa has persisted since the late 1960s, bringing massive suffering to the peoples of the region. The drought is a normal climatic feature: sub-Saharan lands experience wet-dry fluctuations on time scales of decades, centuries and millennia (Glantz, 1987). Research shows that at least the decadal-scale fluctuations relate to global oceanic anomalies (Folland *et al.*, 1986). The drought is of considerable interest because the pattern of drying observed in this region coincides with the patterns predicted by general circulation models of global warming (Bradley *et al.*, 1987).

CONCLUSIONS

This chapter has reviewed the major types of short-term variability experienced in today's climate. These within-year, year-to-year, and decade-to-decade climatic variations are important components of regional climate. Significant forms of short-term climatic variation include: persistent climatic regimes; the 40-50 day oscillation; the Quasi-Biennial Oscillation; teleconnections (including ENSO events); solar cycle-related variations; and decadal-scale circulation regimes.

Short-term climatic variations will be an integral part of future climate, with or without global warming. However, if the planet does warm, scientists are not sure about the form these variations will take. It is possible that new, unique patterns of climatic variability will appear. It is even more likely that familiar forms of climatic variability will continue, but that in some cases their location, temporal character or intensity will change. For instance, some climatic variations may appear more often or less frequently than today, others will appear in new locations and disappear from old haunts, and still others will appear at the same times and places, but will be stronger or weaker than before. Regardless of the regional impact of global warming, climatic variability will continue to be a constant joy, heartache and mystery to future generations.

REFERENCES

Barnston, A.G. and R.E. Livezey. 1989. An operational multifield analog prediction system for United States seasonal temperatures. Part II: Spring, summer, fall, and intermediate three-month period experiments. *Journal of Climate* 2:513-541.

Barnston, A.G. and R.E. Livezey. 1991. Statistical prediction of January-February mean Northern Hemisphere lower tropospheric climate from the 11-year solar cycle and the Southern Oscillation for west and east QBO phases. *Journal of Climate* 4:249-262.

Barnston, A.G., R.E. Livezey and M.S. Halpert. 1991. Modulation of Southern Oscillation-Northern Hemisphere mid-winter climate relationships by the QBO. *Journal of Climate* 4:203-217.

Bradley, R.S., H.F. Diaz, J.K. Eischeid, P.D. Jones, P.M. Kelly and C.M. Goodess.

1987. Precipitation fluctuations over Northern Hemisphere land areas since the mid-19th century. *Science* 237:171-175.

Cook, E.R. and G.C. Jacoby. 1977. Tree ring-drought relationships in the Hudson Valley, New York. *Science* 198:399-401.

Diaz, H.F. and R.G. Quayle. 1980. The climate of the United States since 1895: Spatial and temporal changes. *Monthly Weather Review* 108:249-266.

Fein, J.S. and P.L. Stephens. 1987. *Monsoons*. John Wiley and Sons, New York.

Folland, C.K., T.N. Palmer and D.E. Parker. 1986. Sahel rainfall and worldwide sea temperatures, 1901-85. *Nature* 320:602-607.

Glantz, M.H. 1987. Drought in Africa. *Scientific American* 256(June):34-40.

Glantz, M.H., R.W. Katz and N. Nicholls. 1991. *Teleconnections Linking Worldwide Climate Anomalies*. Cambridge University Press, Cambridge.

Held, I.M. 1983. Stationary and quasi-stationary eddies in the extratropical troposphere: Theory. In: *Large-Scale Dynamical Processes in the Atmosphere*, B.J. Hoskins and R.P. Pearce, editors. London: Academic Press, 127-168.

Holton, J.R. and H. Tan. 1980. The influence of the equatorial Quasi-Biennial Oscillation on the global circulation at 50 mb. *Journal of the Atmospheric Sciences* 37:2200-2208.

Karoly, D.J. 1989. Southern Hemisphere features associated with El Niño-Southern Oscillation events. *Journal of Climate* 2:1239-1252.

Krishnamurti, T.N. and S. Gadgil. 1985. On the structure of the 30 to 50 day mode over the globe during FGGE. *Tellus* 37A:336-360.

Labitzke, K. and H. van Loon. 1988. Association between the 11-year solar cycle, the QBO, and the atmosphere. Part I: The troposphere and stratosphere in the Northern Hemisphere winter. *Journal of Atmospheric and Terrestial Physics* 50:197-206.

Labitzke, K. and H. van Loon. 1989. Association between the 11-year solar cycle, the QBO, and the atmosphere. Part III: Aspects of the association. *Journal of Climate* 2:554-565.

Lamb, P.J. 1987. On the development of regional climatic scenarios for policy-oriented climatic-impact assessment. *Bulletin of the American Meteorological Society* 68:1116-1123.

Lamb, P.J. and R.A. Peppler. 1987. North Atlantic Oscillation: Concept and an application. *Bulletin of the American Meteorological Society* 68:1218-1225.

Leathers, D.J., B. Yarnal and M.A. Palecki. 1991. The Pacific/North American teleconnection pattern and United States climate I: Regional correlations. *Journal of Climate* 4:517-528.

Leathers, D.J. and M.A. Palecki. 1992. The Pacific/North American teleconnection pattern and United States climate II: Interannual to Interdecadal associations. *Journal of Climate* (in press).

Livezey, R.E. and A.G. Barnston. 1988. An operational multifield analog prediction system for United States seasonal temperatures. Part I: System design and winter experiments. *Journal of Geophysical Research* 93:10953-10974.

Madden, R. and P.R. Julian. 1971. Detection of a 40-50 day oscillation in the zonal wind of the tropical Pacific. *Journal of the Atmospheric Sciences* 28:702-708.

Madden, R. and P.R. Julian. 1972. Description of global scale circulation cells in the tropics with a 40-50 day period. *Journal of the Atmospheric Sciences* 29:1109-1123.

Mullen, S.L. 1989a. Model experiments on the impact of Pacific sea surface temperature anomalies on blocking frequency. *Journal of Climate* 2:997-1013.

Mullen, S.L. 1989b. The impact of orography on blocking frequency in a general circulation model. *Journal of Climate* 2:1554-1560.

Namias, J. 1986. Persistence of flow patterns over North America and adjacent ocean sectors. *Monthly Weather Review* 114:1368-1383.

Namias, J. 1988a. Abrupt change in climate regime from summer to fall 1985 and stability in the fall. *Meteorology and Atmospheric Physics* 38:34-41.

Namias, J. 1988b. Basis for the forecast for summer 1987 following a radically different spring. *Weather and Forecasting* 3:162-169.

Namias, J. 1990. Basis for prediction of the sharp reversal of climate from autumn to winter 1988-1989. *International Journal of Climatology* 10:659-678.

National Research Council 1982. *Solar variability, weather, and climate*. National Academy Press, Washington, D.C.

Mooley, D.A. and B. Parthasarathy. 1984. Fluctuations in all-India summer monsoon rainfall during 1871-1978. *Climatic Change* 6:287-301.

Palmer, T.N. and D.A. Mansfield. 1986. A study of wintertime circulation anomalies during past El Niño events using a high resolution general circulation model. Part II: Variability of the seasonal mean response. *Quarterly Journal of the Royal Meteorological Society* 112:639-660.

Philander, S.G. 1990. *El Niño, La Nina and the Southern Oscillation*. Academic Press, New York.

Rex, D.F. 1950. Blocking action in the middle troposphere and its effect on regional climate: II. The climatology of blocking action. *Tellus* 2:275-301.

Rex, D.F. 1951. The effect of blocking action upon European climate. *Tellus* 3:100-111.

Rogers, J.C. 1984. The association between the North Atlantic Oscillation and the Southern Oscillation in the Northern Hemisphere. *Monthly Weather Review* 112:1999-2015.

Ropelewski, C.F. and M.S. Halpert. 1987. Global and regional scale precipitation patterns associated with the El Niño/Southern Oscillation. *Monthly Weather Review* 115:1606-1626.

Ropelewski, C.F. and M.S. Halpert. 1989. Precipitation patterns associated with the high index phase of the Southern Oscillation. *Journal of Climate* 2:268-284.

Tribbia, J.J. 1991. The rudimentary theory of atmospheric teleconnections associated with ENSO. In *Teleconnections Linking Worldwide Climate Anomalies*, M.H. Glantz, R.W. Katz and N. Nicholls, editors. Cambridge University Press, Cambridge, 285-308.

van Loon, H. and K. Labitzke. 1988. Association between the 11-year solar cycle, the QBO, and the atmosphere. Part II: Surface and 700 mb in the Northern Hemisphere winter. *Journal of Climate* 1:905-920.

Wallace, J.M. and D.S. Gutzler. 1981. Teleconnections in the 500 mb geopotential

height field during the Northern Hemisphere winter. *Monthly Weather Review* 109:784-812.
Yarnal, B. 1985. Extratropical teleconnections with El Niño/Southern Oscillation (ENSO) events. *Progress in Physical Geography* 9:315-52.
Yarnal, B. and G. Kiladis. 1985. Tropical teleconnections associated with El Niño/Southern Oscillation (ENSO) events. *Progress in Physical Geography* 9:524-58.
Yarnal, B. and H.F. Diaz. 1986. Relationships between extremes of the Southern Oscillation and the winter climate of the Anglo-American Pacific coast. *Journal of Climatology* 6:197-219.
Yarnal, B. and D.J. Leathers. 1988. Relationships between interdecadal and interannual climatic variations and their effect on Pennsylvania climate. *Annals of the Association of American Geographers* 78:624-641.

Global Climate Change: Implications, Challenges and Mitigation Measures. Edited by S.K. Majumdar, L.S. Kalkstein, B. Yarnal, E.W. Miller, and L.M. Rosenfeld. © 1992, The Pennsylvania Academy of Science.

Chapter Four

CONTEMPORARY GLOBAL WARMING: Are We Sure?

THOMAS R. KARL

National Oceanic and Atmospheric Administration (NOAA)
National Environmental Satellite and
Data and Information Service (NESDIS)
National Climatic Data Center (NCDC)
Global Climate Laboratory
Federal Building
Asheville, NC 28801

ABSTRACT

It can be stated with a reasonably high degree of confidence that since the late Nineteenth Century the average global mean temperature has increased, but we cannot be sure about the exact magnitude of this increase. This uncertainty arises because of inadequate spatial and temporal sampling and changes in observation methods. The Inter-governmental Panel on Climate Change (IPCC) tried to express this uncertainty by giving a range of warming since the late Nineteenth Century (0.3 to 0.6°C). We are certain however, that this increase of temperature has been irregular over both space and time. Considering only global mean temperatures, the 100-year rise of temperature is not inconsistent with the lower range of projections pertaining to a doubling of CO_2 concentrations (or their equivalent in other anthropogenic trace greenhouse gases) which translates to about a 1 to 3°C temperature increase. Such a doubling is likely sometime during the Twenty-First Century.

Closer examination of the thermometric record however, highlights many peculiar characteristics of the observed temperature rise which makes its direct linkage to an enhanced anthropogenic greenhouse effect uncertain. This includes (1) the rate and timing of the temperature increase over the last 100 years relative the natural variability of the climate system, (2) the spatial patterns of the observed temperature change compared to the expected change, and (3) the rise of the nighttime temperature relative to the daytime during the past 30 to 40 years.

INTRODUCTION

Over the past several years the change in the global thermometric record has been the central focus in the debate over the effect of increasing concentrations of greenhouse gases. The reason for such attention is actually quite simple. The clearest signal of an enhanced greenhouse gas concentration in Global Climate Models (GCMs) is reflected in the thermometric response near the earth's surface. For this reason, a considerable amount of effort has been expended in reducing and identifying the uncertainties in the historical temperature record, and relating these data to the projections of climate change expected in an enhanced greenhouse world. In this article we examine some of the reasons why uncertainty persists about the cause/effect relationship between the observed global warming and the greenhouse effect.

DO WE REALLY KNOW HOW TEMPERATURES HAVE CHANGED THIS CENTURY?

There is little doubt that we are living in a warmer world today than did our grandfathers. The exact magnitude of the global temperature increase over the past 100 plus years is still in question, but not the sign of the change. There are many reasons for the uncertainty in the magnitude of the change. Unfortunately, for some regions of the globe, the uncertainty is of the same magnitude as the anticipated change of temperature well into the Twenty-First Century (attributed to anthropogenic greenhouse gases). Fortunately, time series used to derive temperature change over continental scales often (but not always) have smaller random errors and biases than do similar time series over smaller space-scales (Karl and Williams, 1987; Karl and Jones, 1990; Trenberth *et al.*, 1991). This can result in situations whereby a higher degree of confidence is associated with large-scale temperature changes compared with local and regional space-scales. Over the past few years we have come to better appreciate the uncertainties associated with our knowledge of the recent temperature change, but there still remains some problems. These are briefly discussed in the next two subsections.

Uncertainties of Land Temperatures

Inhomogeneities arise because:
1. Station locations have changed,

2. Stations located in the vicinity of human activities are often affected by local modifications of the climate, such as urban heat island effects, that are not representative of large-scale climate variation and change,
3. The spatial coverage of the data is neither globally comprehensive nor optimally spaced and the degree of coverage changes over time, and
4. Changes have occurred in instruments, exposures of thermometers, observing schedules, observing practices, and other factors, including land use, sometimes introducing systematic biases into the record.

All of the uncertainties have been assessed in one way or another in a variety of analyses as reviewed by Karl *et al.* (1989). Even today few of the corrections for these inhomogeneities should be considered definitive.

Changing station locations can seriously impact time series of temperature change to such an extent that they overshadow any climate signal in the record. Jones *et al.* (1986a, b) have addressed this issue in their global data base by visual inspection of time series temperature difference plots between neighboring stations in order to eliminate these unwanted effects. Others have used more quantitative approaches, but only over specific regions (Karl *et al.*, 1987). Other than introducing regional biases into temperature data sets, the only suspected (Karl and Jones, 1990) global-scale biases that may have entered our data base via station relocations relates to the migration of city stations to outlying airport locations. In fact, this may have actually mitigated another bias, the urban heat island bias (Karl and Jones, 1989, 1990).

Another important uncertainty relates to the impact of urban heat islands on regional and global time series of temperature. The most comprehensive analysis on a spatial scale (Jones *et al.*, 1990), indicates that the effect of increased urban heat on the change of global temperatures is quite small, an order of magnitude smaller than the apparent rise of global temperature over the past 100 years. New methods of identifying possible contamination of the temperature records include the use of space-based products (Johnson *et al.* 1991, Gallo *et al.*, 1991). These procedures may lead to more definitive analyses with respect to urban heat island contamination, but only over the past decade and into the future.

Another source of bias in the thermometric climate record pertains to the inadequate and changing spatial coverage of the land stations. Changing spatial coverage in the land-based climate record has been tested by so-called frozen grid experiments, whereby the grid cells (5° x 10°) with observations during the Nineteenth and first half of the Twentieth Century are used to estimate late Twentieth Century observed climate anomalies, 1950s, 1960s, and 1970s. These experiments indicate that the poor spatial coverage in the Nineteenth Century adds significant noise to the global temperature record (up to 0.1°C per decade), but it does not seriously affect trend analysis (Jones *et al.*, 1986a). As a result, these frozen grid experiments suggest that spatial coverage is not a severe problem in our analyses, but it does not completely resolve the issue because even during the period 1951-80 land-based coverage was not complete. Furthermore, what has not been pointed out in these analyses is the recent descrepancy of the temperature trend over the past 13 years (1979-92) between the space-based global Micro-wave Sounding Unit (MSU) lower tropospheric temperatures with the in-situ surface temperatures. In-situ temperatures warm a greater rate compared with MSU. Current analyses using the MSU space-based data (Spencer and Christy, 1990) and the in-situ data indicate that the difference is due to inadequate global coverage. Nonetheless experiments using longer data sets with strong asymmetric hemispheric temperature change (Manabe *et al.*, 1990) indicate that the historical coverage is likely to be accurate within an order of magnitude of the recent 100-year temperature rise. (Karl *et al.*, 1992a,b).

Changes in observing methods also affect our ability to differentiate climate change from observing biases. These practices can have substantial impact on regional space-scales (Karl *et al.*, 1992c), but because the biases are not the same from country-to-country, the effect of these biases on global scale temperatures are likely to be mitigated to an appreciable degree.

A recent analysis (Balling, 1991) has provided evidence to suggest that human-induced changes in land-use, e.g. desertification, has also impacted the temperature record. Over the past Century human-induced desertification has increased, and based on Balling's recent analysis it is likely that some of the observed warming in the climate may be directly related to local changes in land-use, and not attributable to large-scale climate forcing. Again however, the impact is much more important on regional and local space-scales than on global space-scales.

Uncertainties of Ocean Temperatures

Since the oceans cover over 70% of the earth's surface, the uncertainties in the thermometric record over the oceans can dominate the uncertainties about global temperature change. There are two dominant sources of uncertainty to consider. First, the adequacy of spatial coverage over the oceans and second, the effect of changes in the observation methods. The first problem stems from the fact that ships of opportunity are our primary source of information about ocean temperature change since the Nineteenth Century. These ships have followed preferred navigation routes, and consequently leave vast portions of the ocean unsampled. Changes in these routes over the course of the Century are also significant, e.g. the opening of the Panama Canal. The second problem is the result of both systematic observing changes and poor documentation about the changes. These changes have required adjustments to the ocean temperature record as large or larger than the apparent rise of temperature since the late Nineteenth Century.

The issue of spatial sampling over the oceans has been addressed in several analyses (Parker, 1987; Folland and Colman, 1988; and Trenberth *et al.*, 1991). Parker (1987) performs a frozen grid experiment, similar to the land experiment of Jones *et al.* (1986a). Again, recent analyses (Karl *et al.*, 1992a,b) indicate that 100-year trends of temperature are relatively insensitive to the historical inadequacies of spatial sampling since the late Nineteenth Century. Folland and Colman's (1988) eigenvector analysis on the sea-surface temperature anomaly field provides some clues as to why experiments appear rather insensitive to poor spatial sampling. Their analysis indicates that the first eigenvector is positively related to all the oceans, and in fact the time series of the first eigenvector is remarkably similar to the global sea-surface temperature anomaly time series. This suggests that although there may be many important modes large-scale temperature variability on decadal time scales, Century time-scale temperature change is more coherent and reasonably well captured by the historical coverage of the last 100 years. As a consequence, the importance of spatial sampling increases considerably when the time-scale is shortened. A recent analysis by Trenberth *et al.* (1992), using the global-scale coverage from the MSU instrument aboard NOAA satellites, suggests that even in the most recent decades, we cannot reliably estimate southern hemisphere monthly temperature anomalies using in-situ measurements.

There have been many studies which have tried to address the problems associated with the changes in observing procedures associated with both sea-surface temperature measurements and marine air temperature. At the present time the most widely accepted method of trying to correct for the changes in sea-surface temperature observations is the technique developed by Folland and Parker (1988) and applied by Bottomley *et al.* (1990) and Farmer *et al.* (1989). These corrections address the problem of differing heat losses from wooden, metal, and canvas buckets. The correction is dependent on many factors including, air-sea temperature difference, wind speed, solar radiation, ship speed, and the interval of time between the actual drawing of the water sample and the measurement. Since 1942, no corrections have been applied to the temperature measurements because a mix of insulated buckets and ship intake temperatures have been used to measure sea-surface temperatures. This may result in some additional bias, but it is not thought to be too severe on a global scale (less the $0.1°C$). Unfortunately, even if the technique developed by Folland and Parker (1988) to correct for the changing observing methods were perfect, poor documentation regarding the type of measurement used to make the sea-surface measurement, which continues today in the International Marine Tape Exchange, will limit our ability to be very precise.

Another method of correcting the sea-surface temperature record is to compare it to the marine air temperatures or to the land-based record near coastal waters. This has been done by Folland (1984), Jones *et al.* (1986a, b) and others. A major problem with this approach is that it results in global-scale corrections which are not geographically dependent and any errors in the land data, such as urban biases are then transferred to the ocean record as well. Comparisons of marine air temperatures with independently corrected sea surface temperatures have been completed by Farmer *et al.* (1989) and Bottomley *et al.* (1990). Since the turn of the Century, the marine air temperatures and the corrected SSTs are in reasonable agreement (IPCC, 1990) for both types of sea-surface temperature corrections.

Proxy Measurements of Temperature

The litany of uncertainties associated with global temperature change would be enough to make even the most optimistic specialists quite uneasy about our ability to quantify or even identify the sign of the changes of global temperature over the past 100 years if it were not for some independent sources of information. This includes changes in sea-level, mountain glaciers, and comparison of the land-based record with marine air temperatures and sea-surface temperatures using independent corrections. Figure 1 from the IPCC (1990) report shows remarkable agreement between the various thermometric data sets. Furthermore, since the late Nineteenth Century many analyses of sea-level change have been undertaken (IPCC, 1990). These analyses indicate that indeed, sea-level has risen by about 10 cm, of which about half may be directly related to steric rise in sea-level (thermal expansion of the oceans) with the remaining half attributable to melting snow and ice. Additional independent evidence regarding the sign of the temperature change comes from surveys of glacial advance and retreat. Although these surveys lack the important mass balance required for certainty, they suggest that on a global basis, mountain glaciers are in retreat. Lastly, recent analyses of the changes of Northern Hemisphere snow cover and Northern Hemisphere (>30°N) surface air and temperature are in close agreement.

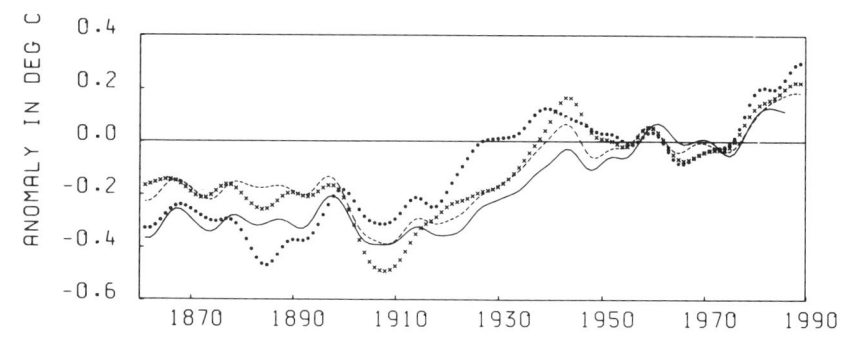

FIGURE 1. IPCC global sea surface, night marine air (crosses) and land air temperature anomalies (dots) 1861-1989, relative to 1951-80. Sea surface temperatures are values from Farmer *et al.* (1989) (solid line). Night marine temperatures from the United Kingdom (UK) Meteorological Office. Land air temperatures are equally weighted averages of data from Jones (1988), Hansen and Lebedeff (1987), and Vinnikov *et al.* (1991).

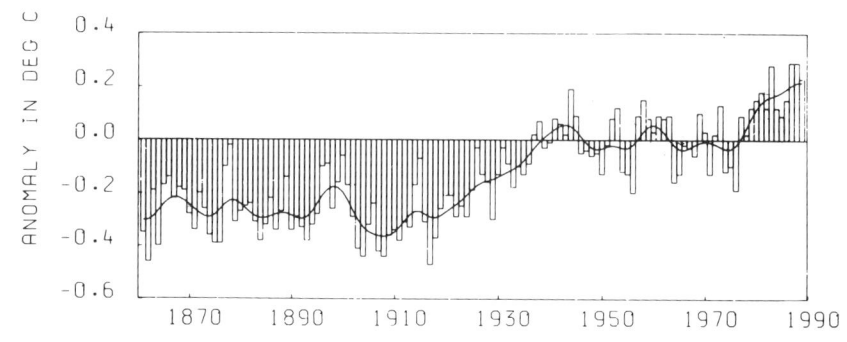

FIGURE 2. IPCC combined land air and sea surface temperatures, 1861-1989, relative to 1951-1980. Land air updated from Jones (1988) and sea surface temperatures from the UK Meteorological Office and Farmer *et al.* (1989). Sea surface temperature component is the average of the two data sets.

THE RATE OF GLOBAL WARMING

The rise of temperature can be, and often has been, summarized as a trend or constant rate of change. Figure 2 indicates that the rate of temperature change has been anything but constant. In fact, the data suggest that all of the warming that we have experienced since the turn of the Century occurred during the 1920s, 1930s, and renewed itself again in the late 1970s. During most of the years since the late Nineteenth Century global temperatures were acutally quite stationary.

One of the major challenges in our quest to understand how temperatures have responded to the global increases of anthropogenic greenhouse gases is related to the timing and abruptness of the observed global warming. This can be illustrated by searching the thermometric record for periods associated with the largest and smallest rates of temperature increase and relating these changes to the radiative equivalent of greenhouse gas increase. The problem is well demonstrated in Figure 3 where the 50-year period associated with the greatest increase of temperature occurred during a period of only modest increases of greenhouse gases. Contrarily, if we search the record for the 50-year period associated with the smallest increase of global mean temperature the rate of increase of anthropogenic greenhouse gases is substantially greater.

There are two logical conclusions that can be drawn from Figure 3. Either the climate system is insensitive to the increases of greenhouse gases that we have experienced over the last 100 or more years, or other forcings must be acting which prevent direct interpretation of evidence such as presented in Figure 3. Global Climate Modeling studies (IPCC, 1990) and considerations of Paleoclimatic evidence (IPCC, 1990) favor the latter conclusion. This implies that our efforts to detect the greenhouse effect, and relate it to the observed climate record will be confounded by other factors such as, volcanic eruptions, solar forcings, other natural climate fluctuations, and even other anthropogenic forcings such as the generation of tropospheric aerosols. This means we will be forced to find the greenhouse fingerprint in a house full of suspects.

THE SPATIAL CHARACTERISTICS OF THE WARMING

Barnett and Schlesinger (1986) outlined a strategy to detect the greenhouse fingerprint in the climate records. Their method consists of comparing the observed spatial patterns of the global temperature change to those predicted from the GCMs with enhanced concentrations of greenhouse gases. Figure 4 depicts the considerable spatial variability of the trends of temperature over the past 40 years. This characteristic of spatial variability is also found in GCMs with enhanced greenhouse concentrations. Barnett and Schlesinger (1986) developed a number of statistics to match the spatial patterns of change projected in the GCMs with the observed spatial climate variability.

In later work, Barnett and Schlesinger (1987) and Barnett (1991) reported that they could not find the expected fingerprint of the greenhouse effect in the surface temperature records. Their analysis was based on quantitative comparisons of the covariances of standardized global temperature anomalies with the expected patterns predicted from GCMs.

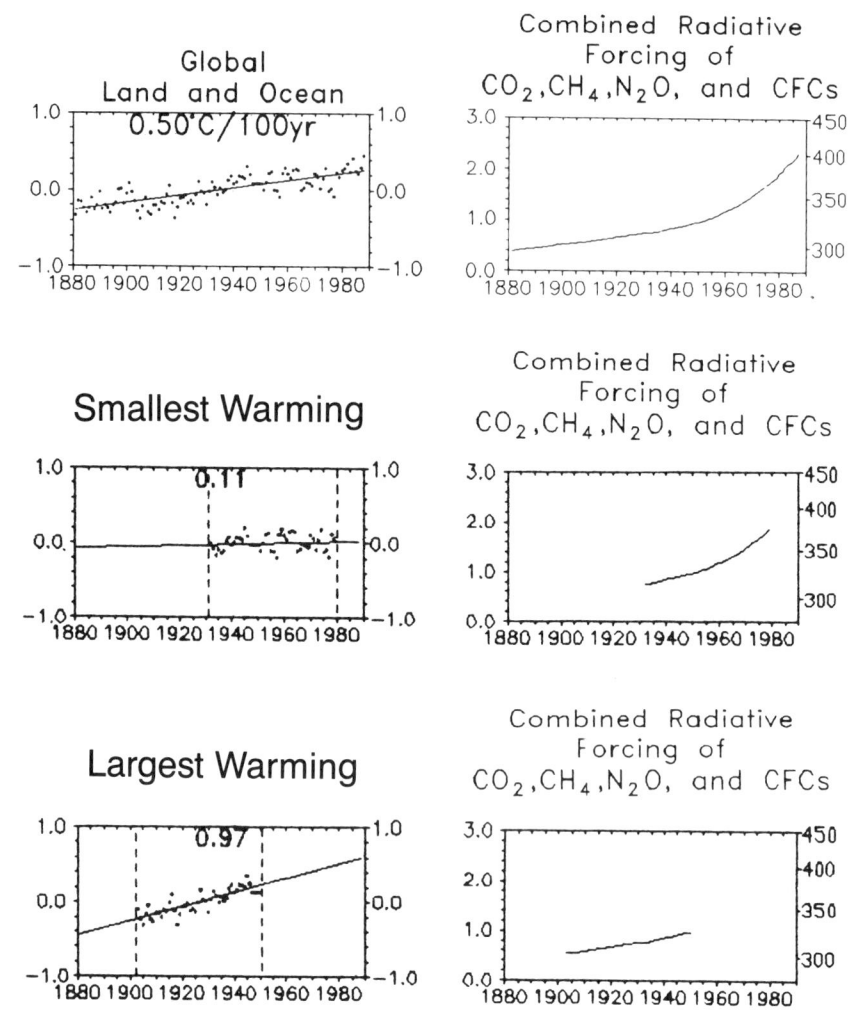

FIGURE 3. Change of global temperature and the radiative equivalent of greenhouse gas forcing (Wigley, 1987). Bottom two panels pertain to the 50-year period with the smallest rate of temperature rise and the associated changes of greenhouse forcing, and similar time series for the 50-year period with the largest rate of temperature increase.

GLOBAL WARMING AND EXTREME TEMPERATURES

One of the more important questions about how temperatures have and will change relates to extremes of temperatures. There have been a dearth of analyses on how changes of extreme temperature or even mean monthly maximum and minimum temperatures have changed over the past 100 or more years. This stems from the difficulty of compiling the records from the daily data. Internationally exchanged climate summaries have not included this type of detail. Recently however, Karl *et al.* (1991) have compiled a data set over the USA, USSR, and

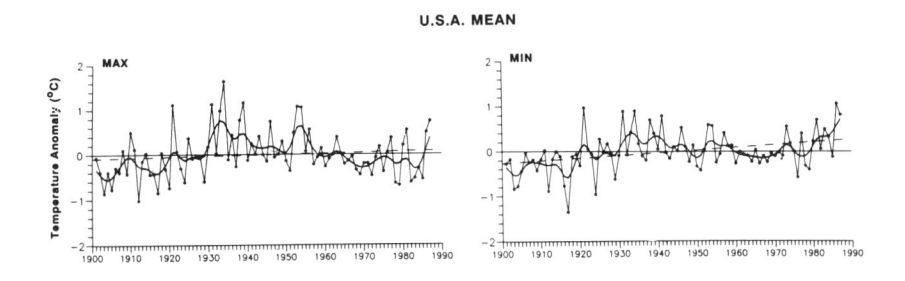

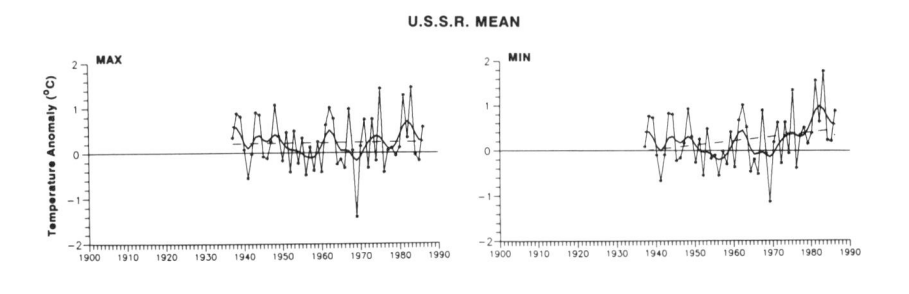

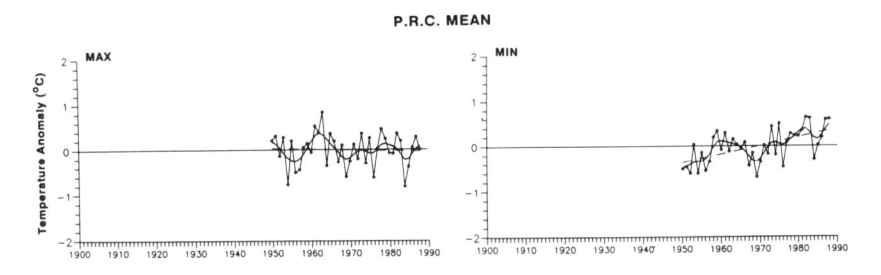

FIGURE 4. Variations of temperature (°C) for mean annual maximum and minimum temperature over the United States (USA), the United Soviet of Socialist's Republics (USSR), and the People's Republic of China (PRC) (from Karl *et al.*, 1991c).

the PRC which includes mean monthly maximum and minimum temperatures. These data are summarized in Figure 5 for changes of the mean maximum and minimum temperature. The data indicate that virtually all of the warming in these three countries, which comprise nearly 40% of the Northern Hemisphere land mass, is due to a rise of the minimum, or nighttime temperature. Similar time series related to the change in the seasonal extremes of temperature reflect a narrowing of the extremes with minimum temperatures increasing, but little change in the extreme maximum.

The cause(s) of the asymmetric diurnal changes are uncertain. There is some evidence to suggest that changes of cloud cover plays a direct role (where increases

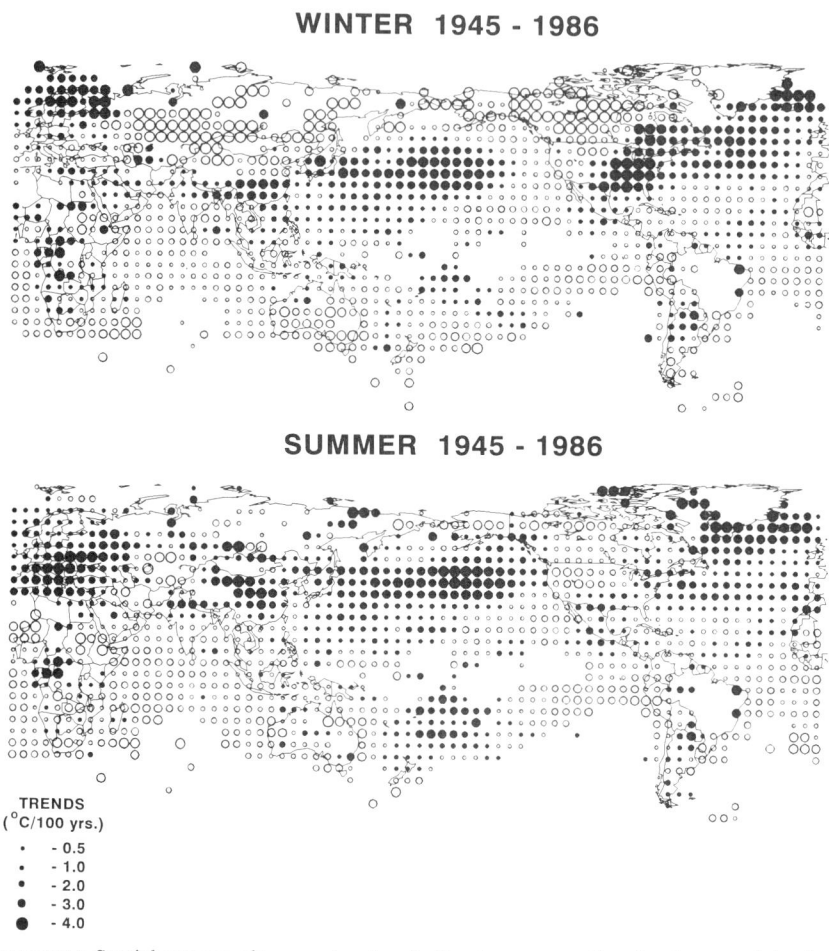

FIGURE 5. Spatial patterns of temperature trends. Trends are proportional to the area of the circle. Negative trends are depicted with solid circles and positive trends with open circles. Land air temperatures are from Jones (1988), and sea surface temperatures from Farmer *et al.* (1989).

in cloudiness result in reduced maximum and higher minimum temperatures). Regardless of the exact cause, these results imply that either: (1) the warming in much of the Northern Hemisphere is grossly affected by non-greenhouse forcing, confirming what we already suspect, or (2) that climate model projections related to the expected change in the diurnal temperature are underestimating (overestimating) the rise of the daily minimum (maximum) relative to the maximum (minimum). GCMs project an almost symmetric increase of the daily minimum temperature relative to the maximum temperature.

CONCLUSIONS

We are quite confident that global mean temperatures have increased since the Nineteenth Century. The exact magnitude is uncertain, but it is believed to be in the range of 0.3 to 0.6°C. We can neither attribute this warming to the greenhouse effect, nor can we dismiss the possibility that it may be a result of enhanced anthropogenic greenhouse gases. The problem arises because of the nature of the warming: its magnitude is smaller than expected, the spatial patterns of the warming do not seem to match model projections, and the character of the warming, nighttime versus daytime, does not seem to match our projections. Nonetheless, because of other forcings, both natural and manmade (tropospheric aerosols) it is possible that the sensitivity of the climate system to greenhouse gases may be undesirably large.

REFERENCES

Balling, R.C., Jr. 1991. Impact of desertification on regional and global warming. *Bull. Amer. Meteor. Soc.*, 72:232-234.

Barnett, T.P. 1986. Detection of changes in global tropospheric temperature field induced by greenhouse gases. *J. Geophys. Res.* 91:6659-6667.

Barnett, T.P., M.E. Schlesinger and X.-J. Jsing. 1991. Greenhouse-Gas Induced Climatic Change: A Critical Appraisal of Simulations and Observations (Ed. M.E. Schlesinger) 511-536, Elseview, Amsterdam.

Barnett, T.P. and M.E. Schlesinger. 1987. Detecting changes in global climate induced by greenhouse gases. *J. Geophys. Res.*, 92:14,772-14,780.

Bottomley, M., C.K. Folland, J. Hsiung, R.E. Newell and D.E. Parker. 1990. Global Ocean Surface Temperature Atlas (GOSTA). Joint Meteorological Office/Massachusetts Institute of Technology Project. Project supported by U.S. Dept. of Energy, U.S. National Science Foundation and U.S. Office of Naval Research. Publication funded by UK Dept. of Energy and Environment, about 20 pp. plus 313 plates.

Farmer, G. T.M.L. Wigley, P.D. Jones and M. Salmon. 1989. Documenting and explaining recent global-mean temperature changes. Climatic Research Unit, Norwith, Final Report to NERC, UK, Contract GR/3/6565.

Folland, C.K., D.E. Parker and F.E. Kates. 1984. Worldwide marine temperature fluctuations 1856-1981. *Nature*, 310:670-673.

Folland, C.K. and A.W. Colman. 1988. An interim analysis of the leading covariance eigenvectors of worldwide sea surface temperature anomalies for 1951-80. LRFC 20, Meteorological Office, UK.

Folland, C.K. and D.E. Parker. 1990. Observed variations of sea surface temperature. Proc NATO Advanced Research Workshop on Climate-Ocean Interaction. Oxford, UK, 26-30, Sept. 1988. Kluwer Academic Press.

Gallo, K.P., J.D. Tarpley, J.F. Brown, A.L. McNab and T.R. Karl. 1991. The use of multi-satellite data and advanced analysis techniques for climatological applications: Assessment of urban heat island temperature bias. Proceedings of the Seventh Conference on Applied Climatology, Salt Lake City, Utah. *American Meteorological Society.*

Hansen, J. and S. Lebedeff. 1987. Global trends of measured surface air temperature. *J. Geophys. Res.* 92:13,345-13,372.

Inter-governmental Panel on Climatic Change (IPCC). 1990. Scientific Assessment of Climatic Change, World Meteorological Organization/United Nations Environmental Programme, Geneva, Switzerland, 366 pp.

Johnson, G.L., J.M. Davis, T.R. Karl, A.L. McNab and J.D. Tarpley. 1991. On the use of TOVS-derived temperatures in estimates of urban bias. IN REVIEW, *J. Climate.*

Jones, P.D., P. Ya. Groisman, M. Coughlan, N. Plummer, W-C. Chang and T.R. Karl. 1990. How large is the urbanization bias in large area averaged surface air temperature trends? *Nature.*

Jones, P.D. 1988. Hemisphere surface air temperature variations: Recent trends and an update to 1987. *J. Clim.* 1:654-660.

Jones, P.D., S.C.B. Raper, R.S. Bradley, H.F. Diaz, P.M. Kelly and T.M.L. Wigley. 1986a. Northern Hemisphere surface air temperature variations, 1851-1984. *J. Clim. Appl. Met.*, 25:161-179.

Jones, P.D., S.C.B. Raper, R.S. Bradley, H.F. Diaz, P.M. Kelly and T.M.L. Wigley. 1986b. Southern Hemisphere surface air temperature variations, 1851-1984. *J. Clim. Appl. Met.*, 25:1213-1230.

Karl, T.R. and C.N. Williams, Jr. 1987. An approach to adjusting climatological time series for discontinuous inhomogeneities. *J. Clim. Appl. Met.*, 26:1744-1763.

Karl, T.R., H.F. Diaz and G. Kukla. 1988. Urbanization: its detection and effect in the United States climate record. *J. Clim.*, 1:1099-1123.

Karl, T.R., J.D. Tarpley, R.G. Quayle, H.F. Diaz, D.A. Robinson and R.S. Bradley. 1989. The recent climate record: What it can and cannot tell us. *Rev. Geophys.*, 27:405-430.

Karl, T.R. and P.D. Jones. 1989. Urban bias in area-averaged surface temperature trends. Bull. Amer. Meteor. Soc., 265-270.

Karl, T.R. and P.D. Jones. 1990. Reply to comments on "Urban bias in area-averaged surface temperature trends." *Bull. Amer. Met. Soc.*, 71:572-574.

Karl, T.R., G. Kukla, V.N. Razuvayev, M.J. Changery, R.G. Quayle, R.R. Heim, Jr., and D.R. Easterling. 1991. Global warming: Evidence for asymmetric diurnal temperature change. *Geophys. Res. Lett.*

Karl, T.R., R.W. Knight and J.R. Christy. 1992a. Global and hemispheric temperature trends: Uncertainties related to inadequate spatial sampling. Seventh Conference on Probability and Statistics in the Atmospheric Sciences. Toronto, Canada, American Meteorological Society.

Karl, T.R., R.W. Knight and J.R. Christy. 1992b. Global temperature trends: Uncertainties related to incomplete spatial sampling. IN REVIEW, *Journal Geophys. Research.*

Karl, T.R., R.G. Quayle and P. Ya. Groisman. 1991b. Detecting climate variations and change: New challenges for operational observing and data management systems. IN PRESS *Journal of Clim.*

Manabe, S., K. Byran and M.J. Spelman. 1990. Transient response of a global ocean-atmosphere model to a doubling of atmospheric carbon dioxide. *J. Phys. Oceanogr.*, 20:722-749.

Spencer, R.W. and J.R. Christy. 1990. Precise monitoring of global temperature trends from satellites. *Science*, 247:1558-1562.

Trenberth, K.E., J.R. Christy and J.W. Hurrell. 1992. Monitoring global monthly mean surface temperatures. IN PRESS *J. Climate.*

Parker, D.E. 1987. The sensitivity of estimates of trends of global and hemispheric marine temperatures to limitations in geographical coverage. LRFC 12, Meteorological Office, UK.

Vinnikov, K. Va., P. Ya. Groisman and K.M. Lugina. 1991. The empirical data on modern global climate changes (temperature and precipitation). *J. Clim.*, 3:662-677.

Wigley, T.M.L. 1987. Relative contributions of different trace gases to the greenhouse effect. *Climate Monitor*, 16:14-29.

Global Climate Change: Implications, Challenges and Mitigation Measures. Edited by S.K. Majumdar, L.S. Kalkstein, B. Yarnal, E.W. Miller, and L.M. Rosenfeld. © 1992, The Pennsylvania Academy of Science.

Chapter Five

HYDROLOGICAL IMPACTS OF GLOBAL WARMING OVER THE UNITED STATES

JOHANNES J. FEDDEMA[1] and JOHN R. MATHER[2]

[1]Department of Geography
UCLA, CA
and
[2]Department of Geography
University of Delaware
Newark, DE 19716

INTRODUCTION

Scientists continue to debate the effect of increasing CO_2 and other trace gases on regional, continental, and global climates. While some feel the issue is fairly well settled, the majority agree that we need to study the problem in considerably more detail to achieve a reasonable understanding of the possible range of future climates. Of possibly greater significance to humans, however, is information on how future climates will influence life on earth. In other words, how will any new climate scenarios influence our nation's water resources, agricultural production, standard of living, or economic activities?

The world has always existed under changing climates. Humans have survived through the "little Ice Age," the so-called "climatic optimum," and the "year without a summer." We are accustomed to wide swings in climatic conditions, both on a short term basis (year to year) and over longer time periods (decades to centuries). While the global warming anticipated by many as a result of increased CO_2 and other trace gases may not be more severe than some of the climate changes witnessed by humans in the past few thousand years, it could, if it occurs, result in significant changes and adjustments in our current living habits and/or way of doing business. Thus, because of the large number of unknowns that presently exist in our estimates of future climates, it behooves us to consider possible future climate scenarios and to begin to develop options that will enable us to prosper no matter what climatic future lies ahead of us.

Wetherald (1989) has provided a simple, but comprehensive review of current efforts to model both present and future climatic conditions resulting from different CO_2 conditions. He discusses the similarities as well as the differences in the leading Global Circulation Models (GCMs) that have been developed over the last decade to study the consequences of increased CO_2. He concludes that while many unknowns still exist, and crude assumptions have had to be employed in a number of cases to approximate atmospheric relationships and feedbacks, the GCMs still may give us the most realistic picture of what future conditions will be like.

We already have a well-tested model that expresses how atmospheric energy (expressed as air temperature) and precipitation influence the water resources of a place or area. The climatic water budget, originally developed by Thornthwaite in the early 1940s and later modified by Thornthwaite and Mather (1955), has been used extensively to provide information on hydroclimatic factors such as soil moisture storage, actual evapotranspiration, water deficit, soil water surplus, water runoff or streamflow, and snow accumulation and melt. Evaluation of water budget computations reveal that if precipitation increases in a region with no change in temperature (and, thus, potential evapotranspiration) or with a decrease in temperature, soil moisture storage and streamflow will increase. An increase in precipitation accompanied by an increase in temperature is more difficult to evaluate because the relative magnitudes of these changes would determine whether soil moisture content and streamflow increase or decrease. A precipitation decrease accompanied by an increase in climatic demand for water should result in a decrease in soil moisture storage and streamflow. Clearly, the seasonal patterns of these changes would strongly influence the actual pattern of increase or decrease in soil moisture conditions, streamflow, or other hydroclimatic factors.

Any attempt to understand future hydroclimatic conditions through the application of temperature and precipitation data derived from one of the global circulation models is fraught with uncertainties that the user must fully understand. There is certainly little agreement concerning the timing and magnitude of changes due to increased CO_2 and other trace gases. Mitchell (1989) has recently summarized the global mean changes in temperature and precipitation assuming a doubling of CO_2 obtained from an evaluation of five different GCMs (Table 1). Agreement on the magnitude of the mean global temperature increase is closer than on mean global precipitation. The influence of climatic warming on precipitation is more in doubt since there are a number of feedback relations that must be considered. These include the effect of the "greenhouse" conditions on the distribution of cloudiness and winds which will have their own impact on precipitation distribution, as well as the possible occurrence of more or fewer storms or drought periods as a result of increased CO_2 amounts. A 1984 EPA study on the potential climatic impacts of increasing CO_2 emphasized in bold type the warning that "the values given for hydrologic conditions in particular grids should not be used for planning purposes." The reason for this warning is that the present models do not necessarily predict present conditions with sufficient accuracy. Built-in errors under present conditions might produce even greater errors under future scenarios. At the same time, available global circulation models provide information for a rather coarse grid of points spaced at either a 4 x 5° or 8 x 10° grid network, depending on the GCM being evaluated. In areas of significant topographic and vegetation diversity, these grid densities provide only rough estimates of what might be expected within the grid area.

TABLE 1
Global Mean Changes in Five Recent CO_2-Doubling Studies (from Mitchell, 1989)

Study	Source	Surface Temperature Change, K	Precipitation Change, %
GISS	Hansen et al. (1984)	4.2	11.0
NCAR	Washington and Meehl (1984)	4.0*	7.1
GFDL	Wetherald and Manabe (1986)	4.0	8.7
UK-MET	Wilson and Mitchell (1987)	5.2	15.0
OSU	Schlesinger and Zhao (1987)	2.8	7.8

* provided through personal communications in 1988.

METHODOLOGY

Based on the output from four different GCMs (GISS, GFDL, OSU, and UK-MET: see Kalkstein, 1991), four different climate scenarios have been created to evaluate predicted climate conditions in the United States under double CO_2 conditions. To evaluate not only changes predicted on the GCM grid scales, but also the effects at the sub-GCM grid level, a half degree by half degree gridded climatology of the U.S. was used as a base for this evaluation (Legates and Willmott, 1990a, 1990b). Due to the recognized difficulty of interpolating between GCM grid points, no interpolation of GCM data was performed. Instead, current climatology data at each half-by-half degree grid point were modified by the calculated changes at the closest GCM grid location.

Using only temperature and measured precipitation climatology values at each half-by-half degree grid value, the double CO_2 conditions were calculated as follows for each GCM:

T(HALF CO_2) = T(HALF CLIMATOLOGY) + (T(GCM CO_2)-T(GCM CONTROL))

where T(HALF CO_2) is the predicted mean monthly temperature at a particular half-by-half degree grid location under double CO_2 conditions, T(HALF CLIMATOLOGY) is the observed mean monthly temperature at the same half-by-half degree grid point, T(GCM CO_2) is the mean monthly temperature at the nearest GCM grid location for the double CO_2 model run, and T(GCM CONTROL) is the mean monthly temperature value at the nearest GCM grid point in the control (single CO_2) run.

The following relation was used to determine precipitation after a doubling of CO_2:

$$P(HALF\ CO_2) = P(HALF\ CLIMATOLOGY) * (P(GCM\ CO_2)/P(GCM\ CONTROL))$$

where $P(HALF\ CO_2)$ is the predicted monthly precipitation total at a particular half-by-half degree grid point under double CO_2 conditions, $P(HALF\ CLIMATOLOGY)$ is the observed monthly precipitation total at the same grid location, $P(GCM\ CO_2)$ is the predicted monthly precipitation total at the nearest GCM grid location in the double CO_2 GCM run, and $P(GCM\ CONTROL)$ is the monthly precipitation for the nearest GCM grid location in the control run. In those few instances where a center of a half-by-half degree grid location fell exactly on the boundary between two GCM grid cells, the half-by-half grid point was consistently modified by the GCM grid location lying to the south or east.

Applying data from the four different GCMs, four double CO_2 scenarios of mean monthly temperature and precipitation values were created at the half-by-half degree grid resolution. To evaluate the impact of CO_2 warming on hydrologic factors over the United States, these temperature and precipitation values were used as input to a water budget algorithm (WATBUG: Willmott, 1977). In this algorithm, potential evapotranspiration estimates are made by the standard Thornthwaite approach, using an assumed 150 mm storage at field capacity and soil moisture depletion curve G (Mather, 1978) at each grid point. This depletion curve permits soil moisture to be removed at the potential rate until the soil moisture content is at 70 percent of field capacity. Below that value, soil moisture is removed at a constantly declining rate as the soil dries. Annual totals of specific water budget variables obtained from the four predicted double CO_2 climates were then compared to annual water budget results obtained from the half-by-half degree observed climate data.

RESULTS

Temperature changes in all models were quite consistent in their spatial distribution, with the smallest temperature increases generally occurring in the southeastern portion of the country. However, there was some variation in the magnitude of the average (mean latitude weighted average of all the grid values) deviation from current climate. The UK-MET model showed the greatest difference (a 6.86 C° increase), while the OSU model was lowest with only a 3.08 C° increase. The GFDL and GISS models showed less variability than the other two, and had average temperature differences of 4.33 C° and 4.60 C°, respectively. In all models, every half-by-half degree grid point is predicted to experience an increase in mean annual temperature under double CO_2 conditions.

Predicted monthly precipitation values under double CO_2 conditions showed considerably more variability particularly within models (Figure 1). In terms of a nationwide average, all GCMs predicted a considerable increase in total annual precipitation, ranging from 34 mm (OSU) to 71 mm (UK-MET) with intermediate increases

54 Global Climate Change: Implications, Challenges and Mitigation Measures

of 43 mm and 54 mm for the GISS and GFDL models, respectively. Except for the OSU model in the Pacific Northwest, all the models predicted a general increase in precipitation throughout the northcentral and northwestern sections of the country. Most of the models also seemed to agree on increased precipitation in the Southeast (particularly Florida) and a drying along the Mississippi River valley. The greatest discrepancy among models is in the southwest and northeast regions, with some models predicting a decrease in precipitation and others an increase in precipitation, resulting in large differences in predicted annual precipitation in these regions. Significant variations in precipitation are also found on the sub-GCM grid scale, because similar percentage change values are being applied to widely varying annual precipitation values at the half-by-half degree grid scale in regions with large precipitation gradients (e.g., the GISS GCM over the Mississippi River valley). These variable inputs of precipitation totals at the sub-GCM grid scale will have a significant impact on the evaluation of hydrologic variables in the same regions.

As expected from use of the Thornthwaite model, changes in potential evapotranspiration show the same regularity as the relative temperature changes (Figure 2). Because all models predict increased temperatures at all grid locations, each of these grid points also shows a significant increase in potential evapotranspiration. However, due to the nonlinear relationship between potential evapotranspiration

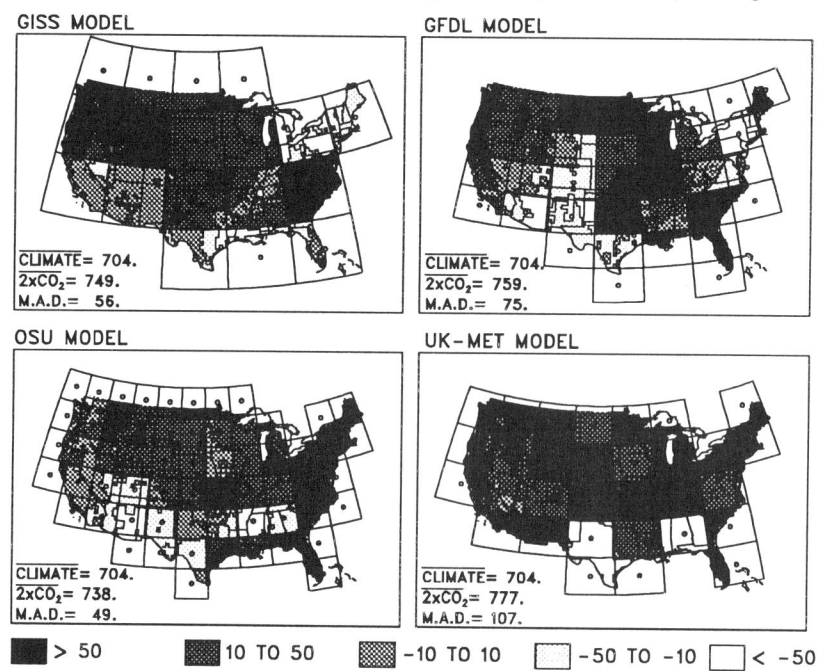

FIGURE 1. Change in annual precipitation (in mm) from present with doubling of CO_2 based on data from four GCMs (Climate is mean current climate value for country, $2 \times CO_2$ is mean value under double CO_2 conditions, M.A.D. is mean absolute difference in values, all based on half-by-half degree grid scale values).

and temperature, the actual observed differences in mm of water show a slightly different spatial distribution from the predicted temperature increases. The southern and southeast parts of the country show the most dramatic increases in potential evapotranspiration, even though the temperature increases were relatively low in the same locations. This seemingly contradictory result is the consequence of temperatures in the southeast approaching the 26.5 C° point of convergence in the Thornthwaite computations of potential evapotranspiration. In regions where initial temperatures already exceed 26.5 C°, such as the Southwest, a similar temperature increase will have less effect on increasing potential evapotranspiration. The only model inconsistent with the general spatial distribution of increased potential evapotranspiration values is the UK-MET model where, due to extremely high temperature increases in the Midwest and Great Lakes regions, there is a dramatic increase in potential evapotranspiration in the same area. In absolute terms, the mean annual increase in potential evapotranspiration ranges from 141 mm for the OSU model to 311 mm for the UK-MET model, with the GFDL and GISS models having an increase of 198 mm on average. As was the case with precipitation, the wide range in temperatures within a given GCM grid cell results in some nonlinear increases of potential evapotranspiration within the grid cells, particularly in mountain regions where there is considerable temperature variability within a GCM grid

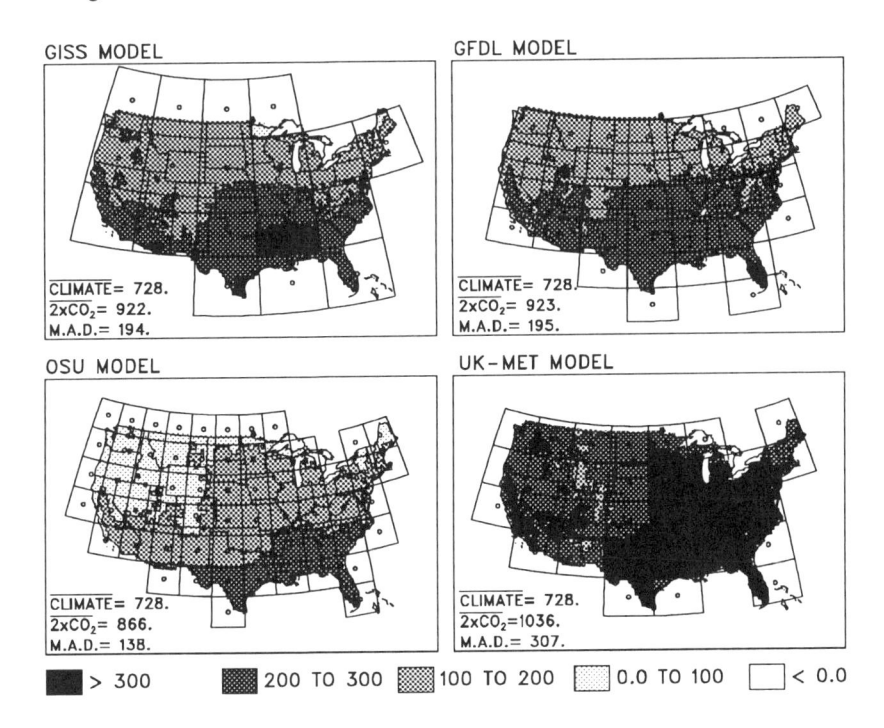

FIGURE 2. Change in annual potential evapotranspiration (in mm) from present with doubling of CO_2 based on data from four GCMs. (See Figure 1 for legend)

cell, and potential evapotranspiration estimates may differ by more than 100 mm within these cells.

Calculating water budget components as a function of potential evapotranspiration and precipitation reveals that under double CO_2 conditions, predicted changes in actual evapotranspiration vary widely (Figure 3). As a whole, actual evapotranspiration changes most closely follow those of precipitation particularly in dry regions where actual evaporation tends to equal precipitation. With a general increase in potential evapotranspiration, actual evaporation will be expected to increase unless limited by precipitation. There is a predicted increase in actual evapotranspiration ranging from 54 mm (GFDL) to 104 mm (UK-MET) with increases of 63 mm and 83 mm for OSU and GISS models, respectively. These increases are considerably more than equivalent increases in precipitation. Thus, the water for such increases has to originate from some other factor in the water budget. In this case, surplus water amounts will be reduced to allow for an increase in actual evapotranspiration. Most of the increase in actual evapotranspiration therefore occurs in the wetter regions of the country, particularly in the moist areas east of the Mississippi River and in the Northwest. The models disagree most significantly in the southeastern region, where precipitation predictions are also

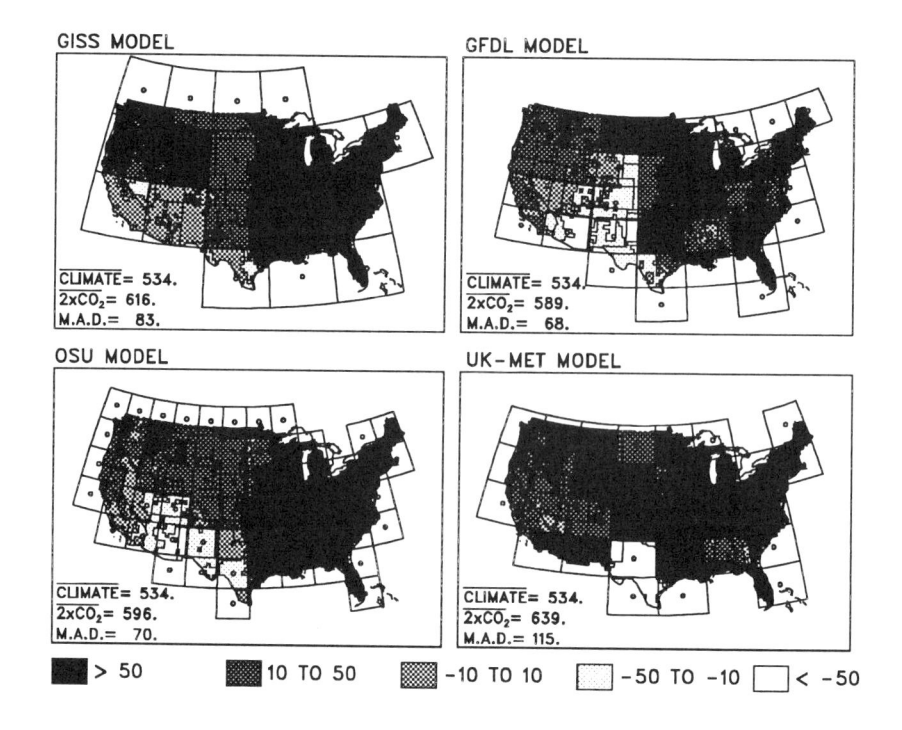

FIGURE 3. Change in annual actual evapotranspiration (in mm) from present with doubling of CO_2 based on data from four GCMs. (See Figure 1 for legend)

quite variable. The increase in actual evapotranspiration might suggest an increase in moisture availability for biological processes, although the much larger increases in potential evapotranspiration at equivalent locations reveal that there would be a relative drying of the climates in most locations. Increased water demands in these regions have not been met by increased supply.

Of particular interest is the finding that in the regions where potential evapotranspiration and precipitation are nearly equal under current conditions, changes in actual evapotranspiration may be of different signs within a given GCM grid cell. Typically, this is most notable near the line running from the Texas Gulf coast to the upper reaches of the Mississippi River valley where precipitation equals potential evapotranspiration. Particularly good examples of this are shown in the GISS and GFDL models in southeastern Texas. In both these models, potential evapotranspiration has increased in these grid cells, showing a greater water demand in the area. Simultaneously, precipitation has decreased in the same area. In the drier western part of the GCM grid cells, this results in a decrease of actual evapotranspiration due to the decrease in water supply, and the lack of a water source for supplementing lower amounts of precipitation (there is no water surplus in this region under current climate conditions). Similarly, in the eastern portions of the GCM grid cells, precipitation also decreases, and potential evapotranspiration is increased.

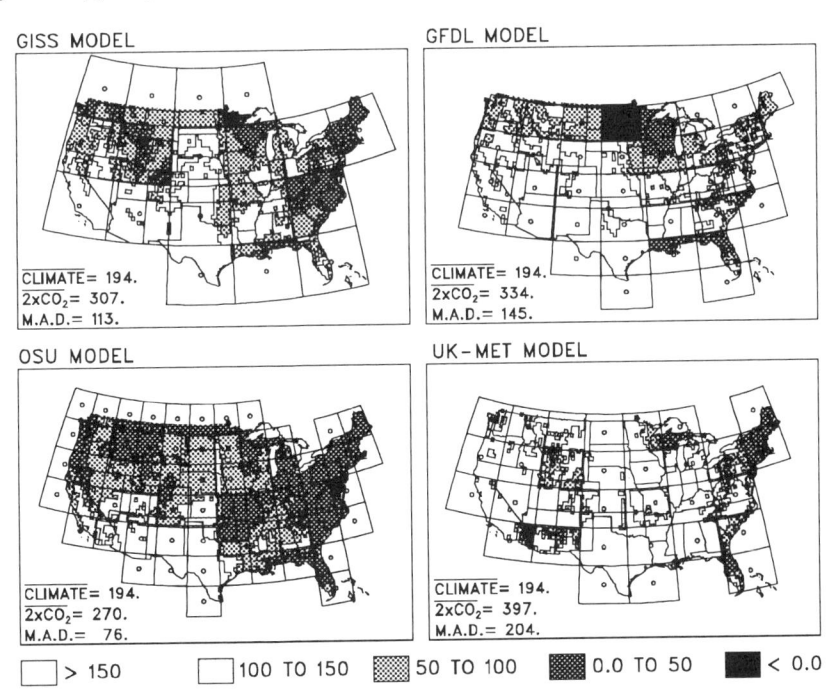

FIGURE 4. Change in annual moisture deficit (in mm) from present with doubling of CO_2 based on data from four GCMs. (See Figure 1 for legend)

However, because the greater potential evapotranspiration occurs partly at a time when there is currently a rainfall surplus, surplus will be diminished or eliminated under double CO_2 conditions, and this quantity of water will be available as actual evapotranspiration, resulting in an overall increase in actual evapotranspiration and agricultural productivity. This illustrates that results based on GCM grid scales have to be interpreted very cautiously in regions of high climatic moisture and temperature gradients, where gross aggregations of climate may be misleading with respect to certain parameters.

The effects of double CO_2 conditions on biological processes is best demonstrated in the distribution of changes in the moisture deficit between the double CO_2 and current condition models (Figure 4). Moisture deficits represent the moisture shortage for optimal crop productivity in a region. Although there is significant variation between models, overall changes show an increase in the total moisture deficit for all models, ranging from 79 mm (OSU) to 207 mm (UK-MET), with 116 mm and 144 mm for GISS and GFDL, respectively. Spatially, most of the models predict the largest increase in the moisture deficit in the southwestern portion of the nation. This area is already extremely dry, so that with little soil moisture and no surplus for most of the year, there is little opportunity for a redistribution in moisture sources. Also, several of the models predict a decrease in precipitation for this area while there is an increase in potential evapotranspiration. The result is the observed increase in moisture shortage. In all the models, there are very few areas where there is a decrease in the moisture deficit, showing that there will be a general drying of the entire country. However, the impact of the increased potential evapotranspiration will be least felt in already moist areas particularly in the region east of the Mississippi. In these areas, there may be some increase in the intensity of a seasonal summer drought, but under normal circumstances there is sufficient surplus water available for at least part of the year so that any increase in potential evapotranspiration is partially compensated by decreasing the surplus and putting some of this moisture towards increasing actual evapotranspiration (Figure 3). Similar results are also found in the moist Northwest. Again, there is evidence that predicted moisture deficit conditions can vary widely at the sub-GCM grid scale, for potential increases in moisture deficit can be offset by a reduction in moisture surplus in moist areas, while this is not possible in dry regions.

Evaluation of predicted moisture surplus conditions (Figure 5) shows that in dry regions there is no change in the moisture surplus, because there is no moisture surplus there under current conditions. This is expected as all precipitation water is immediately lost by actual evapotranspiration. Predicted changes in precipitation are not sufficient to exceed potential evapotranspiration and create a moisture surplus. It is interesting that all models show a very steep gradient between the dry regions having no change in surplus and wet regions where much of the surplus under current conditions will be converted into additional moisture for actual evapotranspiration. Therefore, the largest decreases in surplus moisture will occur in regions where surplus water is used to compensate for increases in potential evapotranspiration. In areas where additional water supply in the form of precipitation may offset relatively small increases in potential evapotranspiration

Hydrological Impacts of Global Warming Over the United States   59

due to low temperatures, regardless of the double CO_2 temperature increase, surplus may actually increase, particularly in the mountains of the extreme east and west. In those models in which precipitation predictions vary widely from the other models another exception may be found. For example, the midwest region shows a remarkable increase in precipitation in the GFDL model, which is not matched in the other models. Surplus, in this region, shows an overall increase because the increase in precipitation is greater than the overall increase in potential evapotranspiration.

To integrate the effect of double CO_2 climates on the moisture regime of the United States, Figure 6 shows the changes in the moisture index as defined in Willmott and Feddema (1992). The index runs from -1 to 1 where negative values indicate a dry climate (precipitation < potential evapotranspiration), and positive values show a moist climate (precipitation > potential evapotranspiration). Model values indicate that generally the country will undergo a significant drying of its climate under double CO_2 conditions. Differences range from a decrease in the moisture index of -0.09 (OSU), -0.11 for both GFDL and GISS, and -0.17 for the UK-MET model. The moist eastern coastal and western mountain regions will benefit from the increases in potential and actual evapotranspiration which are easily compensated for by the excess moisture in these areas. The resultant increase in the

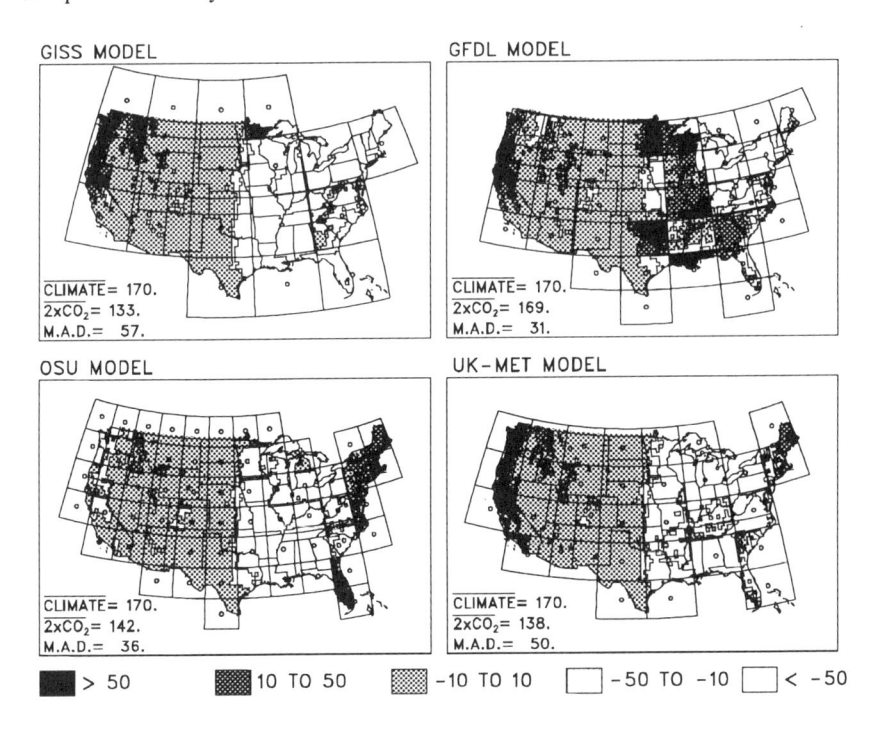

FIGURE 5. Change in annual moisture surplus (in mm) from present with doubling of CO_2 based on data from four GCMs. (See Figure 1 for legend)

growing season and potential productivity of these regions suggests that economically these areas will have the most benefit from predicted global warming scenarios. Regions most adversely affected appear to be those where the difference between precipitation and potential evapotranspiration tend to be very small under current climate conditions. In particular, this encompasses the Colorado/New Mexico plateau region extending into the southwestern desert, and areas in the Midwest and along the Mississippi valley. All these regions would experience a considerable increase in potential evapotranspiration, and most would also experience decreases or very small increases in precipitation resulting in dramatic drying of the climate. From an agricultural perspective, this would mean an increase in dryness in a climate that already barely has enough moisture to meet agricultural water demand. Such changes could have severe repercussions on the types and quantities of crops that might be produced in these regions without large increases in irrigation. At the same time, less water would also be available for irrigation and groundwater recharge in these areas. A similar situation might be expected in the northern Midwest (Dakotas) where three of the four models also agree on a substantial drying of the climate (GFDL is the exception). Most of the rest of the country can expect less dramatic but at least some drying of climate conditions, with the impact on agricultural productivity being determined by the quantity of surplus water

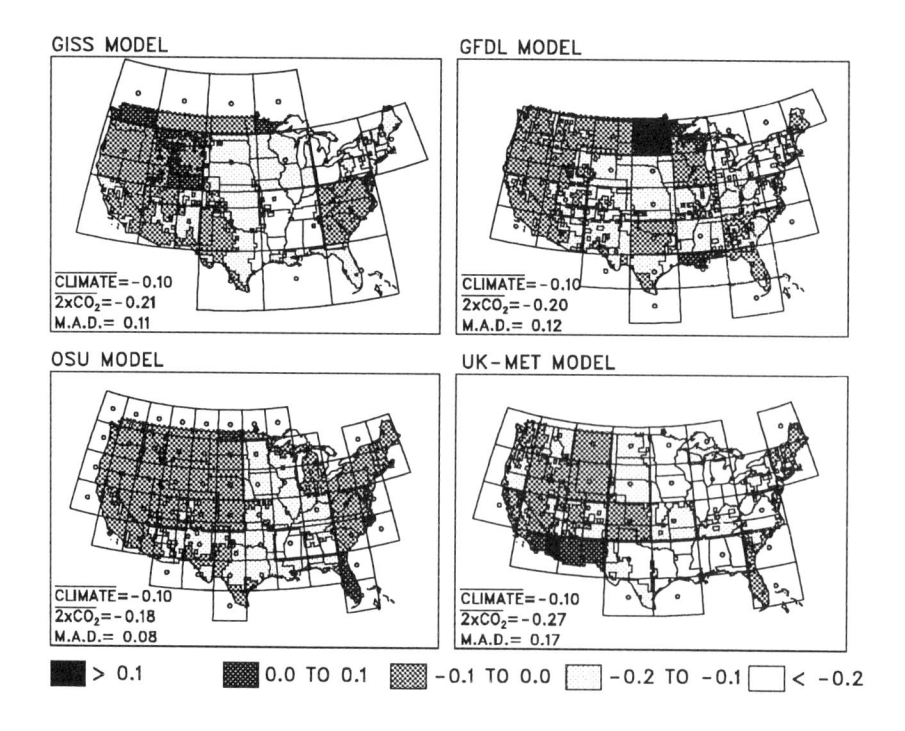

FIGURE 6. Change in annual climatic moisture index from present with doubling of CO_2 based on data from four GCMs. (See Figure 1 for legend)

currently available to compensate for the predicted increases in potential evapotranspiration. Based on predicted changes in the moisture index it would appear that very few regions would benefit from global warming with regard to water resources and agricultural productivity.

CONCLUSIONS

Using four GCM results as a template to show long term changes in mean monthly temperature and monthly total precipitation on a half-by-half degree grid, the United States is projected to experience a significant drying of its climate under double CO_2 conditions. Overall the models agree that all regions of the country may expect a considerable increase in water demand by natural vegetation and agricultural crops. While a general increase in precipitation may be expected at the same time, this increase is not expected to compensate for the increased demand on water resources. The expected result will be a decrease in agricultural productivity under dryland farming conditions especially in those regions where precipitation already is barely adequate to meet the water demand. Regions particularly affected include the Mississippi valley, the New Mexico/Colorado plateau, and the Midwest and Great Plains states. Regions that already have an adequate water supply, may expect a potential increase in summer droughts, but this will be accompanied by increased growing season lengths and most of the water deficit created by the increased moisture demand will be met by utilizing current moisture surpluses available in these areas.

Although the models tend to agree on the overall signal of climate change under double CO_2 conditions, there are several regions where there is considerable disagreement in the direction and magnitude of the observed climate changes, particularly in the southwestern and northeastern portions of the country. Furthermore, it is demonstrated that, from a useable moisture point of view (actual evapotranspiration), there can be quite extreme (and even opposite) variations in predicted climate changes at the subgrid level. These changes are mostly due to the loss of surplus water available under current climate conditions, and the conversion of the same water for use in evaporative processes under a warmer double CO_2 climate with higher potential evapotranspiration demands.

REFERENCES

Environmental Protection Agency, 1984: Potential Climatic Impacts of Increasing Atmospheric CO_2 with Emphasis on Water Availability and Hydrology in the United States. Strategic Studies Staff, Office of Policy Analysis.

Hansen, J. A. Lacis, D. Rind, G. Russell, P. Stone, I. Fung, R. Ruedy and J. Lerner. 1984. "Climate Sensitivity: Analysis of Feedback Mechanisms," in J.E. Hansen and T. Takahasi (eds.), *Climate Processes and Climate Sensitivity,* 130-163. Geophysical Monograph Series, vol. 29, American Geophysical Union, Washington, D.C.

Kalkstein, L.S. (ed.) 1991. *Global Comparisons of Selected GCM Control Runs and Observed Climate Data.* United States Environmental Protection Agency, Office of Policy, Planning, and Evaluation. Climate Change Division.

Legates, D.R. and C.J. Willmott, 1990a. Mean Seasonal and Spatial Variability in Gauge-Corrected Global Precipitation. *International Journal of Climatology,* 10:111-127.

Legates, D.R. and C.J. Willmott. 1990b. Mean Seasonal and Spatial Variability in Global Surface Air Temperature. *Theoretical and Applied Climatology,* 14:11-21.

Mather, J.R. 1978. *The Climatic Water Budget in Environmental Analysis.* Lexington, MA: Lexington Books.

Mitchell, J.F.B. 1989. The "Greenhouse" Effect and Climate Change. *Reviews of Geophysics,* American Geophysical Union, 27(1):115-139.

Schlesinger, M.E. and Z. Zhao. 1987. Seasonal Climate Changes Induced by Doubled CO_2 as Simulated by the OSU Atmospheric OCM/Mixed Layer Model. *Report 70,* Oregon State University Climate Institute, Corvallis, 73 pp.

Thornthwaite, C.W. and J.R. Mather. 1955. The Water Balance. *Publications in Climatology,* Laboratory of Climatology, 8(1):1-104.

Washington, W.M. and G.A. Meehl. 1984. A Seasonal Cycle Experiment on the Climate Sensitivity Due to a Doubling of CO_2 with an Atmospheric General Circulation Model Coupled to a Simple Mixed Layer Ocean Model. *Journal of Geophysical Research,* 89:9475-9503.

Wetherald, R.T. 1991. "Changes of Temperature and Hydrology Caused by an Increase of Atmospheric Carbon Dioxide as Predicted by General Circulation Models," in R.L. Wyman (ed.) *Global Climate Change and Life on Earth,* 1-17. New York: Routledge, Chapman and Hall.

Wetherald, R.T. and S. Manabe. 1986. An Investigation of Cloud Cover Change in Response to Thermal Forcing. *Climatic Change,* 8:5-24.

Willmott, C.J. and J.J. Feddema. 1992. A More Rational Climatic Moisture Index. *Professional Geographers,* 44(1):84-88.

Wilson, C.A. and J.F.B. Mitchell. 1987. A Doubled CO_2 Climate Sensitivity Experiment with a GCM Including a Simple Ocean. *Journal of Geophysical Research.* 92:13,215-13,343.

Wilson, C.A. and J.F.B. Mitchell. 1987. A Doubled CO_2 Climate Sensitivity Experiment with a GCM Including a Simple Ocean. *Journal of Geophysical Research.* 92:13,215-13,343.

Global Climate Change: Implications, Challenges and Mitigation Measures. Edited by S.K. Majumdar, L.S. Kalkstein, B. Yarnal, E.W. Miller, and L.M. Rosenfeld. © 1992, The Pennsylvania Academy of Science.

Chapter Six

THE PHYSICS OF THE GREENHOUSE EFFECT

THOMAS P. ACKERMAN
[1]Department of Meteorology and Earth System Science Center
The Pennsylvania State University
University Park, PA 16802

INTRODUCTION

Although the climate of the earth is complicated and highly non-linear, many of its properties can be illustrated with rather simple models. In this paper, we focus our attention on a few of these simple models and use them to demonstrate the physics of global warming and some of the important feedback processes. In some cases, we even can use the models to come up with back-of-the-envelope estimates of the magnitudes of the warming produced by different physical mechanisms. As we proceed, however, it should be remembered that reality is far more complex than the simple models we use here.

PLANETARY RADIATIVE EQUILIBRIUM

If we consider a planetary atmosphere as a single thermodynamic system, we can apply to it the first law of thermodynamics, which states the change in internal energy of the system is the difference between the rate at which heat is added to the system and work is done by the system. For an ideal gas, the change in internal energy is simply the change in temperature of the gas multiplied by its mass and specific heat capacity at constant pressure. If we assume that the atmosphere under consideration is in equilibrium, i.e., the mean atmospheric temperature is constant, then the heat added must equal the work done. In an ideal gas, work is done by compression or expansion. However, if we average over the entire atmosphere, then from continuity every expansion is balanced by an equal compression and the net work done is zero. Thus, we find that at equilibrium, the net heat added to the atmosphere must also be zero.

If the planetary system is observed from outside the atmosphere, then the heating can consist only of radiation absorbed and emitted. For convenience, we break Q into two components: Q_{SOL}, representing the absorbed solar radiation, and Q_{IR}, representing the thermal radiation emitted by the planet. From our application of the first law,

$$Q_{SOL} + Q_{IR} = 0. \tag{1}$$

This condition is usually referred to as *radiative equilibrium*.

We can use (1) to compute the global mean temperature by balancing the energy absorbed and emitted. The rate of solar absorption is simply

$$Q_{SOL} = S_0(1 - A_P)\pi R_e^2$$

where A_P is the planetary albedo (or reflectivity) and S_0 is the solar irradiance at the radius of the earth. For the earth A_P has a value of about 0.3, due to scattering by molecules (also called Rayleigh scattering), particles, clouds and the earth's surface. For computational purposes in this paper, the solar irradiance is taken to be 1365 W/m². This quantity, sometimes referred to as the solar constant, is actually not constant, but varies slightly with solar activity and possibly with some long-term trend as well. R_e is the radius of the earth. It appears in the equation as the cross-sectional area of the planet because we are interested in the flux absorbed, which is the component of the solar radiation perpendicular to the earth's surface. For the planet as a whole, the intercepting area is its cross-section.

The rate of thermal emission can be written as

$$-Q_{IR} = \sigma T_e^4 (4\pi R_e^2)$$

where σ is the Stefan-Boltzmann constant, and T_e is the radiative equilibrium temperature, i.e., the average temperature at which the planet must radiate in order to balance the absorbed solar radiation. Note that we have written Q_{IR} with a negative sign because it is defined to be the energy absorbed *by* the system. The geometrical factor in this case is the surface area of the planet, since the exiting flux is emitted perpendicular to the planet everywhere. Note that we do not distinguish between the radius of the planetary surface and the radius of the emitting surface since for the earth they vary by less then 10 km on average. Or, to put it more simply, the thickness of the atmosphere (which is not a well-defined quantity) is very small relative to the radius of the earth.

Equating these two expression results in the equilibrium condition:

$$\tfrac{1}{4}(1 - A_P)S_0 = \sigma T_e^4. \tag{2}$$

As discussed, the factor of 1/4 in this equation arises from considering the planetary geometry. If we consider a single average column on the earth, we can argue for the same factor by noting that the sun shines only half the day and the cosine of the incident angle is on average about one half. From (2), we compute a value of the radiative equilibrium temperature of 255 K or $-18°C$.

From observations we can show that the average surface air temperature of the earth is about 286 K, which is significantly higher than the radiative equilibrium temperature. Does this mean that our energy balance expression is incorrect? Actually, no. In fact, the computed value of T_e is very close to the average value observed by satellite. This temperature of 255 K represents the average temperature of the thermal radiation emitted by the surface plus atmosphere. In the earth's atmosphere, this temperature occurs on average at an altitude of about 5 km. Thus, the effective radiating level of the combined system is not at the surface but at approximately the middle (by mass) of the atmosphere. (We will return to this point in Section VII.)

THE ATMOSPHERIC GREENHOUSE EFFECT

As we saw above, the surface air temperature is higher than the radiative equilibrium temperature of the planet. This is presumably due to the thin sheath of gas that surrounds the planet. What is it about the atmosphere that produces this warming of the surface?

Over 99% of the atmosphere is composed of diatomic and triatomic molecules. These molecules can absorb radiative energy over a wide range of the electromagnetic spectrum. However, they are in general much more effective at absorbing energy at infrared wavelengths (from about 1 to 100 μm) than at visible and near-infrared wavelengths (from about 0.3 to 1 μm). The reasons for this difference are complex and beyond the scope of this paper. As a consequence, the bulk of the solar radiation passes through the atmosphere and is absorbed at the surface. This warm surface emits thermal radiation, essentially as a black-body. The thermal radiation is partially absorbed in the atmosphere by a variety of atmospheric constituents such as H_2O, CO_2, and O_3. Although this absorption varies with wavelength, we will for the moment assume that it is spectrally invarient. This assumption is equivalent to stating that the atmosphere absorbs as a grey-body with an emissivity, ϵ_a, that is independent of wavelength (at thermal infrared wavelengths) and less than unity.

Based on this simple interpretation of the physics of the atmosphere, we can construct a one-layer model of the earth-atmosphere system as shown in Figure 1. The atmosphere is represented by a single layer with a uniform temperature, and the earth by another layer with a uniform temperature. It is assumed that no solar radiation is absorbed in the atmosphere. The validity of this assumption will be addressed in the following section. At equilibrium, the energy balance in this simple system can be represented by two equations in two unknowns, the temperature of the surface (T_s) and the temperature of the atmosphere (T_a). These two equations are:

$$\tfrac{1}{4}(1 - A_P)S_0 + \epsilon_a \sigma T_a^4 = \sigma T_s^4 \qquad (3)$$

$$2\epsilon_a \sigma T_a^4 = \epsilon_a \sigma T_s^4 \qquad (4)$$

The first of this pair represents the energy balance at the surface, and the second the energy balance of the one-layer atmosphere. The factor of 2 arises because a layer radiates in two directions, both up and down. Solving these two equations, we find that

$$T_a = \frac{1}{\sqrt[4]{2}}T_s \tag{5}$$

$$\sigma T_s^4 = \frac{\frac{1}{4}S_0(1-A_P)}{(1-\frac{1}{2}\epsilon_a)} \tag{6}$$

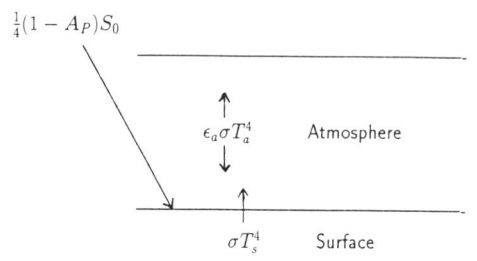

FIGURE 1: A schematic diagram of a one-layer atmospheric model.

Note that in this simple model, the atmospheric temperature depends only on the surface temperature. This result arises from Kirchhoff's law, which states that the absorptivity and emissivity of a black (or grey) body must be equal. The surface temperature is both a function of the solar energy absorbed and the emissivity of the atmosphere because the surface receives energy both from the sun and the thermal emission of the atmosphere. This latter term is the source of what we commonly call the "greenhouse effect". Although a greenhouse operates on quite another principle (suppressing heat loss by preventing convection and turbulent mixing), the name has become ingrained in the popular press and perhaps conveys a useful image. It might be more useful to describe this process as the "insulation effect" since the atmosphere behaves somewhat as insulation, preventing the loss of heat from the surface. Of course, the effects of home insulation are largely produced by preventing convective, rather than radiative, losses so this analogy is also not scientifically accurate.

The surface and atmospheric temperatures computed from (5) and (6) are presented in Figure 2. We can rewrite these equations making use of the relationship $\sigma T_e^4 = \frac{1}{4}(1-A_P)S_0$. Here T_e is the radiative equilibrium temperature found in the previous section. In this case, we have

$$T_s = T_e/(1-\tfrac{1}{2}\epsilon_a)^{1/4} \geq T_e \tag{7}$$

$$T_a = T_e/(2-\epsilon_a)^{1/4} \leq T_e \tag{8}$$

The inequalities arise because the emissivity of the atmosphere in our simple model is constrained to lie between 0 and 1. The lower value represents the surface with no atmosphere and the upper value represents the theoretical limit of a black-body atmosphere. As the emissivity approaches 0, $T_s \rightarrow T_e$, since there is no atmosphere to radiate energy down to the surface. We can derive an atmospheric temperature for this case, but it is physically meaningless. An examination of (4) shows that we are solving an equation in which both sides are 0; hence there is no solution for T_a. As the emissivity approaches 1, we find that $T_a = T_e$. This is due to the fact that no thermal emission emitted by the surface is transmitted by the atmosphere. Thus the energy emitted from the atmosphere must be equal to the radiation absorbed by the planet.

The spectrally-averaged emissivity of the earth's atmosphere is about 0.8. This leads to a surface temperature of about 288 K and an atmosphere temperature of about 244 K. The former is higher than the radiative equilibrium temperature, the latter is lower. In short, the atmosphere warms the surface by about 33 K over the radiative equilibrium temperature. Without this effect, the surface of our planet would be a substantially colder and less hospitable environment.

Before concluding this section, a word of warning should be given. It is often stated that if there were no atmosphere, the surface temperature of the earth would be lower by about 33 K. This is, however, a meaningless statement because the solar absorption would also change if the atmosphere were absent. The plantary reflectivity is determined largely by the presence of clouds and molecules, both of which would be absent if the atmosphere were absent. In addition, the surface of the planet would be quite different in the absence of oceans and vegetation. The earth would then resemble the moon or Mars. For both these bodies, the surface temperature can be predicted quite accurately from (3) with $\epsilon_a = 0$.

CHANGING THE ATMOSPHERIC EMISSIVITY

The emissivity of the earth's atmosphere is determined primarily by the concentrations of H_2O and CO_2, and secondarily by the concentrations of other trace gases that absorb infrared radiation. Thus, altering the concentration of any of these gases will alter the emissivity of the atmosphere, which in turn changes the surface temperature. It is of interest to develop a relationship between the change in surface temperature in our simple model for a given change in atmospheric emissivity.

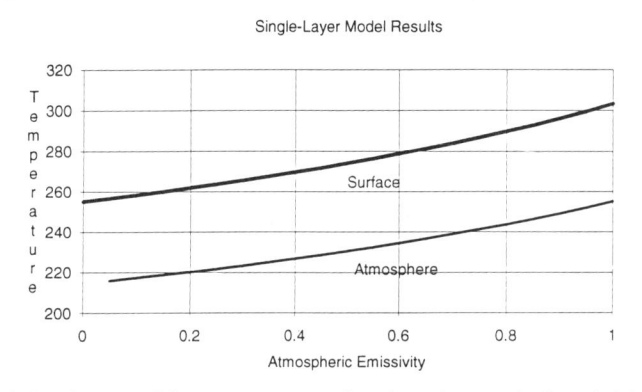

FIGURE 2. One-layer model temperatures as a function of atmospheric emissivity.

This is essentially the slope of the curve shown in Figure 2.

Differentiating (3) with respect to emissivity and rearranging, we find

$$\frac{\Delta T_s}{T_s} = \frac{1}{4} \frac{\frac{1}{2}\Delta\epsilon_a(\sigma T_s^4)}{(1 - \frac{1}{2}\epsilon_a)(\sigma T_s^4)}, \qquad (9)$$

which has an straightforward physical interpretation. The denominator on the right (not including the factor of 4) is the solar energy absorbed in our system. The numerator is the change in energy radiated to the surface due to the change in emissivity ($\Delta\epsilon_a$). (Note that, from (4), $T_a^4 = T_s^4/2$.) Thus the ratio of the change in surface temperature to the surface temperature itself is equal to one-fourth the ratio of the change in emitted energy by the atmosphere to the energy absorbed in the system. The factor of one-fourth arises from the fourth power dependence of blackbody radiation.

Let us now apply this expression to an actual atmospheric problem. From more sophisticated radiative transfer calculations (see for example, Ramanathan and Coakley, 1979), we can determine that doubling the atmospheric CO_2 concentration will increase the downward infrared radiative flux at the surface (the numerator in our expression above) by roughly 4 W/m². For current conditions, the absorbed solar energy amounts to some 240^w/m² and $T_s \approx 288K$. Plugging these values into (13), we find $\Delta T_s = 1.2K$. In other words, for each doubling of CO_2, we expect the surface temperature to increase by about 1.2 K.

At this point, we might conclude that we have solved the greenhouse problem. Given a radiative transfer algorithm sufficiently powerful to calculate the change in downward infrared flux for a particular change in the concentration of some gaseous absorber, we can use our simple model to evaluate the corresponding change in surface temperature.

Is this all there is to the greenhouse problem? Well, not quite! In the course of developing our simple model and solution, we introduced several simplifying assumptions and neglected some other important processes. Five questions come to mind:

1. Is a one-layer atmosphere model adequate?
2. Is solar energy actually deposited only at the surface? If not, what impact does this have on our results?
3. Is the atmosphere grey?
4. What about the response of that other important absorber, water vapor?
5. What about clouds?

In the succeeding sections, each of these questions will be addressed and an attempt made to relate the implications of the answers to our problem of greenhouse warming.

A MULTI-LAYER MODEL

The answer to our first question is an emphatic "no". A simple examination of any observed vertical profile of atmospheric temperature shows that the temperature decreases typically by 60 to 80°C from the surface to the tropopause, defined as the boundary between the relatively well-mixed lower 8 to 12 km of the atmosphere (the troposphere) and the stably stratified stratosphere. Above the tropopause, the temperature increases with height to the top of the stratosphere. This increase is caused by the absorption of solar energy by ozone and oxygen. Although this increase in temperature has a number of important consequences, it has little effect on surface temperature and atmospheric energetics, so we will not discuss it further. The fact that temperature decreases by such a large amount across the tropopause means that our assumption of a single atmospheric layer with a uniform temperature is incorrect.

The shape and magnitude of the tropospheric temperature profile is discussed in a later section. However, it is fairly easy to demonstrate from radiative considerations why the temperature must decrease with altitude in an atmosphere with more than one layer. A schematic diagram of a multi-layer model is shown in Figure 3. Here each layer is labeled with a temperature T_i and an emissivity ϵ_i. If we again assume that solar absorption occurs only at the surface, we find at equilibrium that the temperature of the layers decreases monotonically from the surface to the top layer. While the actual values of the layer temperatures depend on the layer emissivities, the monotonic decrease holds regardless of the actual emissivity values. This smooth temperature decrease with height may not be intuitively obvious, but can be rationalized from simple energetics arguments. The surface is the warmest, just as in our one-layer model, because it receives both solar radiation and downward thermal radiation from each of the layers above. The temperature of any atmospheric layer is determined by balancing the partial absorption of radiation emitted by all the layers above and below it against the radiation emitted by the

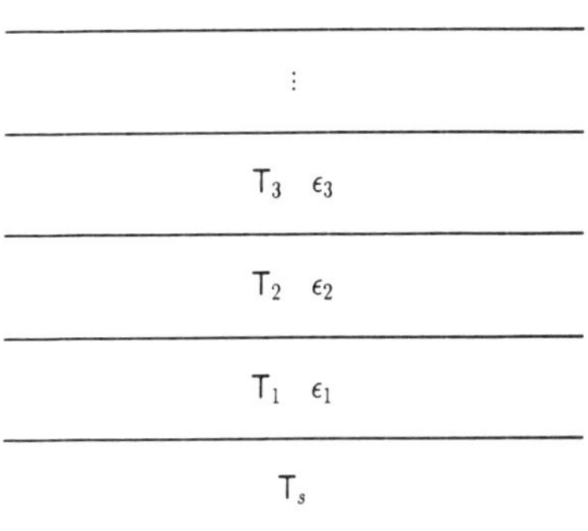

$T_n \quad \epsilon_n$

$\vdots$

$T_3 \quad \epsilon_3$

$T_2 \quad \epsilon_2$

$T_1 \quad \epsilon_1$

T_s

FIGURE 3. A schematic diagram of a multi-layer atmospheric model.

layer itself. The top layer can only absorb radiation emitted from the layers below, although it still radiates in both directions. We can view this problem as each of the layers exchanging thermal radiation with each other, with the most important exchange occurring between adjacent layers. However, the closer the layer is to the upper boundary, the fewer layers there are between that layer and cold space. Thus, the higher the layer, the more energy it radiates to space without receiving any compensating exchange in return. Hence, the layer temperatures decrease with height.

It is a relatively straight-forward problem to consider a two-layer atmosphere. The result is a set of three coupled equations in three unknowns. Interested readers are encouraged to write out this set and solve it as a function of layer emissivities to convince themselves that indeed the temperature decreases with height. It is also instructive to consider a system of n black layers. In this case, the equations simplify considerably because the layers are nontransmitting. This simple model also shows that the temperature decreases with altitude.

Given this behavior of our atmospheric system, it seems quite likely that we need to expand our one-layer model to a multi-layer model in order to do justice to the impact of increasing concentrations of absorbing gases. The implications of this change are considered below.

ATMOSPHERIC RADIATIVE TRANSFER

Answering the second and third questions from our list requires us to first take a small sidetrip and discuss the transfer of solar and infrared radiation in the atmosphere. When a beam of radiation interacts with a volume of atmosphere, three types of processes may occur. First of all, photons may be absorbed by either molecules or particles found in that volume. Absorption by molecules is of course highly dependent on the wavelength of the incident radiation, while absorption by particles tends to be a much weaker function of wavelength. In either case, the absorption is strongly dependent on the specific molecules and particle composition. Thus, we need to be able to specify the composition of the atmosphere in some detail to be able to compute absorption of radiation.

The second process that may occur is scattering, which may be thought of in classical terms as a simple redirection of the photon due to interaction with a molecule or particle. Scattering is a complex phenomenon and an adequate description is beyond the scope of this short paper. Excellent treatments of the various aspects of the problem can be found in books such as van de Hulst (1975) and Bohren and Huffman (1983). For our purpose it is sufficient to note that the effect of scattering is to remove photons from the incident beam of radiation and direct them into some other direction. For particles with dimensions near and larger than the wavelength, the preferential direction of scatter is in the forward cone, i.e., in directions not very far from the original direction of the photon. This is a consequence of the nature of the scattering of electromagnetic waves and quite simply has no analog in classical particle scattering. Thus, thin layers of large particles (for example, a thin cirrus cloud) tend to simply diffuse an incoming beam of radiation slightly about its original direction without causing substantial change to the number of photons traversing the layer.

In a optically thick medium, i.e., one in which a photon may be scattered many times as it traverses the medium, the result is to produce a diffuse field of transmitted and reflected photons. This is the effect of a layer of stratus clouds on solar radiation. Although the cloud is sufficiently thick that relatively few direct-beam photons from the sun penetrate the layer (the solar disc cannot be observed), many photons do penetrate the layer as a diffuse beam (hence, the white light of layer clouds). Of course, many photons are also reflected, leading to the bright appearance of clouds on visible-wavelength satellite imagery. The fact that so many diffuse photons penetrate the cloud is due to the tendency for the photons to be forward scattered. If this were not the case, our world would be very dark whenever we found ourselves under any substantial cloud cover!

The third process of concern is emission. Since our volume must be in thermal equilibrium, any absorption of energy must be compensated for by emission. This emission is of course a function of temperature as specified by Planck's law, and is also uniform in direction.

A consideration of these fundamental processes leads us to make a distinction between solar and thermal radiation. Because atmospheric temperatures are much lower than the temperature of the sun, no atmospheric emission occurs at solar wavelengths (less than about 4 μm). Furthermore, much of the solar radiation occurs at solar wavelengths that are not absorbed by atmospheric gaseous constituents because the energy is too high to be effective in exciting rotational or vibrational transitions, and not sufficiently high to cause electronic transitions or dissociation. Hence, the solar radiative transfer problem can be posed largely as the transfer of an external source of photons through a principally scattering medium, with occasional absorption bands. The thermal radiative transfer problem is in some ways much more complicated since it is an internal source problem, i.e., each volume is a source as well as a sink of photons, with highly variable absorption as a function of wavelength. For a clear atmosphere, scattering may be largely neglected, which simplifies our problem enormously. However, in a cloudy atmosphere, we may be forced to consider all three processes simultaneously!

Considering first the transfer of solar radiation through a cloudfree atmosphere, we obtain the result shown in Figure 4 (taken from Liou, 1980). The upper curve depicts the spectrally-dependent solar intensity at the top of the atmosphere. This curve is closely approximated by a 5500°C black body curve. The curve just beneath the upper curve represents the amount of solar radiation that would reach the surface if only molecular, or Rayleigh, scatter by the atmosphere were considered. Because the molecular scattering coefficient scales with λ^{-4}, the effect of scattering is much more pronounced at short than at long wavelengths. (It is this particular effect that gives rise to the blue color of the clear sky.) The cross-hatched areas indicate areas of gaseous absorption and the particular molecular species responsible for each absorption band is shown on the bar on the horizontal axis. Neglecting for the moment the small but very important role played by ozone at wavelengths shorter than 0.4 μm, the bulk of the solar absorption in the atmosphere is due to water vapor. If we note that about 70% of the water vapor in an atmospheric column is found within 2 km of the ground, it is apparent the majority of the absorption indicated in the figure occurs near the ground. Thus our initial assumption that absorption occurred only at the ground is fairly reasonable.

The one important exception to this assumption is absorption of ultraviolet radiation by ozone and molecular oxygen. This is of course beneficial from the point of the biological community, but it is also responsible for producing the stratosphere, which we shall discuss in more detail below.

Thermal infrared radiative transfer in clear sky conditions is dominated by gaseous absorption and emission. A useful way of visualizing this is shown in Figure 5 (Liou, 1980). The smooth curves on the diagram represent the Planck function at the indicated temperature, while the heavy jagged curve is an interferometric measurement made at nadir view from a satellite over the Pacific Ocean. This measurement is expressed in terms of equivalent radiative temperature, which is simply the temperature that a black-body would have if it were emitting the measured radiance. Several features of interest are immediately apparent. In the region between 8 and 12 μm (8-12 cm^{-1}), the observational curve nearly parallels the Planck curves at a temperature of about 295 K. This is the so-called atmospheric window in which atmospheric gaseous absorption is nearly zero. A weak continuum absorption due to water vapor, which becomes important in tropical atmospheres with large water columns, does occur in this region. Thus, the observed warm

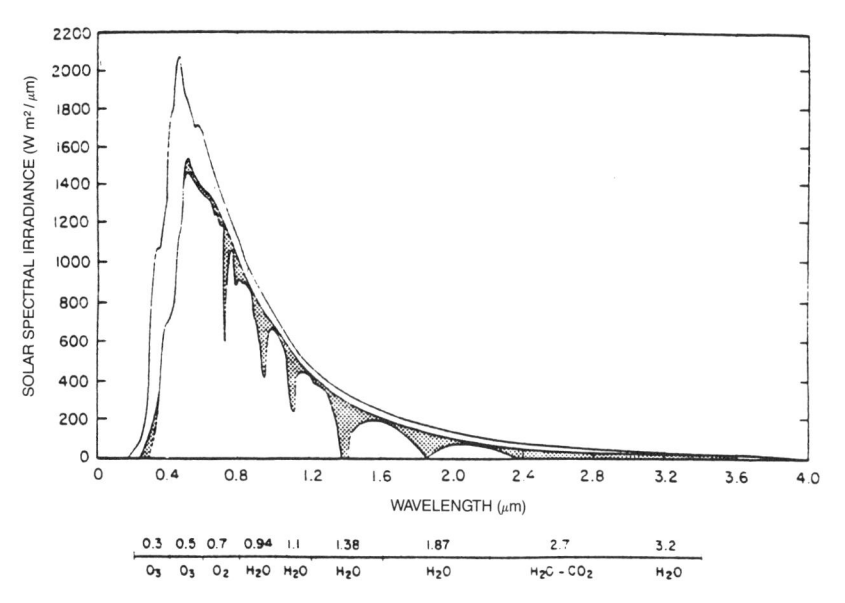

FIGURE 4. Spectral distribution of solar irradiance at the top of the atmosphere (upper curve), and at sea level (lower curve). The shaded areas represent absorption due to various gases (indicated on bar below the horizontal axis); the upper envelope of the shaded areas indicates the reduction of solar radiation due to molecular scattering. (taken from Liou, 1980)

radiation is being emitted from the surface or boundary layer and transmitted to space, thereby cooling the surface and lower atmosphere.

Several regions of the spectrum show radiative temperatures much lower than that of the window. In particular, the absorption feature at 15 μm (667 cm^{-1}) due to the fundamental vibrational band of CO_2 stands out. These very low temperatures associated with this band indicate that emission is occurring high in the atmosphere. Similar low temperatures are associated with the ozone feature at 9.6 μm and the water vapor vibrational feature shortward of about 7.5 μm. Low emission temperatures would also be seen longward of 400 μm due to water vapor rotational features if the interferometer had been capable of measuring at these wavelengths. While it is again beyond the scope of this paper to discuss these processes in detail (the interested reader is referred to the excellent book by Goody and Yung, 1989), the important inference to draw from this figure is that emission occurs at a variety of levels in the earth system because the opacity of the cloud-free atmosphere varies enormously with wavelength.

Returning now to our two questions about radiative transfer, we can provide some answers based on the preceding discussion.

- Absorption of solar radiation occurs principally at the surface or in the lowest kilometer or two or the atmosphere. The most notable exception is O_3 absorption of uv radiation in the upper atmosphere. Therefore, our assumption that absorption occurs only at the surface is moderately good, but will not give a correct stratospheric temperature profile.
- Molecular absorption in the thermal infrared is highly spectrally-dependent. Consequently, infrared absorption and emission are strong functions of altitude, as is the associated radiative cooling of the atmosphere. Therefore, our assumption that the atmosphere is a grey absorber is very poor!

So now what do we do to patch up our model?

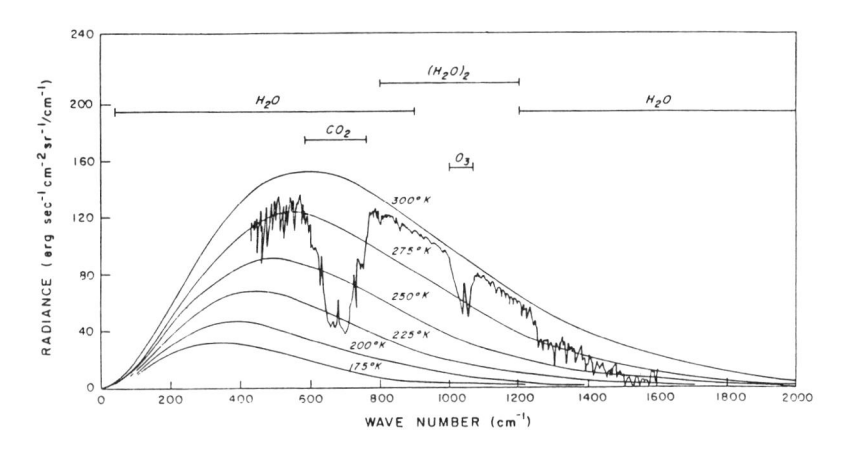

FIGURE 5. The terrestrial infrared spectrum observed at nadir view by the Nimbus IV IRIS instrument near Guam (15.1°N, 215.5°W) on April 27, 1970. The horizontal bars indicate the atmospheric gases responsible for the observed absorption features. The smooth curves are the Planck function calculated at the indicated temperature. (taken from Liou, 1980).

RADIATIVE EQUILIBRIUM IN MULTIPLE LAYERS

The straight-forward solution to fixing the model is to rewrite the condition for radiative equilibrium in terms of a vertical coordinate. This requires the solar heating and infrared emission to cancel each other, not just at the top of the atmosphere, but at each level within the atmosphere. Solving this problem is considerably more complicated than that outlined above for the multi-layer model. First, the solar absorption must be computed in the atmosphere as well as at the ground. In addition, infrared absorption and emission must be treated in spectral detail, rather than with a simple emissivity expression. However, without going into elaborate detail, the solution steps can be described.

We begin by subdividing the atmosphere into discrete layers (Figure 3). The criteria used for the division are not rigorous, but basically are that the spectrally-averaged emissivity for each layer should be considerably less than one, and that, when the equilibrium temperature is found, the temperature change from layer to layer should not be too large. Next, concentrations of H_2O, CO_2, and O_3 are specified for each layer from a given atmospheric profile. Clouds and surface properties are specified as well. Then, an initial temperature profile, $T(z)$, is assumed and Q_{SOL} and Q_{IR} are computed as functions of z. A new $T(z)$ profile is computed from the initial $T(z)$ profile plus the calculated heating rates. The latter procedure is interacted until the condition $dT(z)/dt = 0$ is met everywhere.

The results of a radiative equilibrium calculation for the earth's atmosphere are shown in Figure 6. Note the extremely warm surface temperature of about 345 K or 60°C and the rapid decrease of temperature with height. This is the result of the strong solar heating at the surface and in the lowest layers of the atmosphere. In order to balance this heating with infrared emission, it is necessary for the surface to get very warm. The temperatures in the region from 5 to 15 km are extremely cold because the greenhouse gases emit infrared radiation in this region but there is very little solar absorption. Finally, warming occurs due to solar absorption by O_3 in the region above 15 km.

The radiative-equilibrium temperature profile shown in Figure 6 is obviously incorrect. The surface temperature is far too warm and the lower atmosphere too cold. Once again we have neglected an important physical process in the atmosphere. In this case, it is convection. The atmosphere is a compressible fluid. Thus, as a parcel of air rises in the atmosphere, it expands and does work against its environment, thereby cooling. We define this rate of temperature decrease as the lapse rate = $\Gamma = -dT(z)/dz$. For a dry atmosphere, this rate becomes the *dry adiabatic lapse rate* defined as

$$\Gamma_D = g/c_P = 9.8 K/km$$

In the case of a saturated atmosphere, condensation of water occurs during the lifting process and latent heat is released, thereby reducing the parcel cooling. For this process, we define the *moist adiabatic lapse rate*, Γ_M, which is not constant but is a function of parcel temperature. For our purpose here, we simply note that $\Gamma_M \leq \Gamma_D$ always.

Now, for a dry parcel convection will occur if $\Gamma > \Gamma_D$, while for a saturated parcel convection will occur if $\Gamma > \Gamma_M$. Looking again at our radiative equilibrium profile in Figure 6, we see that the lapse rate in the lower km or so of the atmosphere clearly exceeds these conditions. Simply put, if we heat the surface air to these high temperatures, it will become unstable and rise. The effect of this process is to move heat upward from the surface into the atmosphere. The average atmospheric lapse rate is a result of some complex mixture of dry and saturated processes and, from observations, we find that it is on the order of 6.5K/km. Thus, to complete our model, we make one simple addition assumption: the lapse rate between any two layers is constrained such that $\Gamma \leq 6.5 K/km$. The resulting equilibrium temperature profile produced by solving the equilibrium problem subject to this constraint is shown in Figure 6. This atmospheric state is called *radiative-convection equilibrium* (RCE).

Comparing the two profiles in Figure 6, we see that the RCE surface temperature has dropped relative to radiative equilibrium, while the atmospheric temperatures have risen. A comparison with observations shows that this type of model does a good job of representing the average thermal structure of the atmosphere (as well it might, since we chose our lapse rate based on observations!). More importantly,

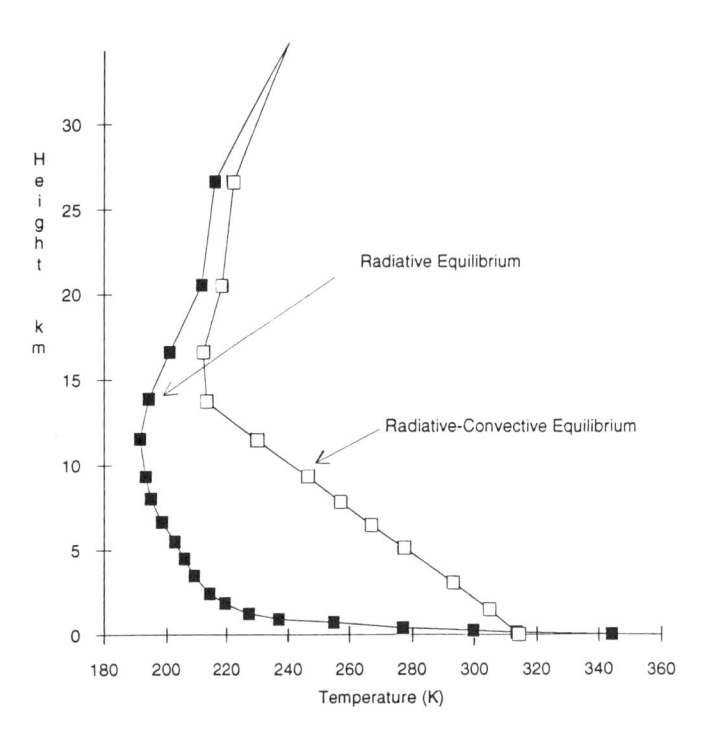

FIGURE 6. Radiative equilibrium (solid markers) and radiative-convective equilibrium (open markers) temperature profile for the earth atmosphere.

however, we have developed what might be called the poor man's Global Climate Model. It can be (and has been) used to investigate the impact of a variety of interesting climate-related processes such as the impact of changes in greenhouse gas concentrations on temperature profiles. Thus, in trying to answer our questions about the nature of radiative transfer in the atmosphere, we have been led to the development of a far more sophisticated and useful model than our simple one-layer emissivity model.

WATER VAPOR

Our fourth question dealt with the response of water vapor concentration in the atmosphere to changes in temperature. The concentration of water vapor in an atmospheric parcel is limited by the saturation vapor pressure, $e_s(T)$, which is, as indicated, a function of temperature. This relationship is given by the Clausius-Clapeyron Equation as

$$e_s(T) = e_{s0} \exp[(L_v/R_w)(1/T_o - 1/T)]$$

Here, e_{s0} is the saturated vapor pressure at a reference temperature T_o, L_v is the latent heat of vaporization for water, and R_w is the gas constant for water. A plot of this equation is shown in Figure 7. Note the exponential rise in the vapor pressure with increasing temperature.

Quite obviously, air parcels are not in general saturated with water vapor. If they were, fog and clouds would be present everywhere. Meteorologists typically define the water vapor concentration of a parcel in terms of the relative humidity (RH), which is the ratio of the parcel water vapor concentration to the saturation vapor pressure (i.e., $RH = e/e_s(T)$). Thus, RH is simply an indicator of how close a parcel is to saturation. On both theoretical and observational grounds we are led to suggest that, averaged over large regions of the globe, RH is approximately constant with time in the atmosphere. Although the reasons behind this fact are fairly complex, we can intuitively understand it by noting that a major share of our planet is covered by water. Hence, there is plenty of water availability at the surface. The limiting process with regard to atmospheric water vapor concentrations then becomes the rate at which water vapor can evaporate[1] into the near-surface atmospheric layer and be transported upward. The evaporation rate is basically controlled by two factors: the near-surface wind speed and the RH of this near-surface layer. As the RH of the layer approaches 1, the evaporation rate becomes smaller and smaller. Thus, for very dry air, evaporation rate is rapid and the RH rises but for

---

[1] Some confusion can occur regarding the usage of this term. In the case of a vapor over a liquid surface, evaporation and condensation are always occuring since molecules are continually moving from one phase to the other. At equilibrium (RH = 1), the two processes are exactly in balance and there is no *net* evaporation, defined as the difference between evaporation and condensation. In meteorology we are predominantly concerned with this net transfer of water from the surface to the atmosphere and hence we simply equate the concept of evaporation with this net transfer.

moist air (RH approaching 1) the evaporation rate slows to near zero. This tends to maintain a high and approximately constant RH in the atmosphere when averaged over large areas and times.

We now consider the issue of water vapor feedback. From the preceeding, we see that if the average temperature of an air parcel should increase (say due to an increasing CO_2 concentration), then the saturation vapor pressure of that parcel would also increase (Figure 7). But if, as we have just argued, the RH is to remain approximately constant, then the water vapor concentration (e) in the parcel must also increase. But, as we noted above, water vapor is the most important greenhouse gas so an increase in water vapor concentration leads to an increase in temperature. In short, we have a positive feedback. As T increases, e increases and as e increases, T increases!

We can use our same simple one-layer model to quantify this effect. We stated earlier that doubling the CO_2 concentration will increase the downward atmospheric radiation by about 4 W/m^2 and raise the temperature by about 1.2 K. If we then compute the additional water vapor that would be put into the atmosphere by this temperature rise (using the Clausius-Clapeyron Equation) and the additional rise in downward atmospheric radiation, we find the increase in forcing to be about a factor of 4. This factor of 4 increase in forcing translates directly into a factor of 4 increase in temperature (Eqn. 9). Thus for a doubling of CO_2, we expect an increase in temperature on the order of 4.8 K, which is in fact very near that predicted by 3-dimensional climate models.

CLOUDS

Our final question dealt with the role of clouds in the greenhouse problem. This is, without a doubt, the most complicated component of the problem and cannot be treated comprehensively in this short article. However, it is possible to summarize briefly the important radiative properties of clouds and their impact on climate. From this, in turn, we can deduce their role in climate change.

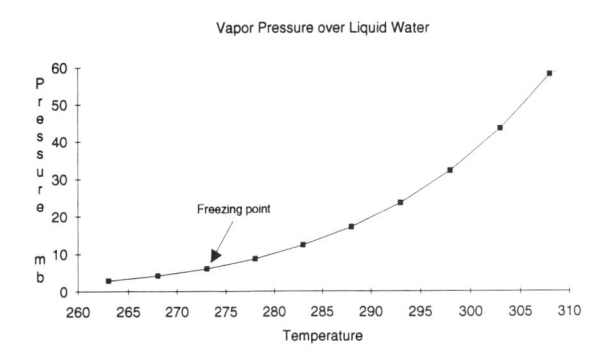

FIGURE 7. Saturation vapor pressure for water as a function of temperature. The freezing point of water is indicated for reference.

We begin by considering the bulk optical properties of a cloud. The most important of these properties is the combined effect of cloud optical thickness and absorption. To a first approximation, absorption of solar radiation by clouds can be neglected. Clouds simply increase the planetary albedo by reflecting incident solar radiation. The amount of sunlight reflected by the cloud increases monotonically with increasing optical thickness to some asymptotic limit. At thermal wavelengths, clouds are absorbers and increasing optical thickness increases the absorption. The absorption process varies approximately exponentially with optical thickness. Thus, an optical thickness of 1 or 3 causes the cloud to be nearly opaque. (These phenomena, and others of interest, are discussed in a lucid and elegant treatment of multiple scattering by Bohren, 1987.)

Because low clouds have relatively high liquid water concentrations, they have large optical thicknesses and are essentially black (non-transmitting) in the thermal infrared region. Thus increasing their optical thickness has little impact on infrared radiation, but does increase their solar reflectivity. High clouds, conversely, have relatively small optical depths, so changing their optical depths has a significant impact on the thermal radiation field.

The second important optical property of clouds has to do with the probability of scattering radiation either forward or backward, i.e., scattered into either the forward or backward hemisphere relative to the original direction of propagation. This particular quantity is a property of the size of the particles that make up the cloud. In general, for a fixed wavelength, the larger the particle, the higher the probability of forward scattering. Because ice particles in cirrus clouds are typically much larger than water droplets, cirrus clouds are more strongly forward scattering. As a consequence, they have a relatively small impact on the planetary albedo. The reflectivity of stratus clouds, however, can potentially be affected by changing the average particle size. Some evidence suggests that the number of drops in a stratus cloud can be increased by adding small particles to the environment. These particles may serve as additional nucleation sites and, for a given amount of condensation, this results in more and smaller drops. This increases the optical depth of the clouds and, since the drops are smaller, also increases the probability of back-scatter. Both effects cause the cloud to be more reflective.

The impact of clouds on climate is complicated. Apart from the difficulty of simply calculating the cloud optical properties, we find that, for any particular cloud, the solar and infrared effects are opposite in sign. Because low clouds primarily affect the solar radiation budget through increased reflectivity, they have an overall tendency to cool the earth. Conversely, high clouds tend to warm the system because they primarily act to reduce the outgoing infrared radiation. Thus, we also need to know the vertical location of any particular cloud and the vertical distribution of clouds in general.

Rather than trying to quantify the locations and properties of all clouds and then compute the impact on climate, an alternative approach to the problem is to try to observe directly the impact of clouds on the earth's climate. At this point, we do not have a completely adequate answer to this question. Satellite observations provide us with a means of measuring the impact of clouds on the planetary radiation budget, but this is by no means a trivial task. The Earth Radiation Budget Experiment is an ongoing effort to quantify from space cloud impacts on the radiation budget. This, however, still does not provide the entire answer because we need to know the impact of clouds on the surface energy budget as well. Surface observation systems are too coarse spatially to provide adequate coverage of the globe and we have not yet been able to devise sufficiently accurate algorithms to measure the surface budget from satellites.

Estimates of the effect of clouds on climate suggest that about a half to two-thirds of the planetary albedo is due to clouds, i.e., the planet would absorb an additional 15 to 20% of the incident solar radiation if no clouds were present. Presumably, most of this energy would be absorbed at or very near the surface. It is more difficult to estimate the greenhouse impact of clouds, but clouds probably reduce the outgoing longwave radiation by a factor of 20 to 40% over what would be emitted by a cloud-free planet with the same thermal structure. Clouds increase the downward thermal radiation at the surface by more than a factor of two.

Given our inability to measure cloud effects accurately, we might choose to compute them instead. Unfortunately, our understanding of the detailed cloud microphysical processes that are important in cloud droplet formation and maintenance is also limited. Furthermore, these calculations need to be carried out on very small spatial grids and very short timescales that are completely incompatible with climate models. The best that we can do in climate models is use empirical relationships that attempt to relate cloud properties to large-scale variables (such as average relative humidity and temperature) predicted by the model. The relationships are not well grounded observationally and are inadequate for the task at hand. The improvement of our understanding of cloud processes and the quantification of this understanding in climate models is the current focus of considerable research in meteorology.

If we have so much uncertainty about the role of clouds in the current climate, then we must have a greater uncertainty about the role of clouds if that climate is changed. The issue of cloud feedback in climate change is currently unresolved and the subject of considerable debate.

SUMMARY

All of the preceding discussion can be summarized in a few major points that cut to the heart of the global warming issue.

- The atmosphere, because it contains gases such as CO_2 and H_2O that absorb infrared radiation, raises the average surface temperature from about 255 K to the observed 288 K. The primary greenhouse gas is water vapor, followed by carbon dioxide and ozone. Clouds also play an important role in maintaining the observed surface temperature.
- Model results suggest that doubling the CO_2 concentration of the atmosphere would raise the surface temperature by about 1.2 K, all other processes remaining constant.

- Because the saturation vapor pressure of water increases with increasing temperature, this increase in CO_2 will be accompanied by a rise in the water vapor concentration. This leads to a further increase in surface temperature. In RCE models, the effect of this water vapor feedback is to increase the CO_2 temperature increase by a factor of about 3 or 4.
- The troposphere is in radiative-convective equilibrium. Convection transports energy upwards from the surface and this energy is radiated to space throughout the atmosphere. Because of this convective linking, the temperature of the entire troposphere will increase if the surface temperature increases.
- The impact of clouds on the current climate is complicated and not well understood. It is possible that, in a changing climate, cloud feedbacks could damp out the temperature changes suggested above. It is also possible (and no less probable!) that cloud feedbacks could increase the temperature changes above.

It is vital that our society comes to grip with the nature of the greenhouse problem and the potential for climate change. We are already in the midst of an unprecedented, inadvertant global climate change experiment; our questions now are how far are we going to carry out this experiment and where will it take us?

REFERENCES

Bohren, C.F. 1987. Multiple scattering of light and some of its observable consequences. *Am. J. Phys.*, 55:524-533.

Bohren, C.F. and D.R. Huffman. 1983. *Absorption and Scattering of Light by Small Particles.* John Wiley & Sons, New York, NY, 530 pp.

Goody, R.M. and Y.L. Yung. 1989. *Atmospheric Radiation, Theoretical Basis.* Oxford University Press, New York, NY, 519 pp.

Liou, K.N. 1980. *An Introduction to Atmospheric Radiation.* Academic Press, New York, NY, 392 pp.

Ramanathan, V. and J.A. Coakley. 1978. Climate modeling through radiative-convective models. *Rev. Geophys. and Space Phys.*, 16:465-490.

van de Hulst, H.C. 1957: *Light Scattering by Small Particles.* Dover Publications, New York, NY, 470 pp.

Global Climate Change: Implications, Challenges and Mitigation Measures. Edited by S.K. Majumdar, L.S. Kalkstein, B. Yarnal, E.W. Miller, and L.M. Rosenfeld. © 1992, The Pennsylvania Academy of Science.

Chapter Seven

GLOBAL WARMING:
A Case for Concern*

STEPHEN H. SCHNEIDER**
National Center for Atmospheric Research***
P.O. Box 3000
Boulder, CO 80307

INTRODUCTION

Why Build A Model

The effect of human pollution on climate is an uncontrolled experiment now being performed on "Laboratory Earth". In order to be anticipatory, we can turn to a surrogate lab, we can build mathematical models of the Earth and perform our "experiments" in computers.

Mathematical models translate conceptual ideas into quantitative statements. There are a range of models, from those that simply treat one or two processes in detail to large scale, multi-process simulation models. Such models usually are not faithful simulators of the full complexity of reality, of course, but they can tell us the logical consequences of explicit sets of plausible assumptions. To me, that certainly is a big step beyond pure conception—or to put it more crudely, modeling is a major advance over purely intuitive "hand-waving" predictions of our climatic future.

---

*Adapted in part from "Prediction of Future Climate Change" *New Scientist*, 128, 49-51.
**Any opinions, findings, conclusions or recommendations expressed in this article are those of the author and do not necessarily reflect the views of the National Science Foundation.
***The National Center for Atmospheric Research is sponsored by the National Science Foundation.

However, even the most sophisticated climatic models still are plagued by great uncertainties, which puts a premium on understanding what they can do well, where they have some credibility and their aspects that are pure speculation.

BASIC ELEMENTS OF MODELS

Modelers speak of a hierarchy of models that ranges from simple earth-averaged, time-independent, temperature models up to high-resolution, three-dimensional, time-dependent models known as general circulation models (GCMs) (e.g., Washington and Parkinson, 1986). While three-dimensional, time-dependent models are usually more dynamically accurate, they are very computer intensive. Thus, it is sometimes necessary to run physically, chemically, and (somewhat) biologically comprehensive models in lower resolution. Choosing the optimum combination of factors is an intuitive art that trades off completeness and (the modelers hope) accuracy for tractability and economy. Moreover, the theoretical feasibility of long-range simulation must also be evaluated; in other words, some problems are inherently unpredictable. These include detailed forecasts of weather events past a few weeks, the precise time and space evolving patterns of climatic anomalies (e.g., Chervin 1981), global scale stochastic fluctuations (e.g., Hasselmann 1976) or chaotic behavior (e.g., Lorenz 1968). For those problems where predictability is not ruled out in principle, such a trade-off between accuracy and economy is not "scientific" per se, but rather is a value judgment, based on the weighing of many factors. Making this judgment depends strongly on the problem the model is being designed to address.

To simulate the climate, a modeler needs to decide which components of the climatic system to include and which variables to involve. For example, if we choose to simulate the long-term sequence of glacials and interglacials, our model needs to include as explicitly as possible the effects of all the important interacting components of the climatic system operating over the past million years or so. Besides the atmosphere, these include the ice masses, upper and deep oceans, and the up and down motions of the earth's crust. Even life influences the climate and thus must be included too; plants, for example, can affect the chemical composition of the air and seas as well as the reflectivity (i.e., albedo) or water-cycling character of the land or air. These mutually interacting subsystems form part of the internal components of the model. On the other hand, if we are only interested in modeling very short-term weather events—say, over a single week—then our model can ignore any changes in the glaciers, deep oceans, land shapes, and forests, since these variables obviously change little over one week's time. For short-term weather, only the atmosphere itself needs to be part of the model's internal climatic system.

The slowly varying factors such as oceans or glaciers are said to be external to the internal part of the climatic system being modeled. Modelers also refer to external factors as boundary conditions, since they form boundaries for the internal model components. These boundaries are not always physical ones, such as the oceans, which are at the bottom of the atmosphere, but can also be mass or momentum or energy fluxes across physical boundaries. An example is the solar radiation impinging on the earth. Solar radiation is often referred to by climatic modelers as a *boundary forcing function* of the model for two reasons: the energy output from the sun is not an interactive, internal component of the climatic system of the model; and the energy from the sun forces the climate toward a certain temperature distribution.

We could restrict a model to predict only a globally averaged temperature that never changes its value over time (i.e., is in equilibrium). This very simple model would consist of an internal part that, when averaged over all of the atmosphere, oceans, biosphere, and glaciers, would describe two characteristics: the average reflectivity of the earth and its average greenhouse properties. The boundary condition for such a model would be merely the incoming solar energy. Such a model is called *zero dimensional*, since it collapses the east-west, north-south, and up-down space dimensions of the actual world into one point that represents some global average of all earth-atmosphere system temperatures in all places. It also collapses all three dimensional processes into a global average, which may in some cases such as heat transport by winds, completely neglect that process. If our zero-dimensional model were expanded to resolve temperature or winds at different altitudes, it could produce vertical temperature profiles as a function of atmospheric composition (the classical example being Manabe and Wetherald 1967). If it were further expanded to include latitudes and longitudes and heights, then it would be three-dimensional. The *resolution* of a model refers to the number of dimensions included and to the amount of spatial detail with which each dimension is explicitly treated.

THE PROBLEMS OF PARAMETERIZATIONS AND CLIMATIC FEEDBACK MECHANISMS

Clouds, being very bright, reflect a large fraction of sunlight back to space, thereby helping control the earth's temperature. Thus, predicting the changing amount of cloudiness over time is essential to reliable climate simulation. But most individual clouds are smaller than even the smallest area represented by the smallest resolved element (i.e., "grid box") of a global climate or weather prediction model. A single thunderstorm is typically a few kilometers in size, not a few hundred—the size of many "high-resolution" global model grids. Therefore, no global climate model available now (or likely to be available in the next few decades) can explicitly resolve every individual cloud. These important climatic elements are therefore called *sub-grid-scale phenomena*. Yet, even though we cannot explicitly treat all individual clouds, we can deal with their collective effects on the grid-scale climate. The method for doing so is known as *parameterization*, a contraction for "parametric representation". Instead of solving for sub-grid-scale details, which is impractical, we search for a relationship between climatic variables we do resolve (for example, those whose variations occur over larger areas than the grid size) and those we do not resolve. For instance, climatic modelers have examined years of data on the humidity of the atmosphere averaged over large areas and have related these values to cloudiness averaged over that area. It is typical to choose an area the size of a numerical model's grid—a few hundred kilometers on a side. While it is not possible to find a perfect correspondence between these averaged variables, reasonable relationships have been found in a wide variety of circumstances. These relationships typically require a few factors, or *parameters*, some of which are derived empirically from observed data, not computed from first principles. The parameterization method applies to almost all simulation models, whether dealing with physical, biological or even social systems. The most important parameterizations affect processes called *feedback mechanisms*. This concept is well known outside of computer-modeling circles. The word feedback is vernacular. As the term implies, information can be "fed back" to you that will possibly alter your behavior.

So it is in the climate system. Processes interact to modify the overall climatic state. Suppose, for example, a cold snap brings on a high albedo snow cover which tends to reduce the amount of solar heat absorbed, subsequently intensifying the cold. This interactive process is known to climatologists as the *snow-and-ice/albedo/temperature feedback mechanism*. Its destabilizing, *positive-feedback* effect is becoming well understood and has been incorporated into the parameterizations of most climatic models. Unfortunately, other potentially important feedback mechanisms are not usually as well understood. One of the most difficult ones is so-called *cloud feedback*, which could be either a positive or negative feedback process, depending on circumstances. (e.g., Cess *et al.* 1989, Schneider *et al.* 1978). Another is the extent to which changes in soil temperature or soil moisture from climate change could alter the microbial decomposition of dead organic matter into heat trapping gases such as carbon dioxide or methane (e.g., Andreae and Schimel 1989).

MODEL VERIFICATION

Given the uncertainties associated with model parameterizations and feedbacks, then, can we have confidence in model predictions? At least several methods can be used, and none by itself is sufficient. First, we must check overall model-simulation skill against the real climate for today's conditions to see if the control experiment is reliable. The seasonal cycle is one good test. Figure 1, from the work of Syukuro Manabe and R.J. Stouffer at the Geophysical Fluid Dynamics Laboratory at Princeton, N.J., shows how remarkably well a three-dimensional global circulation model can simulate the regional distribution of the seasonal cycle of surface air temperature—a well understood climate change that is, when averaged over the Earth, larger than ice age-interglacial changes! The seasonal-cycle simulation is a necessary test, of what can be called "fast physics" like cloud formations but it doesn't tell us how well the model simulates slow changes in ice cover or soil organic matter or deep ocean temperatures, since these variables do not change much over a seasonal cycle, though they do influence long-term trends.

A second method of verification is to test in isolation individual physical sub-components of the model (such as its parameterizations) directly against real data and/or more highly resolved process models of specific processes (e.g., evapotranspiration from soils and forests). This still is not a guarantee that the net effect of all interacting physical subcomponents has been properly treated, but it is an important test. For example, the upward infrared radiation emitted from the planet to space can be measured from satellites or calculated in a climate model. If this quantity is subtracted from the emitted upward infrared radiation at the earth's surface, then the difference between these quantities, G, can be identified

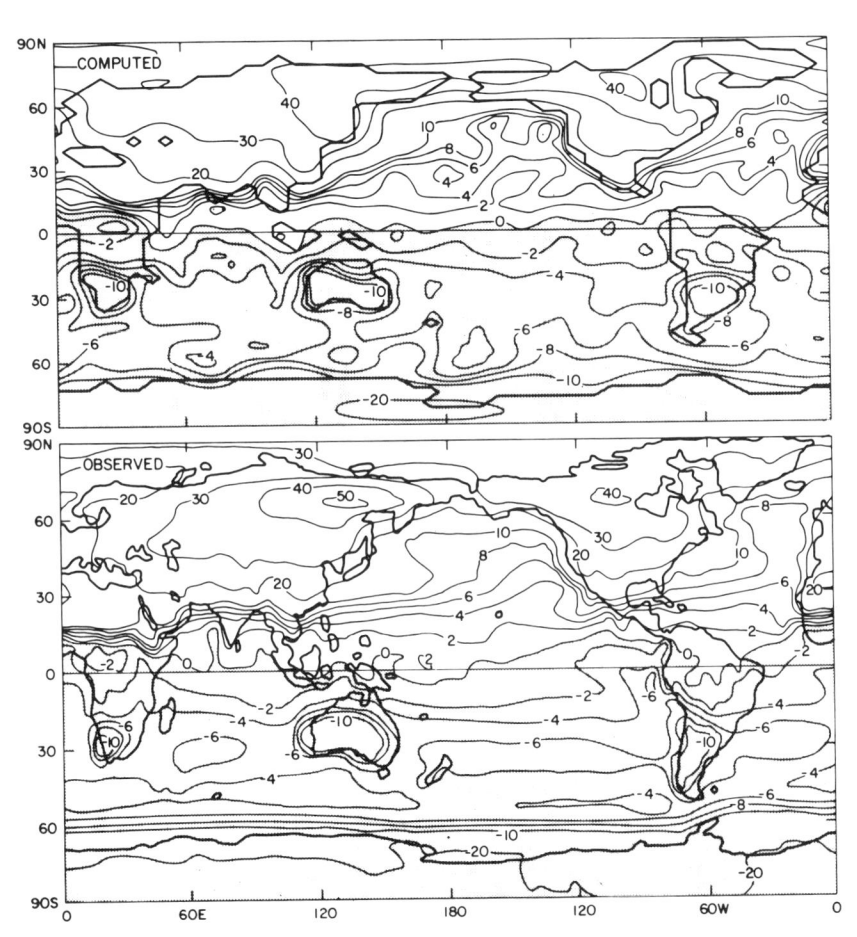

FIGURE 1. A three-dimensional climate model has been used to compute the winter to summer temperature differences all over the globe. The model's performance can be verified against the observed data shown below. This verification exercise shows that the model quite impressively reproduces many of the features of the seasonal cycle. These seasonal temperature extremes are mostly larger than those occurring between ice ages and interglacials or for any plausible future carbon dioxide change (Source: S. Manabe and R.J. Stouffer, 1980, Sensitivity of a global climate model to an increase of CO_2 concentration in the atmosphere, *J. Geophys. Res.* 85:5529-5554.)

as the "Greenhouse Effect", as on Figure 2 from Raval and Ramanathan (1989). Although radiative processes may not be the only ones operative in nature or in the general circulation model (GCM) results, the close agreement among the satellite results (thick line labeled ERBE on Figure 2), a GCM (thick dashed line labeled CCM) and line-by-line radiative transfer calculations (thin dashed line) on Figure 2 all give strong evidence that this physical subcomponent—known as the "water-vapor/greenhouse feedback"—is well modeled at grid scale. Another validation test is to compare model-generated and observed statistics of grid-point daily variability (e.g., Mearns *et al.* 1990, Rind *et al.* 1989). Some variables, (i.e., temperature) are well modeled whereas others (such as relative humidity) yield poorer simulations of observed variability.

All three methods must constantly be used and reused as models evolve if we are to improve the credibility of their predictions. And to these we can add a fourth method: the model's ability to simulate the very different climates of the ancient earth or even those of other planets.

SENSITIVITY AND SCENARIO ANALYSIS

Although the ultimate goal of any forecasting simulation may be to produce a

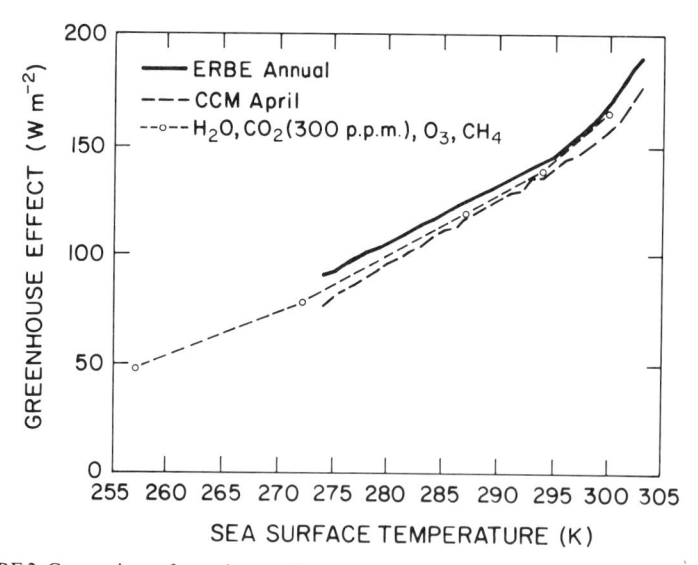

FIGURE 2. Comparison of greenhouse effect, heat trapping parameter, G, and surface temperature, obtained from three sources: bold line: ERBE annual values, obtained by averaging April, July and October 1985 and January 1986 satellite measurements; thick dashed line, three-dimensional climate-model simulations for a perpetual April simulation (National Center for Atmospheric Research Community Climate Model); thin dashed line, line-by-line radiation-model calculations by Dr. A. Arking using CO_2, O_3 and CH_4. The line-by-line model results come close to the CCM and the ERBE values. (Source: Raval, A. and V. Ramanathan 1989, Observational determination of the greenhouse effect, *Nature* 342, 758).

single, accurate time-series projection of some evolving variable, a lesser goal may still be quite useful and certainly is more realizable: to specify plausible scenarios of various uncertain or unpredictable variables and then to evaluate the sensitivity of some predicted variable to either different scenarios or different model assumptions.

Key internal factors such as the cloud feedback or vertical oceanic mixing parameters, can be varied over a plausible range of values in order to help determine which internal processes have the most importance for the sensitivity of the climate to, say, CO_2 buildup. Even though one cannot be certain which of the simulations is most realistic, sensitivity analyses can (1) help set up a priority list for further work on uncertain internal model elements and (2) help to estimate the plausible range of climatic futures to which society may have to adapt over the next several decades. Given these plausible futures, some of us might choose to avoid a low-to-moderate probability, high consequence outcome associated with some specific scenario. Indeed public policy often seeks to avoid plausible, high cost scenarios. So too, do people purchase insurance. On the other hand, more risk-prone people might prefer extra scientific certainty before asking society to invest present resources to hedge against uncertain, even if plausible, climatic futures. At a minimum, cross-sensitivity analysis, in which the response of some forecast variables to multiple variations in uncertain internal and/or external parameters, allows us to examine quantitatively the differential consequences of explicit sets of plausible assumptions.

In any case, even if we cannot produce a reliable single foecast of some future variables, we might be able to provide much more credible sensitivity analyses, which can have practical applications in helping us to investigate a range of probabilities and consequences of plausible scenarios. Such predictions may simply be the best "forecasts" that honest natural or social scientists can provide to inform society on a plausible range of alternative futures of complex systems. How to react to such information, of course, is in the realm of values and politics.

Let us proceed then to several examples of this process.

CLIMATIC MODEL RESULTS

The most important question surrounding the greenhouse gas controversy is simply: What will be the regional distribution of climatic changes associated with significant increases in CO_2 and other trace greenhouse gases. Other trace gases such as chlorofluorocarbons, methane, nitrogen oxides, or ozone can, all taken together, have a comparable greenhouse effect to CO_2 over the next century (e.g., Dickinson and Cicerone 1986). To investigate such possibilities one needs a model with regional resolution. It needs to include processes such as the hydrologic cycle and the storage of moisture in the soils—since these factors are so critical to both climatic change and its impacts on agriculture and water supplies. To investigate plausible climatic scenarios, modelers have typically run "equilibrium simulations" in which instantaneously increased values of carbon dioxide are imposed at an initial time and held fixed while the model is allowed to approach a new equilibrium.

Syukuro Manabe and R.J. Stouffer (1980) for example, in one of the most widely quoted results, find a summer "dry zone" in the middle of Eurasia and North America, as well as increased moistness in some of the monsoon belts. All of this was from a doubling of CO_2 held fixed over time. The model was allowed to run several decades of simulated time in order to reach equilibrium. But Manabe's team used an artificially constructed ocean that consisted of a uniform mixed layer depth of 70 meters with no heat flow to or from the deep ocean below. While this shallow "sea" allows the seasonal cycle to be satisfactorily simulated (see Figure 1), a purely mixed layer ocean does not include the important processes whereby water is transported horizontally from the tropics toward the poles or vertically between the mixed layer and the abyssal depths. The latter processes slow the approach toward thermal equilibrium of the surface waters, and certainly would affect the transient evolution of the surface temperature changes in the ocean due to the actual time-evolving increase of trace greenhouse gases.

Thus, during the transient phase of warming the surface temperature increases from latitude to latitude and land to sea could have a different pattern than at equilibrium. This, in turn, could cause significantly different climatic anomalies during the transient phase than would be inferred from equilibrium sensitivity tests to fixed increases in trace gases.

To answer this very uncertain transient response question more reliably, it has been proposed that three-dimensional atmospheric models be coupled to fairly realistic three-dimensional oceanic models. To date, only a handful of such model experiments have been run (e.g., Washington and Meehl 1989, Stouffer *et al.* 1989, IPCC, 1992), but none over the century or two time scale needed to adequately address this important issue. One reason for detailing the CO_2 transient/regional climate anomaly example here is to exemplify the need for various sensitivity experiments across a hierarchy of models. Such methods are especially essential during the development phase of modeling. In 1980, Starley Thompson and I ran quasi-one-dimensional models to examine the potential importance of this transient issue (e.g., Schneider and Thompson 1981). However, these models could not provide reliable simulations because of their physical simplicity. But the one-dimensional models were economically efficient enough to be able to be run over the century time span needed to explore the CO_2 transient issue. Our findings suggested that it is imperative to run high resolution, coupled atmospheric, oceanic, cryospheric and land surface sub-model models if regional, time-evolving scenarios of climatic changes are to have any hope of credibility. On the other hand, the more complex three-dimensional general circulation models are not yet at an adequate phase of development to have trustworthy coupled atmosphere/ocean models that are both well verified and economical enough to be run and rerun over the hundred years needed for greenhouse gas-transient simulations. For example, a "low-resolution" atmospheric general circulation model with nine vertical levels and a horizontal grid size of 4.5°/at x7.5° long takes some 10 hours of CRAY X-MP time to generate a year of weather maps in 30 minute time steps. To run a 50 year transient experiment is thus some 500 hours of super-computer time. And, to run at more desirable higher resolution we must remember that each time the grid size is cut in half the computer time required goes up by an order of magnitude. Thus, the simple or low resolution model helps to identify and bound potentially important problems, provide some quantitative sensitivity studies, and help set priorities for three-dimensional coupled model research or needed observational programs over the next few decades. Since the agricultural and other environmental impacts of increasing greenhouse gases depend on the specific regional and seasonal distribution of climatic change, resolution of the transient climate response debate, among others, is critical for climatic impact assessment and ultimate policy responses to the advent or prospect of increasing greenhouse gases.

THE CURRENT GLOBAL WARMING DEBATE: SCIENCE OR MEDIA HYPE?

However, before discussing any potential management responses to such projections, let me first point out that not all knowledgeable scientists are in agreement as to the probability that such changes will occur. In fact, if one has followed the very noisy, often polemical debate in the media recently, one might get the (I believe false) impression that there are but two radically opposed schools of thought about global warming: (1) that climatic changes will be so severe, so sudden and so certain that major species extinction events will intensify, sea-level rise will create tens of millions of environmental refugees, millions to perhaps billions of people will starve and devastated ecosystems are a virtual certainty or, alternatively, (2) there is nothing but uncertainty about global warming, no evidence that the twentieth century has done what the modelers have predicted, and the people arguing for change are just "environmental extremists"—thus there is no need for any management response to an event that is improbable and in no case should any such responses interfere with the "free market" and bankrupt nations (for example, see the December 25, 1989 cover story of *Forbes Magazine*, Brookes, 1989). Unfortunately, while such a highly charged and polarized debate makes entertaining opinion page reading or viewing for the ratings-dominated media, it provides a very poor description of the reality of the actual scientific debate or the broad consensus on basic issues within the scientific community. In my opinion, the "end of the world" or "nothing to worry about" are the two *least likely* cases, with almost any scenario in between having a higher probability of occurrence.

Figure 3 shows a projection of global warming possibilities into the 21st century drawn by an international group of scientists that was convened in the mid-1980's by the International Council of Scientific Unions. This assessment is very similar to that in the IPCC (1990) report. Figure 3 shows warming from a very moderate additional half degree C up to a catastrophic 5°C or greater warming before the end of the next century. I do not hesitate to call the latter extreme catastrophic because that is the magnitude of warming that occurred between about 15,000 years ago and 5,000 years ago, the end of the last ice age to our present interglacial epoch. It took nature some 5 to 10 thousand years to fully accomplish that warming, and it was accompanied by a 100 meter or so rise in sea level, thousands-of-kilometers migration of forest species, radically altered habitats, species extinctions, species evolution, and other major environmental changes. Indeed, the ice age to interglacial transition revamped the ecological face of the planet. If the mid-to-upper part of the scenarios occurred as rapidly as they are projected on Figure 3, then indeed they would justify the substantial concern that many scientists have over the prospect of global warming. On the other hand, the many unknown factors which make most scientists, myself included, hesitant to make anything other than qualitative or intuitive probabilistic kinds of forecasts could well suggest a 21st century climate in which global temperature change would be only a degree or so. While even that seemingly small change might be serious for certain life forms (for example, some plant or animal species living near the tops of mountains that would be driven to extinction with even a small warming), by and large changes of less than a degree or so taking place over a century or more would clearly add less stress to natural (and certainly to human) systems than would changes of several degrees taking place in 50 years or less. Indeed, the rate of change may be a most critically important factor to the adaptive capacity of both humans and natural systems, particularly the latter, since ecosystems don't have the option of planting new seeds to match the new climate the way our farmers do [e.g., Peters and Lovejoy (in press)].

Critics of immediate policy responses to global warming are quick to point out the many uncertainties that could reduce the average projections made by climate

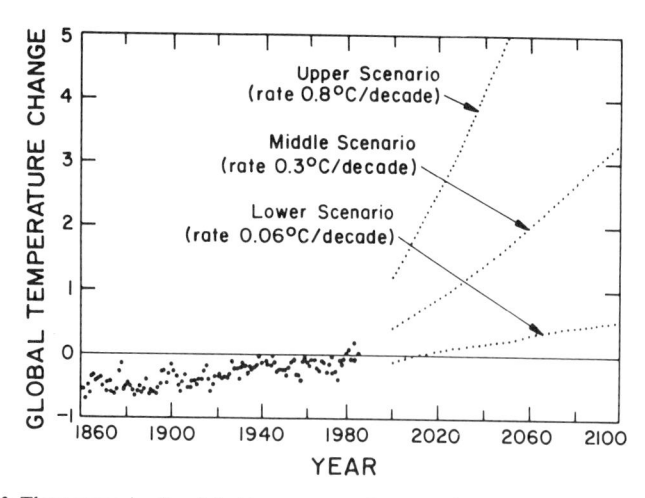

FIGURE 3. Three scenarios for global temperature change to the year 2100 derived from combining uncertainties in future trace greenhouse gas projections with uncertainties of modeling the climatic response to those projections. Sustained global temperature changes beyond 2°C would be unprecedented during the era of human civilization. The middle to upper range represent climatic change at a pace ten to one hundred times faster than typical long-term, natural global average rates of change. [Source: Jaeger, J. April 1988, Developing Policies for Responding to Climatic Change: A Summary of the Discussions and Recommendations of the Workshops Held in Villach 28 September to 2 October 1987 (WCIP-1, WMO/TD-No. 225)].

models (such as the middle line on Figure 3). Many critiques (for example, see George C. Marshall Institute 1989) somehow forget to stress that the sword of uncertainty has two blades: that is, uncertainties in physical or biological processes which make it possible for the present generation of models to have overestimated future warming effects, are just as likely to have caused the models to have understimated change. It is well known that the 25% increase in carbon dioxide that is documented since the industrial revolution, the 100% increase in methane since the industrial revolution and the introduction of man-made chemicals such as chlorofluorocarbons (also responsible for stratospheric ozone depletion) since the 1950's should have trapped about two extra watts of radiant energy over every square meter of earth. That part is well accepted by most climatological specialists. However, what is less well accepted is how to translate that two watts of heating into "X" degrees of temperature change, since this involves assumptions about how that heating will be distributed among surface temperature rises, evaporation increases, cloudiness changes, ice changes and so forth. The factor of 2 to 3 uncertainty in global temperature rise projections as cited in typical U.S. National Academy of Sciences reports or the IPCC reflects a legitimate estimate of uncertainty held by most in the scientific community. Indeed, recent attempts by a British group to mimic the effects of cloud droplets halved their model's sensitivity to doubled CO_2—but is still well within the often-cited 1.5-4.5°C range. However, the authors of the study at the British Meteorological Office, wisely pointed out that "although the revised cloud scheme is more detailed, it is not necessarily more accurate than the less sophisticated scheme." (Mitchell, Ingram and Senior 1989). I have never seen this forthright and important caveat quoted by any of the global warming critics who almost always cite the British work as a reason to lower our concern by a factor of two or so.

Finally, as stated earlier, prediction of the detailed regional distribution of climatic anomalies, that is, where and when it will be wetter and drier, how many more floods might occur in the spring in California or forest fires in Siberia in August, is simply highly speculative, although some plausible scenarios can be given. Some such scenarios are given in Table 1, from the National Academy of Sciences, 1987, assessment.

In an effort to shed some light on these questions, Schneider, Gleick and Mearns (1991) offered a set of "forecasts" on changes in some important meteorological variables, over a range of temporal, spatial, and statistical scales. I believe that carefully qualified, explicit scenarios of plausible future climatic changes are preferable to impact speculations based on implicit or casually formulated forecasts. Therefore, Table 2 was prepared to provide impact assessment specialists with ranges of climate changes that reflected their interpretation of state-of-the-art modeling results. These projections were based on an analysis of then available results and provide what were believed to be plausible estimates about the direction or magnitude of some important anthropogenic climatic changes over the next 50 years or so—a typical time estimate for an equivalent doubling of carbon dioxide—together with a simple high, medium, or low level of confidence for each variable. (By "equivalent doubling" it is meant that carbon dioxide together with other trace

TABLE 1

Possible Climate Changes From Doubling of CO_2
(Source: National Academy of Sciences, 1987)

Large Stratospheric Cooling (virtually certain)	Reduced ozone concentrations in the upper stratosphere will lead to reduced absorption of solar ultraviolet radiation and therefore less heating. Increases in the stratospheric concentration of carbon dioxide and other radiatively active trace gases will increase the radiation of heat from the stratosphere. The combination of decreased heating and increased cooling will lead to a major lowering of temperatures in the upper stratosphere.
Global-Mean Surface Warming (very probable)	For a doubling of atmospheric carbon dioxide (or its radiative equivalent from all the greenhouse gases), the *long-term* global-mean surface warming is expected to be in the range of 1.5 to 4.5°C. The most significant uncertainty arises from the effects of clouds. Of course, the *actual* rate of warming over the next century will be governed by the growth rate of greenhouse gases, natural fluctuations in the climate system, and the detailed response of the slowly responding parts of the climate system, i.e, oceans and glacial ice.
Global-Mean Precipitation Increase (very probable)	Increased heating of the surface will lead to increased evaporation and, therefore, to grater global mean precipitation. Despite this increase in global average precipitation, some individual regions might well experience decreases in rainfall.
Reduction of Sea Ice (very probable)	As the climate warms, total sea ice is expected to be reduced.
Polar Winter Surface Warming (very probable)	As the sea ice boundary is shifted poleward, the models predict a dramatically enhanced surface warming in winter polar regions. The greater fraction of open water and thinner sea ice will probably lead to warming of the polar surface air by as much as three times the global mean warming.
Summer Continental Dryness/Warming (likely in the long term)	Several studies have predicted a marked long-term drying of the soil moisture over some mid-latitude interior continental regions during summer. This dryness is mainly caused by an earlier termination of snowmelt and rainy periods, and an earlier onset of the spring-to-summer reduction of soil wetness. Of course, these simulations of long-term equilibrium conditions may not offer a reliable guide to trends over the next few decades of changing atmospheric composition and changing climate.
High-Latitude Precipitation Increase (probable)	As the climate warms, the increased poleward penetration of warm, moist air should increase the average annual precipitation in high latitudes.
Rise in Global Mean Sea Level (probable)	A rise in mean sea level is generally expected due to thermal expansion of sea water in the warmer future climate. Far less certain is the contribution due to melting or calving of land ice.

TABLE 2

Phenomena	Projection of probable global annual average change (1)	Distribution of change — Regional averge	Distribution of change — Change in seasonality	Distribution of change — Interannual* variability	Significant transients	Confidence of projection — Global average	Confidence of projection — Regional average	Estimated time for research that leads to consensus (years)
Temperature**	+2 to +5 C	−3 to +10 C	yes	down?	yes	high	medium	0 to 10
Sea level	0 to 80 cm***	(2)	no	???	yes***	high	medium	5 to 20
Precipitation	+7 to +15%	−20 to +20%	yes	up?	yes	high	low	10 to 40
Direct solar radiation	−10 to +10%	−30 to +30%	yes	???	possible	low	low	10 to 40
Evapotranspiration	+5 to +10%	−10 to +10%	yes	???	possible	high	low	10 to 40
Soil moisture	???	−50 to +50%	yes	???	yes	???	medium	10 to 40
Runoff	increase	−50 to +50%	yes	???	yes	medium	low	10 to 40
Severe storms	???	???	(3)	???	yes	???	???	10 to 40

(1) For an "equivalent doubling" of atmospheric CO_2 from the preindustrial level.
(2) Increases in sea level at approximately the global rate except where local geological activity prevails or if changes occur to ocean currents.
(3) Some suggestions of longer season and increased intensity of tropical cyclones as a result of warmer sea surface temperatures.
??? No basis for quantitative or qualitative forecast.
* Inferences based on preliminary results for the U.S. from Rind et al. (1989).
** Based on three-dimensional model results. If only trace gas increases were responsible for 20th century warming trend of about 0.5 degrees C, then this range should be reduced by perhaps 1 degree C.
*** Assumes only small changes in Greenland or West Antarctic ice sheets in 21st century. For equilibrium, hundreds of years would be needed and up to several meters of additional sea level rise could be accompanied by centuries of ice sheet melting from an equilibrium warming of 3 or more degrees C.

Source: Modified (by Schneider) from Schneider, S.H., P. Gleick, and L.O. Mearns in P.E. Waggoner, 1990. Climate Change and U.S. Water Resources. New York: John Wiley & Sons, 41-73.

greenhouse gases have a radiative effect equivalent to doubling the pre-industrial value of carbon dioxide from about 280 to 560 ppm.) As another measure of the nature of the uncertainties, a subjective estimate is included of the time that may be necessary to achieve a widespread scientific consensus on the direction and magnitude of the change. In some cases—such as the magnitude and direction of changes in global, annual-averaged temperature and precipitation—such a consensus has virtually been reached. In other cases, such changes in the extent of cloud cover, time-evolving patterns of regional precipitation, or the daily, monthly, or inter-annual variance of many climatic variables, the large uncertainties surrounding present projections will only be reduced with decades of considerably more research. (IPCC 1990, Figures 11.1 and 11.2, suggest 10-20 years for major research progress as well.)

Let us consider, for example, the first row on Table 2: temperature change. The global average change of $+2$ to $+5°C$ is typical of that in most national and international assessments for an equivalent doubling of greenhouse gases, neglecting transient delays. While no probability bounds are usually given, Schneider, Gleick and Mearns intuitively suggest this 2-5°C range as a ± 1 standard deviation estimate. The neglect of transients means that the range given is based on the assumption that trace gases have been increased over a long enough period for the climate to come into equilibrium with the increased concentration of greenhouse gases. In reality, as noted earlier, the large heat capacity of the oceans will delay realization of most of the equilibrium warming by at least many decades. This implies that at any specific time when we reach an equivalent CO_2 doubling (by say 2030), the actual global temperature increase may be considerably less than the ± 2 to $\pm 5°C$ listed in Table 2. However, this "unrealized warming" (Hansen *et al.*, 1985) will eventually be experienced when the climate system thermal response catches up to the greenhouse gas forcing. However, additional factors such as the emission of sulfur oxides can create particles which both contribute to acid rain and brighten the atmosphere, thereby offsetting, for a few decades at least, some of the anticipated global warming (e.g. Charlson *et al.*, 1991).

Forecasts of regional or watershed-scale changes in temperature, evaporation, or precipitation are most germane to impact assessment. But, as Table 2 suggests, such regional forecasts are much more uncertain than global equilibrium projections. Regional temperature ranges given in Table 2 are much larger than global changes, and even allow for some regions of negative change. This regional uncertainty is reinforced by suggestions that SO_2 pollution from industrial activities (largely in industrial areas of the Northern Hemisphere) might increase cloud albedos near the pollution sources (Wigley 1989). Higher northern latitude surface temperature increases are up to several times larger than the global average response, at least in equilibrium. Because of the importance of regional or local impact information, techniques need to be developed to evaluate smaller-scale effects of large-scale climatic changes. [For example, Gleick (1987) employed a regional hydrology model driven by large-scale climate change scenarios from various GCM inputs and Giorgi *et al.* (1990) developed a method to embed a mesoscale model into a GCM.].

Even more uncertain than regional details, but perhaps most important, are estimates for measures of climatic variability such as the frequency and magnitude of severe storms, enhanced heat waves, or reduced frost probabilities (Mearns *et al.* 1984 and 1990; Parry and Carter 1985). For example, some modeling evidence suggests that hurricane intensities will increase with climate changes (Emanuel 1987). Such issues are just now beginning to be considered and evaluated from equilibrium climate-model results, and will, of course, have to be studied again for realistic transient cases to be of maximum value to impact assessors.

Although climatic models are far from fully verified for future simulations, the seasonal and paleoclimatic simulations are strong evidence that state-of-the-art climatic models already have considerable skills. An awareness of just what models are and what they can and can't do is probably the best we can ask of the public and its representatives. Then, the tough policy problem is how to apply the society's values in choosing to face the future given the possible outcomes that climatic models foretell. Modelers will continue to develop and refine new models by turning to larger computers to run them and more observations to improve and verify them. We must ask the indulgence of society to recognize that immediate, definitive answers are not likely, as coupling of higher resolution atmosphere, ocean, i.e., land surface and chemistry submodels will take a decade or more to develop. In essence, what climate models and their applications typify is a growing class of problems not unique to climate but also familiar in other disciplines: nuclear waste disposal, safety of food additives or drugs, efficacy of strategic defense and so forth. These are problems for which guaranteed "scientific" answers cannot be obtained—except by performing the experiment on ourselves.

If the public is totally ignorant of the nature, use, or validity of climatic (or many other kinds of) models, then public policy-making based on model results will be haphazard at best. In this case, the decision-making process tends to be dominated by special interests or a technically trained elite.

ADAPTATION OR PREVENTION?

Public responses to the advent or prospect of global climate change typically come in two categories: adaptation and prevention. With regard to *adaptation*, flexibility of adaptive measures needs to be considered now. Indeed, if the entire global warming debate comes out closer to the views of the present critics and only small changes occur, what would be lost, for example, by improving the flexibility of water supply systems? After all, nature will continue to give us wet and dry years. The 100-year flood will happen sometime, as will the 100-year drought—with or without global warming, which, of course, could change the odds of such extremes. Therefore, increasing management flexibility will pay dividends (e.g., see Waggoner 1990) even in (what I and the IPCC believe to be) the unlikely event that global warming proves to be minimal. This situation provides a metaphor to buying insurance. Only a foolish or desperate person would over- or under-invest in insurance. However, unlike the insurance metaphor, where a premium is "wasted" if one doesn't collect on any damages, here we actually can get benefits for our investment in flexibility as we buy insurance against the prospect of rapid climate change at the same time. Of course, no one gets a return on investment without making an investment, and that is true not just in the private sector, but obviously in the public sector as well.

The other category of management response is *prevention*. That simply means slowing down the rate at which the gases that are injected into the atmosphere that can modify the climate are produced. The principal way to do that falls primarily in the jurisdiction of energy use and production managers. However, since deforestation is an important component of CO_2 production, halting deforestation can help to reduce the atmospheric buildup of CO_2. Thus, prevention is also in the purview of such land management agencies. For example, domestic animals are a major source of methane, and such livestock graze extensively on public lands. Therefore, solutions to methane emission controls will involve consideration of the use of public lands for this purpose. As another example, microbial communities in soils decompose dead organic matter into greenhouse gases, such as carbon dioxide, methane or nitrous oxide. Since these microbes typically increase their metabolic activity when the soil warms, deforestation or grazing which removes vegetation cover and results in warmer soils could enhance the production of these greenhouse gases (e.g., Lashof 1989). These issues are contentious, but nonetheless need to be examined for their potential effectiveness as emission control strategies.

The best strategies should have high leverage—that is, help to solve more than one problem with a single investment. It is my personal view that such high leverage or "tie-in" strategies are the best approach to dealing with prevention, or at least delaying the rate of the build-up of greenhouse gases, and thus the prospect of rapid global warming (e.g., Schneider 1990). It is obvious that the things to do first are the things that make the most sense regardless of whether global warming materializes. Energy efficiency, as is often mentioned, is the single most important "tie-in" strategy, for using energy efficiently will not only reduce the prospect of rapid climate change, but also reduce acid rain (itself a threat to many natural lands), will reduce air pollution in cities, will reduce balance of payment deficits in energy importing countries by reducing dependence on foreign supplies of energy, and in the long run can improve product competitiveness by reducing the energy components of manufactured products, which, for example, are about twice as high for the United States as for her more efficient competitors, Japan, Italy or West Germany (e.g., Chandler 1988). Enforcing the London extension of the Montreal Protocol to ban virtually all CFC use would not only lower the 20% role CFCs play as global heat trapping gases, but also would slow stratospheric ozone depletion.

A number of studies have now shown that the U.S. can lower its greenhouse gas emissions by 10-40% at low to negative (i.e., we actually save money relative to current inefficient practices!) costs (e.g., OTA 1991, NAS 1991), and that a mix of regulatory and market incentive policies are needed to create such an outcome. I believe that flexibility for adaptation and high leverage strategies for prevention are not premature policy actions. Although global warming critics are right when they note that current uncertainties *could* render global warming effects to be small, so too could such uncertainties make current "best guess" estimates too modest. In short, no one knows now whether such uncertainties will reduce or exacerbate present estimates. To be prudent, I believe that if we are to manage effectively a future with increasing uncertainty in environmental conditions at a global scale, then it is imperative that we practice prevention by slowing down the rate at which we force global changes and also that management flexibility be increased to ease adaptation. *Ideology is the enemy of flexibility.* It is incumbent on all of us—individuals, businesses and governments—to rethink any of our ideologically rigid positions in order to fashion ways in which to enhance the flexibility of adaptive management. This can give us the opportunity to manage more effectively the potential consequences of serious climate change, as well as improve the capacity of present systems to deal with the natural variability of the environment, itself a well demonstrated threat to many of our abilities or resources.

REFERENCES

Andreae, M.O. and D.S. Schimel, (eds.). 1989. *Dahlem Workshop on Exchange of Trace Gases between Terrestrial Ecosystems and the Atmosphere*, John Wiley and Sons, New York, 347 pp.

Brookes, W.T. 1989. "The Global Warming Panic". *Forbes*, 25 December 1989:96-102.

Cess, R.D., G.L. Potter, J.P. Blanchet, G.J. Boer, S.J. Ghan, J.T. Kiehl, H. Le Treut, Z.X. Li, X.Z. Liang, J.F.B. Mitchell, J.J. Morcrette, D.A. Randall, M.R. Riches, E. Roeckner, U. Schlese, A. Slingo, K.E. Taylor, W.M. Washington, R.T. Wetherald, I. Yagai. 1989. Interpretation of Cloud-Climate Feedback as Produced by 14 Atmospheric General Circulation Models. *Science* 245:513-516.

Chandler, W.U., H.S. Geller and N.R. Ledbetter. 1988. *Energy Efficiency: A New Agenda*, GW Press, Springfield, VA, 76 pp.

Charlson, R.J., J. Langor, H. Rodhe, C.B. Leory and S.G. Warren. 1991. Perturbation of the northern hemisphere radiative balance by backscattering from anthropogenic sulfate aerosols. *Tellus*, 43ab, 152-163.

Chervin, R.M. 1981. On the Comparison of Observed and GCM Simulated Climate Ensembles. *J. Atmos. Sci.* 38:885-901.

Dickinson, R.E. and R.J. Cicerone. 1986. Future Global Warming from Atmospheric Trace Gases. *Nature* 319:109-115.

Emanuel, K.A. 1987. The Dependence of Hurricane Intensity on Climate. *Nature* 326:483-485.

George C. Marshall Institute. 1989. *Scientific Perspectives on the Greenhouse Problem,* Washington, D.C., 37 pp.

Giorgi, F., M.R. Marinucci and G. Visconti. 1990. Use of a limited area model nested in a general circulation model for regional climate simulation over Europe. *J. Geophys. Res.* 95:18,413-18,431.

Gleick, P.H. 1987. Regional Hydrologic Consequences of Increases in Atmospheric CO_2 and Other Trace Gases. *Climatic Change* 10:137-160.

Hansen, J., G. Russell, A. Lacis, I. Fung, D. Rind. 1985. Climate Response Times: Dependence on Climate Sensitivity and Ocean Mixing. *Science* 229:857-859.

Hasselmann, K. 1976. Stochastic Climate Models. Part I. Theory. *Tellus XXVIII* 6:473-485.

Intergovernmental Panel on Climate Change (IPCC), (2nd Draft) March 1990. *Scientific Assessment of Climate Change,* World Meteorological Organization, Geneva.

Intergovernmental Panel on Climate Change (IPCC) 1992. Climate Change 1992: The Supplementary Report to the IPCC Scientific Assessment. J.T. Houghton, G.J. Jenkins, and J.J. Ephraums, (eds.). Cambridge University Press, Cambridge.

Lashof, D.A. 1989. The Dynamic Greenhouse: Feedback Processes That May Influence Future Concentrations of Atmospheric Trace Gases and Climatic Change. *Climatic Change* 14:213-242.

Lorenz, E.N. 1968. Climate Determinism. *Meteorol. Monogr.* 8, No. 30:1-3.

Manabe, S. and R.J. Stouffer. 1980. Sensitivity of a global climate model to an increase in CO_2 concentration in the atmosphere. *J. Geophys. Res.* 85:5529-5554.

Manabe, S. and R.T. Wetherald. 1967. Thermal Equilibrium of the Atmosphere with a Given Distribution of Relative Humidity. *J. Atmos. Sci.* 24:241-259.

Mearns, L.O., S.H. Schneider, S.L. Thompson and L.R. McDaniel. 1990. Analysis of climate variability in general circulation models: Comparison with observations and changes in variability in $2xCO_2$ experiments. *J. Geophys. Res.* 95:20,469-20,490.

Mearns, L.O., R.W. Katz and S.H. Schneider. 1984. Extreme High Temperature Events: Changes in their Probabilities and Changes in Mean Temperature, *J. Climate Appl. Meteor.* 23:1601-1613.

Mitchell, J.F.B, W.J. Ingram and C.A. Senior. 1989. CO_2 and Climate: A Missing Feedback? *Nature* 341:132-134.

National Academy of Sciences. 1991. *Policy Implications of Greenhouse Warming,* National Academy Press, Washington, D.C., 127 pp.

Office of Technology Assessment. 1991. *Changing by Degrees, Steps to Reduce Greenhouse Gases.* Washington, D.C.

Parry, M.L. and T.R. Carter. 1985. The Effects of Climate Variations on Agricultural Risk. *Climatic Change* 7:95-110.

Peters, R. (ed.). Proceedings of the Conference on the Consequences of the Greenhouse Effect for Biological Diversity, Yale University Press, New Haven (in press).

Rind, D.R. Goldberg and R. Ruedy. 1989. Change in Climate Variability in the 21st Century. *Climatic Change* 14:5-37.

Schneider, S.H. 1990. *Global Warming: Are We Entering the Greenhouse Century?,* Vintage Books, New York, New York, 343 pp.

Schneider, S.H. and S.L. Thompson. 1981. Atmospheric CO_2 and Climate: Importance of the Transient Response. *J. Atmos. Sci.* 37:895-900.

Schneider, S.H., P.H. Gleick and L.O. Mearns. 1991. Climate-Change Scenarios

Schneider, S.H., P.H. Gleick and L.O. Mearns. 1991. Climate-Change Scenarios for Impact Assessment. In *Proceedings of Conference on the Consequences of the Greenhouse Effect for Biological Diversity.* R. Peters, Ed., Yale University Press, New Haven, (in press).

Schneider, S.H., W.M. Washington and R.M. Chervin. 1978. Cloudiness as a climatic feedback mechanism: Effects on cloud amounts of prescribed global and regional surface temperature changes in the NCAR GCM. *J. Atmos. Sci.* 35, 12:2207-2221.

Stouffer, R.J. S. Manabe and K. Bryan. 1989. Interhemispheric Asymmetry in Climate Response to a Gradual Increase of Atmospheric CO_2. *Nature* 342:600-662.

Waggoner, P.E. (ed.). 1990. *Climate Change and U.S. Water Resources*, John Wiley and Sons, New York, 496 pp.

Washington, W.M. and C.L. Parkinson. 1986. An Introduction to Three-Dimensional Climate Modeling. University Science Books, Mill Valley, CA and Oxford University Press, New York, NY, 422 pp.

Washington, W.M. and G.A. Meehl. 1989. Climate Sensitivity Due to Increased CO_2: Experiments With a Coupled Atmosphere and Ocean General Circulation Model. *Climate Dynamics* 4:1-38.

Wigley, T.M.L. 1989. Possible Climate Change Due to SO_2-Derived Cloud Nuclei. *Nature* 339:365-367.

Global Climate Change: Implications, Challenges and Mitigation Measures. Edited by S.K. Majumdar, L.S. Kalkstein, B. Yarnal, E.W. Miller, and L.M. Rosenfeld. © 1992, The Pennsylvania Academy of Science.

Chapter Eight

GLOBAL WARMING:
Beyond the Popular Vision

PATRICK J. MICHAELS
[1]Department of Environmental Sciences
University of Virginia
Charlottesville, VA 22903

INTRODUCTION

The critical scientific question on global environmental change may not be "how much will the globe warm", but rather "how will it warm". While the scientific community is virtually unanimous that there will be some warming, its real effect will be expressed by its regionality, seasonality and diurnality. As a result of these factors, there are now several compelling lines of evidence that indicate the chance of an ecologically or economically disastrous global warming is becoming more remote. These findings are at considerable variance to what might be referred to as the current "Popular Vision" of global warming: a global temperature change of approximately 4°C, enhanced warming at the poles resulting in melting of major areas of sea ice, and starvation and civil strife resulting from ecological chaos.

While this Vision is accomodated by more lurid interpretations of the recent Intergovernmental Panel on Climate Change (IPCC) Policymakers Summary[1], it is not characteristic of the median range of the climate impacts suggested in that report. Rather, the Vision became prominent in the United States because of two synergistic events. The first was the publication of the penultimate generation of general circulation model (GCM) climate simulations in the mid to late 1980's: These predicted a mean global warming of 4.2°C with winter warming of as much as 18°C for the north polar regions[2-6]. The second was the Congressional testimony of June 23, 1988 that there was a "high degree of cause and effect" between current temperatures and human greenhouse alterations[7].

Nowhere in that testimony, and to my best knowledge, nowhere else did NASA's James Hansen ever state that the anomalously warm summer of 1988 in the United States was caused by human alterations of infrared-absorbing trace gases. Nonetheless, the press and the public concluded otherwise: 70% of the respondents to a subsequent CNN poll agreed with the statement that the 1988 drought was in fact caused by "the greenhouse effect"[8]. There is little doubt that such unanimity in fact does reflect a Popular Vision.

As a result, elected officials have been compelled to respond. Senator Gore has compared the current situation to that of 1930's Germany, where several events such as Kristallnacht presaged the holocaust[9]. Those that do not recognize this were labelled modern-day Neville Chamberlains. Draconian interventionist legislation has been submitted to the U.S. Senate[10], and "global warming" is now a touchstone of U.S. foreign policy[11].

The combined effect of this popular sentiment and the political response has been to create a substantial amount of legislation. Are such deep emotional and political committments justified? Is the likelihood of very disruptive climatic change considerable?

In that light, we examine here the problems of (1) Global and Hemispheric temperature histories and trace gas concentrations, (2) Artificial warming from urban heat islands, (3) High latitude and diurnal temperatures, (4) "Popular Vision" climate models and their successors, (5) Possible negative feedbacks in the pollution system.

TRACE GAS CONCENTRATIONS AND TEMPERATURE HISTORIES

While there are several infrared-absorbing trace gases that have increased as a result of anthropogeneration, almost all of the current radiative forcing is associated with (in descending order), Carbon Dioxide, Methane, Nitrous Oxide, and Chloroflourocarbons.

Continuous instrumental records of CO_2 concentration date from the late 1950's at Mauna Loa Observatory, where the 1958 annual average was 315ppm. The concentration now is near 354ppm. "Pre-industrial" (circa 1800) concentrations were initially assumed to be in the range of 295ppm[12], giving a change between 1800 and the mid-1980's (level: 335ppm) of 14%. Assuming that change alone accounts for a change in total radiative forcing of approximately 0.5 w/m^2, which is consistent with both energy balance and GCM forcing, and a climate sensitivity of 1.0°C/Wm^{-2}, the observed warming of approximately 0.5°C in the last 100 years certainly seemed consistent with the changes in trace gases that were assumed in the early 1980's.

Since then, estimates of the background CO_2 concentration have dropped somewhat, and the calculated contributions by other anthropogenerated infrared-absorbing gases have increased dramatically. Initially ice-core studies gave a background of 270-290ppm, with a most likely value of 279ppm[13]. Another analysis obtained a lower bound of 260ppm[14]. The Soviet/French work on the long Vostok core seems to corroborate the lower values[15], which if assumed to be 270ppm, results in an anthropogenerated rise of 31% to date. The observed warming then becomes less than that expected from equilibrium models.

Background CH_4 concentrations, again calculated from ice cores, appear to be around 800ppb[16], compared to a current value of 1700ppb[17]. Sources include rice paddy agriculture, an increasing fraction from deforestation[18], and bovine flatulence, all of which are not likely to decrease in the near future. Further, the rate at which methane is microbially resequestered from the atmosphere may be declining as a result of high intensity agriculture in the industrialized world and "green revolution" culture in developing nations[19]. The current climate forcing of methane is nearly 40% of the effect of increased CO_2[20].

Precise knowledge of the sources and sinks of N_2O is unavailable at this time, and background concentration estimates are much less reliable than those for CO_2 and CH_4. The background level of 285ppb should therefore be taken with some advisement, as should future projections[17]. The current concentration is 298-308ppb.

Virtually all Chloroflourocarbons (CFC's) are anthropogenerated. 1950 concentrations are estimated at 0.001ppb for CFC-11 and 0.005 for CFC-12[3]. Current values are 0.22ppb for CFC-11 and 0.38 for CFC-12. The net warming effect of CFC's is over an order of four magnitudes greater than CO_2 on a per-molecule basis[1].

Wigley's estimate that the radiative effect of the non-CO_2 trace gases is 80% of that caused by a change in concentration from 279-350ppm[21] gives a current *effective* CO_2 concentration of 412ppm or 148-158% of the background range of 260-279ppm. The IPCC states that "Greenhouse gases have increased since pre-industrial times...by an amount that is radiatively equivalent to about a 50% increase in carbon dioxide"[1], which gives a current effective value of 390-420ppm. *We have already gone half-way to an effective doubling of the pre-industrial CO_2 concentration.*

While it is customary to present the history of global temperature as "at least not contradictory to"[22] popular vision climate model projections, those models produce an expected equilibrium warming for this trace gas change of approximately 2.0°C[21], while a least-squares fit of the entire set of Jones and Wigley global data gives a rise of 0.45 ± 10°C in the last 100 years[23]. Further, Figure 1 demonstrates that much of the warming of the southern hemisphere and virtually *all* of the warming of the northern hemisphere was prior to the major postwar emissions of the trace gases[24]. This discrepancy is further compounded by the fact that the "water" (southern) hemisphere, which should warm up least and slowest, in fact shows the more "greenhouse-like" signal.

Least-squares analyses of the Jones and Wigley global record demonstrate that fully 90% of the warming of the global record took place prior to 1945[23]; see Figure 2. At this time the actual CO_2 concentration was in the range of 310ppm and the effective concentration was approximately 325ppm. This is only one-third of the total forcing to date, and implies that there has been very little warming during the period in which the majority of the radiative forcing has taken place.

Oceanic thermal lag may not in fact account for the difference. Wigley calculated, using liberal estimates of the lag, that the hemisphere should still have warmed a degree celsius[21] (primarily after 1950), and the net global warming since then is in fact approximately 0.25°C. Sea surface temperature anlyses of Newell et al.[25] and Oort et al.[26] demonstrate that there is very little lag between that record and land readings. This is corroborated by the oxygen isotope-derived temperatures in the Himalayan ice cap calculated by Thompson[27] which is more a proxy for sea surface temperatures where the moisture originated than it is a true land record.

Thus we are left with a globe that appears to have warmed up two to four *times* less than suggested by straightforward interpretation on the Popular Vision models.

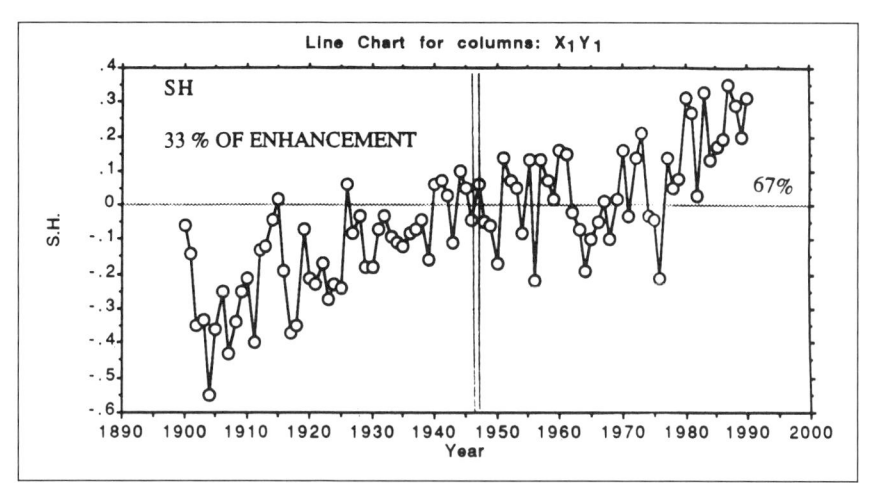

FIGURE 1A. Combined land and ocean annual mean surface air temperatures for the southern hemisphere through 1990[24].

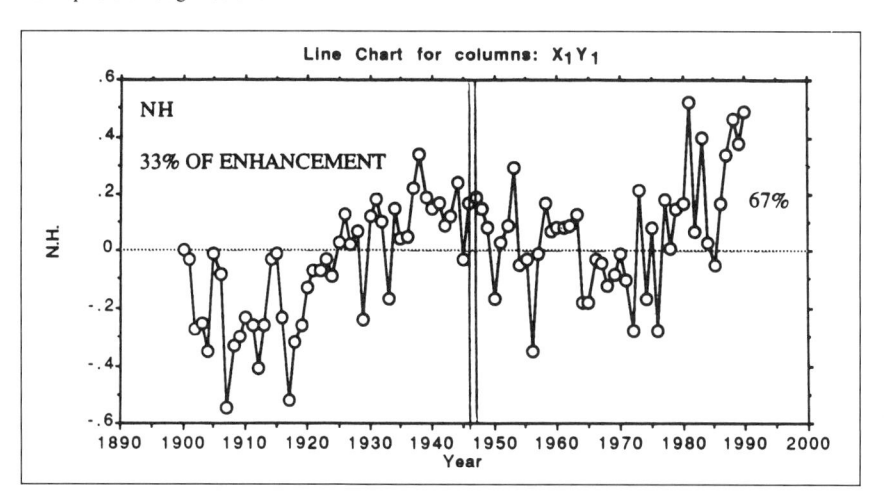

FIGURE 1B. Combined land and ocean annual mean surface air temperatures for the northern hemisphere through 1990[24].

In fact, these discrepancies are well known to those directly involved in research, and it is therefore mystifying why they are so rarely emphasized in public discussion.

Stratospheric temperatures, which should fall in a trace-gas enriched atmosphere, have dropped considerably more in the southern hemisphere than in the north, with the greatest declines in the polar zone that may be associated with the magnified end-of-winter ozone minima. There is no statistically significant trend in northern hemisphere stratospheric temperatures[28].

URBAN HEAT ISLANDS AND LONG-TERM SITE BIAS

Localized warming related to urban buildup has long been recognized as a confounding factor in regional and global temperature analyses. While causes include changes in the surface energy balance, boundary layer thickness and amount of open sky, precise calculations on the various components of the "urban effect" are not available. However, Karl *et al.* have found in the United States that the effect is statistically detectable at populations levels of 2,500 and up[29], and Balling and Idso note the effect at populations as low as 500[30]. The temperature record is further confounded by the fact that exponential population increases suggest the warming bias should be of the same functional form as enhanced greenhouse forcing.

Karl *et al.* have created a deurbanized "Historical Climate Network" (HCN) of 492 stations that have been checked for instrumental and location changes, and adjusted for population[31]. While this record only applies to the coterminous United States and is therefore not necessarily representative of global trends (i.e. it shows a slow but unsteady cooling for the last fifty years), it serves as an important check on other global records because they can be compared over this limited region.

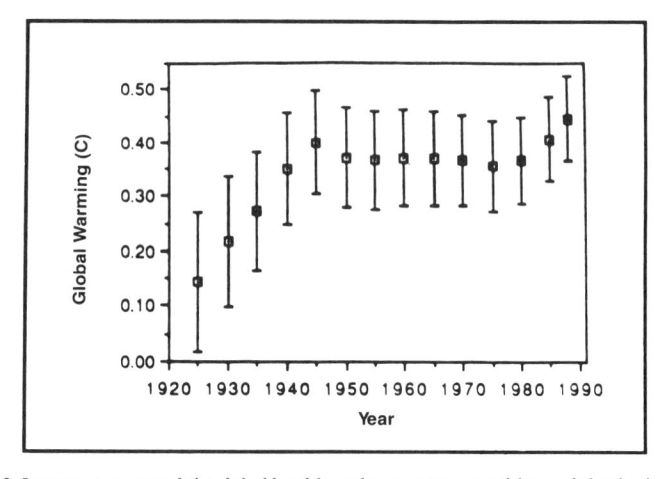

FIGURE 2. Least squares trends in global land-based temperatures, with trends beginning in 1885[23].

Karl *et al.* found an artificial warming of 0.10-15°C in both the NASA and Jones and Wigley (illustrated here) records over the United States. Curiously, the bias appears greatest in the U.S. record. A similar analysis of an analogous Soviet Union record shows no overall urban bias[32], although urban stations are in fact *cooler* than surrounding rural ones from 1953-67; there is no difference between the two through 1980 and then there is some apparent urban warming after 1980. A similar negative urban effect is observed in the 1960's in mainland China[33]. It therefore seems that the global urban bias is currently not appreciably more than 0.10°C, but it should be noted that it will be more pronounced at the end of the record, which shows the warmest temperatures.

In an attempt to determine the reliability of the various long-term climate records, Spencer and Christy correlated satellite-microwave sensed temperatures through late 1989 with Karl's United States HCN, and the hemispheric records of Jones and Wigley and Hansen *et al.*[34] The HCN is the most correspondent (R^2 = .77), Jones and Wigley's record was almost as reliable (R^2 = .72 and .66) for the northern and southern hemisphere, while the NASA record of Hansen and Lebedeff in the southern hemisphere was much less reliable (R^2 = .27). This last record, which is the "warmest" of the global records, has the majority of its warming in the southern hemisphere[35]. The coefficient of determination (R^2) between the U.S. HCN and the satellite data is now .86. The implication is that the methodology for the removal of the urban effect—where temperatures are actually adjusted for population—is most reliable in the temperature histories of Karl *et al.*[31]

An additional, and more insidious problem may be introduced into long-term climate records by a *general* site bias. Over what was the "developing world" of the late 19th century, the longest-standing records have tended to originate at points of commerce. Because of the predominance of water power during that era, it is likely that the longest records are in fact at sites that are preferential for cold air damage. Such locations may be buffered from nighttime warming.

Several attempts have been made to compensate both for population and general biases, usually involving temperatures in the free atmosphere. Angell[39], using the 850-300mb layer, finds a net warming of 0.24°C in the last thirty years after adjusting for El Nino warming and volcanic cooling. Michaels *et al.*[37] used observed 1000-500mb thickness values back to 1948 (the beginning of daily 500mb records) and a proxy record that extended back to 1885 over the U.S. and southern Canada and concluded that the observed secular variability in surface temperatures calculated from this record was approximately three times larger than that measured from ground-based thermometers. The rise in the first half of the 20th century and the subsequent fall through the mid 1970's was in fact concurrent with that of the ground based record, although the magnitude was again greater. Notably, the rise in early 20th century temperatures calculated by this method (of 1.8°C) does not differ appreciably from the expected regional greenhouse warming calculated by GCM's for *next* century. It is unknown whether this magnitude is related to noise in the statistical transfer functions between the cyclone frequency record used to calculate the 500mb height from 1885-1948, or whether it truly reflects free-atmosphere temperature variability.

Weber[38] also examined 1000-500mb thickness values and found warming in the

Northern Hemisphere tropics and subtropics, but cooling in the higher latitudes, with no overall hemispheric warming or cooling between the 1950's and the 1980's.

HIGH LATITUDE AND DIURNAL TEMPERATURES

In Manabe and Wetherald[2], which is quite representative of the penultimate generation of climate models[2-6] the difference between 2 X CO_2 and 4 X at high northern latitudes is twice the difference between 1 X and 2X, (8°C) suggesting that for the currently effective 1.5X there should be considerable Arctic warming, even with oceanic thermal lag. Nonetheless, in the Elsaesser *et al.* compendium of temperature records[39] the data indicate a rapid rise in temperature prior to the majority of trace gas emissions and concurrent with the rise in the Northern Hemisphere average. In most records the rise is followed by a decline, from the 1930's to the mid-70's, of similar proportion; See Figure 3.

Virtually all climate models suggest that the polar warming will be magnified in winter. Nonetheless, there has been a substantial secular *decline* in winter temperatures over the Atlantic Arctic since 1920[40], and there has been no change in polar night temperatures at the south pole (Figure 3b)— a station that surely

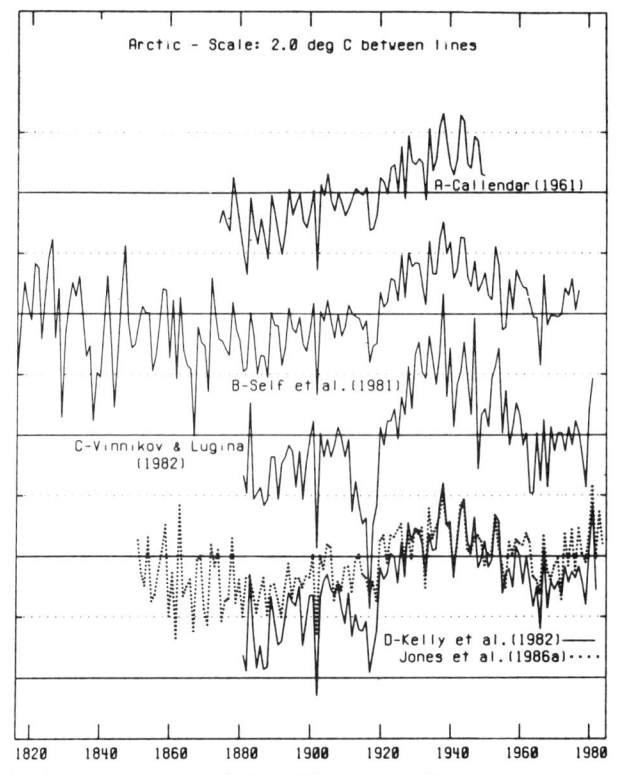

FIGURE 3. Arctic temperature records from Ellsaesser *et al*[39].

has no urban warming or cold air drainage[42]. Perhaps the depth of the winter anticyclone is sufficient to buffer that station from anthropogenic alteration of infrared absorption. There is no reason such a buffering should be confined to the South Pole.

The Popular Vision is not supported by recent studies on diurnal temperatures. Karl *et al.* examined maximum and minimum values from the U.S. HCN[31] and found that the daily range (difference between the two) has declined precipitously since 1950, and is now *averaging* two standard deviations below the mean for the century. In that record, maximum values have actually declined while minimum values have risen (Figure 4).

This behavior is consistent with an enhanced greenhouse combined with the increases in cloudiness (3.5%) and reduced sunshine that have been documented across the country[43]. Weber[44] has also documented a decline in sunshine in Germany, and notes that the effect is enhanced at high elevation, which implicates low-level stratocumulus—clouds whose net radiative effect is surface cooling—as the cause. Warren *et al.*[45] have found increasing cloudiness at most locations around the globe, although the shipboard observations used are subject to substantial observer and scale biases. Nonetheless, the cloud type that shows the most increase is the low level stratocumulus in the northern oceans. The finding of reduced UVB flux at the low elevations[46] in combination with increased values at heights greater than 10 KM[47] is also consistent with an increase in low level cloudiness.

Subtraction of maximum and minimum temperature curves for the other Northern Hemisphere locations in the IPCC chapter on observed temperature[60] gives the same result: a considerable narrowing of the daily range since 1950 that is mainly a result of a rise in night temperatures (Figure 5). Rural data for the Soviet Union, not included in the report[24] exhibit the same behavior.

If anthropogenerated warming takes place primarily at night, the Popular Vision is simply wrong. Evaporation rate increases, which are a primary cause of projected increases in drought frequency[49] are minimized. The growing season is longer, because that period is primarily determined by night low temperatures. Finally, many plants, including some agriculturally important species, will show enhanced growth with increased moisture efficiency[50] because of the well-known "fertilizer" effect of CO_2. Night warming also minimizes polar melting because mean temperatures are so far below freezing during winter that the enhanced greenhouse is insufficient to induce melting. Thus "how" the world is warming appears in fact to be beneficial. If this is indeed the expression of the altered greenhouse, the earth will have to reverse a course it has already embarked upon to get us to the greenhouse disaster.

Curiously, *none* of the analyzed Southern Hemisphere temperature ranges are declining[24,51]. Such an important interhemispheric differential in one of the prime components of the diurnal energy could be very important, as the most likely cause would be an increase in Northern Hemisphere cloudiness with less change in the Southern Hemisphere. One prime candidate for causation is therefore anthropogenerated sulfate, which is both hygroscopic and reflective, and is produced in much lower volume in the Southern Hemisphere.

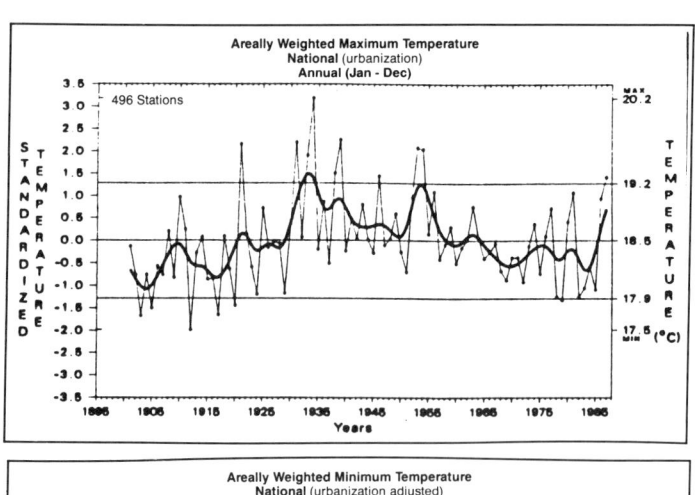

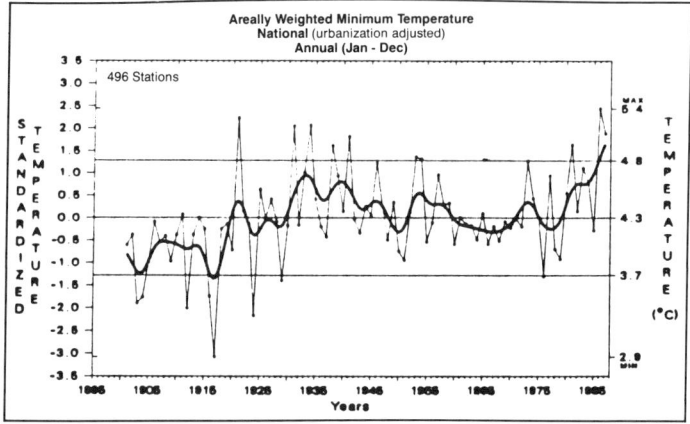

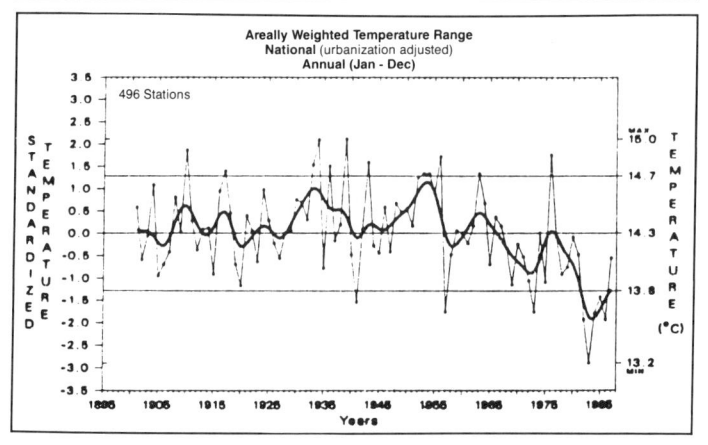

108 Global Climate Change: Implications, Challenges and Mitigation Measures

FIGURE 4. Maximum, minimum and daily temperature range in the U.S. Historical Climate Network[30].

"POPULAR VISION" AND NEW GENERATION CLIMATE MODELS

The penultimate generation of GCM's calculated a mean equilibrium warming for a doubling of atmospheric CO_2 of 4.2°C, with maximum warming of as much as 18°C during north polar winter, and less significant warming (approximately two degrees) over tropical oceans[2-6]. Figure 6 is from Washington and Meehl[6].

Major shortcomings in these calculations included unrealistic ocean dynamics and ocean-atmosphere coupling, inadequate cloud parameterization (prior to the ERBE analysis even the *sign* of the temperature forcing by the earth's cloud cover was unknown and debatable; it is now thought to be negative[52]), and the usual use of stepwise (instantaneous) doubling of CO_2, rather than the low-order exponential increase that exists in the real world.

With a modified ice-water interaction between clouds, projected warming in the United Kingdom Meteorological Office (UKMO) dropped from 5.2° to 1.9°C[53]. With a coupled ocean and climate general circulation model, the net warming in the National Center for Atmospheric Research (NCAR) model dropped to 1.6°C for thirty years after a shock doubling, compared to 3.7° in an earlier equilibrium calculation[54]. Manabe[55] reports a net equilibrium warming for a doubling with a

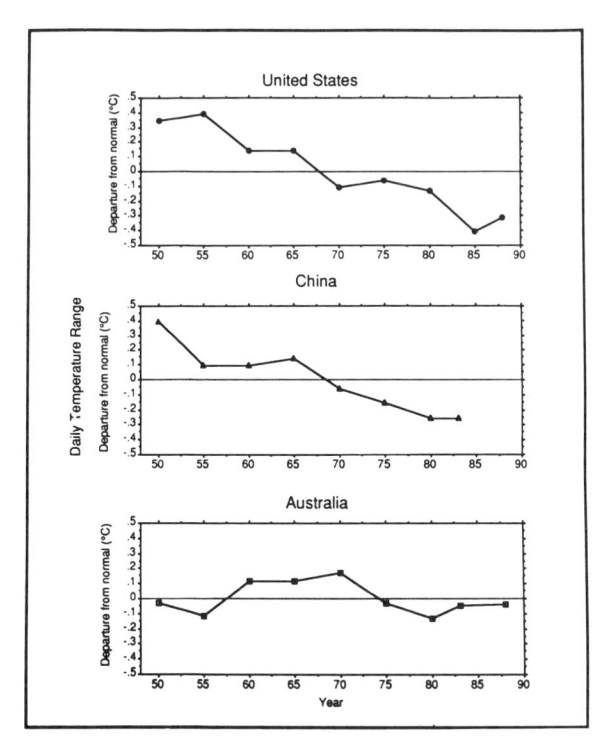

FIGURE 5. Mean daily temperature range at five-year intervals (and including the last year of record) for the U.S., Mainland China, and Australia; Calculated from reference 60.

coupled atmosphere-ocean model of 1.8°C, compared to 4.3° in a more primitive version. The net warming projected for the southern hemisphere is approximately 1.0°C, with 2.5° projected for the Northern Hemisphere. The observed behavior seems inconsistent with this difference unless there is some mitigation of northern hemisphere warming by other means.

In each of these models, estimates of global warming for a doubling of CO_2 have been considerably reduced. The mean of these three runs is 2.0°C, down over two degrees from the previous mean. This substantially reduced warming is calculated *either* with altered cloud or ocean parameters. It therefore seems logical that a combination of the two will produce even less net warming.

Figure 7 details the NCAR calculation with the coupled land and ocean GCM for thirty years after a shock doubling. Areas of warming of greater than 4°C have been dramatically reduced (compare to Figure 6), and in fact there are none in either hemisphere for June-August.

Because substantial warming in all GCM's tends to concentrate at high latitudes, use of this Mercator or related projections presents a highly distorted view of model results that serves to overemphasize areas of warming. In terms of actual area of the globe, on an annual basis, the area of greater than 4°C of warming in the illustrated model is less than five percent. *Virtually all of that warming is projected either for polar twilight or polar night,* when the temperature is far below freezing.

It is tempting to view the unanimity of these newer models as a sign of reliability, but that is not the case. In a transient version of the NCAR model, using an increase in radiative forcing that is quite close to what has actually occurred since 1950, calculations based upon its publication show that the current land temperature should be 1.4°C above the 1950 mean[56]; in fact the rise has been on the order of 0.3°. Thus, even the most conservative transient model appears to have grossly overestimated land warming. Also, even this version predicts very unrealistic regional temperatures for the current greenhouse enhancement. Figure 8 details the distribution of December-February temperature in the five year average of years 26-30 in the thirty year transient run in which trace gas forcing was increased in a more

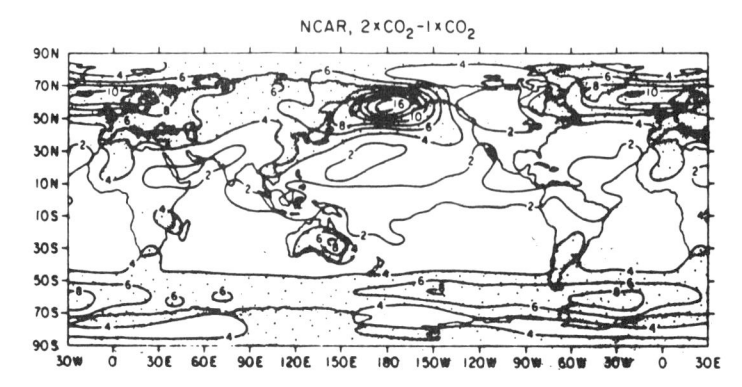

FIGURE 6. Winter temperature changes for a doubling of CO_2 calculated by Washington and Meehl[6] in 1983.

realistic manner than the step-change runs. Roughly speaking, the map should be analogous to a five-year aggregate in the period 1975-85 (Years 26-30 beginning in 1950). The most significant projected anomalies are the 2-4° warming of the northern half of North America and the 3-6° *cooling* of the North Atlantic. Neither occurred, although Atlantic temperatures may have dropped some. This projected cold anomaly does not appear in later years of this transient model[41].

POSSIBLE NEGATIVE FEEDBACKS IN THE POLLUTION SYSTEM

It is clear that human activity, besides enhancing the greenhouse effect, also produces substances that can serve to counter that effect. These include particulates, which serve to scatter radiation, and sulfur dioxide molecules, which is their oxidized sulfate state can serve as cloud condensation nuclei. It is this human negative feedback component of the pollution system that drastically compromises the applicability of temperature/CO_2 correlations in Antarctic ice cores[15] to the current greenhouse excursion.

Mayewski *et al.*[57] recently demonstrated that the anthropogenerated sulfate load in the Northern Hemisphere atmosphere now is equivalent to the maximum loading from the Tambora and Krakatoa volcanoes, both of which were associated with pronounced cooling of short duration. Hansen and Lacis[58] suggest current global forcing from aerosols may be as much as -1.5 w/m². Because anthropogenerated sulfates are primarily produced and reside in the northern hemisphere, we may therefore be overcompensating for current greenhouse forcing (1.5-2.5 w/m², depending upon estimate) with actual negative forcing in the hemisphere that contains most of the world's population.

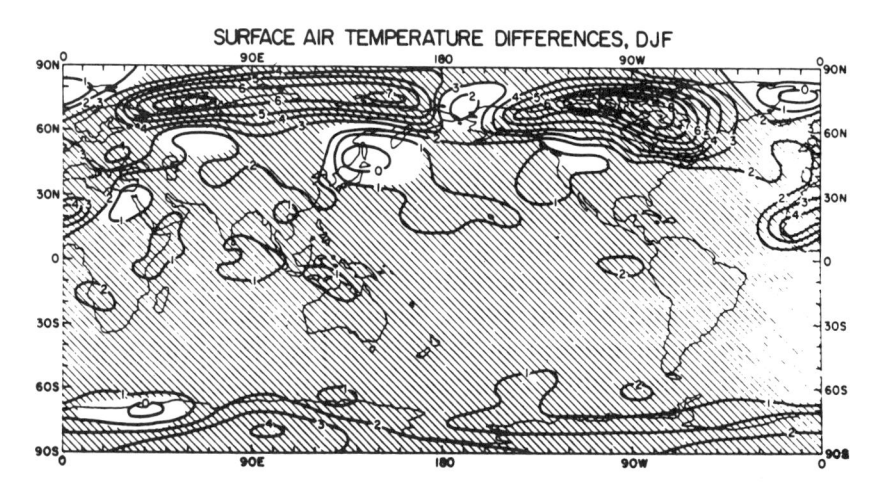

FIGURE 7. Winter temperature changes for a thirty year after a shock doubling of CO_2 calculated by Washington and Meehl[54] in 1989.

Is there any evidence for this? If anthropogenerated sulfates were in fact mitigating enhanced greenhouse warming, we would expect to see the following:

1) Night warming from an increase in both clouds and greenhouse forcing,
2) A counteraction of daytime warming by the greenhouse forcing because of an increase in clouds,
3) A decrease in daily temperature range,
4) Concentration of these effects in the industrial (Northern) hemisphere, and
5) Enhanced brightening of low-level clouds near sulfate source regions.

Previous sections of this paper detail evidence for the first four. With regard to, 5), Cess[59] analyzed a limited set of satellite data and found an increase in brightness in ocean-surface stratocumulus that was heightened near the source regions of Asia and North America. The effect, in which integrated albedo was increased by as much as eight percent, persisted for thousands of miles downstream from the source regions. A "clean" swath of the South Pacific ocean served as a "control", and showed no brightening. Finally, German sunshine analyses[44] and the diminution of UVB, at low altitudes[46] coupled with enhancement at high elevation[47] argue for some type of increase in a low altitude reflectant.

CONCLUSION

The Popular Vision of the world influenced by an enhanced greenhouse is one of ecological disaster resulting from rising temperature, evaporation rates, and sea

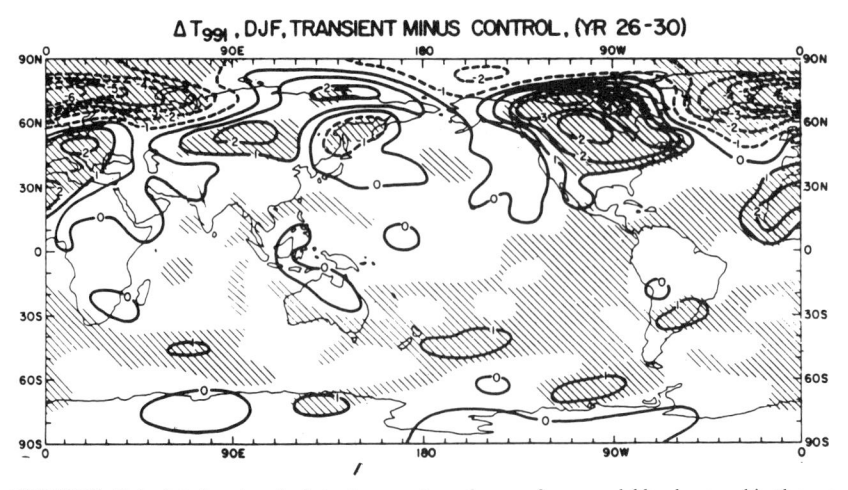

FIGURE 8. Calculated regional winter temperature changes from model background in the year 26-30 average of a model that realistically mimics anthropogenerated greenhouse forcing since 1950[54].

level. That vision persists even though there are now several lines of observational and model evidence that indicate such a "carbon dioxide in/disaster out" scenario is becoming more remote.

In fact, that vision evolved in light of some remarkable inconsistencies. The Northern Hemisphere, which should warm first and most, shows no net change in the last half-century. Repeated measurements now show relative warming at night, which may in fact be beneficial. The amount of global warming is clearly less than it should be according to the earlier GCM's, given the fact that we are already half way to an effective doubling of CO_2. High latitude temperatures, which should be most sensitive, show no obvious greenhouse signal. Other anthropogenerated compounds may be mitigating the warming. New climate models partition almost all of the warming of more than 4°C to polar twilight or night, which will have a minimal effect on ice melting.

If this is the course the earth has embarked upon in response to human insults of the atmosphere, that response is primarily benign. And if human sulfate emissions are indeed the cause of this benign greenhouse, what are the implications for environmental policy?

A PERSONAL PROJECTION

I believe that both the observed and theoretical evidence for mitigation of greenhouse warming by anthropogenerated sulfates is compelling and that it certainly serves to explain several disparate measurements that seem counter to the simplistic greenhouse effect. If this is indeed true, the dimensions of the policy debate on global change have changed dramatically.

Two arguments should now surface. One group will argue, as did Hansen and Lacis,[58] for an advanced timetable to drastically reduce or eliminate fossil fuel combustion, because the much longer residence time of CO_2 means that a substantial warming pulse already has been hidden by sulfates, and that that pulse will emerge soon afer SO_2 emissions are reduced.

The counterargument will be that we may have inadvertently discovered a control technology against greenhouse warming. The acid rain effects of that technology can be minimized by concentrating fossil-fuel based power generation on the eastern margin of continents, while using nonfossil dense energy sources inland.

Is a sulfate-mitigated warming reason to limit fossil fuel combustion immediately, or does it argue that coal, whose reserves can last for centuries, should continue as a primary fuel? The former point of view now acknowledges *both* economic and ecological disruption as usage is curtailed, while the later viewpoint will continue to increase the CO_2 content of the atmosphere dramatically. Will we purposefully produce a significant near-term disruption, or will we hope that the intellectual capital generated by nondiminished economic activity will, over the next 200 years, provide a nondisruptive solution to the global warming problem? That is the environmental question of the 1990's.

REFERENCES AND NOTES

1. Intergovernmental Panel on Climate Change (IPCC). 1990. *Policymakers Summary of the Scientific Assessment of Climate Change.* World Meteorological Organization, United Nations Environment Programme. 39pp.
2. S. Manabe and R.T. Wetherald. 1980. *J. Atmos. Sci. 37*, 99.
3. Hansen, J.E., A. Lacis, D. Rind, G. Russell, P. Stone, I. Fung, R. Ruedy and J. Lerner. 1984. *Geophys. Mon. Ser 29*, 130.
4. M.E. Schlesinger. 1984. *Adv. Geophys. 26*, 141.
5. J.F.B. Mitchell. 1983. *Quart. Jour. Royal Meteor. Soc. 109*, 113.
6. Washington, W.M., and G.A. Meehl. 1983. *J. Geophys. Res 88*, 6600.
7. Testimony of James E. Hansen to the U.S. House of Representatives. 6/23/88.
8. Cable News Network ran a "yes/no" poll on June 25, 1988 in which the question was "is the current drought caused by the greenhouse effect?"
9. A.V. Gore, Jr., 3/18/89. "An Ecological Kristallnach. Listen". New York Times.
10. S.333, by Patrick Leahy (D-VT) calls for a 50% reduction in 1985 CO_2 emissions by 2000.
11. S.M. Goshko. 1/31/89. *The Washington Post.*
12. M.E. Schlesinger in *Carbon Dioxide Proliferation: Will the Icecaps Melt* (IEEE Power Engineering Society, 1982) pp.9-18.
13. A. Neftel, E. Moor, H. Oeschger, B. Stauffer. 1985. *Nature 315*, 45.
14. Raynaud, D., J.M. Barnola. 1985. *Nature 315*, 309.
15. C. Lorious, J. Jouzel, C. Ritz, L. Merlivat, N.I. Barkov, Y.S. Korotkevich, V.M. Kotlyakov. 1987. *Nature 329*, 516.
16. G.I. Pearman, D. Etheridge, F. deSilva, P.D. Fraser. 1986. *Nature 320, 248.*
17. R.A. Rasmussen and M.A. K. Kahlil. 1986. *Science 232*, 1623.
18. H. Craig, C.C. Chou, J.A. Wehlan, C.M. Stevens, A. Engelmeier. 1988. *Science 242*, 1535.
19. P.A. Studler, R.D. Bowden, J. M. Melillo, J.D. Aber. 1989. *Nature 341*, 314.
20. H. Craig and C.C. Chou. 1982. *Geophys. Res. Let. 9*, 1211.
21. T.M. L. Wigley. 1987. *Clim. Monitor 16*, 14.
22. M.C. MacCracken and G.J. Kukla. 1985. In *Detecting the Climatic Effects of Increasing Carbon Dioxide,* U.S. Department of Energy DOE/ER-1235. pp. 163-176.
23. R. Balling, S.B. Idso, *in press.*
24. P.D. Jones. 1988. *J. Climate* 1, 654-657. Data extended through 1990 from figures communicated from Jones to R.W. Spencer of NASA-Huntsville.
25. R.E. Newell and J. Hsiung. 1984. *Sea Surface Temperature, Atmospheric Carbon Dioxide, and the Global Energy Budget: Some Comparisons between Past and Present.* In: W.A. Morner (ed.): Climatic Changes on a Yearly to Millenial Basis, Reidel, Dordrecht, 533-561.
26. A.H. Oort, Y.H. Pan, R.W. Reynolds, C.F. Ropelewski. 1989. *Climate Dynamics 2*, 29.
27. L.G. Thompson, E. Mosley-Thompso, M.E. Davis, J.F. Bolzan, J. Dai, T. Yao,

N. Gundestrup, X. Wu, L. Klein, Z. Lie. 1989. *Science 246*, 474.
28. J.K. Angell. 1986. *Mon. Wea. Rev. 107*, 1922.
29. T.R. Karl, H.F. Diaz, J. Kukla. 1988. *J. Climate 1*, 1099.
30. R. Balling and S.B. Idso, *in press*.
31. T.R. Karl, R.G. Baldwin, M.G. Burgin. 1988. *Historical Climatology Series 4-5,* National Climatic Data Center, Asheville, NC 28801. 107pp.
32. P.D. Jones, *Nature,* in press.
33. W.C. Wang and T.R. Karl, *Geophysical Research Letters,* in press.
34. R.W. Spencer, J.R. Christy. 1990. *Science 247,* 1558.
35. Satellite data update through 1990, personal communication from R.W. Spencer.
36. J.K. Angell. 1990. *Geophys. Res. Let. 118*, 1093.
37. P.J. Michaels. 1990. *Int. J. Envi. Stud. 36,* 55.
38. G.-R. Weber. 1990. *Int. J. Climatol. 10*, 3.
39. H.W. Elsaesser, M.C. MacCracken, J.J. Walton, S.L. Grotch. 1986. *Rev. Geophys. 24*, 745.
40. J.C. Rogers. 1989. Proc. 13th Annual Climate Diagnostics Workshop, NOAA; Available from NTIS.
41. Personal communication from G.A. Meehl, National Center for Atmospheric Research 9/13/90: Years 26-30 of the transient GCM of Washington and Meehl do show cooling in this region, however it did not persist in succeeding pentads.
42. J. Sansom. 1989. *J. Climate 2*, 1164-1172.
43. J.K. Angell. 1990. *J. Climate 3*, 296-306.
44. G.-R. Weber. 1990. *Theor. Appl. Climatol. 41*, 1-9.
45. S.G. Warren, C.J. Hahn, J. London, R.M. Chervin, R.L. Jenne. 1988. *Global Distribution of Total Cloud Cover and Cloud Type Amounts over the Ocean,* U.S. Department of Energy Publication DOE/ER-0406, 42pp + Maps.
46. J. Scotto, G. Cotton, F. Urbach, D. Berger, F. Fears. 1988. *Science 239*, 762-763.
47. C. Bruhl, P.J. Crutzen. 1989. *Geophys. Res. Let. 16*, 703.
48. Personal Communication from T.R. Karl. 9/90. National Climatic Data Center, Asheville, NC 22980.
49. D. Rind, R. Goldberg, J. Hansen, C. Rosensweig, and R. Ruedy. 1990. *J. Geophys. Res. 95*, 9983-10004.
50. N. Sionit., H. Hellmers, B.R. Strain. 1980. *Crop. Sci. 20*, 687-690.
52. V. Ramanathan, R.D. Cess, E.F. Harrison, P. Minnis, B.R. Barkstrom, E. Ahmad, D. Hartmann. 1989. *Science 243*, 53-67.
53. J.F.B. Mitchell. 1989. *Nature 341*, 132-134.
54. W.M. Washington, G.A. Meehl. 1989. *Clim. Dyn. 2*, 1-38.
55. S. Manabe, K. Bryan, M.J. Spelman, *J. Phys. Ocean.* in press.
56. Washington and Meehl (54) state that "The transient experiment . . . shows a difference of about 0.7°C after thirty years for surface air temperature and 0.3 to 0.4° for the global average sea surface temperature difference". The implied land warming is therefore 1.4°C; since 1950 land areas of the globe have warmed between 0.2 and 0.3°C.

57. P.A. Mayewski, W.B. Lyons, M.J. Spencer, M.S. Twickler, C.F. Bock, S. Whitlow. 1990. *Nature* 346, 554-556.
58. J.E. Hansen, A.A. Lacis. 1990. *Nature 346*, 713-718.
59. R.E. Cess. 1989. Presentation to Department of Energy Research Agenda Workshop 4/25/89.
60. C.K. Folland, T.R. Karl and K. Ya. Vinnikov. 1990. *Observed Climate Variations and Change,* U.N. Intergovernmental Panel on Climate Change, Working Group 1, Section 7, 29 pp.

Global Climate Change: Implications, Challenges and Mitigation Measures. Edited by S.K. Majumdar, L.S. Kalkstein, B. Yarnal, E.W. Miller, and L.M. Rosenfeld. © 1992, The Pennsylvania Academy of Science.

Chapter Nine

THE CARBON CYCLE AND THE CARBON DIOXIDE PROBLEM

BERRIEN MOORE III[1]*
and
ROBERT KAUFMANN[2]**

[1]Institute for the Study of Earth, Oceans, and Space
University of New Hampshire
Durham, NH 03824-3525
and
[2]Center for Energy and Environmental Studies
Boston University
Boston, MA 02215

INTRODUCTION

This paper considers both the global carbon cycle and the human use of fossil fuels, which is changing a critical aspect of that cycle: the concentration of atmospheric carbon dioxide. In treating both topics in a single paper, we hope to make connections across academic disciplines. In this light, we have not treated the many nuances and subtle features associated with carbon cycle investigations or energy analysis. We have, however, been honest with respect to the difficulties that are contained within such studies.

---

*Supported in part by a grant from the National Aeronautics and Space Administration #NAGW-2669.
**Acknowledges the support of the staff at the USSR Academy of Sciences during preparation of the manuscript.

The focus in Part I is on the principal sink for the carbon dioxide that is being added to the atmosphere by human activity. This sink is the world's oceans. In Part II, we consider the primary source for carbon dioxide: the burning of fossil fuels.

PART I

The present average concentration of carbon dioxide in the atmosphere is about 355 parts per million, which is equivalent to about 755 Pg. C (10^{15} grams carbon). This is up from less than 600 Pg. C which, based on ice core records, we believe was present in the atmosphere in the middle of the last century (Figure 1). What is of equal importance is that if we examine deeper (and hence older) cores, we find that the atmospheric concentration of CO_2 was relatively constant for several thousand years before 1700. This relative constancy is important because it implies that one can focus upon the perturbation problem and net fluxes rather than worry about gross fluxes. If this were not the case, then one could seriously question whether or not the atmospheric increase of CO_2 could be associated with fossil fuel use and deforestation (Figure 2), since most of the gross fluxes within the global carbon cycle (Figure 3) are far larger than the human induced fluxes associated with fossil fuel burning and deforestation. We should mention that one can also link fossil fuel use to the atmospheric increase of CO_2 through measurements of the atmospheric concentration of carbon-14 CO_2, which is diluted when fossil fuels are burned.

Estimates of the amount of carbon in living organic matter on land vary between 450 and 600 Pg. C. Consequently, there is more carbon in the atmosphere than there is in all the world's forest—a rather startling fact (Figure 3). There are, of course, other "pools" of terrestrial carbon. For instance, there is two to three times more carbon in the surface soils of the world than in vegetation.

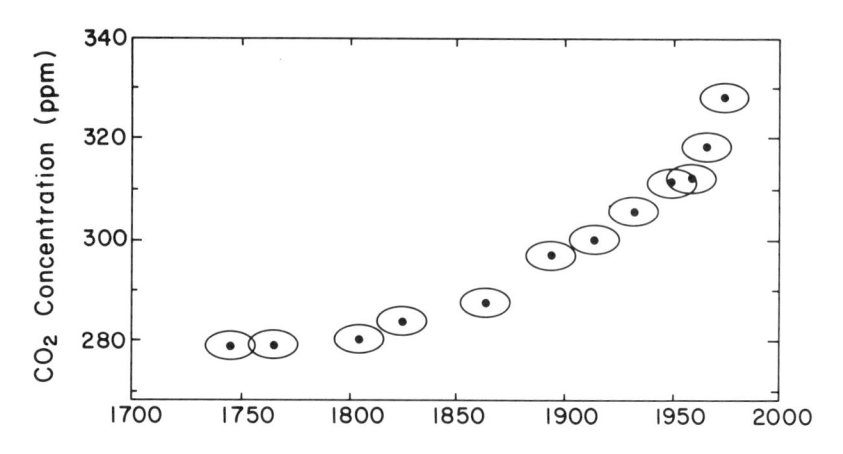

Atmospheric CO_2 concentrations measured in glacier ice formed during the last 200 years.

FIGURE 1. The Siple ice core record with an indication of uncertainty in both time and concentration. From *Trends '90: A Compendium of Data on Global Change.*

The Carbon Cycle and the Carbon Dioxide Problem   119

The amount of carbon released globally from vegetation and soils as a result of deforestation remains uncertain, though much progress has been made in recent years. The difficulties in calculating this release center on two general but important factors: 1) the rate of land-use change and 2) the response of biota to disturbances. It is surprising and frustrating that, even today, estimates of the current

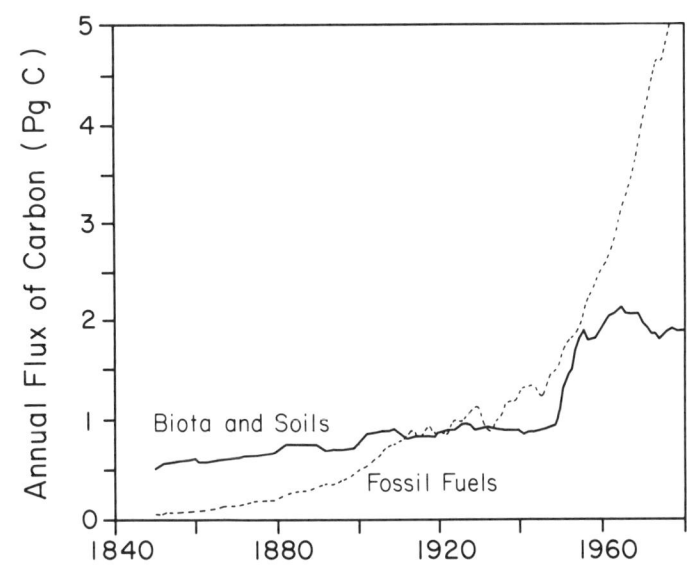

FIGURE 2. The fossil fuel emissions record (in dashes) and an estimate of the biotic release due to land use change (in solid). The fossil fuel release is by Rotty and Marland and taken from *Trends '90: A Compendium of Data on Global Change*. The biotic release is from Houghton et al. 1983.

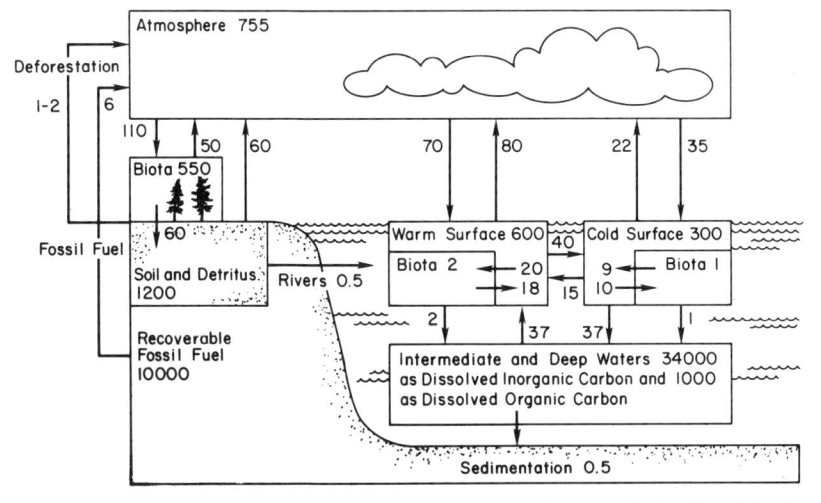

FIGURE 3. The global carbon cycle (from Moore, 1985). The boxes are in Pg. C and the fluxes (arrows) are in Pg. C per year as CO_2.

rate of conversion of closed canopy tropical forests to agricultural land vary from 70,000 to 100,000 square kilometers per year; the combined area of Massachusetts, New Hampshire, and Vermont is approximately 70,000 square kilometers. The analysis of the biotic response to such disturbances has proven even more difficult to determine accurately.

Calculations of the net carbon loss from the global biotic inventory, which attempt to take into account the uncertainties concerning disturbance rates and biotic response, indicate that from 1700 to 1990 the total loss was approximately 120-150 Pg. C, which is roughly 1/2 the amount of carbon dioxide that the combustion of fossil fuels has contributed since 1860. Currently, approximately 5.7 Pg. C are emitted annually in the burning of fossil fuels and another 1-1.5 Pg. C from land deforestation (Figure 2). We shall not consider further the deforestation issue but rather refer the reader to the literature (e.g. Moore *et al.*, 1981; Houghton *et al.*, 1983; Houghton *et al.*, 1985; Palm *et al.*, 1986; Houghton *et al.*, 1987; Houghton *et al.*, 1991; Detwiler and Hall, 1988; Melillo *et al.*, 1988; Houghton, 1990; Houghton and Skole, 1990; and Skole, in press). We should mention, however, that an estimate of about 120 Pg. C of carbon from deforestation is consistent with other calculations that use ratios of carbon isotopes, as found in ice cores and in tree rings, to infer how much CO_2 has come from deforestation; however, the temporal patterns that result from these two types of calculations do differ somewhat.

The oceans are by far the largest active reservoir of carbon. Recent estimates of the total amount of dissolved inorganic carbon, ΣC, establish an amount of about 34,000 Pg. C. Only a small fraction is carbon dioxide (0.5 percent); the bicarbonate ion, making about 90 percent, and the carbonate ion, at slightly less than 10 percent, are the major forms of dissolved inorganic carbon. There is far less dissolved organic carbon, about 1,000 Pg. C, and even less particulate organic carbon.

Although the oceans are the largest active reservoir of carbon and cover almost 70 percent of the globe, the total marine biomass contains only about 3 Pg. C of carbon or just about 0.5 percent of the carbon stored in terrestrial vegetation. On the other hand, total primary production of marine organisms is 30 to 40 Pg. C per year, corresponding to 30 to 40 percent of the total primary production of terrestrial vegetation (see Figure 3). However, only a relatively small portion of this production results in particulate organic carbon, which sinks and decomposes in deeper layers or is incorporated into sediments.

The net flux of atmospheric carbon dioxide into the oceans remain uncertain, although most current models suggest that between 2 and 3 Pg. C per year are taken up by the oceans. If these calculations reflect actual oceanic uptake, then it is inconsistent with the magnitude of the fossil fuel emissions, biospheric source for CO_2, and the observed rate of atmospheric increase in CO_2. Simply stated, the annual budget (Table 1) does not balance, and as yet unknown sinks, on land or in the oceans must play a role. The imbalance is diminished by reductions in the estimate of the rate of deforestation or increases in the regrowth, or if the oceanic uptake is underestimated, or if there are significant shorter term natural variations in the concentration of carbon dioxide in the atmosphere that override the imbalances. The question of which combination of these possibilities is more likely is both important and fascinating. At present, our ability to interpret the carbon cycle (Figure 3) and, thus, predict future carbon dioxide concentrations in the atmosphere and hence how climate may change, is confounded by these unresolved imbalances in the carbon budget.

The change in atmospheric CO_2 concentration resulting from fossil-fuel combustion, land-management activities, and other human-induced disturbances of the global carbon cycle is strongly governed by the CO_2 exchange between the atmosphere and ocean. The ocean is believed generally to be the largest sink for atmospheric CO_2; however, there is some (though highly controversial) evidence that the temperate biosphere may have been a comparable net, annual sink during the last decade (Tans *et al.*, 1990). Acknowledging this and other uncertainties about the carbon cycle, it is still true that future atmospheric concentrations of CO_2 will depend on the rate of industrial CO_2 emissions, the net exchanges between the atmosphere and the terrestrial biosphere, and the rate of uptake by the oceans.

In order to better appreciate the ocean's role in all of this, let us consider briefly the particular biological, chemical, and physical processes that govern the largest sink for CO_2: the ocean.

The Role of the Oceans in the Global Carbon Cycle.

The rate of carbon uptake by the oceans is controlled by sea water temperature, surface chemistry and biology, and by the various patterns of mixing and circulation which determine the amount of carbon transported from surface waters to the deep ocean. The actual exchange of CO_2 between the sea surface and the atmosphere is by diffusion at the air-sea interface and hence, is governed by the CO_2 partial pressure difference between the atmosphere and the sea surface, the wind velocity at the sea surface, and state of the sea surface. The essential uncertainties are the distribution of the partial pressure of CO_2 in the ocean surface and the factors controlling this distribution.

For simplicity, one can consider the partial pressure of CO_2 in sea water as a linear function, depending on temperature, salinity, and alkalinity, of the concentration of dissolved inorganic carbon in sea water. Specifically, for a reasonable range of temperatures, salinities, and alkalinities, and for relatively small changes in dissolved inorganic carbon, ΣC, the major aspects of the current relationship are captured by the linear function:

TABLE 1

Annual Carbon Budget

Input of Carbon as Carbon Dioxide into Atmosphere		
	Fossil Fuel:	5 Pg. C per year
	Deforestation minus regrowth:	1 Pg. C per year
	6 Pg. C per year	
Uptake of Carbon Dioxide		
	Atmospheric Increase:	2.5 Pg. C per year
	Oceanic Uptake:	2.5 Pg. C per year
	Fertilization Effects	?
	5 + ? Pg. C per year	

$$P = (10P_0/\Sigma C_0) \Sigma C - 9 P_0 \tag{1}$$

where P_0 and ΣC_0 are the preindustrial CO_2 partial pressure in the sea surface and the dissolved inorganic carbon concentration in the sea surface, respectively, and P and ΣC are the contemporary values.

Rewriting this linear expression as a normalized ratio of the change in P against the change in ΣC yields as a useful expression:

$$\{(P - P_0)/P_0\}/\{(\Sigma C - \Sigma C_0)/\Sigma C_0\} = 10 \tag{2}$$

Since the atmospheric concentration of CO_2 has increased by approximately 25% over the last 125 years since 1860; then, in light of Equation 2, the increase in surface concentration of total dissolved inorganic carbon would be only 2.5% (assuming the surface layer remains in equilibrium with the atmosphere). Since this sea surface mixed layer is about the same size, with respect to total inorganic carbon, as the atmosphere (the 900 GTs. in Figure 3 reflect a somewhat deeper mixed layer and not just the well-mixed sea surface layer), this would imply that if there are no other processes effecting the concentration of dissolved inorganic carbon in the sea surface that the ocean has only taken up a small amount of the fossil fuel carbon. There are, of course other processes such as biological activity and ocean mixing, and it is these processes that must have removed from surface waters whatever additional carbon dioxide entered the ocean.

There are five interlocked biological and physical features that are central to the oceanic uptake of carbon dioxide: 1) the consumption of CO_2 in primary production in biologically active surface waters; 2) the enrichment of the deep water in CO_2 due to the decomposition and dissolution of detrital matter that originates from biological processes in the surface waters; 3) the sinking of water in polar regions, particularly in the North Atlantic, taking CO_2 with it followed by a general bottom water flow toward the equator; 4) the upwelling of this water in equatorial regions with a corresponding outgassing of CO_2 to the atmosphere and a general poleward flow of surface waters, and 5) accompanying the meridional circulation is the general turbulent mixing processes; whereby, the carbon-rich water at intermediate depths are continuously being exchanged (mixed) with water of less carbon content in the surface layers.

The first two features taken together are often referred to as the "biological pump"; the biology pumps carbon to the bottom. The most obvious pumping is the incorporation, in tissue or as carbonate in shells, into living organism of inorganic carbon that is dissolved in surface waters, lowering the partial pressure of CO_2, followed by the "shipping" of some of this carbon to the bottom "packed" in the remains of dead marine organisms. Where active, the biological pump lowers the partial pressure of CO_2 in surface waters and increases the partial pressure in deep water not in contact with the atmosphere. It is as if the "biological pump" moves the partial pressure around in a way that allows carbon dioxide to work its way into the ocean. However, since the pre-industrial carbon cycle was in quasi-steady state (the net exchange was zero) and since there is little reason to believe that the biological pump has changed over the last 300 years, its direct role in the perturbation problem that has been induced by human activities is perhaps minimal. Its role in a changed climate is an open and important question.

The role of the oceans in the carbon cycle is much dependent on their rate of overturning (the meridional circulation) and mixing. In polar regions, ice formation leaves much of the salt "behind", still in solution. The result is an increase in density; furthermore, in the North Atlantic, evaporation exceeds precipitation. As a consequence, these cold, saline, and hence, dense North Atlantic surface waters sink, and thereby they have the potential to form, in effect, a pipeline or conveyor belt (Figure 4; this term and more importantly the concept are due to Broecker—see Broecker and Ping, 1982) for transferring atmospheric CO_2 to the large reservoirs of abyssal waters which have long residence times. In addition to the bottom water formation in polar regions, there is water exchange between surface waters and intermediate waters due to vertical turbulent exchange in association with the surface ocean currents like the Gulf Stream.

In order to sort out the effects of these many processes that govern the exchange of CO_2 between the atmosphere and the sea, it is necessary to use models. Such models range from complex formulations of the general circulation of the ocean to simpler box models that seek to capture net effects of the various processes. In keeping with the interdisciplinary spirit of this paper, we shall explore briefly four classical box models that treat the oceanic uptake of carbon dioxide. We note in passing that one of these models, the Box Diffusion Model (see below), was used by Intergovernmental Panel on Climate Change (IPCC) as the basis for determining future oceanic uptake of carbon dioxide

Simple Ocean Carbon Models

The penetration of carbon from the surface ocean into deeper regions of the sea is an important limitation or control on the rate of carbon uptake by the ocean. Oeschger *et al.* (1975) sought to capture this vertical structure by developing a layered, diffusion-based simple ocean model (The Box-Diffusion Model (BD); Figure 5a). Siegenthaler (1983) incorporated into the Box-Diffusion Model the "pipe line" concept of Broecker and Ping by simply allowing deeper layers to rise to the

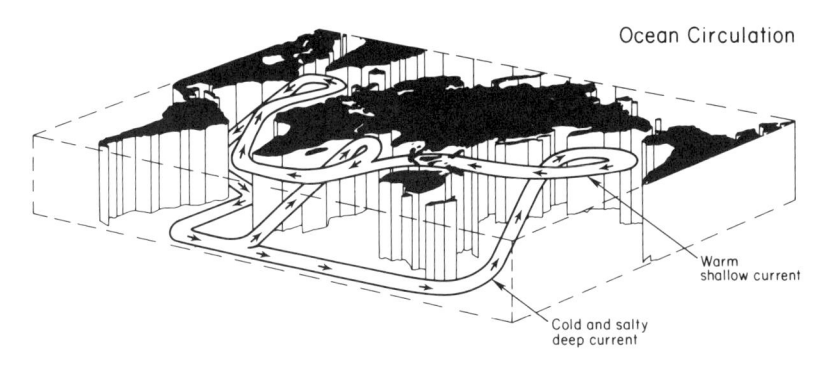

FIGURE 4. The broad meridional circulation of the ocean adopted from Broecker and Ping, 1982.

surface in the "polar" regions (The Out-Crop Model (OC); Figure 5b). Bjorkstrom (1979) also incorporated cold polar regions which communicate directly with deep layers as distinct from warmer regions which did not. This simple, low dimensional box model of the global ocean also distinguished intermediate waters from deeper ones and included advection as well as eddy diffusion (The Advective-Diffusive Model (AD); Figure 6). This model also contained biological processes (organic and carbonate) as well as a full treatment of ocean surface carbon chemistry. Bolin *et al.* (1983) employed five tracers and an inverse calculation to develop a coarse resolution 12 box model for the global ocean (The 12 Box Model (12B); Figure 7).

Box-diffusion Model
(Oeschgen and colleagues)

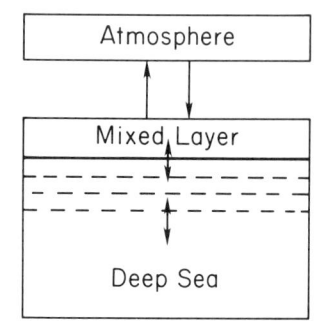

FIGURE 5a. Box-Diffusion Model ("BD"): the turnover of carbon below 75m is represented by a diffusion equation. A constant coefficient of diffusivity is estimated to match an idealized profile of natural carbon-14 (Oeschger *et al.*, 1975).

Box-diffusion Model with Polar Outcrops
(Siegenthaler and colleagues)

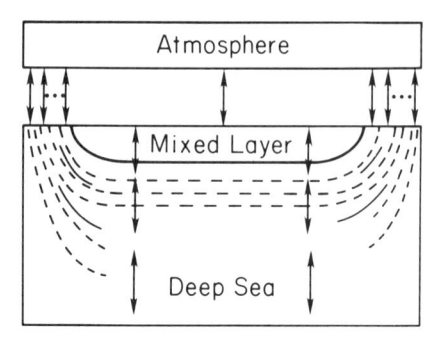

FIGURE 5b. Outcrop-Diffusion Model ("OD"): allows direct ventilation of the intermediate and deep oceans at high latitudes by incorporating outcrops for all sublayers into the box-diffusion formulation (Siegenthaler, 1983).

The four models have, naturally, much in common: they are all diagnostic rather than prognostic; each uses carbon-14 in the parameterization process (in fact, carbon-14 is the basic clock for all of the models and hence controls much of their response); each includes some form of ocean carbon chemistry and some form of ocean mixing. There are, however, major differences: ocean biology is explicitly included only in two (AD, 12B); whereas, it is simply "part of" the diffusive process in both the BD and OC models; deep water formation is not explicitly considered in the box-diffusion models; ocean chemistry varies in complexity; in the box-diffusion and outcrop-diffusion models all of the physics is captured by a single constant eddy diffusivity term, whereas, in the Bjorkstrom model the system is primarily advective (eddy diffusivities can be incorporated); the Bolin model has both advection and eddy diffusivities; and, perhaps most importantly the geometrical configurations are quite different. In addition, there are a host of smaller differences between ocean volume and surface area, parameterization

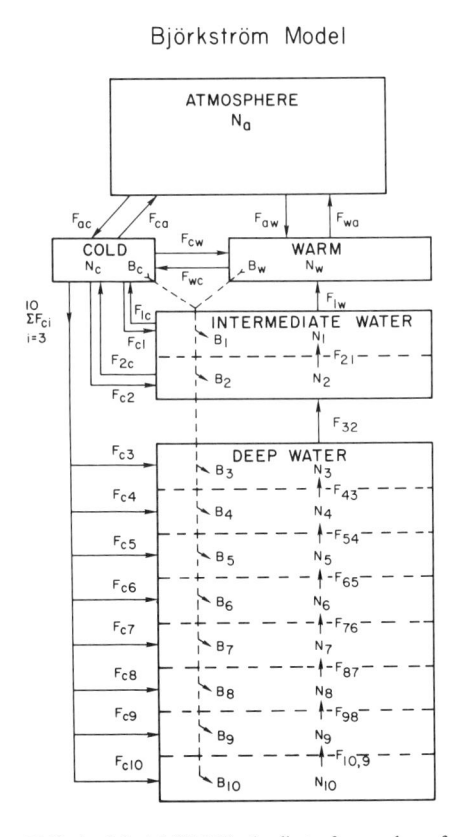

FIGURE 6. Advective-Diffusive Model ("AD"): the first of a number of models whose structures are meant to be more realistic than the simple diffusive assumption. The surface ocean is divided into cold and warm compartments. Water downwells directly from the cold surface compartment into intermediate and deep layers (Bjorkstrom, 1979).

procedures, carbon-14 profiles, initial conditions and atmospheric residence time.

As a result of these large and small differences, we find (in the next section) a range of responses; briefly, the outcrop-diffusion model is the most efficient in taking up CO_2; the box-diffusion model, the Bjorkstrom model, and the Bolin model all tend toward less carbon uptake. The OC model is more efficient because of the infinitely rapid connection of the atmosphere to the deeper layers; the reason the three other models are so similar is the over-arching importance of 14-carbon in setting the basic rates within the models, none of which have a direct link to deeper layers.

Basic Results

We consider three future records or scenarios simply to capture the general characteristics of the response of just the ocean-atmosphere system to different CO_2 forcing. The initial two emission scenarios do not reflect any particular energy future; rather they are chosen to reflect a range of responses. The third reflects a *highly simplified* policy-motivated scenario. Again, we emphasize that these three scenarios are not policy driven but rather reflect simply the dynamics of the models.

The first two forcing functions (Energy scenarios) are:

1. A Gaussian-like fit (see Bacastow and Bjorkstrom, 1981) to the actual fossil fuel record (i.e., Rotty-Marland Record for the period 1860-1986 as supplied by the Carbon Dioxide Information Analysis Center at the Oak Ridge National Laboratory) to 1986 such that a) it has the same total input for the period 1860-1986; b) it passes through the 1986 data point; and c) the total integral is eight times the amount of CO_2 as in the pre-industrial atmosphere (see Figure 8).

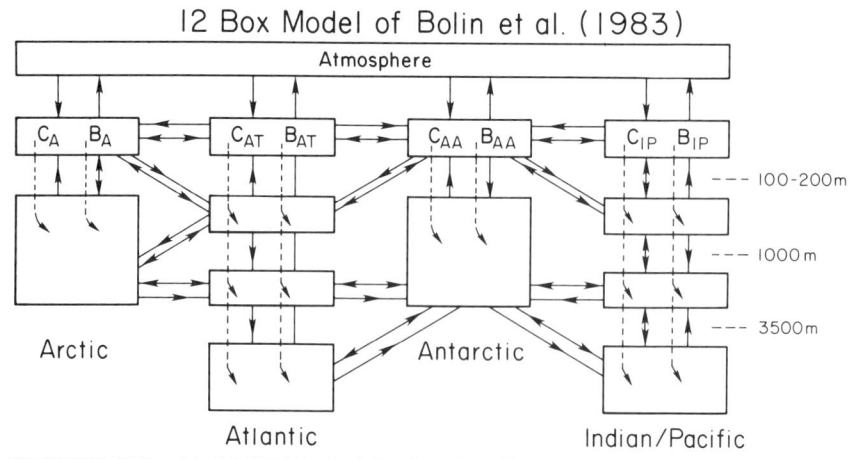

FIGURE 7. 12-Box Model ("12B"): the Atlantic and Pacific-Indian Oceans are each divided into surface, intermediate, deep and bottom water compartments. The Arctic and Antarctic Oceans are divided into surface and deep water compartments. The model is calibrated against multiple tracer distributions (Bolin *et al.*, 1983).

2. The actual fossil fuel record to 1986 and then the best fit straight line to the 1977-1986 data for the period 1987-2200 (see Figure 9).

We use the four models discussed earlier and actually include two parameterizations for the OC model. One uses steady-state ^{14}C to parameterize the dynamic

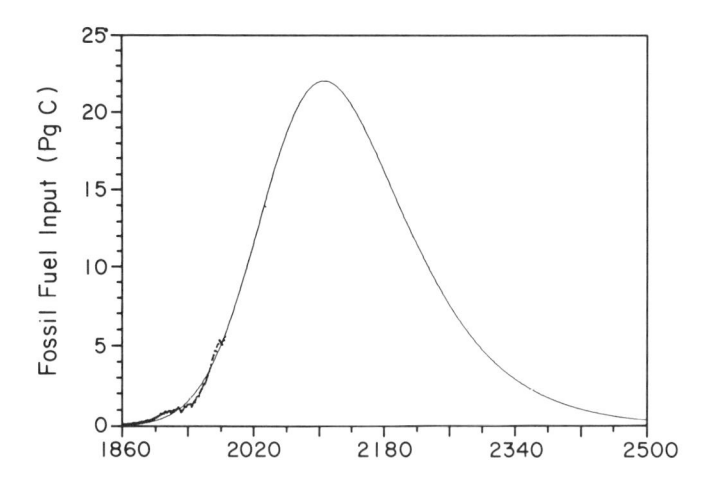

FIGURE 8. A Gaussian-like fit to the fossil fuel record for the period 1860-1986 a) it has the same total input for the period 1860-1986; b) it passes through the 1986 data point; and c) the total integral is eight times the amount of CO_2 as in the pre-industrial atmosphere.

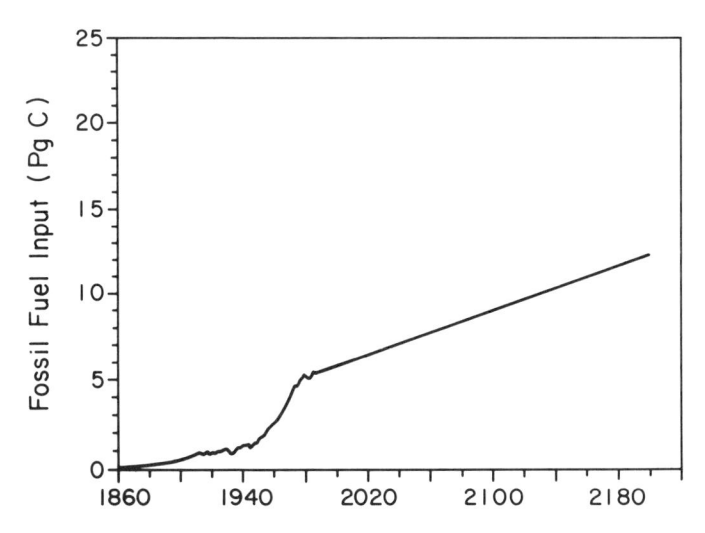

FIGURE 9. The fossil fuel record to 1986 and then the best fit straight line to the 1977-1986 data for the period 1987-2200.

128 Global Climate Change: Implications, Challenges and Mitigation Measures

processes in the model, and the other uses bomb ^{14}C (see Siegenthaler, 1983). Hence, for each forcing term, there are actually five responses. The results are shown in Figures 10-11. In using the Gaussian-like input functions, we extend the time frame to 2500 in order to show how the atmosphere-ocean system begins to relax as the forcing term goes to zero. Because of the 'infinitely' fast transfer of CO_2 to deep layers, the OC model, in either parameterization, is by far the most efficient in taking up carbon dioxide.

This range of models certainly includes what is the currently accepted role of

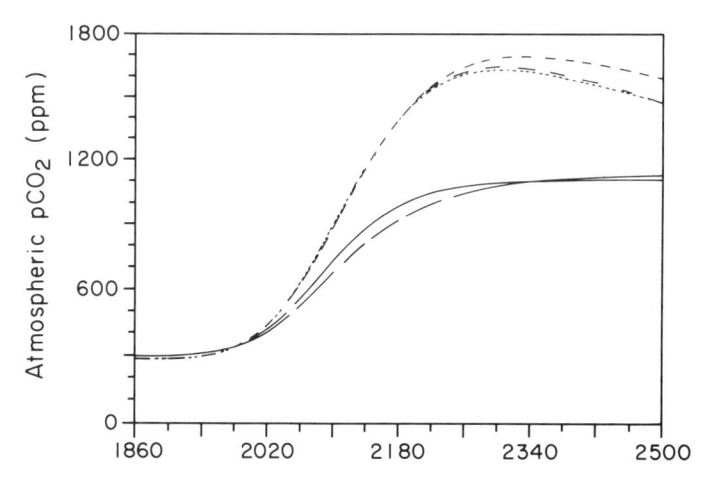

FIGURE 10. The response of the different ocean-atmosphere models to the Gaussian forcing (See Figure 8). The solid represents two versions of the OC model; the short dashed is the AD model; the medium dashed is the BD model, and the long dashed is the 12B model.

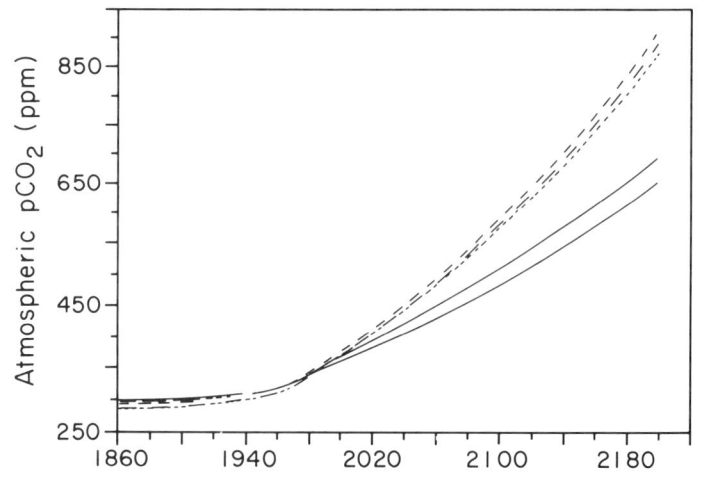

FIGURE 11. The response of the different ocean-atmosphere models to the fossil fuel forcing followed by the linear forcing (Figure 9). The solid represents two versions of the OC model; the short dashed is the AD model; the medium dashed is the BD model, and the long dashed is the 12B model.

the ocean in the global carbon cycle; however, it does not include a model that is sufficiently "efficient" so as to allow for the uptake of the currently accepted net flux of CO_2 from deforestation (Figure 2 again). Therefore, either 1) the current understanding of the oceans is insufficient, or 2) the current estimates of land use-derived CO_2 are in error, or 3) there are important missing processes. It is likely that all three are true.

Finally, in light of a recent editorial by Firor (1988), we tested the four models using an emission scenario investigated by Maier-Reimer and Hasselman (1987), the results of which are referenced by Firor. In this scenario, fossil fuel use decreases at 2% per year until it reaches half the 1986 level of 5.56 Pg. C per year; a graph of this forcing function is given in Figure 12. In his editorial, Firor suggests that society may be able to "come close to stabilizing the atmospheric burden of CO_2 with a 50% reduction in fossil fuel use."

The responses of the four models to this emissions scenario are given on the graph of Figure 13. In using these models, one sees that after the input function becomes constant, the rise in atmospheric CO_2 concentration averages between 0.4 ppm/yr (for the outcrop model) and 0.5 ppm/yr (for the box-diffusion model). Admittedly, this rise is considerably less than the approximately 1 ppm/yr rise that is presently occurring, but it does not constitute a stable atmosphere. Tests of the responses of the four models to similar emissions scenarios indicate that emissions would have to be cut to approximately one-tenth of the present level to bring about something close to stabilization (a 0.05 ppm/yr rise).

This topic and other related questions raised by Firor (1988) have been the subject of additional investigations (Harvey 1989), and the differences between these studies where atmospheric stabilization is achieved and the current findings need further investigation. This is but one area that simple models of the ocean-atmosphere carbon system might continue to prove useful. Two other interesting issues are: 1) to define better the character of the terrestrial sink, and 2) to determine well the deforestation source.

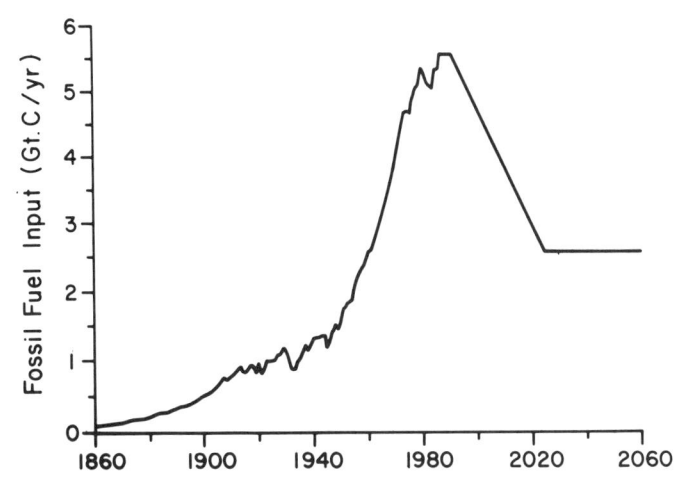

FIGURE 12. A plot of fossil fuel CO_2 emission to 1986 followed by a five year period at the 1986 level, and then decreasing at 2% per year until it reaches half the 1986 level of 5.56 Pg. C per year.

Let us now turn to obtaining a better appreciation of determining logical energy futures rather than purely simple mathematical "forcing functions."

PART II

The predominance of the fossil fuel component of carbon emissions and the complex coupling of fossil fuel use to economic activity warrants our focus. This part of the chapter describes this source of carbon dioxide in some detail. The discussion is broken into three sections. In the first section, we describe how economic activity drives the combustion of fossil fuels. In the second section, we review the sources of emissions since the 1970s and describe general agreements among forecasters regarding future emissions. These agreements are small compared to the disagreement. The third section reviews the uncertainties which will have a large effect on future emissions.

The Relation Between Economic Activity and the Combustion of Fossil Fuels

From a physical perspective, economic production is a work process. Relatively diffuse natural resources are organized into useful goods and services. The process of organization opposes the second law of thermodynamics, which states that the universe as a whole tends towards a greater state of entropy (disorganization). Energy is required to oppose this tendency; therefore, every economic activity uses some energy.

The type of energy used to do work has changed tremendously during human history. For most of our existence, human muscles did all economic work. As technical capabilities increased, humans subsidized their effort with work derived from animals, falling water, and burning wood. Concurrent with the start of the industrial revolution, humans supplemented their efforts with increasing quantities of coal

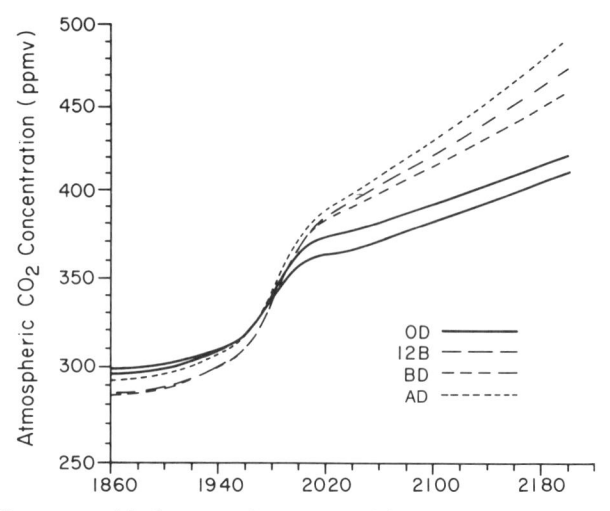

FIGURE 13. The response of the four atmosphere-ocean models to the historic fossil fuel forcing followed by a declining forcing as shown in Figure 12. The solid represents two versions of the OC model; the short dashed is the AD model; the medium dashed is the BD model, and the long dashed is the 12B model.

oil, and natural gas. Although it is not possible to assign a direction of causality between energy use and economic activity (Kraft and Kraft, 1978; Arkarca Long, 1980; Yu and Hwang, 1984) the tremendous increase in economic production during the industrial revolution would not have been possible without fossil fuels.

In spite of the threat to global climate posed by the CO_2 emitted by fossil fuels, currently these energies are preferred over existing alternatives. Fossil fuels can be obtained from the environment for relatively little cost, as measured by the energy return on investment (EROI), which compares the amount of energy obtained with the amount of energy used by society to obtain them (see Hall *et al.*, 1986). The EROI for fossil fuels is much larger than existing alternatives and in a form which humans can use efficiently (Cleveland *et al.*, 1984). At present, humans generally can use a heat unit of coal, oil, or natural gas to do more work than a heat unit of wood, sunlight, or failing water (Odum, 1971).

The general relation of the amount of carbon dioxide emitted to the use of fossil fuel to do economic work and the amount of economic output generated is summarized by

$$CO_2 \text{ Emitted} = (\text{Economic Activity})_i * (\text{Energy Intensity})_{ij} * (\text{Emission Rate})_j \quad (3)$$

in which i is a type of economic activity and j is a type of energy such as natural gas. Economic activity is measured by the real value-added in producing sectors such as oil refining or by the final consumption of a particular good, such as clothing. Energy intensity is measured by the quantity of energy that is associated with the production or consumption of a particular good or service. Emission rate is measured by the CO_2 emitted per heat unit of fuel burned.

Changes in economic activity affect emissions via the income and mix effect. *Ceteris paribus*, the emission of carbon dioxide, rises in proportion to the level of economic activity. Changes in types of goods and services produced and consumed affect emissions through the mix effect. Variations exist in the amount of fossil fuel associated with the production or consumption of a good or service. In general, goods require more fossil fuel to produce than services; therefore, a shift in the pattern of consumption or production from goods to services tends to increase emissions even if real GDP remains constant.

The energy intensity for a particular good or service varies over time and/or among nations. This variation is possible because the amount of energy used by society to do economic work is much greater than the minimum that is dictated by the second law of thermodynamics (Gibbons and Chandler, 1981). As a result, technical change affects emissions via the intensity effect by changing the amount of fossil fuel used to produce a good or service.

Finally, differences in emission rates affect the amount of carbon dioxide that is released by economic activity. Emission rates vary greatly among fossil fuels. Coal emits more carbon per heat unit than the other fuels. On average, coal emits 25.1 grams of carbon per BTU (gC/BTU), oil emits 20.3 gC/BTU, and natural gas emits 14.5 gC/BTU (Manne and Richels, 1991). We stress average because emission rates vary by grade. These variations are relatively small for oil and natural gas, but can be large for coal. For example, ranks of coal with a relatively low heat content such as lignite emit can emit 10 - 30 percent more carbon per heat unit than high quality coals.

The Past, Present, and Future of Carbon Emissions

The emission of carbon dioxide from the combustion of fossil fuel started in earnest with the industrial revolution. Historical analyses indicate that emissions have been increasing steadily since 1860. This rate accelerated rapidly after World War II and is associated with the unprecedented period of economic expansion (see again Figure 2).

The increase in emissions reflects several changes. One important change is the share of emission from different fuels. Since the 1970s, emissions from coal and natural gas have increased relative to oil (Figure 14). This change is associated with the large increase in oil prices initiated by OPEC in fall 1973.

The geographic origin of emissions also has changed. Since 1970, developing nations have increased their emissions relative to OECD nations which is decreasing (Figure 15). This change is associated with the rapid rates of economic activity that accompany economic development. This development has been most pronounced in Asia, which has increased its share of emissions rapidly since the 1960s.

The mix effect also is changing emissions. The electricity sector is increasing its emissions relative to other sectors in all regions (Figure 16). On the other hand, the fraction of emissions from energy end-use sectors varies among regions. The fraction of emissions from the industrial sector in OECD nations is decreasing while the fraction of emissions from the industrial sector in developing nations is increasing

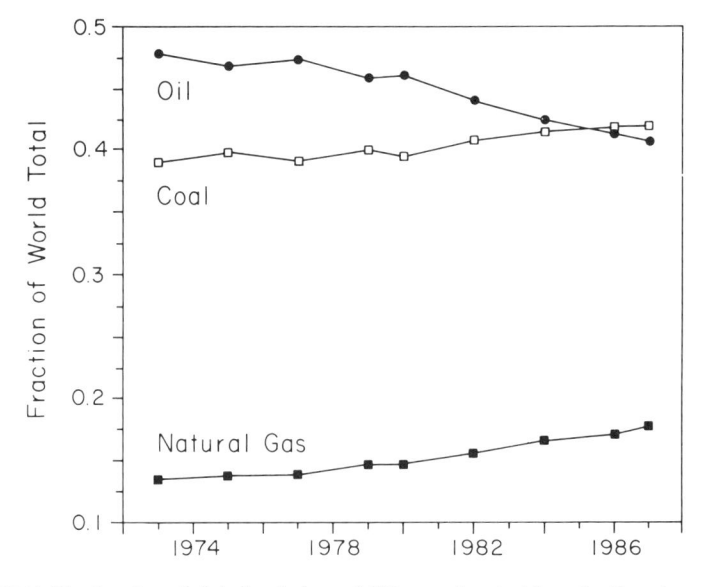

FIGURE 14. The fraction of global emissions of CO_2 associated with coal, oil, and natural gas.

slightly (Figure 17). Finally, the fraction of emissions from the transportation sector is increasing in OECD nations (Figure 18), which is clearly associated with the high degree of mobility in these societies.

Many models have been developed to forecast how these changes may or may not continue. Although there is no consensus, some areas of agreement have emerged. Most forecast emissions to increase steadily over the next 50 to 100 years if no

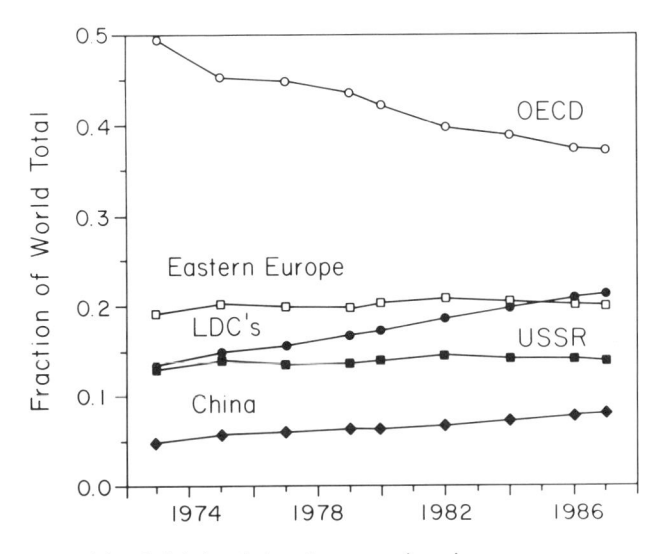

FIGURE 15. The origin of global emissions by economic regions.

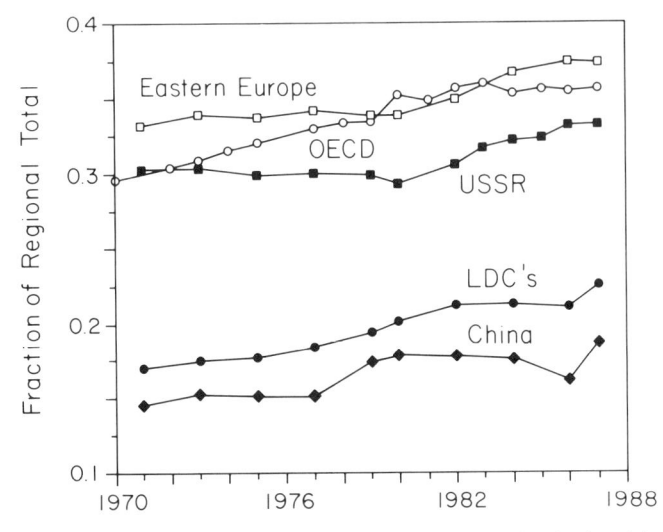

FIGURE 16. The fraction of a region's total emissions that are released by the electricity generating sector.

policies to slow emissions are adopted. For example, Manne and Richels (1991) forecast CO_2 fossil fuel emissions to increase from 6 to 25 Pg. C per year between 1990 and 2100. Most forecast the share of carbon emitted by developed countries will decrease relative to developing nations. For example, Sathaye and Ketoff (1991) forecast that emissions from developing nations will increase several-fold by 2025; whereas, many forecasts for developed countries double at most by 2025.

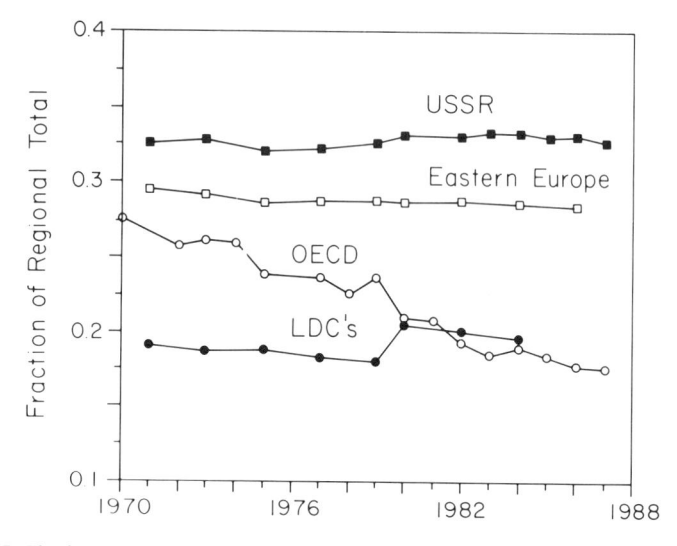

FIGURE 17. The fraction of a region's total emissions that are released by the industrial sector.

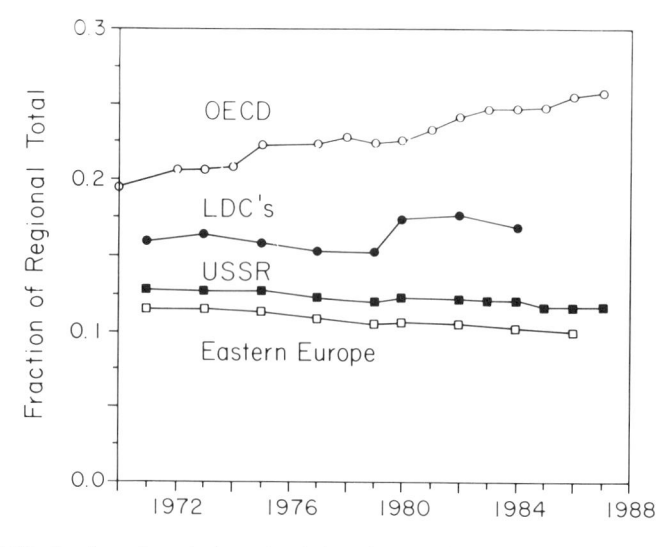

FIGURE 18. The fraction of a region's total emissions that are released by the transportation sector.

Uncertainties Regarding Future Emissions of Carbon Dioxide From Fossil Fuels

The only certainty regarding the forecast(s) for emissions of carbon dioxide is that they will not be correct. This expectation does not reduce the usefulness of building models and making forecasts. Indeed, one of the most important uses of models is to identify uncertainties that have a large impact on system behavior. Towards this end, differences in forecasts and the assumptions used to derive them identify several areas of uncertainty regarding future rates of emission. We divide this uncertainty among three coupled categories: energy prices, energy supply, and energy demand.

a. Prices. The importance of energy prices introduces considerable uncertainty in the forecast because economists are not able to describe OPEC behavior using standard models. OPEC is not a price taker; decisions concerning output do affect market prices. Conversely, OPEC does not have sufficient control of the market or discipline among members to qualify as a cartel.

Most forecasts for emissions of carbon dioxide assume, implicitly or explicitly, a modicum of OPEC cooperation. Under these conditions, the "base case" shows a slow but steady increase in real oil prices. But a wide continuum of OPEC behaviors is possible, and extremes on either end have a dramatic effect on oil prices. At one extreme, a collapse in cooperation among OPEC members would reduce prices until price equalled the marginal cost of production plus resource rent. The price of coal and natural gas would follow the lowering of oil prices (Hamilton, 1983). Although information needed to calculate marginal cost and rent for OPEC oil is not available, analysts estimate that this sum lies between $5 and $10 per barrel (Adelman, 1989).

A large decline in real oil prices would increase emissions through several mechanisms. Low energy prices speed economic activity, which would increase emissions through the income effect. The price decline would reduce the price of energy intensive goods and thereby increase their consumption via the mix effect. Finally, the decline in prices would reduce incentives to reduce energy intensity through conservation and thereby increase emissions via the intensity effect.

At the other extreme, OPEC may find a way to maintain a high degree of discipline. Although this extreme is less likely, the potential to increase price was demonstrated in 1973-1974 and 1979-1981. Higher prices probably would reduce emissions through income, mix, and intensity effects by depressing economic activity, discouraging consumption of energy intensive goods, and encouraging producers to reduce energy intensity, respectively.

Because we are dealing with an economic issue, we can not resist saying, "On the other hand," the rise in oil prices may increase emissions by stimulating users to substitute coal for oil. Higher oil prices would increase the bill for imported oil. To reduce the loss of foreign exchange, consuming nations may substitute indigenous resources, which often is coal. Because more than one heat unit of coal is needed to replace a single heat unit of oil and because a heat unit of coal emits more carbon than oil, this increase in emissions could offset some or all of the reduction generated by the income, mix, and intensity effects. For example, Kaufmann (in preparation) finds that 1.1 to 1.67 units of coal are needed to replace one unit of oil and that 1.11 to 8.83 units of coal are needed to replace one unit of natural gas. These rates of substitution imply that using coal instead of oil increases emissions 36 to 106 percent while using coal instead of natural gas increases emissions 80 to 1429 percent.

b. Supply. Several aspects of energy supply introduce significant amounts of uncertainty to the forecast for emissions. One of the largest uncertainties, the types of fuels that will be consumed in the future, is generated by one certainty: the world is depleting its conventional supply of crude oil. Although estimates for recoverable supply are far from unanimous, they have narrowed considerably over the last 20 years. Consensus indicates that 2,000 to 2,400 billion barrels of crude oil ultimately will be recovered from conventional sources worldwide. As of 1988, 610.1 billion barrels had been produced, which leaves another 1197 to 1867 billion barrels (Masters *et al.*, 1991) to be consumed. Under a relatively wide range of assumptions about real oil prices and economic activity, the remaining supply of conventional crude oil will be exhausted by the middle of the next century (EMF, 1991).

Coal and natural gas are the leading replacements. Although natural gas is a more desirable fuel than coal from an economic and environmental perspective, coal may replace oil because of uncertainty about the size and distribution of the natural gas. Uncertainty regarding supply is sufficient to justify scenarios in which the gas resource base is 4 times larger than commonly assumed. Even if natural gas remains when conventional sources of crude oil are exhausted, the distribution may limit gas consumption. Natural gas fields are highly concentrated and often are located far from consumers. Natural gas cannot be moved inexpensively by ship, therefore, high transportation costs may prevent natural gas from replacing oil.

Coal has many characteristics that may enable it to replace oil. Proved reserves could satisfy world demand for several hundred years. Furthermore, coal deposits are scattered and constitute the only significant domestic energy resource in many nations. Local availability is most important in areas that are expected to undergo rapid economic growth over the next 100 years such as China and India.

Finally, oil from conventional sources may be replaced by oil from unconventional sources such as oil shale and tar sands. Although the exact size of the resources base in uncertain, identified deposits of oil shale are larger than conventional oil resources and expected deposits could make this resource several times larger than conventional crude oil (Marland, 1979). In spite of their abundance, several uncertainties and problems may limit unconventional resources. One of the most important uncertainties is the cost of production. To measure this, analysts calculate the price for crude oil would "trigger" their use. A recent study by the National Research Council (1990) estimates that unconventional sources of oil could be viable at $29 to $81. This threshold is exceeded by many forecasts for oil prices, which implies a significant contribution from unconventional resources. But this type of analysis may overstate their potential. In the past, the price of crude oil has exceeded estimates for trigger prices for unconventional sources of crude oil several times, but they have not been recovered without considerable government subsidy (Fearon and Wolman, 1982). One reason for the failure of unconventional resources to enter the market may be their low EROI. Because energy is such an important component of their cost, the cost of producing oil from unconventional resources rises with the price of oil. Thus, trigger prices always are slightly higher than the price of crude oil from conventional sources, regardless of its absolute level. Another obstacle to unconventional sources of crude oil is their high emission rate. In general, unconventional sources of crude oil emit about 60 percent more CO_2 (40.8 gC/BTU) than a similar quantity of coal (Manne and Richels, 1990).

Nonfossil energies are an attractive replacement because they emit no carbon dioxide. Because the technical, economic, and political viability of alternative fuels is not well understood, the presence of these "backstop" fuels introduces significant uncertainty to the forecast of emissions. The noncarbon alternatives for generating electricity that are currently economically feasible, hydropower, nuclear power, and geothermal are not expected to increase in importance because of limits associated with the resource base and environmental impacts.

Geothermal sites are limited and cannot provide significant increases in electricity production. A similar limit exists on expansion of hydropower. In nations where a significant increase in hydropower is possible, such as Canada and Brazil, large environmental and social impacts may limit growth. For example, expansion of hydropower in these two nations could drown large areas of forest and destroy native peoples. Uncertainty regarding environmental, political, and economic issues diminish the potential for nuclear power to lower emissions. Real or not, fear associated with accidents in the US, USSR, and Japan have caused the public to rethink their support for nuclear power. Serious issues associated with the disposition of spent fuel have not been solved. Finally, the economic advantage of nuclear power relative to coal fired capacity is not clear. Even in France, where nuclear power supplied 70 percent of electricity in 1988, the cost advantage of nuclear power relative to coal disappears if the subsidies that keep the price of coal mined in France above the import prices are removed (Virdis and Reiber, 1991).

Uncertainty also exists surrounding the costs and availability of fuels that could replace oil in the transportation sector. The physical requirements of fuels for the transport sector have focused attention on alcohol derived from biomass. Uncertainties associated with the costs and sustainable supply of biomass cloud the potential to replace oil with alcohol derived from biomass. Some analyses indicate that large quantities of biomass can be converted to alcohol fuels or gasified at relatively little cost (Williams, 1990). Other analyses indicate that the EROI for alcohol fuels is small or less than one (Hopkinson and Day, 1980) and that a large scale effort to produce alcohol from biomass could have serious environmental impacts (Pimentel *et al.*, 1981).

c. Demand. On the demand side, uncertainty regarding the relation between energy prices and the quantity of energy consumed and uncertainty about economic growth cloud the forecast for emissions. The relation between energy prices and energy use is summarized by the economic concept of price elasticity. The own-price elasticity of demand for a given energy type measures the percent change in energy consumption that is associated with a one percent change in the real price of that energy. The cross-price elasticity of demand for energy measures the percent change in the consumption of one type of energy that is associated with a one percent change in the price of an alternative fuel.

Uncertainty regarding the relation between energy prices and energy intensity is summarized by the wide range in empirical estimates for own- and cross-price elasticities of demand for energy (Bohi and Zimmerman, 1987). The wide range in values reflects uncertainty regarding the potential for energy saving technical change and the potential for substitution. The resolution of this uncertainty will have a large effect on the types of policies implemented to slow emissions.

One of the most important foci of uncertainty regarding energy prices and energy intensity is the presence and size of autonomous increases in energy efficiency (AIEE). AIEE is defined as a reduction in energy intensity that is not caused by a change in absolute or relative prices. AIEE reflects energy saving technical changes that allow producers or consumers to maintain the same level of economic activity with less energy, and consequently assumptions regarding AIEE have a large effect on the forecast of emissions.

Manne and Richels (1990) use a value of 0.4, which reduces energy intensity 0.4 percent per year. This value for AIEE is an approximate mean chosen from a relatively wide range (Hoeller *et al.*, 1990). The range reflects a great deal of uncertainty regarding the existence and size of AIEE. Williams (1990) argues that the AIEE could be significantly greater than 1.0 percent per year. This larger value is based on analyses that show a steady decline in energy intensity. For example, an analysis by Howarth (1991) indicates that the energy intensity of the U.S. manufacturing sector decreased steadily between 1958 and 1985 and that the rate of decline was not affected by the 1973-1974 OPEC price increase for oil.

On the other hand, other analysts, some of whom are associated with the biophysical school of economic analysis, claim that AIEE has little or no effect on energy intensity. The argument for the unimportance of AIEE is based in part on statistical analyses of the energy/real GDP ratio. Analyses of this ratio in the big five nations indicate that changes in the types of fuels used, energy prices, and the mix of goods and services produced and consumed account for nearly all the change in amount of energy used to produce a real dollar of GDP (Gever *et al.*, 1986; Kaufmann, in press).

To date, there is no resolution regarding the potential for energy saving technical change to reduce emissions. One obstacle is the differences in the system boundaries used by analysts to measure AIEE. Analyses that generate large values for AIEE examine only the energy used on-site. Analyses that indicate small values for AIEE include the energy used off-site to support the activity. Pending the resolution of the "correct" value for AIEE, the forecast for emissions could change significantly.

The other aspect of uncertainty regarding the relation between energy prices and energy intensity is the potential for conservation. An increase in energy prices causes producers and consumers to substitute nonenergy factors. For example, higher prices may induce consumers to purchase insulation and producers to purchase fuel efficient machinery. The energy saved by substituting capital, labor, or some other raw material is termed conservation.

The potential for conservation to reduce emissions is determined by the market's ability to induce the economically efficient level of conservation. When the price of energy rises, the conservation measures adapted depend on a balance between the cost of the energy saved and the marginal cost of energy savings, which is the economic cost of reducing energy use. The market should induce adaptation of all measures for which the cost of reducing energy use is equal to or less than the cost of the energy saved.

The conservation opportunities available are represented by a supply schedule, which arrays opportunities such that the marginal cost of a unit of energy saved increases with the total amount of energy saved. Uncertainty exists regarding the slope and intercept of this schedule, but uncertainty about the intercept has generated the most controversy regarding future emissions. Some analysts claim that the intercept is negative. That is, energy conservation would reduce total costs at current energy prices. Analysts term such opportunities the "free lunch factor."

The presence and size of the "free lunch factor" is the focus of much debate. Some analysts claim that the "free lunch factor" is quite large. For example, many advocates of energy conservation claim that the U.S. could save money and reduce emissions by installing compact fluorescent light bulbs. The initial costs of these bulbs are quite high, but that the savings on electricity bills pay back this investment rapidly such that the total cost of lighting is reduced. On the other hand, many economists argue that the claims for free lunch technologies are exaggerated and that a thorough economic analysis indicates that they are not economically feasible at current energy prices. For example, the economic viability of the compact fluorescent bulb depends greatly on assumptions regarding the hours per day that the bulb is used and the interest rate that is used to discount the reductions in the electric bill.

The other determinant of energy demand is the level of economic activity. Uncertainty regarding population and economic growth rates in developing nations, developed nations, and formally centrally planned economies (CPE's) has a tremendous effect on the forecast for emissions. Developing nations are expected to increase their share of total emissions steadily but the uncertainty regarding economic development could accelerate or retard this trend significantly. Even setting aside the uncertainties about the future possible levels of economic growth in the developing world, there remain significant uncertainties about energy consumption.

The effects of economic growth on emissions could be accelerated by structural changes associated with economic development; the rapid building of infrastructure and the shift from noncommercial fuels (e.g. biomass, fuel wood, charcoal) to commercial fuels (e.g. fossil fuels). Rapid accumulation of infrastructure is a relatively energy intensive process because it requires the production of large amounts of energy-intensive raw materials such as metals, minerals, and chemicals.

Most of the nations that are now considered developed passed through a period in which their energy intensity rose rapidly. For example, the United Kingdom increased its energy intensity between 1700 and 1880. The peak in energy intensity coincided with the peak in the construction of infrastructure, such as the building of railroads (Humphrey and Stanislaw, 1979). Analysts are uncertain whether developing nations will repeat this increase in energy intensity. There are two reasons to believe that this increase can be reduced or eliminated. Coal, a relatively inefficient fuel, was the predominant source of energy when the industrial nations developed. If energies other than coal are used to power economic development, some of the increase in intensity and emissions can be avoided.

The degree of technology transfer between industrial and developing nations also will affect the increase in energy intensity associated with economic development. When industrial nations developed, they invented basic technologies such as the steam engine and the steel making process. Since then, the energy efficiency of these processes has been improved significantly. If developing nations are able to "leap frog" the old technologies by obtaining energy efficient technologies from industrial nations, this transfer could reduce or eliminate the increase in energy intensity.

The forecast of economic growth in industrial nations contains significantly less uncertainty than the other group of nations. Nevertheless, changes in the structure of developed nations may reduce energy intensity and hence demand; thereby offsetting some or all of the increase in emissions that is generated by economic growth. Deindustrialization is changing the structure of many industrial nations from reliance on energy intensive goods to reliance on information (but not energy) intensive services. The effect of deindustrialization on global emissions of carbon emission from fossil fuel is unclear. If the shift in final demand for energy intensive goods parallels the shift in production, then it will reduce emissions. If final demand does not parallel the shift in production, we will simply have relocated the sites of production and not reduced emissions. An analysis of U.S. imports and exports indicates that some portion of the decrease in energy intensity of the U.S. economy has been accomplished by importing energy intensive goods rather than producing them domestically (Hannon, 1982).

The greatest uncertainty regarding economic growth and energy demand is associated with the former CPE's of Eastern Europe and the USSR, and China. The forecast of emissions for Eastern Europe and what formerly was the USSR depends on the transition to a market economy. Presently, the economies of these nations are in free fall, shrinking 10 to 20 percent in 1991. Few analysts are willing to state with any certainty when these nations will be able to halt this decline. Uncertainty regarding the structure that will evolve in these nations also will have an important effect on their energy intensity. The energy/GDP ratios in China, the USSR, and the nations of Eastern Europe are among the highest in the world.

Some of this inefficiency is associated with energy prices that are set well below those set by market economies. If these nations change to a market economy, higher fuel prices and access to western technology will induce them to reduce intensity.

Another cause for the high energy demand/real GDP ratio in the former CPE's is their dependence on low quality coal. This coal is produced locally and is produced only because the state subsidizes such efforts. If the former CPE's shift to a market economy and end subsidies to coal producers, these nations may replace low quality coals with oil, natural gas, or higher quality coals.

SUMMARY

As we have seen, there is significant uncertainty about the global carbon cycle and the future forcing that human activity will place on this cycle; however are aspects of the CO_2 problem that are *not* uncertain. Carbon dioxide is a greenhouse gas; its concentration in the atmosphere was essentially constant for several thousand years before 1700; since 1700 the concentration has steadily increased, slowly at first and more rapidly with the advent of fossil fuel use, and finally this change will alter the heat balance of the planet.

What is not certain is the nature of the changes that will be associated with this alteration; how fast these changes will occur, and finally, what humans may do in the face of these uncertainties. What is also certain is that we better not put our heads in the sand.

LITERATURE CITED

Adelman, M.A. 1989. Problems in Modelling world oil supply. Energy Modelling Forum. September 28, 1989.

Arkarca, A.T. and T.V. Long. 1980. On the relationship between energy and GNP: A reexamination. J. Energy Development. 5:326-331.

Bacastow, R.B. and A. Björkström. 1981. Comparison of ocean models for the carbon cycle, In: *Carbon Cycle Modelling*, B. Bolin (Ed.). John Wiley and Sons, Chichester, England, pp. 29-79.

Baes, C.F. 1982. Effects of ocean chemistry and biology on atmospheric carbon dioxide. In: *Carbon Dioxide Review: 1983*, W.C. Clark (Ed.). Oxford University Press, New York, pp. 189-204.

Björkström, A. 1979. A model for CO_2 interaction between atmosphere, oceans, and land biota. In: *The Global Carbon Cycle*, Bolin, E.T. Degens, S. Kempe and P. Ketner (Eds.). John Wiley and Sons, New York, pp. 403-458.

Bohi, D.R. and M.B. Zimmerman. 1987. An update on econometric studies of energy demand behavior. Annual Review of Energy 9:105-154.

Bolin, B. 1970. The carbon cycle. Scientific American 223:124-132.

Bolin, B., E.T. Degens, S. Kempe and P. Ketner, (Eds.). 1979. *The Global Carbon Cycle*. New York: John Wiley and Sons. 491 pp.

Bolin, B. 1983. Changing global biogeochemistry. In: *The Future of Oceanography*. 50th Anniversary Volume, Woods Hole Oceanographic Institution, (Ed.) P. Brewer, Berlin: Springer-Verlag. pp. 305-326.

Bolin, B. 1986. How much CO_2 will remain in the atmosphere? In: *The Greenhouse Effect, Climate Change, and Ecosystems*, B. Bolin, B. Döös, Jäger, and R. Warick (Eds.). John Wiley and Sons, Chichester, England, pp. 93-155.

Bolin, B. and H. Stommell. 1961. On the abyssal circulation of the world ocean-IV. Deep-Sea Research, 8:95-110.

Bolin, B., A. Björkström, K. Holmen and B. Moore. 1983. The simultaneous use of tracers for ocean circulation studies. *Tellus*, 35B, 206-236.

Broecker, W.S. 1974. *Chemical Oceanography*. Harcourt Brace Jovanonich, New York, 214 pp.

Broecker, W.S. and T.S. Peng. 1982. *Tracers in the Sea*. Lamont-Doherty Geological Observatory, Columbia University, New York, p. 690.

Carbon Dioxide Assessment Committee, National Research Council. 1983. Changing Climate. Washington, D.C.: National Academy Press Trabalka, J.R., (Ed.). 1985. Atmospheric Carbon Dioxide and the Global Carbon Cycle. 315 pp. United States Department of Energy, Department of Commerce, Springfield, Virginia 22161.

Cleveland, C., R. Costanza, C.A.S. Hall, R. Kaufmann. 1984. Energy and the United States economy: A biophysical perspective. Science 225:890-897.

Craig, H. 1957. The natural distribution of radiocarbon and the exchange time of carbon dioxide between atmosphere and sea. *Tellus*, 9, 1-17.

Detwiler, R.P. and C.A.S. Hall. 1988. Tropical forests and the global carbon cycle. Science 239:42-47.

Energy Modelling Forum. 1991. International Oil Supplies and Demands. EMF Report 11.

Fearon, J.G. and M.G. Wolman. 1982. Shale oil: Always a bridesmaid? Forty years of estimating costs and prospects. Energy Systems and Policy 6:63-96.

Firor, J. 1988. Public policy and the airborne fraction, guest editorial. Climatic Change, 12, 103-105.

Frieden, B.J. and K. Baker. 1983. The market needs help: The disappointing record of home energy conservation. Journal of Policy Analysis and Management 2:432-448.

Gever, J., R. Kaufmann, D. Skole and C. Vorosmarty. 1986. *Beyond Oil: The Threat to Food and Fuel in the Coming Decades*. Ballinger Press, Cambridge, Mass.

Gibbons, J.H. and W.U. Chandler. 1981. *The Conservation Revolution*. Plenum Press, New York.

Hall, C.A.S., C. Cleveland and R. Kaufmann. 1986. *Energy and Resource Quality: The Ecology of the Economic Process*, John Wiley & Sons, New York.

Hamilton, J.D. 1983. Oil and the macroeconomy since World War II Journal of Political Economy. 91:228-248.

Hannon, B. 1982. Analysis of the energy cost of economic activities: 1963-2000. Energy Systems and Policy 6(3):249-278.

Harvey, D. 1989. Managing atmospheric CO_2. Climatic Change, 15, 343-381.

Hausman, J.A. 1979. Individual discount rates and the purchase and utilization of energy-using durables. The Bell Journal of Economics 10:33-54.

Hoeller, P.A. Dean, and J. Nicolaisen. 1990. A survey of studies of the costs of reducing greenhouse gas emissions. OECD Working Paper.

Hopkinson, C.S. and J.W. Day. 1980. Net energy analysis of alcohol production from sugarcane. Science 207:302-303.

Houghton, R.A. 1990. The future role of tropical forests in affecting the carbon dioxide concentration of the atmosphere. Ambio 19:204-209.

Houghton, R.A., R.D. Boone, J.M. Melillo, C.A. Palm, G.M. Woodwell, N. Myers, B. Moore and D.L. Skole. 1985. Net flux of CO_2 from tropical forests in 1980. Nature 316:617-620.

Houghton, R.A., R.D. Boone, J.R. Fruci, J.E. Hobbie, J.M. Melillo, C.A. Palm, B.J. Peterson, G.R. Shaver, G.M. Woodwell, B. Moore, D. L. Skole and N. Myers. 1987. The flux of carbon from terrestrial ecosystems to the atmosphere in 1980 due to changes in land use: Geographic distribution of the global flux. *Tellus* 39B:122-139.

Houghton, R.A., J.E. Hobbie, J.M. Melillo, B. Moore, B.J. Peterson, G.R. Shaver and G.M. Woodwell. 1983. Changes in the carbon content of terrestrial biota and soils between 1860 and 1980: A net release of CO_2 to the atmosphere. Ecological Monographs 53:235-262.

Houghton, R.A. and D.L. Skole. 1990. Carbon. In: *The Earth As Transformed by Human Action*. B.L. Turner, W.C. Clark, R.W. Kates, J.F. Richards, J.T. Mathews, and W.B. Meyer (Eds.). Cambridge University Press, Cambridge, U.K. pp. 393-408.

Houghton, R.A., D.L. Skole and D.S. Lefkowitz. 1991. Changes in the landscape of Latin America between 1850 and 1980. II. A net release of CO_2 to the atmosphere. Forest Ecology and Management 38:173-199.

Howarth, R.B. 1991. Energy use in U.S. manufacturing: The impacts of the energy shocks on sectoral output, industry structure, and energy intensity. The Journal of Energy and Development 14(2):175-191.

Humphrey, W.S. and J. Stanislaw. 1979. Economic growth and energy consumption in the UK. Energy Policy 7(1):29-42.

Kaufmann, R.K. In review. A biophysical analysis of the energy/real GDP ratio: implications for substitution and technical change. Ecological Economics.

Kaufmann, R.K. In preparation. The relation between marginal product and price: an analysis of energy markets.

Keeling, C.D. 1973. The carbon dioxide cycle: Reservoir models to depict the exchange of atmospheric carbon dioxide with the oceans and land plants. In: *Chemistry of Lower Atmosphere*, S.I. Rasool (Ed.). Plenum Press, New York, pp. 251-329.

Keeling, C.D. 1982. The oceans and the terrestrial biosphere as future sinks for fossil fuel CO_2. In: *Interpretation of Climate and Photochemical Models, Ozone and Temperature Measurements*, American Institute of Physics, New York.

Keeling, C.D. and B. Bolin. 1967. The simultaneous use of chemical tracers in oceanic studies. I: General theory of reservoir models. *Tellus*, 19, 566-581.

Kraft, J. and A. Kraft. 1978. On the relationship between ernergy and GNP. J. Energy Development 3:401-403.

Maier-Reimer, E. and K. Hasselman. 1987. Transport and storage in the ocean - an inorganic ocean-circulation carbon cycle model. Climate Dynamics, 2, 63-90.

Manne, A.S. and R.G. Richels. 1991. Global CO_2 emission reductions - the impacts of rising energy costs. The Energy Journal. 12(1):87-107.

Marland, G. 1979. Shale oil: US and world resources and prospects for near term commercialization in the United States. Oak Ridge Associated Universities. Oak Ridge, Tenn. ORAU/IEA 78-8(R).

Masters, C.D., D.H. Root, E.D. Attansi. 1991. Resource constraints in petroleum production potential. Science 253:146-152.

Melilo, J.M., J.R. Fruci, R.A. Houghton, B. Moore and D.L. Skole. 1988. Land-use change in the Soviet Union between 1850 and 1980: causes of a net release of CO_2 to the atmosphere. *Tellus* 40B:116-128.

Moore, B. 1985. The oceanic sink for excess atmospheric carbon dioxide. In Duedall, I., D.R. Kester and P.K. Park (Eds.), *Wastes in the Ocean, Volume IV: Energy Wastes in the ocean*. John Wiley & Sons.

Moore, B., R.D. Boone, J.E. Hobbie, R.A. Houghton, J. M. Melillo, B.J. Peterson, G.R. Shaver, C.J. Vorosmarty, G.M. Woodwell. 1981. A simple model for analysis of the role of terrestrial ecosystems in the global carbon budget. In B. Bolin (Ed.), Carbon Cycling Modelling, SCOPE 16, John Wiley & Sons, New York. pp. 365-385.

Nordhaus, W.D. and G.W. Yohe. 1983. Future carbon dioxide emissions from fossil fuel. In: *Climate Change*, Carbon Dioxide Assessment Committee, National Academy of Sciences, Washington, D.C., pp. 87-153.

Odum, H.T. 1971. *Environment, Power, & Society.* Wiley-Interscience, New York.

Oeschger, H., U. Siegenthaler and A. Gugelman. 1975. A box diffusion model to study the carbon dioxide exchange in nature. *Tellus,* 27, 168-192.

Palm, C.A., R.A. Houghton, J.M. Melillo and D.L. Skole. 1986. Atmospheric carbon dioxide from deforestation in Southeast Asia. Biotropica 18:177-188.

Pimentel, D., M.A. Moran, S. Fast, G. Weber, R. Bukantis, L. Balliett, P. Boveng, C.J. Clelevand, S. Hindman, and M. Young. 1981. Biomass energy from crop and forest residues. Science 212:1110-1115.

Sathaye, J. and A. Ketoff. 1991. CO_2 emissions from developing countries: Better understanding the role of energy in the long term. The Energy Journal 12(1): 161-196.

Siegenthaler, U. 1983. Uptake of excess CO_2 by an outcrop-diffusion model of the ocean. Journal of Geophysical Research, 88, 3599-3608.

Skole, D.L. in press. Acquiring global data on land cover change. In: *Land Cover and Land Use Change.* Turner et al. (Eds.), Proceedings of the 4th Global Change Institute, OIES, UCAR.

Takahashi, T. 1977. Carbon dioxide chemistry in ocean water. In: Proceedings of the Workshop on the Global Effects of Carbon Dioxide from Fossil Fuels, Conference Publication 770385-UC-11, W.P. Elliot and L. Machta (Eds.). U.S. Department of Energy, Washington, D.C., pp. 63-71.

Tans, Pieter P., Inez Y. Fung and Taro Takahashi, Observational Constraints on the Global Atmospheric CO_2 Budget. Science 247:1431-1438.

Trends '90: A Compendium of Data on Global Change. 1990. T.A. Boden, P. Kanciruk, and M. P. Farrell (Eds.) Carbon Dioxide Information Analysis Center, Environmental Sciences Division, Oak Ridge National Laboratory, Oak Ridge, Tenn. U.S.A.

Virdis, M.R. and M. Reiber. 1991. The cost of switching electricity generation from coal to nuclear fuel. The Energy Journal 12 (2):109-134.

Williams, R.H. 1990. Low cost strategies for coping with CO_2 emission limits. The Energy Journal 11(3):35-59.

Yu, E.S.H. and B.K. Hwang. 1984. On the relationship between energy and GNP: Further results. Energy Economics 6:186-190.

Global Climate Change: Implications, Challenges and Mitigation Measures. Edited by S.K. Majumdar, L.S. Kalkstein, B. Yarnal, E.W. Miller, and L.M. Rosenfeld. © 1992, The Pennsylvania Academy of Science.

Chapter Ten

CLIMATE MODEL DESCRIPTION AND IMPACT ON TERRESTRIAL CLIMATE

ROY L. JENNE
National Center for Atmospheric Research
Box 3000
Boulder, CO 80309

INTRODUCTION

The impact of a greenhouse warming on the terrestrial climate is discussed. The main focus is on precipitation and temperature because these have the largest social and economic effects. The intent of this text is to provide the scientific user with a feeling for the attributes and responses of the climate models without detailing their full complexity. Some of the simulated climate changes over the U.S., Sahel, and China are presented. Climate model output has been made available from several models; the data can be readily used for a variety of assessment studies.

Data from these climate models were used to support studies sponsored by EPA in which about 35 PIs across the U.S. studied (during Oct. 1987- June 1988) the effect of changes in climate. The studies included the effect of climate change on crop yields, irrigation, agricultural economics, water resources, power generation, forestry, and changes in estuaries and beaches due to sea-level changes. A set of assessment studies, also sponsored by EPA, and involving many countries, started about October 1989. These were for agriculture (25 countries), rivers (10 countries), and forests (10 countries).

A group was needed to provide model data inputs and information to many PIs who mainly used personal computers for the studies. Therefore, NCAR[1] took on the task of operating a data center to provide data for assessment studies. This involves extracting data from the models' global archives, and putting them into one simple format with appropriate documentation. Data are available on tape at NCAR from more model simulations than are described here.

WHAT IS A CLIMATE MODEL?

A climate model is a very sophisticated set of computer programs that solve the equations that describe, for example, how winds blow and how pressure and temperature change in the atmosphere. The models simulate radiation and clouds; they evaporate water from the soil and use excess precipitation to build up soil moisture. When it snows, the model will change the reflectance of the surface. The details of a thunderstorm cannot be directly handled by these low resolution models, but the statistical effects of them over a broad region are included.

At least a simple ocean and its sea ice must also be included in any climate model whose purpose is to tell us something about how the climate will differ when a major change (such as double CO_2) is put into the model. The effects of ocean heat storage and sea ice are too large to be ignored.

The reader might ask why one should believe a climate model when a weather forecast model (run at better resolution) does not have good skill beyond a few days. The reason that climate models are still useful for long time periods is that they do not have to forecast the exact timing and shape of each weather system as a forecast model does. They only need to tell us the right statistics about all these events. The models should be able to tell us whether the main belt of westerly storms would shift in latitude or intensity when more CO_2 is added to the atmosphere. If the model correctly handles clouds, surface evaporation, and the radiative effects of gases, for example, it should then be able to give us an insight about the earth's climate. If we change something in the earth system (or in the model) such as the amount of carbon dioxide or the land surface properties, the model should be able to give us information about how the earth's climate will respond.

The climate models are still simple compared with the complexity of the real climate system. The outputs from several climate models have been used in EPA studies so that the differences among the models would give some feeling for how well they can be trusted. The models still require a lot of further development.

MODEL RESOLUTION

Climate model resolution has been relatively low because of the computer time needed to run these models. The 9-layer GISS model run in 1982 had a horizontal resolution of about 900 km. The various models run during 1984-88 usually had

---

[1]National Center Atmospheric Research, sponsored by the National Science Foundation (NSF).

about 9 levels and a resolution of 500 to 550 km. In 1989-90 some models provide higher resolution data (350-450 km), (Table 1).

DATA AVAILABLE FOR ASSESSMENT STUDIES

The data most used by PIs who studied crops, forests, and rivers, were long-period mean model data (about 10 years) for each month for the present climate (1x CO_2) and for the 2x CO_2 climate. In the case of the transient runs, means for each decade were given.

NCAR prepared the data that was most needed into a simple common format. For most models we have only long-term monthly averages of the most important 7 to 20 variables. The primary variables include temperature, precipitation, surface incident solar radiation, surface air moisture, total cloud cover, surface average wind speed (if available), evaporation, and sea level pressure. NCAR also has data for individual months for some models, and for all 170 fields in the model. Several years of daily model data are also present.

SOLAR CONSTANT USED IN THE MODELS

The solar constant gives the average intensity of solar radiation at the earth, but before it enters the atmosphere.

The solar constant is about 1366 to 1367 W/m² (based on satellite measurements). The earth is now closest to the sun about Jan. 4; the intensity then is nearly 7% greater than in July. All of the models change the intensity through the year in a proper way. A solar constant of 1367 W/m² is equivalent to an average of 342 watts on each square meter of the earth (because the area of a sphere is four times the area of an inscribed circle).

TABLE 1: CHARACTERISTICS OF SELECTED CLIMATE MODELS
All models are global in extent, have a smoothed topography that varies between models, and have an annual cycle. All models (except the transients) give data for the present climate (1x CO_2) and double CO_2 climate (2x CO_2). The EPA studies, based on these models, were made between October 1987 and June 1988.

	When Calculated	Model Resolution (lat x lon)	Model Levels	Diurnal Cycle	Base 1x CO_2 (ppm)	$\triangle$T for Double CO_2	Increase In Global Precip
GISS	1982	7.83 x 10°	9	yes	315	4.2°C	11.0%
GISS Transients	1984-85	7.83 x 10°	9	yes	315*	—	—
OSU	1984-85	4.00 x 5.0°	2	no	326	2.84°C	7.8%
GFDL**	1984-85	4.44 x 7.5°	9	no	300	4.0°C	8.7%
GFDL** (Better Ocean)	Feb. 1988[A]	4.44 x 7.5°	9	no	300	4.0°C	8.3%
GFDL R30**	May 1989[A]	2.22 x 3.75°	9	no	300	4.0°C	8.3%
UK	June 1986[A]	5.00 x 7.50°	11	yes	320	5.2°C	15.0%
Canada (CCC)	Nov. 1989[A]	3.75 x 3.75°	10	yes	330	3.5°C	3.8%

*(in 1958)
[A]These are the completion dates of the model run.
**This is a spectral model that has 15 waves, the R30 has 30 waves. The other models are gridpoint models with resolution as given.

The energy flux of the direct beam of solar radiation on a clear day near sea level is very close to 1000 watts per m^2 (measured perpendicular to the beam). The atmosphere screens out the rest.

The intensity on a horizontal surface is less; also the sun only shines for part of the day, and clouds reflect part of the radiation. Thus, the average intensity on a horizontal surface in mid latitudes is about 60 W/m^2 in winter and 260 W/m^2 in summer. The climate models include clouds, water vapor, and radiation. Thus, they attempt to simulate the processes described above.

AMOUNT OF CO_2 USED IN THE CONTROL RUNS OF THE GENERAL CIRCULATION MODELS (GCMs)

The amount of CO_2 in the atmosphere was measured to gradually increase from 315 ppm in 1958 to 342 ppm in 1983. The amount of CO_2 used in models for the present climate (1xCO_2) is as low as 300 ppm in some models, and as high as 330 ppm in others (Table 1). Since models are tuned to provide an adequate simulation of the present climate, these differences in the amount of CO_2 used for the control run probably are not significant. The amount of CO_2 used for the 2x CO_2 run is always double the 1x run, as one might expect. When the concentration of CO_2 in air was 342 ppm, then there were 722 giga-tonnes (10^9 metric tons) of carbon in the CO_2 molecules in the atmosphere. The world now releases over 5 Gtonnes of carbon from fossil fuel burning each year. We also probably release about 1.0 Gtonnes of carbon from biomass changes, but this number is not nearly as reliable as the fossil fuel amount, which is well known. During 1978-85, there was a net increase of about 2.9 Gtonnes of carbon in the atmosphere each year; this is accurately measured. Most of the difference between the amount released each year and the increase in atmospheric carbon goes into the ocean.

EARTH'S ELEVATION FOR CLIMATE MODELS

The actual detailed elevation of the earth cannot be used directly in a climate model. It has to be smoothed so that the resolution of the topography is similar to that of the model. In the western United States, the highest model elevations are about 1800 meters (5900 feet). This compares with actual elevations of about 11,000 to 12,500 feet for many mountain ridgelines. Topography causes precipitation to change rapidly near hills and mountains. We can't expect low-resolution models to reproduce those effects, but they can given us insight into what happens over broad regions as the greenhouse gases (such as CO_2) increase.

IS THERE A DIURNAL CYCLE IN THE MODEL?

At most latitudes the sun rises in the morning and sets in the evening. Many models do not yet include this full cycle. Rather, they have solar input all the time, but at a lower intensity than its real noontime value. The GISS, UK, and Canadian models have a diurnal cycle (Table 1).

GISS made model test runs (with and without a diurnal cycle) to determine what effect the diurnal cycle has on the climate produced by the model. The sea-surface temperature was held constant for both runs. The cycle was the only change in the model. Some of the climate changes that took place for the run without a diurnal cycle, as compared to one with it were:
- Low-latitude temperatures in winter were about 10°C warmer over land without a cycle; high latitude temperatures were a few degrees warmer;
- The precipitation over the warmer land (without a diurnal cycle) increased by up to a few mm per day;
- Near the equator, there was 50% increase in low clouds over land for no cycle. For example, 30% coverage of clouds would increase to 45%. This is probably because the diurnal cycle helps to "burn off" the low clouds in the daytime.

The reader should not expect the climate models without a diurnal cycle to have this much difference from a model with a cycle. The reason is that models are individually tuned to try to obtain a better simulation of the present climate. Also, when the model output is used for assessment studies, it is the climate differences from present CO_2 to double CO_2 that are important. To develop these differences, the same model is always used for both computer runs. The following information about model sensitivity for added CO_2 was from David Rind (GISS, Feb. 1988).

GCM SENSITIVITY

The sensitivity of a CO_2 climate model is often measured by how much the temperature of the surface air warms up, on the global average, when CO_2 is doubled. Most present models warm up by 3.0° to 4.2°C when CO_2 is doubled.

In all these models the surface warming due to a doubling of CO_2 by itself is only about 1.2°C. Any additional heating is from various feedback processes such as from water vapor and changes in clouds. All of the model runs that are used for these assessment studies permit clouds to vary in the model. Variable clouds are necessary to achieve a proper climate simulation.

GISS used a one-dimensional model to determine the components of their model's 4.2°C increase in surface air temperature with the doubling of CO_2. The changes in water vapor and clouds from the full GCM were used in the 1-D model to make these calculations of sensitivity. The following explains nearly all of the temperature changes:

a. Temperature change at surface due to doubling CO_2 alone	1.2°C
b. Water vapor increased 33% in the total atmospheric column, and it was located at higher levels	1.6°
c. There was a decrease of 1.7% in cloud amount, with some increase in cirrus clouds. Cloud tops became somewhat higher.	0.8°C
d. Changes in the surface albedo, mostly due to a reduction of sea ice	0.4°
	4.0°C
Actual temperature increase in the full GCM	4.2°C

Also, note that the vapor feedback, (a large 1.6°C), is associated with the total temperature change of 4°C. If the overall temperature changes were less, this feeback would also be reduced.

EFFECTS OF CLOUDS ON CLIMATE

For a low cloud, the cloud-top temperature is about the same as the land temperature, so that the cloud loses a lot of infrared radiation to space just as the land would. The low cloud reflects sunlight, which keeps the lower atmosphere cooler. The lower clouds have a cooling effect on surface temperature, but they do keep the nighttime temperature somewhat warmer.

An increase in high clouds warms the climate. The clouds do reflect more solar radiation, which is a cooling effect. However, in the infrared they radiate at a much lower temperature than either the earth's surface or low clouds. The net effect of more high clouds is a warming.

GFDL made a model run to test for the sensitivity to clouds. The 2x CO_2 run was 3.0°C warmer (for the world) than the control run when both had the same fixed clouds. When the clouds in both were allowed to vary, the corresponding increase in temperature was 4.0°C.

Climate models developed in the 1970s showed about a 2 to 3°C warming for a doubling of CO_2. Recent GCM studies generally have a corresponding warming of about 4.0°C. Ramanathan (1988) says the main reason for this increase in model warming is the inclusion of interactive clouds in the model. Since clouds are handled primitively in GCMs, he notes that it is still too early to make reliable estimates of the exact amount of warming to be expected.

The GFDL and Canadian models both show a drop in overall cloudiness with a warmer earth (Table 2). This tends to make the earth darker, and therefore, it absorbs more solar energy. A warmer earth also has less snow and ice, which again makes it darker. It may seem peculiar that a warmer earth, with more precipitation, should have fewer clouds. It all depends on what happens to the frequency of fog and stratus clouds in addition to other clouds. Note that two models (Table 2) show that the planetary albedo is over 1% less (darker) for the warmer earth. For the Canadian model, the albedo stayed the same (and that model did not heat up as much for double CO_2).

INCREASE IN PRECIPITATION

All the models show an increase in the average global precipitation as the earth heats up with double the CO_2. The increase is from 4 to 15% as shown in Table 1. The change in precipitation is greater when the model warms up more.

There are regions over the U.S. where the models do not show an increase in precipitation when CO_2 doubles. When the temperature increases, evaporation will generally increase (often faster than model precipitation). Thus, variables like soil moisture, runoff, and lake leves do not necessarily increase when precipitation increases. One must model the details of the hydrology to determine what will occur.

THE OCEAN

Each of the models has a fairly simple ocean called a slab ocean, usually about 60 meters deep. The models also form sea ice and melt it. It is desirable to include the effects of ocean currents in a climate model. In a real atmosphere-ocean, more heat is moved out of the tropics by ocean currents than by the atmosphere. The annual northward transports of energy in the atmosphere-ocean system are given in Table 3, from Carissimo, *et al* (1985).

TABLE 2: CLOUDS AND RADIATION IN CLIMATE MODELS

Global clouds, albedo, and IR are given for several models. The temperature sensitivity of the model is also given (change from 1x to 2x CO_2). For comparison, the earth's present albedo, based on measurements from satellites is between 28 and 30%.

Model	Temp Change	Run	Total Global Clouds	Planetary Albedo	Outgoing IR (W/M^2)
GISS (1982)	4.2°C	1x CO 2x CO_2		30.24% 28.80% −1.44%	233.0 237.5 +4.5
GFDL (Feb. 1988 Qflux)	4.0°C	1x CO_2 2x CO_2	50.93% 50.59% −.34%	33.83% 32.66% −1.17%	
CCC (Nov. 1989)	3.5°C	1x CO_2 2x CO_2	51.82% 50.68% −1.14%	32.8% 32.8% −0.0%	

TABLE 3: ANNUAL NORTHWARD TRANSPORTS OF ENERGY
IN THE ATMOSPHERE-OCEAN SYSTEM (units 10^{15} watts)

	Atmosphere	Ocean	Total
30°N	2.0	3.6	5.6
20°N	1.4	3.5	4.9
10°N	1.0	1.9	2.9
Equator	0.0	0.3	0.3
10°S	−0.8	−1.6	−2.4
20°S	−1.6	−2.7	−4.3
30°S	−3.1	−2.2	−5.3

THE Q-FLUX PROCEDURE (written late 1987)

A model with a primitive slab ocean can not simulate ocean water temperature and sea ice very accurately. The main reason is that they do not include heat transport by ocean currents. New climate model runs now are often made in two stages. During the first run a local heat flux (called Q-flux) is calculated that should be added at each ocean location in order to simulate the actual SST and sea ice for the present climate. This is called a Q-flux procedure. It effectively includes heat flux by ocean currents.

On the second model run, the Q-fluxes from the first run are imposed. In the case of the control run, the result will be a better simulation of the present climate, for both ocean and land areas. For the 2x CO_2 run, the same Q-fluxes are inserted as for the control run. We know that the 2x CO_2 fluxes should actually be different. However, this procedure is likely to give better sea water temperatures for 2x CO_2 than a run with no pseudo fluxes. For 2x CO_2 the Q-flux procedure finally gives a different SST and a different amount of sea ice, because the climate forcings are different. GISS first used the Q-flux procedure for the 1982 model runs.

HOW MUCH DOES THE OCEAN SLOW DOWN A CLIMATE WARMING?

Consider a warming of the atmosphere by 1 or 2 degrees (delta T) that is imposed on the ocean. With a slab ocean 60m thick, the e-folding time is about 12 years. This means that the slab would have warmed by about 65% of delta T in 12 years and by about 85% of delta T in 25 years. In high middle latitudes, the ocean is mixed to much deeper depths than such a slab ocean. These regions would further retard an overall atmospheric warming. The actual delays are still in the process of being sorted out. This information was largely based on a conversation with Mike Schlesinger, OSU (Feb. 88). Also, see Schlesinger (1988) and Ramanathan (1988).

GODDARD INSTITUTE MODEL

The GISS model was used for a very early simulation (1982) that is still relatively competitive, (Hansen, *et al,* 1983). It has a diurnal cycle. It has low horizontal resolution (7.83°lat. x 10°long.), but it has 9 levels. It represents a very nice pioneering effect. The group originated the Q-flux procedure to improve the ocean and sea ice simulation. Ocean water temperatures and ice cover are computed based on hour-by-hour energy exchange with the atmosphere, the specified pseudo ocean heat transports (Q-flux), and the ocean mixed layer heat capacity. It is a slab ocean, not over 65 m deep for the 1x and 2x CO_2 runs, and with some variation of mixed depth over the seasonal cycle. For example, the depth is shallower in mid-latitudes in summer than in winter. The transient model cases are similar, but can have a deeper mixed depth when appropriate.

MODEL RUNS AT GFDL

The GFDL Model uses a slab ocean 68, deep. It has had R15 resolution (4.4° x 7.5°), (Manabe and Wetherald, 1987). The version completed in 1985 did not have ocean currents or a pseudo heat flux to improve the water temperatures for the present climate. A new model run at GFDL was completed in Feb. 1988. It was still largely the same as before except that it includes Q-flux procedure to essentially reproduce the SST and the sea ice extent of the present climate. GFDL calls this their Q-Flux model. The sensitivity of the previous model run was 4.0°C (global change from 1x to 2x CO_2). The new run was expected to be less sensitive (smaller temperature change, perhaps 3.5°), but it came out almost exactly the same at 4.0°C (within 0.05°C). Too much sea ice had been forming in the N. Hemisphere in the simulation of the present climate. When this melted in the 2x simulation it gave a large warming compared to the 1x run. With this problem fixed, the sensitivity for the N. Hemisphere was less. But this was balanced by a reverse effect in the S. Hemisphere. There, the simulation of sea ice was not as great as is now observed. With this fixed, the warming from 1x to 2x CO_2 in the S. Hemisphere increased. The global average stayed the same.

R. Wetherald at GFDL inspected the simulation of temperature and precipitation over N. America and Eurasia from this new model. He saw significant improvements, but noted that we are never going to see a huge improvement with models having very low resolution (such as R-15 used by GFDL, about 550 km). The Q-flux model run (1988) shows a smaller increase in summer dryness compared to the previous model (Wetherald and Manabe, 1992). Since the polar amplification of temperature is smaller, the drying effect due to evaporation is less.

Better resolution at GFDL: Wetherald made a model run with twice the resolution (run at R30 instead of R15, still 9 levels). There are 96x80 points for the world instead of 48x40 points. A Q-flux procedure is included. There is no diurnal cycle. These simulations start with the results from the R15 runs and process ten more years of simulations (for both 1x and 2x CO_2). Five years was not enough for a stable climatology. The R30 run was completed about May 1989.

With the better resolution, the rain is much better. The soil moisture is better, probably because the precipitation is better. The sensitivity of the R-30 run was still 4.0°C, the same as for R-15. With more intense systems (because of higher resolution), the humidity criteria has to be changed to create clouds, or else there would be too few clouds. They changed it for the R-30 case and ended up with more clouds than in the R-15 run (about 60% global cloud cover compared to 51% in the R-15 case). With the same humidity criteria as before, the R-30 would have had only about 45% cover.

CANADIAN CLIMATE MODEL (MAY 1990)

A model has been developed by the Canadian Climate Centre (CCC) and run for 1x and 2x CO_2 conditions (Boer, *et al,* 1991). The model has a higher resolution

(T32; 3.75 x 3.75° than many climate simulations. There are ten levels. It has a 50m-deep slab ocean, and uses a Q-flux procedure. The model has a typical relative humidity scheme to develop clouds. The soil moisture bucket is variable; for example, the bucket is deeper where there are forests. It has a diurnal cycle. The climate sensitivity is lower than many models: it shows a temperature increase of 3.5° and a 3.8% increase in precipitation when CO_2 doubles (Table 1). I believe that the lower numbers for sensitivity are more likely to be correct. Data to use for global assessment studies were released for general use in May 1991.

UK METEOROLOGICAL OFFICE (UKMO) MODEL RUNS

John Mitchell, in UKMO, reviewed the best of what had been done in other groups and defined the UK 1x and 2x CO_2 runs about Nov. 1985. The model runs were finished about June 1986, (Wilson and Mitchell, 1987). It is a grid point model with a resolution of 5° x 7.5°. There are 11 levels (on sigma surfaces), plus the surface. There is a diurnal cycle. It has a slab ocean that is 50m thick, similar to other models. Mitchell used a Q-flux procedure to correctly simulate the present ocean water temperature and sea ice. They use a model spin-up period of about 20 years, and then collect 15 years of data.

The model climate warms by 5.2°C for CO_2 doubling. Most other models show only about 4°C. Some thought that this UK model gets more warming because it has more layers which permits better development of the important high clouds. I think that most people now believe that this sensitivity is too high.

We will discuss some of the UK model experiments with a smaller amount of warming. In the standard model above, cloud cover is a function of relative humidity (RH). Another version of the model was constructed with an explicit cloud-water variable (CW); in this case, a balance is maintained between cloud water content and water vapor. Ice particles fell out quickly (probably too fast) with a fixed speed of 1 ms^{-1}. On replacing the RH cloud scheme with the CW scheme, the global annual average surface warming (for double CO_2) changed from 5.2°C to 2.7°C (Mitchell et al., 1989). With the RH scheme, cloud reductions (for 2x CO_2) occur throughout the mid-latitude troposphere. With CW, the main difference is near the freezing level; in the warmer doubled-CO_2 simulation, ice cloud which is depleted rapidly by precipitation, is replaced by water cloud which is depleted slowly. Therefore the amount of clouds increases, which gives a smaller temperature increase.

Since 1 ms^{-1} may be too large a fall rate for ice particles, an experiment was run in which these rates were reduced to about 0.2 ms^{-1} for high clouds and to 0.7 ms^{-1} near the freezing level. The resulting global warming was 3.2°C compared to 2.7° above. Another experiment was run in which cloud radiative properties depended on cloud water content. For double CO_2, the global temperature increased by only 1.9°C compared to 2.7°C in the related experiment above. These reductions in CO_2 warming caused by changing the cloud algorithm are very large. The magnitude of the change surprised everyone when it happened. A consensus seems to be developing that the temperature change to expect from doubling of CO_2 is probably about 2.5°C. The main model runs now all show a greater temperature change than this.

The UK completed a run with double the resolution (2.5 x 3.75°), in Nov. 1989. This run includes water content of clouds, more realistic cirrus dissipation, but not variable optical properties. The sensitivity is 3.2°C. NCAR does not yet have data from this simulation.

OREGON STATE UNIVERSITY MODEL

This OSU model (1985 version) was one of the three models used for the first round of EPA assessment studies, for which most of the work was done from Oct. 1987-June 1988. The model was completed in 1985 (Schlesinger and Zhao, 1988). It has just two levels. It does not have a diurnal cycle. It has a slab ocean that is 60m deep. (It was only 5m deep during the 40-year spin-up period). It does not have currents. Therefore, the ocean sea surface temperature finally comes into a balance determined by the local surface energy budget. In general, the model gets too warm in the tropics. It does not seem to get too cold in polar areas or develop too much ice. It melts a little too much ice.

GREENHOUSE CHANGES SHOWN BY CLIMATE MODELS

The next few sections will give some indication of the results of simulations from a selection of climate models. Space is not available to show very many results or very many regions. More information (and the digital data) is available from NCAR.

The climatology that was used for observed surface air temperatures was from Taljaard, et al, 1969 and from Crutcher and Meserve, 1970. These tapes at NCAR also have many upper-air fields. The climatological precipitation was from the RAND set, where the precipitation came from Möller, 1951.

CHANGES IN US ANNUAL TEMPERATURE

The OSU model increases US annual temperatures by 2 to 3.6°C when CO_2 doubles (Figure 1). The other models have a greater global warming for the CO_2 doubling, and also show more warming for the US region: GISS (3.8 to 5.3°C), UK-Met (5.1 to 9.5°C), GFDL (3.6 to 5.5°C, not shown), GFDL higher resolution (3.6 to 5.7°C). The Canadian model gives 3.2 to 6.9°, not shown in the figure.

CHANGES IN US ANNUAL PRECIPITATION

Changes in simulated annual precipitation for the US as CO_2 doubles, are given for five model runs in Figure 2. The UK Met model and the high resolution simulation from GFDL show more precipitation in the new climate than given by the other models. Two simulations from the GFDL model are shown. In general, there is a much larger increase over the US in the most recent higher resolution run (GFDR30), with 2.2° by 3.7° cells. The Canadian Model shows some significant drying over the Great Plains.

156 Global Climate Change: Implications, Challenges and Mitigation Measures

```
10.8    8.0    6.9    6.0    5.5    4.5    4.1    4.3
12.7   10.0    8.9    8.5    9.3    9.3    8.9    9.8
14.4   11.7   12.5   11.1   13.2   13.7   13.3   18.0
15.8   14.8   17.8   15.1   17.4   17.8   17.8   19.8
17.3   16.7   20.3   19.4   20.9   21.5   22.2   23.5
19.3   18.9   22.0   21.7   22.2   24.7   24.6   24.8
```

1A   ANN   ATMOSPHERIC TEMPERATURE   C   CLIMATE

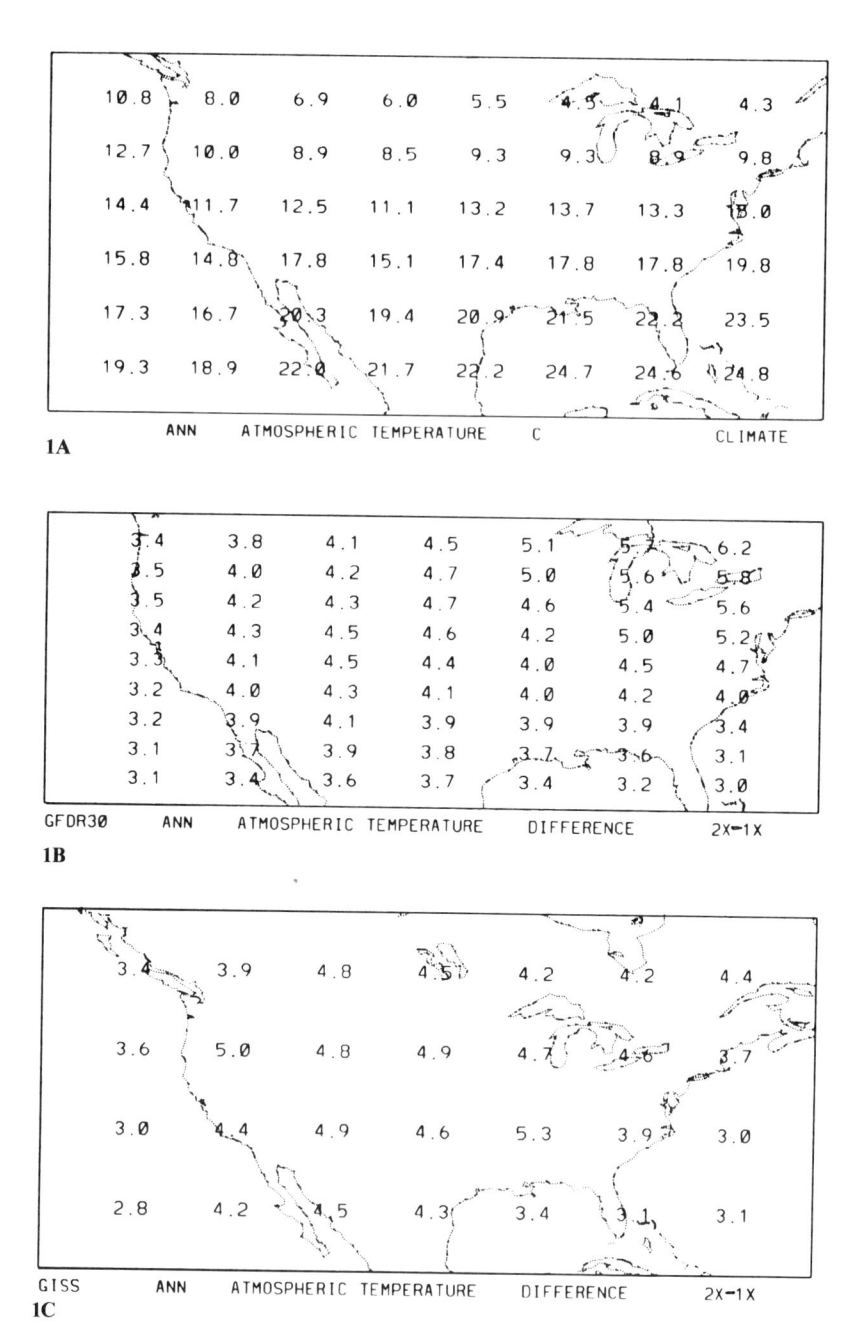

```
3.4    3.8    4.1    4.5    5.1    5.2    6.2
3.5    4.0    4.2    4.7    5.0    5.6    5.8
3.5    4.2    4.3    4.7    4.6    5.4    5.6
3.4    4.3    4.5    4.6    4.2    5.0    5.2
3.3    4.1    4.5    4.4    4.0    4.5    4.7
3.2    4.0    4.3    4.1    4.0    4.2    4.0
3.2    3.9    4.1    3.9    3.9    3.9    3.4
3.1    3.1    3.9    3.8    3.7    3.6    3.1
3.1    3.4    3.6    3.7    3.4    3.2    3.0
```

GFDR30   ANN   ATMOSPHERIC TEMPERATURE   DIFFERENCE   2X−1X
1B

```
3.4    3.9    4.8    4.5    4.2    4.2    4.4
3.6    5.0    4.8    4.9    4.7    4.6    3.7
3.0    4.4    4.9    4.6    5.3    3.9    3.0
2.8    4.2    4.5    4.3    3.4    3.1    3.1
```

GISS   ANN   ATMOSPHERIC TEMPERATURE   DIFFERENCE   2X−1X
1C

FIGURE 1. Change of annual temperature over the US for double CO_2 as given by four climate models. The present annual mean temperature is given in the upper plate. (FIGURE 1 continues on the following page).

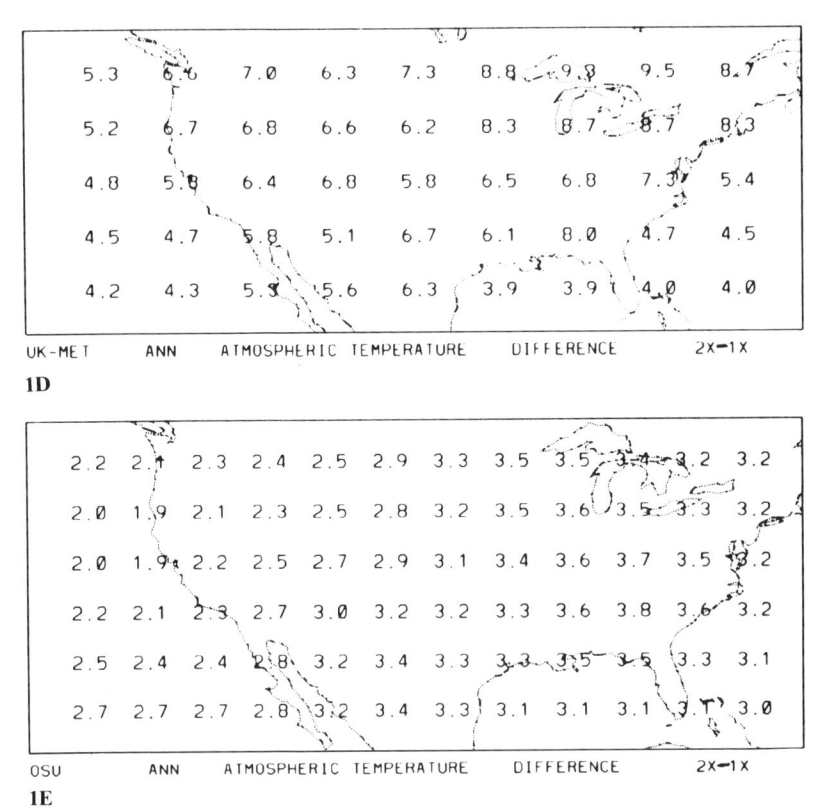

```
 5.3   6.6   7.0   6.3   7.3   8.8   9.8   9.5   8.7
 5.2   6.7   6.8   6.6   6.2   8.3   8.7   8.7   8.3
 4.8   5.8   6.4   6.8   5.8   6.5   6.8   7.3   5.4
 4.5   4.7   5.8   5.1   6.7   6.1   8.0   4.7   4.5
 4.2   4.3   5.3   5.6   6.3   3.9   3.9   4.0   4.0
```

UK-MET   ANN   ATMOSPHERIC TEMPERATURE   DIFFERENCE   2X−1X

1D

```
 2.2   2.1   2.3   2.4   2.5   2.9   3.3   3.5   3.5   3.4   3.2   3.2
 2.0   1.9   2.1   2.3   2.5   2.8   3.2   3.5   3.6   3.5   3.3   3.2
 2.0   1.9   2.2   2.5   2.7   2.9   3.1   3.4   3.6   3.7   3.5   3.2
 2.2   2.1   2.3   2.7   3.0   3.2   3.2   3.3   3.6   3.8   3.6   3.2
 2.5   2.4   2.4   2.8   3.2   3.4   3.3   3.3   3.5   3.5   3.3   3.1
 2.7   2.7   2.7   2.8   3.2   3.4   3.3   3.1   3.1   3.1   3.1   3.0
```

OSU   ANN   ATMOSPHERIC TEMPERATURE   DIFFERENCE   2X−1X

1E

It is a curious fact that many of the models do not do a good job of simulating the present annual rainfall over S.E. US, (for Arkansas, Mississippi and Alabama). Often only 30 to 50% of the real amount is simulated. One reason is that the model topography is too flat, but that doesn't explain all of the problem. Space does not permit us to show the results in a figure here.

AFRICAN SAHEL ANNUAL PRECIPITATION

For the Sahel region of Africa, we only have space to show the ratio of the simulated present-day precipitation climate to the observed climate (Figure 3). One plate of the figure shows the precipitation for the present climate (note that 3 mm/day is equal to about 100 cm of water per year). Several of the models carry precipitation further into the Sahara desert than is observed; therefore, we see ratios such as 1.5 to 4.0 near the Northern boundary. However, the amounts are so small that this is not a concern. Four of the five models only simulate about half of the observed rainfall for the countries just south of the Sahel region. The actual climatology of precipitation over the ocean is not known well, so be cautious of numbers there.

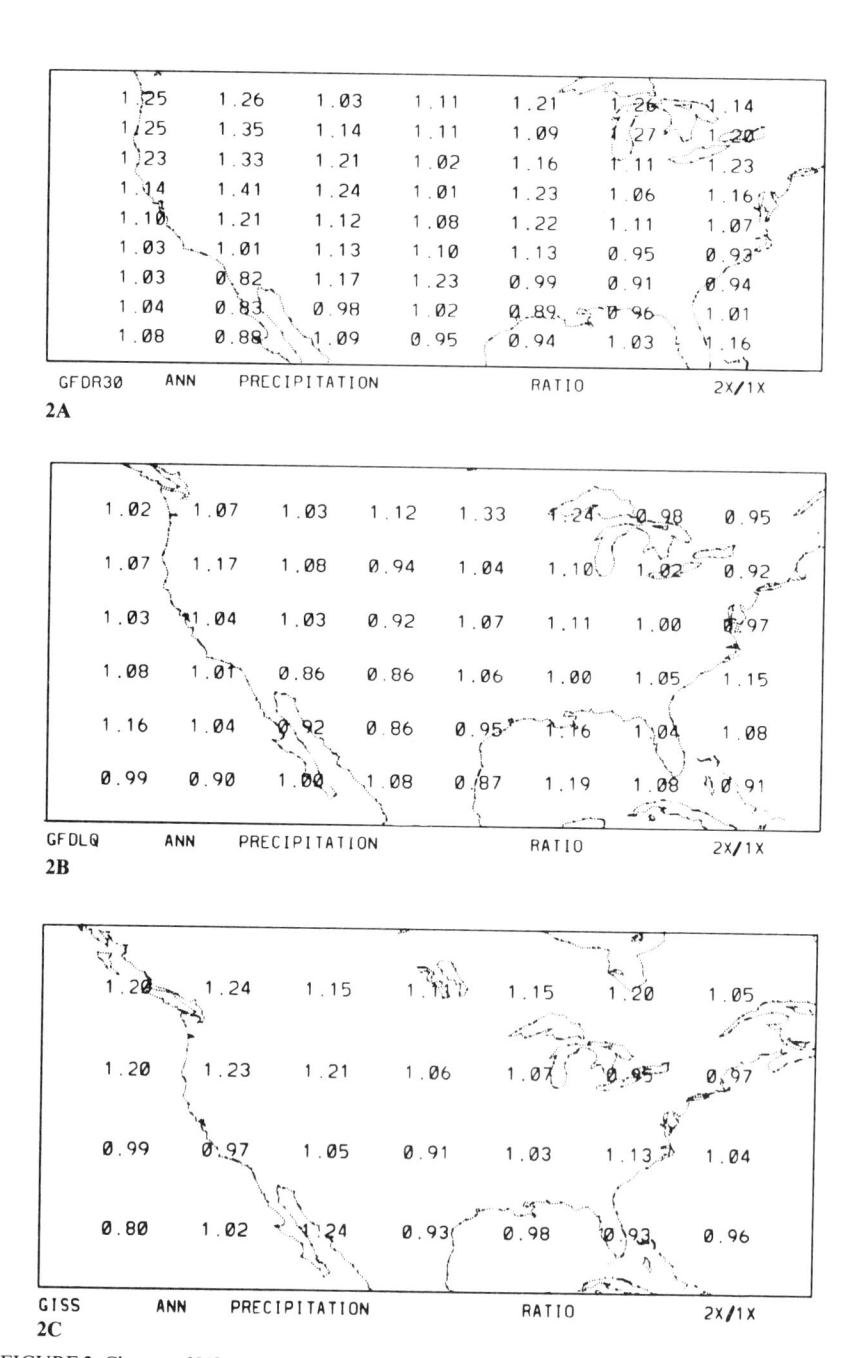

158 Global Climate Change: Implications, Challenges and Mitigation Measures

```
1.25  1.26  1.03  1.11  1.21  1.26  1.14
1.25  1.35  1.14  1.11  1.09  1.27  1.20
1.23  1.33  1.21  1.02  1.16  1.11  1.23
1.14  1.41  1.24  1.01  1.23  1.06  1.16
1.10  1.21  1.12  1.08  1.22  1.11  1.07
1.03  1.01  1.13  1.10  1.13  0.95  0.93
1.03  0.82  1.17  1.23  0.99  0.91  0.94
1.04  0.83  0.98  1.02  0.89  0.96  1.01
1.08  0.88  1.09  0.95  0.94  1.03  1.16
```
GFDR30   ANN   PRECIPITATION   RATIO   2X/1X
2A

```
1.02  1.07  1.03  1.12  1.33  1.24  0.98  0.95
1.07  1.17  1.08  0.94  1.04  1.10  1.02  0.92
1.03  1.04  1.03  0.92  1.07  1.11  1.00  0.97
1.08  1.01  0.86  0.86  1.06  1.00  1.05  1.15
1.16  1.04  0.92  0.86  0.95  1.16  1.04  1.08
0.99  0.90  1.00  1.08  0.87  1.19  1.08  0.91
```
GFDLQ   ANN   PRECIPITATION   RATIO   2X/1X
2B

```
1.20  1.24  1.15  1.11  1.15  1.20  1.05
1.20  1.23  1.21  1.06  1.07  0.95  0.97
0.99  0.97  1.05  0.91  1.03  1.13  1.04
0.80  1.02  1.24  0.93  0.98  0.93  0.96
```
GISS   ANN   PRECIPITATION   RATIO   2X/1X
2C

FIGURE 2. Change of US annual precipitation as given by five climate models. The numbers given are the precipitation in 2x CO_2 climate divided by the 1x control run. For example, 1.17 means 17% more precipitation in the double CO_2 climate. (FIGURE 2 continues on the following page).

```
  1.10   1.07   1.05   1.11   1.10   1.14   1.10
  1.19   1.15   1.01   0.95   1.07   0.99   0.98
  1.27   1.27   1.05   0.84   0.90   0.93   0.95
  1.31   1.11   1.05   0.99   0.93   0.95   0.98
  1.44   0.99   0.89   0.89   0.99   0.90   0.97
  1.29   0.97   0.97   0.86   0.92   0.95   0.97
```

CANADA   ANN   PRECIPITATION   RATIO   2X/1X
2D

```
 1.18  1.16  1.17  1.21  1.06  1.10  1.27  1.29  1.28
 1.19  1.17  1.19  1.28  1.12  1.01  1.14  1.17  1.20
 1.22  1.35  1.14  1.11  1.20  1.12  1.11  1.02  1.18
 1.27  1.20  1.28  1.28  0.82  1.03  0.80  1.17  1.23
 1.22  1.23  1.13  1.40  0.87  0.91  1.08  1.16  1.33
```

UK-MET   ANN   PRECIPITATION   RATIO   2X/1X
2E

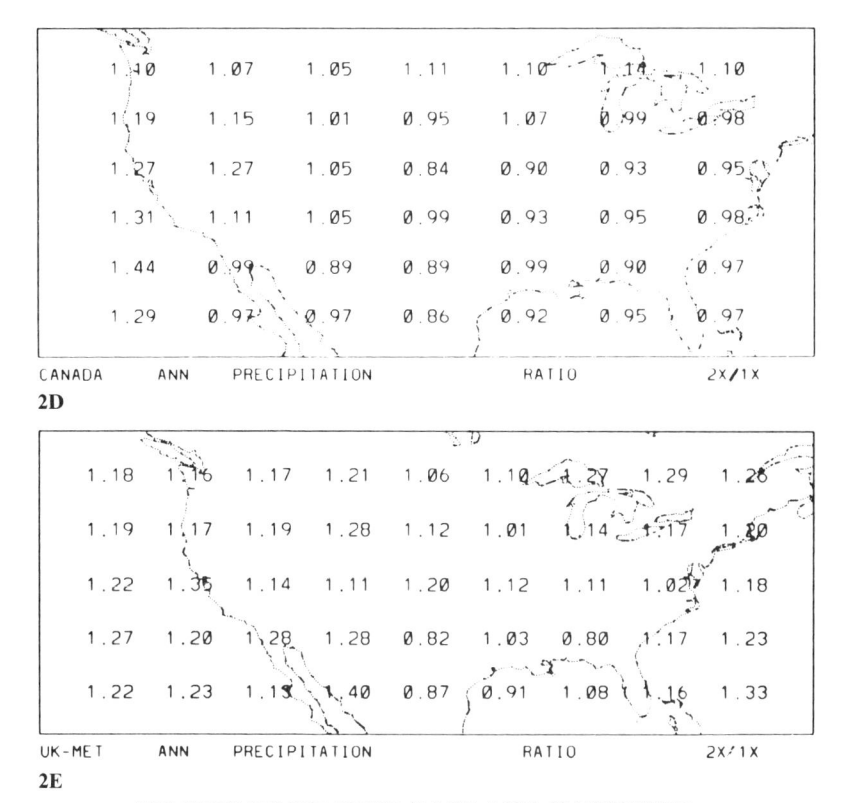

TEMPERATURE OVER EAST ASIA IN WINTER

The warming of the winter climate over East Asia in the double CO_2 model runs is shown in Figure 4. The higher resolution GFDL model warms the region by about 4°C compared with the control run. This is similar to the older GFDL run (not shown), but somewhat less. The Canadian model warms this region by 4-9°C, which is much more than the overall sensitivity of the model (only 3.5°). GISS warms it by about 3-7°C, and UK-Met by 4-8°C. These are very large changes. For the high resolution models, not all points are printed (GFDR30 and Canadian CCC).

COMPUTER REQUIREMENTS FOR CLIMATE MODELS

The development of climate models and their execution has been retarded by the lack of computer time. In any climate model run, for the present climate or double CO_2, the atmosphere and ocean must first be "spun-up" to their equilibrium state. This spin-up usually requires 5 to 25 simulated years, depending on how it is done. Then the model has to be run for 10 years to obtain an adequate sample of weather situations to form a proper climatology.

A total of about 50 simulated years is needed for a climate simulation (includes 1x and 2xCO_2 runs) which required about 250 CPU hours on computers like a

3A

1.60	1.43	1.08	0.91	0.68
4.12	4.18	3.01	2.14	1.61
5.32	5.19	3.86	4.87	3.56
3.08	2.99	2.54	4.78	4.80
0.79	0.80	0.91	2.48	4.25

ANN PRECIPITATION MM/DY CLIMATE

3B

0.32	0.42	0.66	0.93	1.46	2.14	1.61	1.10
0.25	0.39	0.57	0.67	0.90	1.28	1.05	0.98
0.25	0.37	0.49	0.40	0.48	0.61	0.77	0.77
0.30	0.35	0.31	0.24	0.29	0.27	0.43	0.56
0.49	0.46	0.34	0.33	0.46	0.34	0.34	0.56

CANADA ANN PRECIPITATION RATIO 1X/CLIM

3C

1.30	0.32	1.49	1.96	2.15
1.05	0.44	0.96	1.34	2.06
0.69	0.77	1.00	1.00	1.18
0.70	0.84	1.34	1.25	1.11
2.49	1.66	1.94	1.16	1.01

UK-MET ANN PRECIPITATION RATIO 1X/CLIM

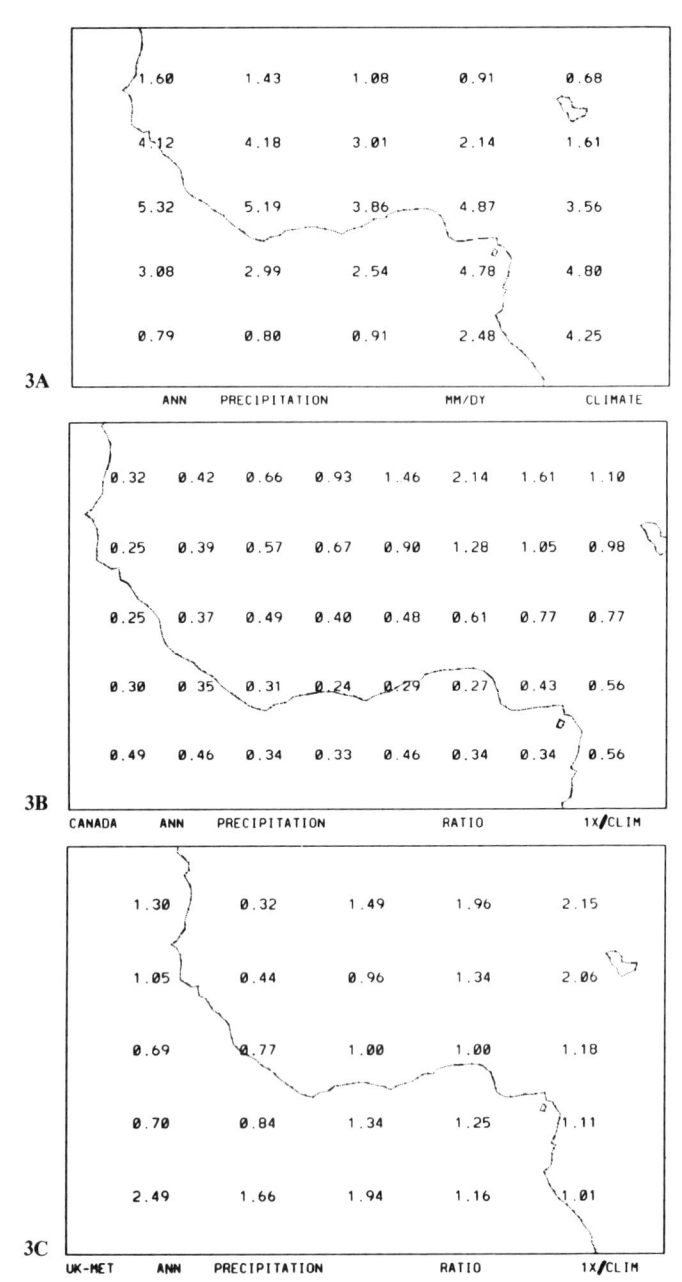

FIGURE 3. Simulation of African annual precipitation by five models compared with climatology. Example: 1.15 means that the model simulates 15% more precipitation for the present than is given by climatology. The chart depicted by 3A is the present climatology. (FIGURE 3 continues on the following page).

3D

0.35	0.24	0.46	1.18	1.26	1.00	1.43	1.35
0.46	0.69	0.64	0.83	1.43	1.56	1.38	0.86
0.84	1.50	0.99	0.76	1.30	1.89	1.61	1.01
0.82	1.45	1.09	0.90	1.07	0.83	1.30	1.69
0.58	0.62	0.44	0.63	0.76	0.34	0.77	2.18
0.50	0.29	0.22	0.36	0.45	0.31	0.58	1.28
0.34	0.26	0.28	0.28	0.30	0.32	0.49	0.77

GFDR30 ANN PRECIPITATION RATIO 1X/CLIM

3E

0.66	0.96	2.00	1.18	1.10
0.48	0.58	0.66	0.91	1.11
0.47	0.54	0.60	0.70	1.07
0.65	0.69	0.97	0.64	0.80
1.45	0.94	1.10	0.72	0.69

GFDLQ ANN PRECIPITATION RATIO 1X/CLIM

3F

1.88	0.60	0.21	0.17	0.36	0.50	0.66	0.17
1.01	0.55	0.26	0.13	0.26	0.45	0.51	0.78
0.85	0.75	0.50	0.41	0.56	0.42	0.36	0.40
1.73	1.71	1.41	1.43	1.50	0.78	0.37	0.30

OSU ANN PRECIPITATION RATIO 1X/CLIM

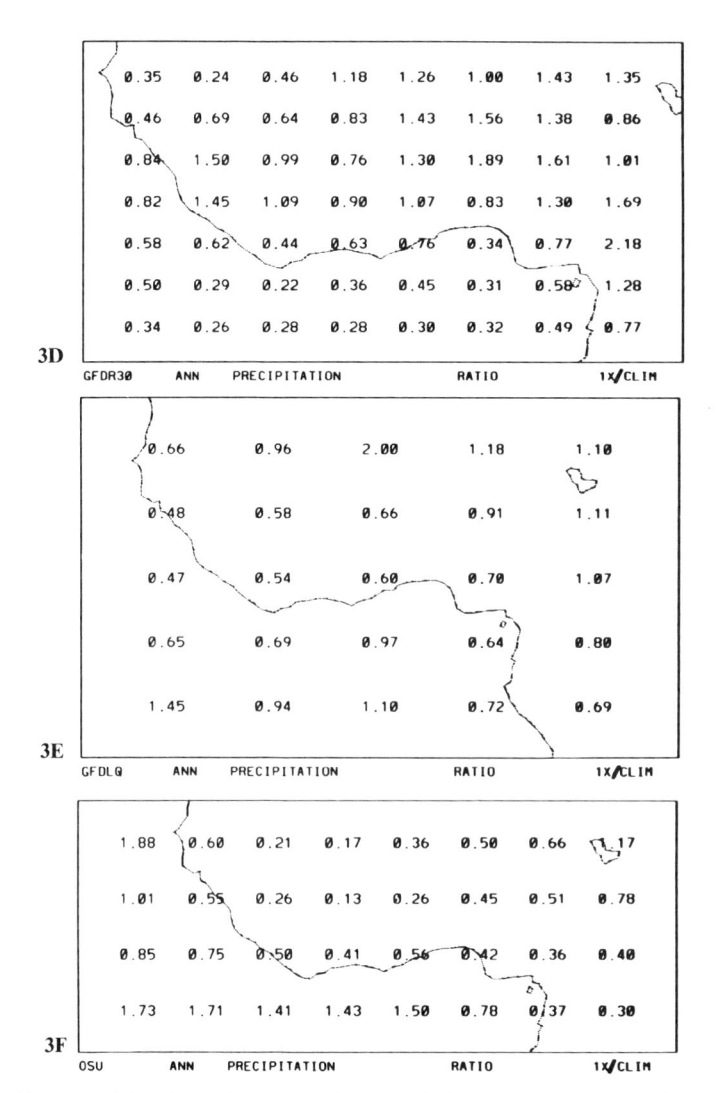

CRAY-1A or CDC 205. These were the world's fastest computers during 1977-82. A Cray could delivery about 15 hours of CPU time in a day; therefore, the model run above required about 17 days of dedicated computer time. A model experiment was actually run over a period of months. These times are for a model with about 9 levels and a resolution of 550 km. If the horizontal resolution of the model is doubled (same number of levels), the running time increases by a factor of six or seven.

ACKNOWLEDGEMENTS

It has been a pleasure to work with the modeling groups to make these data

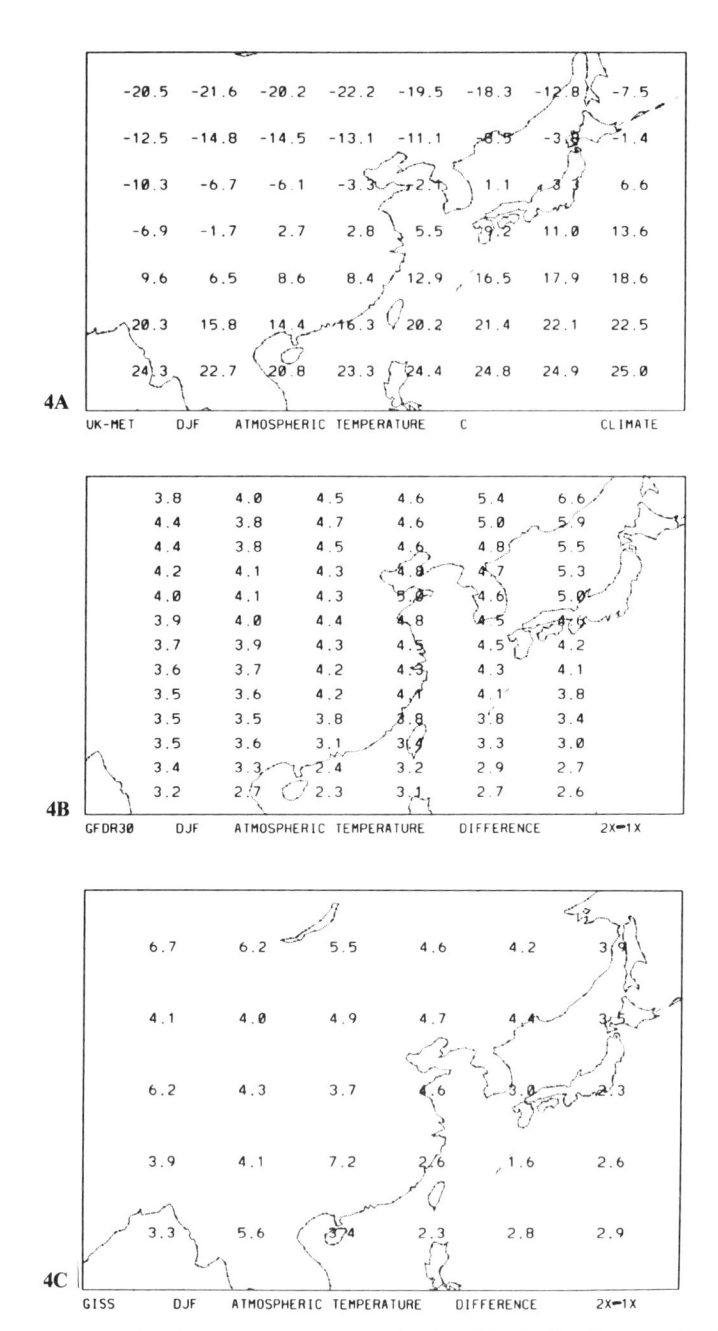

162 Global Climate Change: Implications, Challenges and Mitigation Measures

-20.5	-21.6	-20.2	-22.2	-19.5	-18.3	-12.8	-7.5
-12.5	-14.8	-14.5	-13.1	-11.1	-8.5	-3.8	-1.4
-10.3	-6.7	-6.1	-3.3	-2.1	1.1	4.3	6.6
-6.9	-1.7	2.7	2.8	5.5	9.2	11.0	13.6
9.6	6.5	8.6	8.4	12.9	16.5	17.9	18.6
20.3	15.8	14.4	16.3	20.2	21.4	22.1	22.5
24.3	22.7	20.8	23.3	24.4	24.8	24.9	25.0

4A
UK-MET   DJF   ATMOSPHERIC TEMPERATURE   C   CLIMATE

3.8	4.0	4.5	4.6	5.4	6.6
4.4	3.8	4.7	4.6	5.0	5.9
4.4	3.8	4.5	4.6	4.8	5.5
4.2	4.1	4.3	4.8	4.7	5.3
4.0	4.1	4.3	5.0	4.6	5.0
3.9	4.0	4.4	4.8	4.5	4.6
3.7	3.9	4.3	4.5	4.5	4.2
3.6	3.7	4.2	4.3	4.3	4.1
3.5	3.6	4.2	4.1	4.1	3.8
3.5	3.5	3.8	3.8	3.8	3.4
3.5	3.6	3.1	3.0	3.3	3.0
3.4	3.3	2.4	3.2	2.9	2.7
3.2	2.7	2.3	3.1	2.7	2.6

4B
GFDR30   DJF   ATMOSPHERIC TEMPERATURE   DIFFERENCE   2X-1X

6.7	6.2	5.5	4.6	4.2	3.9
4.1	4.0	4.9	4.7	4.4	3.5
6.2	4.3	3.7	4.6	3.0	2.3
3.9	4.1	7.2	2.6	1.6	2.6
3.3	5.6	3.4	2.3	2.8	2.9

4C
GISS   DJF   ATMOSPHERIC TEMPERATURE   DIFFERENCE   2X-1X

FIGURE 4. Increase in winter temperature over Asia when CO_2 doubles. For example, 5.6 means that the model temperature increased by 5.6°C. The chart depicted by 4A gives normal present-day winter temperature. (FIGURE 4 continues on the following page).

4D

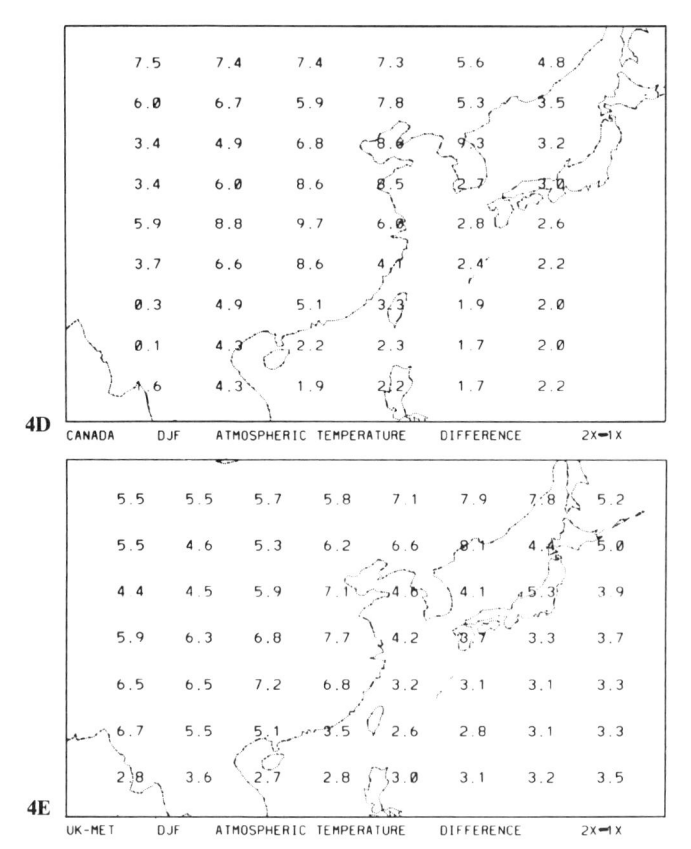

7.5	7.4	7.4	7.3	5.6	4.8
6.0	6.7	5.9	7.8	5.3	3.5
3.4	4.9	6.8	8.0	9.3	3.2
3.4	6.0	8.6	8.5	2.7	3.0
5.9	8.8	9.7	6.0	2.8	2.6
3.7	6.6	8.6	4.1	2.4	2.2
0.3	4.9	5.1	3.3	1.9	2.0
0.1	4.3	2.2	2.3	1.7	2.0
.6	4.3	1.9	2.2	1.7	2.2

CANADA   DJF   ATMOSPHERIC TEMPERATURE   DIFFERENCE   2X—1X

4E

5.5	5.5	5.7	5.8	7.1	7.9	7.8	5.2
5.5	4.6	5.3	6.2	6.6	8.1	4.4	5.0
4.4	4.5	5.9	7.1	4.6	4.1	5.3	3.9
5.9	6.3	6.8	7.7	4.2	3.7	3.3	3.7
6.5	6.5	7.2	6.8	3.2	3.1	3.1	3.3
6.7	5.5	5.1	3.5	2.6	2.8	3.1	3.3
2.8	3.6	2.7	2.8	3.0	3.1	3.2	3.5

UK-MET   DJF   ATMOSPHERIC TEMPERATURE   DIFFERENCE   2X—1X

available. They have been patient with my questions about the models. David Rind was the main contact for the GISS model, and for some sensitivity studies. R. Ruedy at GISS did a considerable amount of work to summarize data and make it rapidly available. At GFDL, Richard Wetherald provided the data and answered questions. The expert on the OSU model is Michael Schlesinger, who provided much information. John Mitchell provided data and information for the UK model. George Boer has provided similar data from the Canadian model.

Dennis Joseph, at NCAR, prepared the computer graphics from the models for use in the figures. Janice Powell helped with the preparation of this text.

REFERENCES

1. Boer, G.J., N. McFarlane and M. Lazare, 1991. Greenhouse gas-induced climatic change simulated with the CCC second-generation GCM. Accepted for publication in the *Journal of Climate*.
2. Carissimo, B.C., A.H. Oort, T.H. VonderHaar, 1985. Estimating the meridional energy transports in the atmosphere and ocean. *J. of Phys. Ocean.* 15:No.1, 82-91.

3. Crutcher, H.L., and J.M. Meserve, 1970. Selected level heights, temperatures and dew points for the northern hemisphere. NAVAIR 50-1C-52 (revised), Chief of Naval Operations, Washington, D.C.
4. Hansen, J., G. Russell, D. Rind, P. Stone, A. Lacis, S. Lebedeff, R. Ruedy and L. Travis, 1983. Efficient three-dimensional global models for climate studies: models I and II. April *Monthly Weather Review.* III: 609-662.
5. Jaeger, L., 1976. Monatskarten des niederschlags für die ganze erde. *Berichte des Deutschen Wetterdienstes,* 139.
6. Manabe, S. and R.T. Wetherald, 1987. Large-scale changes in soil wetness induced by an increase in carbon dioxide. April *J. Atmos. Sci.* 44:1211-1235.
7. McFarlane, N.A., G.J. Boer, J.-P. Blanchet and M. Lazare, 1991. The CCC second generation GCM and its equilibrium climate. Submitted to the *Journal of Climate.*
8. Mitchell, J.F.B., C.A. Senior and W.J. Ingram, 1989. CO_2 and climate: A missing feedback? *Nature.* 341:132-134.
9. Möller, F. Vierteljahrskarten des niederschlags für die ganze erde. 1951. *Petermanns Geographische Mitteilungen*, Justus Perthes, Gotha. pp 1-7.
10. Ramanathan, V., 1988. The greenhouse theory of climate change: A test by an inadvertent global experiment. *Science.* 240:293-299.
11. Schlesinger, Michael E., 1988. Model projections of the climatic changes induced by increased atmospheric CO_2. Symposium on Climate and Geo-Sciences, 22-27 May 1988. Reidel Publ. Co., Dordrecht, Holland, (in press).
12. Schlesinger, M.E. and Z.-c. Zhao, 1988. Seasonal climate changes induced by doubled CO_2 as simulated by the OSU atmospheric GCM/mixed-layer ocean model. CRI Report and *J. Climate* (submitted).
13. Taljaard, J.J., H. van Loon, H.L. Crutcher, and R.L. Jenne, 1969. Climate of the upper air: southern hemisphere. 1: Temperatures, dewpoints, and heights at selected pressure levels." NAVAIR 50-1C-55, Chief of Naval Operations, Washington, D.C., pp 135.
14. Wetherald, R.T. and S. Manabe, est. 1992. A reevaluation of CO_2-induced hydrologic change as obtained from low and high resolution versions of the GFDL general circulation model. (July 1990-91, in preparation). Part of the text was published in the 1991 IPCC Report.
15. Wilson, C.A. and J.F.B. Mitchell, 1987. A Doubled CO_2 Climate Sensitivity Experiment with a Global Climate Model Including a Simple Ocean. *JGR.* 92: D11-11/20 13,315-13,343.

Global Climate Change: Implications, Challenges and Mitigation Measures. Edited by S.K. Majumdar, L.S. Kalkstein, B. Yarnal, E.W. Miller, and L.M. Rosenfeld. © 1992, The Pennsylvania Academy of Science.

Chapter Eleven

MONITORING GLOBAL TEMPERATURE CHANGES FROM SATELLITES

JOHN R. CHRISTY
Atmospheric Science Program
University of Alabama in Huntsville
Huntsville, AL 35899

INTRODUCTION

The 1988 summer heat wave over the central and eastern United States was fresh on the minds of a group of atmospheric scientists later that Fall taking a lunch break at a conference in New Hampshire. Their discussion (and concern) centered around the amount of publicity given to this one hot summer and how it was being associated with predictions of global warming due to the predicted enhanced greenhouse effect. This publicity had caused the average person to believe that a global warming catastrophe had begun, and that every succeeding year was going to be warmer than normal.

Dr. Richard McNider (University of Alabama in Huntsville) expressed a common sentiment found among scientists when he pointed out the tremendous uncertainty inherent in the surface temperature record on which these dire predictions were based. "There must be some way to monitor the atmosphere from space so that there would be no question about lack of global coverage or whether urban heat islands are affecting the trend." Dr. James Dodge (NASA Headquarters) pointed out that the satellite record was too short to give the long time period values needed for the determination of a trend. However, he tossed a question anyway to Dr. Roy Spencer (NASA Marshall Space Flight Center), a microwave expert, "Is there any potential for the microwave temperatures from the polar orbiters to be of use on this issue?" Spencer thought a moment and indicated that the Microwave Sounding Units (MSUs) have been orbiting since 1978 and because of their external calibration might be helpful.

Following the conference, Spencer returned to Huntsville, Alabama where he began the investigation of the MSU data. When initial research revealed a promise of high precision in the results, he asked me to contribute to the effort by developing a technique to merge data from the six different satellites into one consistent time series covering the last 12 years. Our research resulted, among other things, in a record of the fully global distribution of atmospheric temperatures for the layer from the surface to about 10 km. As will be shown below, this temperature can be highly correlated with the near-surface temperature where we as humans live. The coverage, consistency and precision of the data were such that the MSU temperatures were to provide valuable information on issues of global warming and climate change.

CLIMATE MONITORING WITH MSU TEMPERATURES

The readings from microwave emissions are nominally called "brightness temperatures" because they measure the electromagnetic radiation of molecular oxygen in the air as it absorbs and emits in the microwave part of the spectrum. This is similar to the way our eye detects different levels of visible light emitted (not reflected) by a source such as a lightbulb. The brighter the source, the more intense is the radiation. Just as our eyes respond to varying intensities of visible light, the MSU is able to record different levels of microwave emissions from the oxygen in the air. This degree of "brightness" is strongly related to the temperature of the oxygen molecules, and therefore to the temperature of the atmosphere as a whole since oxygen is a well-mixed gas in the atmosphere.

The MSUs aboard the six satellites known as Tiros-N, NOAA-6, -7, -9, -10 and -11 have monitored the atmosphere for the past twelve years. These satellites were launched at roughly 2-year intervals with a life expectancy of about four years. As such, two satellites usually were in operation on any given day. Specifically, the MSUs recorded emissions at four frequencies near the 60 gHz oxygen absorption band. In general, as the frequency departs from 60 gHz, the emission of greater amounts of oxygen is measured. In other words, the sensor "sees" deeper into the atmosphere at 58 gHz than at 59 gHz. The two satellites flying on a given day have different orbits, one which makes a northbound pass over the equator around 7:30 a.m. local time and one around 2:30 p.m. local time. Each orbits the earth from pole to pole approximately 14 times per day with one-day coverage as shown in Figure 1. Note that the earth is rotating on its axis beneath the satellite, so every time the satellite crosses the equator going north this is 1/14 of an earth rotation (day), giving about 100 minutes per orbit.

These satellites contain many instruments only one of which is the MSU. This device measures the brightness temperature at a particular microwave frequency through a sensor which views a scene reflected in a rotating mirror. As the earth comes into view on the mirror, the sensor measures "microwave counts" of eleven "footprints" or scenes over the earth as the mirror revolves. The mirror also points to "cold-space" and a "hot target" on board whose brightness temperature is monitored by platinum resistence thermometers. So, in one rotation (scan) of the mirror, there are thirteen microwave readings composed of eleven earth-atmosphere scenes, one of cold space and one of the hot target.

The atmospheric values of brightness temperature are calculated through the relationship determined by comparing the cold and hot target microwave counts with their known temperatures (2.7 K for cold space and the measured reading from the hot target). Using this relationship of counts to temperature, the atmospheric microwave readings are converted into temperature.

The MSU is an "externally calibrated" instrument because it is checked against two known temperatures (cold space and hot target) outside of the sensor itself. So, even if the sensor were to degrade over time, it would do so for the known hot and cold targets in the same manner as for the atmospheric footprints. Though the relationship between microwave counts and temperature would change, the magnitude of the temperatures so computed would be stable. This is one of the powerful characteristics of the MSU instruments, especially for long-term climate monitoring.

Two of the channels have shown remarkable stability since 1978 and are therefore candidates for long-term studies. These are channel 2 (MSU2) at 53.74 gHz and channel 4 (MSU4) at 57.95 gHz. Briefly stated, the MSU2 measures emissions in the broad vertical layer from the surface (around 1000 mb) to about 100 mb (about 15 km) or 90% of the atmosphere's mass. Most of the energy measured by the channel 2 instrument originates below 300 mb. MSU4 views a layer above 200 mb, primarily between 100 and 50 mb. A detailed description of the weighting functions and the sensor capabilities, including small effects of non-thermal related emissions (e.g. thunderstorm cores, high elevation surfaces, ice, etc.) is provided in (1).

The original purpose of the MSUs was to aid in producing vertical temperature profiles at specific pressure levels for assimilation into operational weather forecast

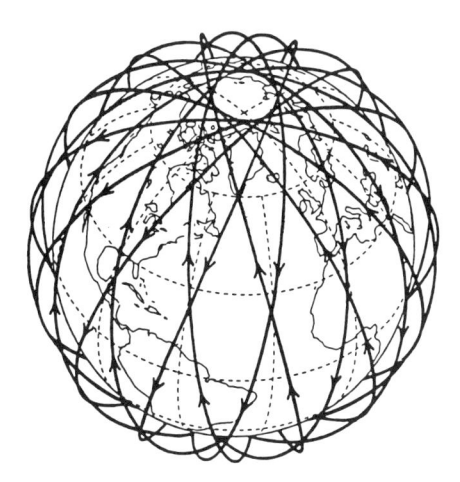

FIGURE 1. Polar orbiting satellites circle the earth 14 times per day as shown above (courtesy Dr. Stanley Kidder, University of Alabama in Huntsville).

models. The process of converting broad atmospheric emissions from infrared and microwave frequencies into specific level data is extremely complicated, relying on considerable theory and several statistical relationships. The algorithms for these satellite-produced profiles have been changed through the years and along with operational adjustments in the generation of the weather maps, has made the maps of only marginal use for precision climate studies.

Spencer took a different approach with the MSU, analyzing the raw radiances which represent deep-layer intergrals of temperature and found a level of precision that was phenomenal. This precision was due to a combination of the external calibration, mentioned above, and the large volume (over 50,000 km^3) of air measured at each of the footprints. The standard error of measurement (s.e.) for one satellite has been determined by comparing two simultaneously operating satellites. For MSU2 and MSU4, the s.e. for monthly gridpoint (2.5° of latitude) anomalies is about 0.05°C in the tropics and 0.1°C in midlatitudes. Global mean s.e. values for daily and monthly anomalies are near 0.05°C and 0.01°C respectively. (Most of the time, two satellites were in orbit so that the s.e. for these periods would be 0.7 multiplied by the values given above.)

One of the crucially important features of the MSU data is that they can be verified against completely independent measurements from ground-based devices. One of these devices is a radiosonde which is an instrument package tied to a helium-filled balloon released twice a day in many locations worldwide. The package contains a transmitter which beams information about the atmosphere back to an earth station. One quantity produced from radiosonde observations is known as the "thickness" or depth of portions of the air column through which the balloon ascends. This thickness is proportional to the temperature of the column since warm air occupies a greater volume than cold air.

Comparisons of monthly anomalies between MSU2 and 1000-200 mb thicknesses (the height of the lowest 80% of the atmosphere's mass) measured from balloons have been excellent both in many-station averages and at grid-point levels (Figure 2). Though the temperature of this deep layer may seem of little value to the operational forecaster who needs specific level data, it is a vital quantity to monitor because minute changes and trends may be accurately determined. It is in this deep-layer value that indications of global climate change may be most unambiguously observed.

A key point to realize in Figure 2 is that the method by which the satellite data were merged from one to the next is shown to have been quite good. There are no obvious errors over the ten years of comparisons and no significant differences in the decadal trends. Because nothing special was done to the MSU2 grid points used in this comparison study, this means the entire globe (10,000 gridpoints) is represented by highly accurate, deep layer MSU2 temperatures. A detailed description of these comparisons is found in (2) and (3).

MSU2 represents the thermal variability measured primarily in the troposphere, (the atmosphere below 12 km). Due to the equivalent barotropic nature of the "monthly-mean" atmosphere, much of what happens at the surface is also characteristic of the deep-layer temperature. However, there is considerable influence on MSU2 by emissions from the upper troposphere and lower stratosphere. Even though the atmospheric mass which contributes to that part of the signal is only about 15% of the total, the magnitude of the anomalies in the lower stratosphere is quite large during periods of vulcanism or sudden stratospheric warmings or coolings.

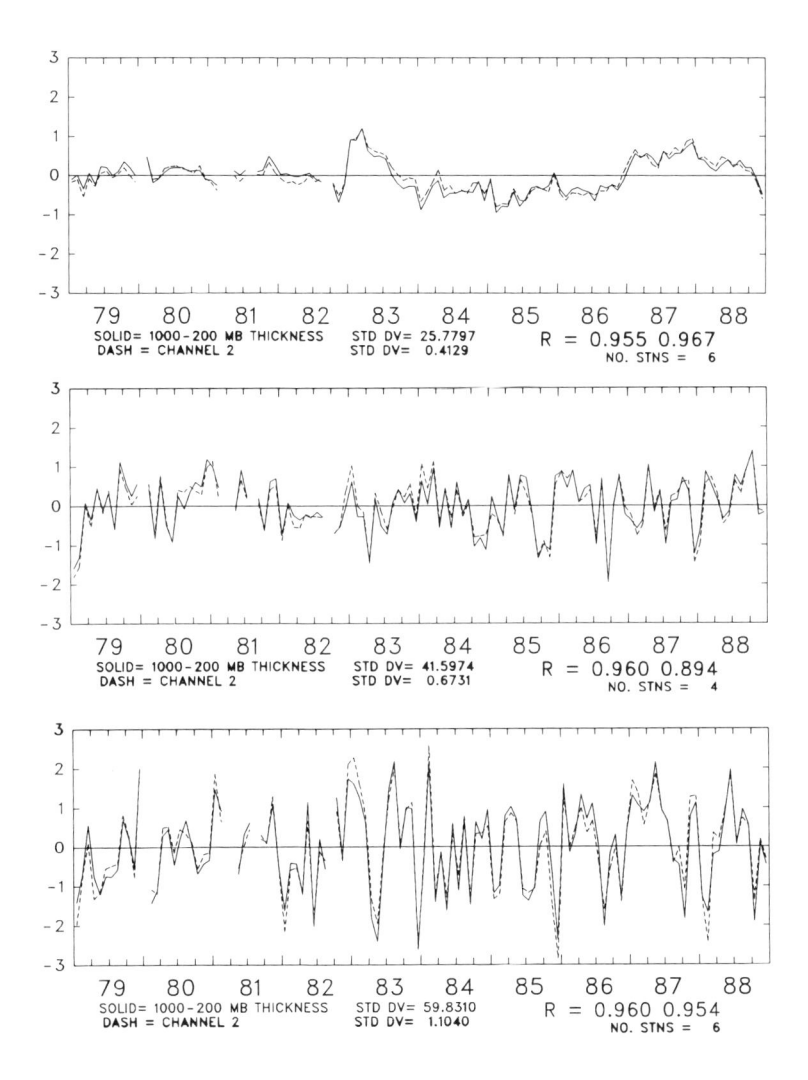

FIGURE 2. Comparisons between MSU2 temperatures (°C, solid) and the thickness (converted to °C) of the 100°-200 mb atmospheric layer measured from balloon soundings (dashed). Results for three climate regions (top to bottom) are for 6 Caribbean, 4 U.S. West coast and 6 Great Lakes stations. The monthly correlations (R) are 0.96 for all three areas and annual correlations are .97, .89 and .95 respectively.

170 Global Climate Change: Implications, Challenges and Mitigation Measures

In fact, the tropospheric and stratospheric anomalies are often of opposite sign, so that a natural cancellation of the tropospheric value is possible for the MSU2. This cancellation is most noticeable in higher latitudes where the portion of atmosphere above the tropopause (the boundary between the troposphere and the stratosphere) is much greater than in the tropics.

A tropospheric retrieval technique has been developed which uses a linear combination of near-limb and near nadir view angles to eliminate the higher altitude influence on MSU2. [Recall that there are eleven footprints. The middle or sixth is directly downward (nadir) progressing to angled views seen in footprints one and eleven on either side. As the angle of view is further from nadir, the path of oxygen observed is at a higher altitude.] The resulting quantity is a tropospheric retrieval ($MSU2_R$) which represents the temperature of a layer from the surface to about 250 mb, but with a peak in the weighting function in the lower troposphere around 700 mb (about 3 km). Tests show that $MSU2_R$ is most highly correlated with the 1000-500 mb thickness (lowest 50% of atmosphere by mass) and has no stratospheric influence (3). The effect of the lower weighting function is seen in Figure 3 which compares the surface temperature anomalies (T_{sfc}) from the Intergovernmental Panel

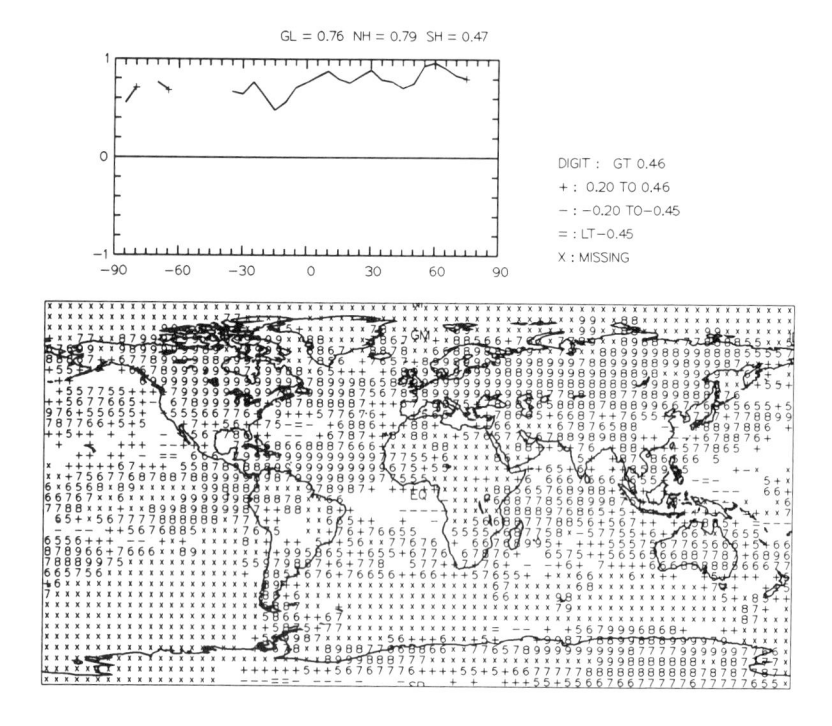

FIGURE 3. Correlations between $MSU2_R$ and the surface temperature data set for annual anomalies. Digits are correlations multiplied by ten and rounded. A " + " represents correlations between +.25 and +.45 and " − " the opposite. An " = " indicates correlations below −.46. Blank regions have correlations between −.25 and +.25. The surface data are missing at grids marked "x". The upper curve represents correlations of monthly zonal mean anomalies.

of Climate Change (IPCC) gridded data and that of MSU2$_R$. In many land and oceanic areas the correlations are quite high.

Figure 3 raises questions about the relation of MSU2$_R$ and the surface data. Anomalies for the rectangular area covering the United States and Canada were produced for both the IPCC T$_{sfc}$ and the MSU^2$_R$. These were normalized and plotted in Figure 4 (upper) for annual and monthly values for the eleven years 1979-1989. The differences are shown immediately below these time series. The comparison is not only remarkable, it is probably as close to the theoretical limit of agreement one can achieve with these two different atmospheric quantities (surface versus the deep layer). The decadal trend (normalized to the IPCC variance) is +0.44°C in both with annual and monthly correlations of 0.98 and 0.95 respectively. Other well-monitored continental regions have high correlations as well.

Similar calculations for the globe (Figure 4, lower) are quite different, with the decadal trends (normalized to the IPCC variance) for the MSU2$_R$ and IPCC at −0.05°C and +0.15°C respectively. The annual and monthly correlations were only 0.76 and 0.63. *These descrepancies certainly beg for an answer since the anticipated global warming trend of +0.3°C per decade is at the level of disagreement between the MSU2$_R$ and the IPCC T$_{sfc}$.*

Why do we not see strong positive correlations world-wide? One reason is that in some regions of the world there is large interlayer variability between T$_{sfc}$ and the troposphere above. For example, over subtropical oceanic areas there is generally cool surface water being transported equatorward which creates a relative cold near-surface air layer. Above this layer is the warm, subsiding air of the subtropical high pressure system creating a very stable situation. In such regions, an especially cool surface air layer, about one kilometer thick, is decoupled from the very warm air above. As a consequence, changes in the near-surface or mid-troposphere temperatures need not have a strong positive relationship (3). In fact, one may actually find negative correlations here because a surface perturbation of warm air will likely mix upward and adiabatically cool the tropospheric air.

Other reasons for the low correlation certainly relate to the very unsystematic manner in which the T$_{sfc}$ is recorded, and the level of noise thus produced. These characteristics are addressed in (4) where it is shown that for most of the earth's oceans, the T$_{sfc}$ measurements are inadequate for the generation of accurate monthly anomalies. Errors which enter the T$_{sfc}$ record over the oceans from ships are due to the following reasons:

—Changes in instrumentation and measurement techniques over time
—Constantly changing set of ships reporting for a given grid-square
—Errors in individual observations
—Ignoring diurnal (24-hour day) variations
—Ignoring natural variations occuring within an average month
—Assuming a point measurement within a 5°C box represents the entire box

The estimated monthly error for most of the world's oceans is at least 0.2°C for 5°x5° boxes, which is more than twice the error of the MSU2 measurements and approaches the actual standard deviation of the surface signal in the tropical and

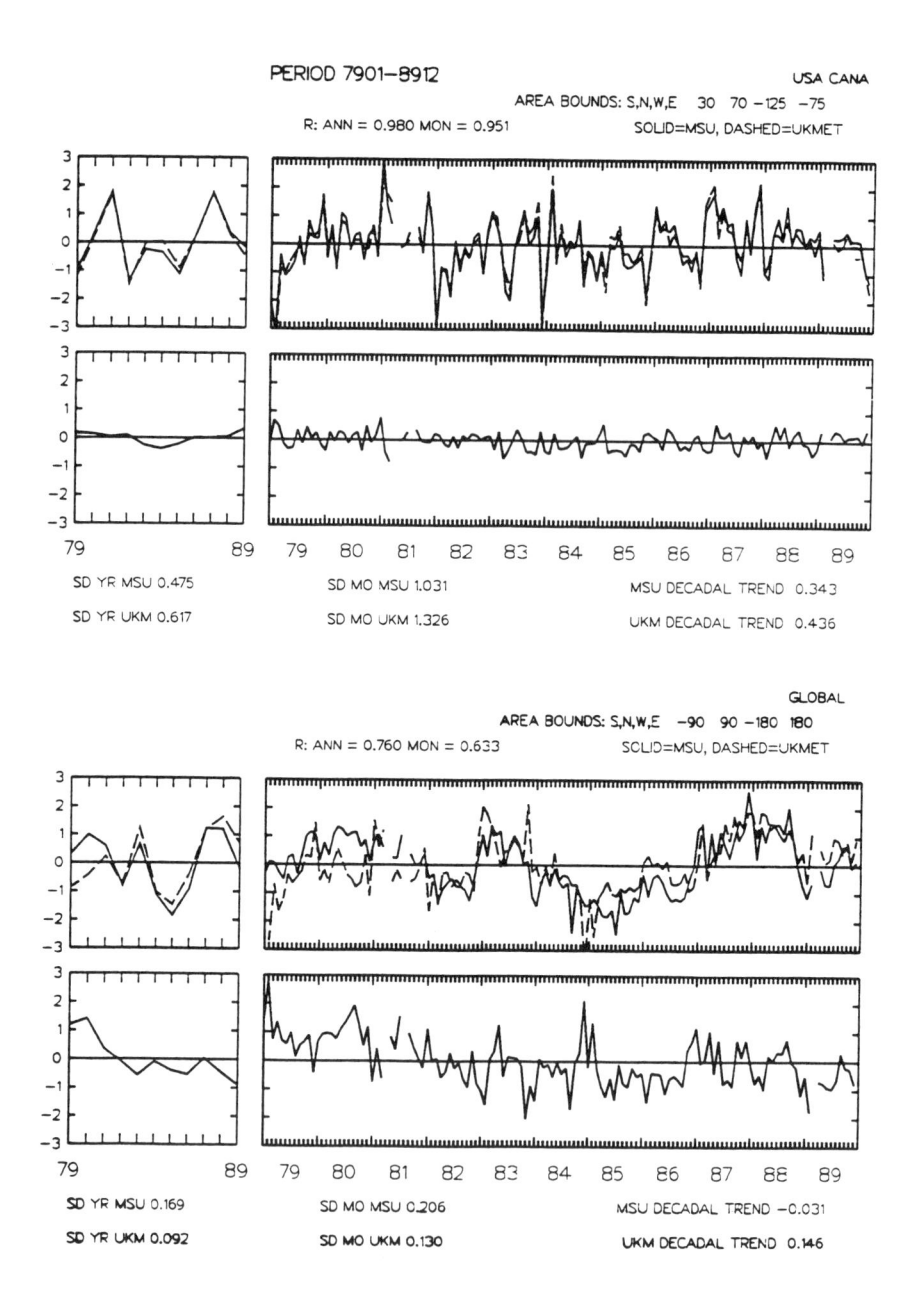

172 Global Climate Change: Implications, Challenges and Mitigation Measures

FIGURE 4. Upper: Time series of the average temperature over the U.S. and Canada for annual anomalies (left) and monthly anomalies (right) for $MSU2_R$ and the surface data (here as UKM or UKMET). The differences are plotted immediately beneath. Units are in standard deviations. The annual and monthly correlations are 0.98 and 0.95 respectively. Lower: As above but for all grids (i.e. quasi-global). Annual and monthly correlations are 0.76 and 0.63.

subtropical areas. The ability of the T_{sfc} record to provide a surrogate for the true global temperature is at least open to question as a result of the comparisons with MSU2 and MSU2$_R$.

A clear example of the problem with the T_{sfc} record is seen in the anomalies for March 1990. The surface temperatures, lacking accurate values over much of the globe (at least 30% had no reports at all), indicated this month was by far the warmest "global" anomaly in the 100+ years of record (James Hansen, Philip Jones, personal communication). Examination of the MSU2$_R$ distribution of anomalies (Figure 5) indicates that, indeed, the northern mid-latitude continents were extremely warm. For the globe as a whole, however, MSU2$_R$ March 1990 was only at the *70th percentile* for the last twelve years of data.

Compounding the situation is the fact that March T_{sfc} anomaly was so warm that it influenced the annual mean for 1990 to be the highest annual value for the entire period of record, and considerable publicity was devoted to this result. The MSU2$_R$ anomaly for all of 1990 was only fourth warmest since 1978. For a global mean quantity, one should not expect perfect agreement between these two sets. However, one should see matchups between extreme months and years as well as some agreement in the trends. The MSU2$_R$ 1990 anomaly was considerably below that of 1987, showing clearly it could not have been the warmest year in the troposphere.

Figure 6 provides the filtered daily global, Northern and Southern hemispheric (NH and SH) anomalies of MSU2$_R$ since 1979. Overall, there has been a slight warming globally of +.03°C per decade in these 12+ years. The NH has experienced several warm spells beginning in 1987 and that is represented by an upward trend of about 0.1°C per decade. This is partially offset by a slight downward trend in the SH. What is clear from these time series is remarkable variability from one

MARCH 1990

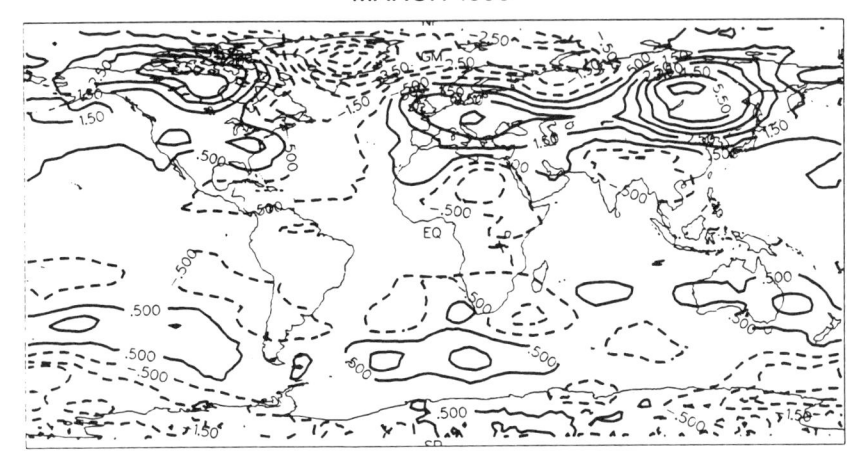

FIGURE 5. Anomalies in tropospheric temperature (MSU2$_R$) for March 1990. Though surface reports claimed this month was the all-time warmest in the past 100 years, the satellites show it to be cooler than fully 30% of the months monitored since 1978. Contours are (°C) −4.5, −3.5, −2.5, etc.

year to the next. These large fluctuations remind us that fitting trends to these data must be done with caution, especially when the result is so near to zero. For example, when the global trend in MSU2$_R$ is calculated for just one year less than the total (i.e. 1979-89) the result is actually negative. The important point to remember here is that since the global trend is not significantly different from zero, its value may change sign with the addition of just a few months worth of new data as it did from 1989 to 1990.

We wish to point out that those aspects of the T$_{sfc}$ observations that climatologists find undesireable (unsystematic measurement methods grid to grid, lack of consistency in instruments and sites through time, non-global coverage, urbanization, lack of independent verification) do not contaminate the MSU recored. Of course, it is imperative that the surface record not only be maintained, but expanded and standardized because surface processes are crucial for many human and climate problems. We do not want to minimize their value in any way. The disagreement between anomaly magnitude and decadal trends presented here could be a real phenomenon of interlayer variability about which we know little in the global sense. As such, it points out the uncertainty which faces the scientific community in understanding the climate system and in determining in which direction the temperature may be going.

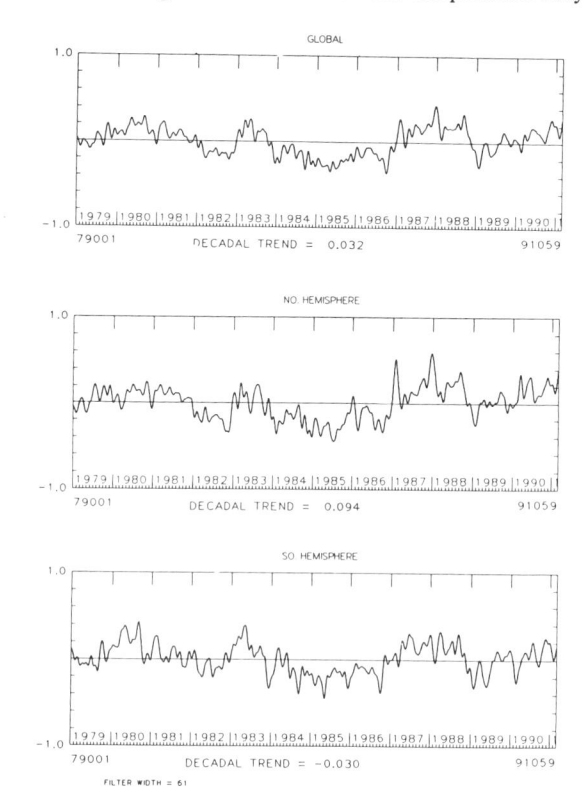

FIGURE 6. From top: Global, Northern and Southern Hemispheric daily filtered anomalies of MSU2$_R$ (°C). Note the large fluctuations that occur during periods of a month or two.

LOWER STRATOSPHERE

The temperatures for the lower stratosphere (MSU4) are given in Figure 7. The obvious global signal is the heating in 1982 caused by the volcanic eruptions in Africa and of El Chichon in Central America. Evidently these eruptions had such force and orientation that considerable aerosol was injected into the stable stratosphere. This material absorbed radiative energy and caused the air to experience a warming for about a year. These aerosols eventually settled out of the stratosphere, leaving the global temperature fairly constant since 1984.

If the stratosphere experiences an increase in carbon dioxide or other greenhouse gases, it is probable that this region would steadily cool down. The reasoning is that greenhouse gases are effective at radiating their thermal energy away, and at these altitudes, that energy would be lost to space. Trying to detect a downward trend in MSU4 due only to greenhouse gas concentrations is impossible because the volcanic impact swamps the smaller signals. However, since 1984 there has not been

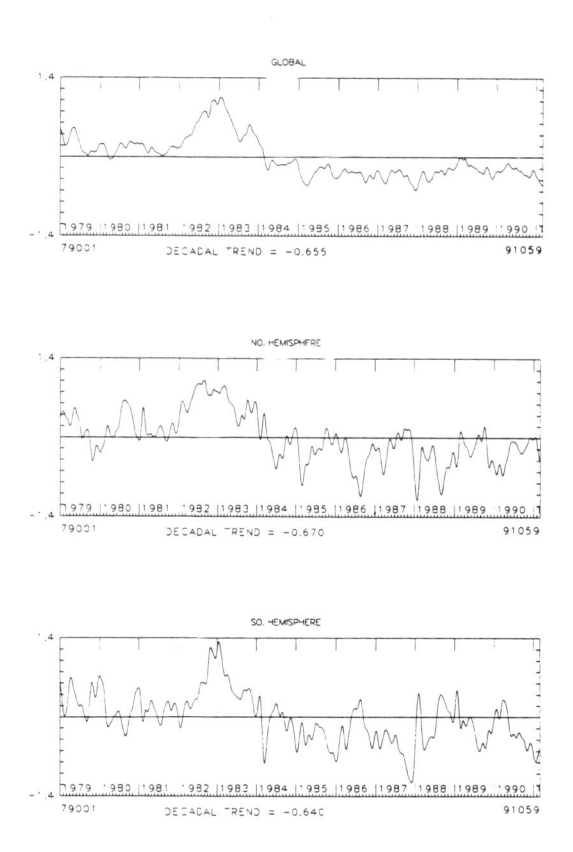

FIGURE 7. As in Figure 6 but for MSU4. The large positive anomaly in 1982 was due to heating from radiatively-active volcanic material injected into the stratosphere.

a downward trend, though we insist such a short record should not be used to deduce conclusions of longer-term trends.

The stratosphere is home to the ozone layer which absorbs solar ultraviolet radiation. Such absorption generates heat and higher temperatures. There is generally a strong positive correlation between ozone, near 24 km altitude, and MSU4, near 20 km, which allows one to monitor ozone changes with the MSU. In Figure 8 one sees the anomalies of MSU4 for five sections of the globe. The very cold anomaly in October 1987 in the South polar area (lowest time series) is due in part to the massive ozone depletion observed that year. Note also, however, that the warmest temperatures in the austral spring occurred the very next year. An examination of the NH polar region also shows large fluctuations (top section Figure 8). This indicates the very large variability that occurs in the polar stratospheric regions.

CONCLUSIONS

What is the ultimate value of the MSU data? Only those in the discipline of atmospheric science can appreciate meteorological data which are global in extent, consistent in measurement technique and stable for long periods. Actually, there are other global data sets in our community, but these do not have a consistently accurate base world-wide. When someone detects a change in a global variable,

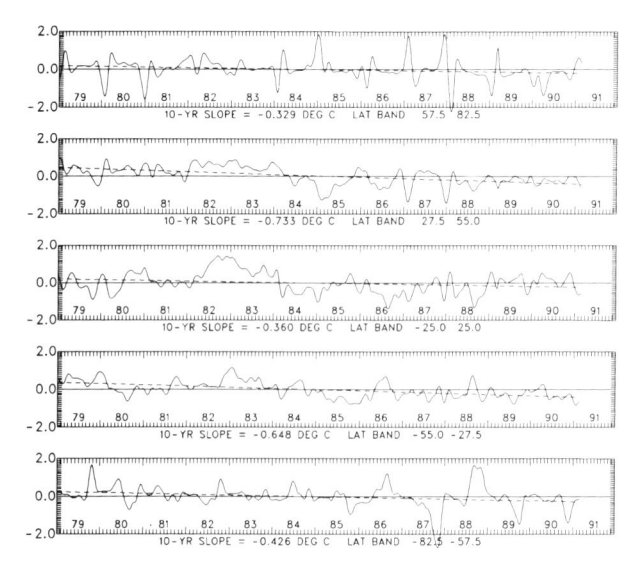

FIGURE 8. MSU4 temperature anomalies (°C) for five sections of the globe: north polar, north midlatitude, equatorial, south midlatitude and south polar. Note the massive cooling in October 1987 in the south polar region from ozone depletion.

say temperature in these data sets, the cause is often determined to be a spurious artifact due to a newly implemented analysis procedure. The MSU data provide accurate, global, deep-layer temperatures of that portion of the atmosphere from which should come the first signs of global warming (troposphere) and cooling (stratosphere). To date, the 12 + year $MSU2_R$ record has not shown a warming trend significantly different from zero. (We will be more confident in these trends when another decade is monitored.) Before governments launch their assaults on fossil fuel emissions (remember, carbon dioxide is a benign gas) they must weigh such evidence as provided from the satellites. There may be many worthwhile reasons for reducing carbon emissions; it is quite unclear, however, whether global warming is scientifically among those reasons.

A particularly important application of the MSU data is in comparisons with output from General Circulation Models (GCMs). These models attempt to mimic the global atmosphere by applying rules of physics that require the atmosphere to respond in a specific way to specific types of forcing. Do these GCMs replicate the real atmosphere well enough so that one may have confidence in a GCM prediction?

One experiment completed in 1990 compared the global tropospheric temperature produced from a GCM with that provided by MSU2 for nine years. The GCM was given actual sea surface temperatures (SSTs) month by month just as the real atmosphere would have perceived. The results indicated that when the SSTs became warm by natural fluctuations considerable heat was transferred to the model atmosphere, just as was observed in the real atmosphere. However, after the SSTs cooled down again, the real atmosphere lost its heat quite rapidly (see Figure 4) while the model continued to retain it. Why did the model not relinquish this heat? No doubt the answer relates to the clouds and radiation processes described very crudely in the model equations.

It is precisely problems such as this that may lead to completely erroneous results in an experiment as subtle as increasing the concentration of carbon dioxide in the model atmosphere. The GCMs have very strong rules which require the atmosphere to heat up when carbon dioxide is added. However, as demonstrated above, a GCM may not have the correct rules that allow the atmosphere to cool off as is observed in the real world.

This real world of ours is an interesting and very complex system. The way the atmosphere responds to increasing greenhouse gases (and no one doubts these gases are increasing) certainly will hold some surprises for us. Perhaps we will never be able to extract a greenhouse signal from the natural variability of the climate system. Perhaps the present levels of carbon dioxide will prevent the climate from descending into an ice age . . . who knows? The questions about global climate change are unanswered and enormously challenging. The answers will come from the examination of precision global data. In fact, most of the real advances in any science spring from investigations involving new and accurate data. The MSU dataset is one of several powerful resources being produced today just at a time in which we are asking these global environmental questions.

REFERENCES

1. Spencer, Roy W., John R. Christy and Norman C. Grody. 1990. Global atmospheric temperature monitoring with satellite microwave measurements: Method and results 1979-84. *J. Climate* 3:1111-1123.
2. Spencer, Roy W. and John R. Christy. 1992a. Precision and radiosonde validation of satellite gridpoint temperature anomalies, Part I: MSU channel 2. *J. Climate, 5*, in press.
3. Spencer, Roy W. and John R. Christy. 1992b. Precision and radiosonde validation of satellite gridpoint temperature anomalies, Part II: A tropospheric retrieval and trends during 1979-90. *J. Climate, 5*, in press.
4. Trenberth, Kevin E., John R. Christy and James Hurrell. 1992. Monitoring global monthly mean surface temperatures. *J. Climate, 5*, in press.

Global Climate Change: Implications, Challenges and Mitigation Measures. Edited by S.K. Majumdar, L.S. Kalkstein, B. Yarnal, E.W. Miller, and L.M. Rosenfeld. © 1992, The Pennsylvania Academy of Science.

Chapter Twelve

THE URBAN HEAT ISLAND: Contaminant to the Global Temperature Record?

ROBERT C. BALLING, JR.
Office of Climatology and
Department of Geography
Arizona State University
Tempe, Arizona 85287-1508

INTRODUCTION

The search for the global warming signal is complicated by a number of potential contaminants to the historical temperature records. Instrument changes, calibration errors, station relocations, observation errors, time of observation, microclimatic changes near the instruments are among the well-known problems that can produce spurious patterns in the temperature trends (Mitchell, 1953, 1958; Karl *et al.*, 1986; Karl and Williams, 1987; Karl and Quayle, 1988; Baker and Ruschy, 1989; Karl *et al.*, 1989). However, while each of these problems must be carefully considered in the search for the greenhouse signal, they are dwarfed somewhat by the potential impact on the temperature record caused by anthropogenic-induced warming associated with urban heat islands. Recognizing the problem, a number of investigators have attempted to disentangle the urban heat island effect from the historical land-based temperature records of the globe. The purpose of this chapter is to review the literature on how much of the "global warming" of the past century can be attributed to the heat island effect.

PHYSICAL CAUSES OF URBAN HEAT ISLANDS

The physical causes for localized warming in urban areas have been studied vigorously for more than a century (e.g., Howard, 1833), and these causes are reasonably well understood. Outstanding reviews of the urban heat island processes include, among many others, Oke (1973, 1979, 1982), Garstang et al. (1975), Landsberg (1981), Goward (1981), Lee (1984), Cayan and Douglas (1984), Brazel (1987), and Karl et al. (1988). Although the city size, morphology, relief, elevation, and regional climate will determine the intensity, persistence, and configuration of the heat island, several generalizations can be made regarding the causes of the localized warming.

Possibly the principal reason for the development of the heat island is the waterproofing of the urban surface. In many cities, the natural vegetation is largely removed and surface is covered by nearly impervious materials. Precipitation quickly runs off the urban surface into underground storm sewers, and the resulting surface and near-surface moisture in the city is minimized. Given reduced quantities of moisture, large portions of the net radiation at the surface go toward heating the surface and air as opposed to evaporating or transpiring water. The result is a substantial increase in local temperatures.

A second major reason for the development of the heat islands is the lower albedo (reflectivity) of the urban landscape. Although the physical characteristics of the urban mosaic may be extremely complex (Marotz and Coiner, 1973), the urban structures typically replace natural surfaces of higher albedos. The $\approx 10\%$ decrease in the albedo of the city compensates for a reduction of incoming radiation from the sun that may be lost due to locally high concentrations of atmospheric pollution. In addition, the complex urban geometry can create a greater absorption of incoming shortwave radiation due to the variety of exposures relative to the rays of the sun.

The thermal properties of building materials in urban areas may also provide an explanation for the increase in local temperatures (Goward, 1981). Some urban interface materials are able to store more heat during the daytime than would be stored by dry soils. This heat that is stored in the daytime is then released during the nighttime period providing the energy for maintaining higher nocturnal temperatures. However, because water has a particularly high heat capacity (and thermal inertia), the availability of surface moisture in the urban and rural environments can dominate the role of thermal properties in explaining the existence of an urban heat island.

The geometry of the urban mosaic provides yet another explanation for the high surface and near-surface air temperatures. The urban canyons reduce the loss of long-wave terrestrial radiation by narrowing the sky-view factor (Arnfield, 1982). The buildings of the city increase the surface friction thereby reducing wind speeds; this reduction in wind speed creates a convergence of sensible heat within the urban environment. Other causes for the heat island include the actual release of heat from a variety of anthropogenic activities within the city and the trapping of long-wave terrestrial radiation by low-level atmospheric aerosols. These, and other, mechanisms for creating the observed urban heat islands are discussed in considerable detail by Oke (1979) and Brazel (1987).

THE PHOENIX, ARIZONA URBAN HEAT ISLAND

Phoenix, Arizona provides a tremendous example of the climate impact of rapid urbanization. Phoenix population has grown very quickly in recent decades going from approximately 700,000 in 1960 to nearly two million residents in 1990. Unlike many other cities, this period of rapid growth took place during a time when excellent weather observations were taken throughout the expanding metropolitan area. In addition, much of the city is growing into once irrigated areas thereby reducing the surface and sub-surface moisture levels. The particularly strong solar loads of the desert setting accentuates the impact of the varying thermal properties of the city. Accordingly, Phoenix has produced a classic heat island that has been studied intensely over the past few decades.

A plot of mean annual temperatures from 1960 to 1990 reveals the large rise in temperature that has occurred during this period of rapid population growth (Figure 1). As can be seen there, temperatures in Phoenix show a rise of $0.042°C$ yr^{-1} over the entire 1931-1990 time period; however, the plot clearly shows that the significant warming began in the early 1960s and continued to the present. The linear rise in temperature from 1960-1990 is $0.125°C$ yr^{-1} yielding a linear increase of $3.75°C$ over that time period. The background temperature, as measured by a number of rural locations in Arizona, has shown only a small warming of less than $0.5°C$ over the recent period of rapid growth in Phoenix (Balling and Idso, 1989, 1990).

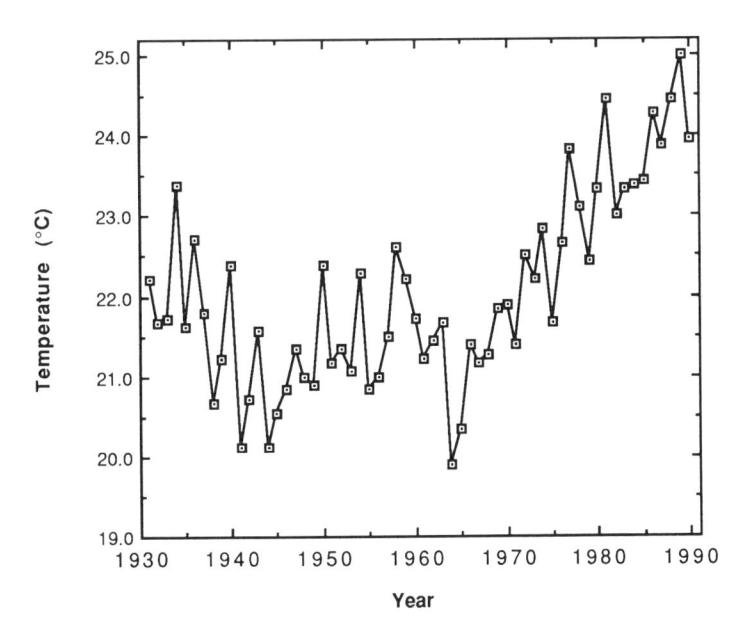

FIGURE 1. Phoenix, Arizona Mean Annual Temperatures (°C), 1931-1990.

Indeed, maps and transects of the Phoenix heat island produced with temperature records from around the city and satellite-based sensor systems (Hsu, 1984; Balling and Brazel, 1987c, 1988, 1989) show the warming to be largely confined to the growing metropolitan area. Therefore, the bulk of the warming shown in Figure 1 most likely comes from the local urban heat island effect in Phoenix.

A number of papers have been written providing details on the changes in temperature observed in the Phoenix area. Balling and Brazel (1986a, 1986b) showed that the bulk of the heat island induced warming in Phoenix occurs at night; the rise in temperature at 0200 LST was found to be 3.5 times larger than the rise in temperature observed at 1400 LST. The temperature increases were as much as five times larger in the summer season when compared to temperature increases in the winter season (Balling and Cerveny, 1987). While temperatures were rising, the dew point levels were falling at a significant rate as irrigated areas were eliminated in the expanding metropolitan area (Brazel and Balling, 1986). Not surprisingly, the significant decline in dew points began in the early 1960s when population growth accelerated in the Phoenix area. The decline in dew points, coupled with the rise in temperatures, produced a particularly large drop in the relative humidity levels recorded within the city.

The impact of the urban growth has many other observable effects in the Phoenix area. The stronger temperature gradient between the city center and the surrounding rural area appears to produce a significant increase in local wind speeds despite the increase in the surface friction. Throughout much of the year, early morning (0500 LST) wind speeds increased by more than 50% during the period 1948-1985 (Balling and Cerveny, 1987). Given the increase in temperatures, increase in wind speeds, and decrease in dew points and relative humidity levels, evaporation rates in the city increased by 30% when a period 1969-1985 was compared to a period 1917-1968 (Balling and Brazel, 1987a). The heat island also appeared to significantly affect the movement, development, and maintenance of summertime convective storms in the Phoenix area (Balling and Brazel, 1987b). Like nearly every other city of the world, urban growth in Phoenix has produced a substantial impact on the local weather and climate; in particular, the growth of this city has produced an enormous increase in local temperatures.

HEAT ISLANDS AND GLOBAL CHANGE

The example from Phoenix rather clearly illustrates how growth in a city can create a rise in temperature that may be confused as a greenhouse warming signal. Given the explosive growth in the cities in many parts of the world, investigators have attempted to disassociate the heat island warming from the land-based temperature records from any warming that may be resulting from the build-up of greenhouse gases. Fortunately, researchers are approaching an agreement on the level of heat-island contamination to the global temperature record.

Kukla *et al.* (1986) analyzed the temperature records over the period 1941-1980 for 34 carefully selected urban-rural "pairs" in the United States. They concluded that an urban warming bias of 0.12°C/decade was present in the data; this led the authors to speculate about the level of contamination that may be present in hemispheric and/or global temperature trends. However, the authors did not attempt to quantify the heat island signal in these hemispheric or global data bases.

Hansen and Lebedeff (1987) recognized the urban heat island problem in the preparation of their global temperature data base. They calculated the trend in global temperature with all records, then they calculated the trends without any cities of more than 100,000 residents. The difference in the trends revealed a bias of 0.1°C over a 100-year period. Hansen and Lebedeff "subjectively estimated" another 0.1°C heat island bias in the remaining data; they conceded that their data contained an urban heat island contamination of approximately 0.2°C per century. Their unadjusted data showed a 0.7°C trend over the past 100 years — the adjustment for urban warming yields a true global warming signal of approximately 0.5°C over the past century.

Karl et al. (1988) extended the basic analyses of Kukla et al. (1986) to evaluate the degree of urban warming bias contained within the United States Historical Climatology Network (HCN) of over 1200 stations (Quinlan et al., 1987). They computed trends in the temperature differences between rural and urban locations and related the differential trends to the population within the urban center. Using a 1901-1984 time period, Karl et al. (1988) found an urban heat island bias of between 0.05°C and 0.10°C in the HCN data. Because most of the HCN stations are in rural locations (the 1980 median population of the network locations was 5,832), the urban bias was considerably less than the bias determined in the Kukla et al. (1986) study. Karl et al. (1988) also showed that the urban bias was much greater for the minimum temperatures and almost negligible for the maximum temperatures.

Balling and Idso (1989) performed a similar analysis on the HCN data. Using the period 1920-1984, they calculated a linear temperature trend for each station for annual and seasonal maximum, minimum, and mean temperatures. Recognizing that the *change* in population, and not the total population, will initiate the differential temperature trend, they calculated the difference in population between 1920 and 1980 for each of the HCN locations. Stepwise regression analysis was used to link the spatial variance in the temperature trend with geographic variables (e.g., latitude, longitude) and the proportional change in population, defined as $\Delta P = (P_{1980} - P_{1920})/P_{1920}$. Their stepwise linear regression model showed that the ΔP term is statistically significant in explaining spatial variance in temperature trends; they concluded that a 0.05°C correction was required for the largely rural HCN annual temperature trends. The ΔP term was not significant for maximum temperatures, but highly significant for minimum temperatures. For example, the urban heat island bias in the annual minimum temperatures was 0.11°C over the 1920-1984 period and 0.16°C for winter minimum temperatures.

Balling and Idso (1989) linked their findings to the Hansen and Lebedeff (1987) global data base. By comparing the population-adjusted temperature trends from the eastern United States to the temperature trends from the Hansen and Lebedeff data for the identical area, a bias of 0.37°C over the 1920-1984 period was noted in the Hansen and Lebedeff time series. Balling and Idso (1989) concluded that the Hansen and Lebedeff (1987) temperature data must have a significant urban heat island bias that could account for much of the observed "global warming."

In the same year, Jones et al. (1989) published an article comparing the Karl et al. (1988) urban-adjusted data to the widely-used Jones et al. (1986a, 1986b, 1986c) hemispheric and global gridded temperature data. By comparing the gridded data for the United States to the Karl et al. (1988) adjusted data the contiguous United States, Jones et al. (1989) computed an urban bias in their data of 0.08°C over the period 1901-1984. Arguing that no more than 40% of the Northern Hemispheric landmass is likely to have a similar bias, Jones et al. (1989) concluded that the level of heat island bias in the Jones et al. (1986a) temperatures for the Northern Hemisphere is probably between 0.01°C and 0.10°C; a value near 0.05°C was considered most reasonable.

Karl and Jones (1989) performed a similar analysis for both the Jones et al. (1986a) data and the Hansen and Lebedeff (1987) data. Again, the Karl et al. (1988) urban-adjusted data for the United States were used as a baseline. For the period 1901-1984, the Jones et al. (1986a) temperature data revealed a 0.11°C bias while the Hansen and Lebedeff (1987) data showed a 0.32°C bias. Given a global trend in temperature of approximately 0.5°C over the same time period, the urban heat island bias appeared to produce an important part of the overall trend. However, the work was based on the United States data, and many questions remained about the overall representativeness of urbanization in this highly developed, mid-latitude landmass.

The Karl et al. (1988) analysis of matched urban-rural pairs was extended for other parts of the globe by Jones et al. (1990). Their results showed a heat island bias of 0.15°C for the United States over the period 1901-1984. However, in western U.S.S.R., the bias was essentially 0°C for the 1901-1987 period but near 0.12°C for 1930-1987. Eastern Australia showed a bias of 0.04°C for 1930-1988 and eastern China actually revealed a "bias" of -0.04°C from 1954-1983. They concluded that the urban bias is probably near 0.05°C for the entire globe for this century, and they noted that this value is fully an order of magnitude less than the observed temperature trend over the same time period.

CONCLUDING REMARKS

The debate (e.g., Wood, 1988; Wigley and Jones, 1988) regarding urban heat island contamination in the global temperature records is likely to continue into the immediate future. However, at this point, one could conclude that (a) the Hansen and Lebedeff (1987) temperature data are seriously affected by urban warming, (b) the Jones et al. (1986a, 1986b, 1986c; Jones, 1988) data set has a global urban warming bias somewhere between 0.01°C and 0.10°C, with the most likely value near 0.05°C, and (c) the heat island bias in the Jones et al. data is likely to be an order of magnitude less than the overall trend. The recent data set introduced by Vinnikov et al. (1990) appears to be similar to the Jones et al. data.

With respect to the greenhouse debate, one may argue that a 0.05°C urban warming bias in this century is relatively unimportant given a total global trend near

0.5°C over the same time period. However, the 0.05°C urban-related warming in the global temperature record represents only one portion of the overall trend. Balling (1991) has shown that desertification may be producing another 0.05°C to 0.15°C non-greenhouse related warming signal in the land-based temperature data. Others (e.g., Wu *et al.*, 1990) have shown that atmospheric turbidity can explain more than 0.1°C of the temperature trend of the past century; the role of solar variability in explaining any significant portion of the trend is still debated (e.g., Gilliland, 1982; Newell *et al.*, 1989; Hansen and Lacis, 1990; Kelly and Wigley, 1990; Wigley and Raper, 1990; Wu *et al.*, 1990). Many other factors, including the potentially important moderating role of aerosol sulfates (Charlson *et al.*, 1987, 1990; Schwartz, 1988; Wigley, 1989), must be resolved. Nonetheless, it is relatively clear that when the "global warming" of the past century is partitioned in these various categories, the urban warming portion remains very important and may be on the same order of magnitude as the warming trend uniquely caused by the build-up of greenhouse gases.

REFERENCES

Arnfield, A.J. 1982: An approach to the estimation of surface radiative properties and radiation budget of cities. *Physical Geography,* 3, 97-122.

Baker, D.G. and D.L. Ruschy, 1989: Temperature measurements compared. *The State Climatologist,* 13, 2-5.

Balling, R.C., Jr., 1991: Impact of desertification on regional and global warming. *Bulletin of the American Meteorological Society,* 72, in press.

Balling, R.C., Jr. and S.W. Brazel, 1986a: 'New' weather in Phoenix? Myths and realities. *Weatherwise,* 39, 86-90.

Balling, R.C., Jr. and S.W. Brazel, 1986b: Temporal analysis of summertime weather stress levels in Phoenix, Arizona. *Theoretical and Applied Climatology,* 36, 331-342.

Balling, R.C., Jr. and S.W. Brazel, 1987a: The impact of rapid urbanization on pan evaporation in Phoenix, Arizona. *Journal of Climatology,* 7, 593-597.

Balling, R.C., Jr. and S.W. Brazel, 1987b: Recent changes in Phoenix, Arizona summertime diurnal precipitation patterns. *Theoretical and Applied Climatology,* 38, 50-54.

Balling, R.C., Jr. and S.W. Brazel, 1987c: Time and space characteristics of the Phoenix urban heat island. *Journal of the Arizona-Nevada Academy of Science,* 21, 75-81.

Balling, R.C., Jr. and S.W. Brazel, 1988: High-resolution surface temperature patterns in a complex urban terrain. *Photogrammetric Engineering and Remote Sensing,* 54, 1289-1289.

Balling, R.C., Jr. and S.W. Brazel, 1989: High-resolution nighttime temperature patterns in Phoenix. *Journal of the Arizona-Nevada Academy of Science,* 23, 49-53.

Balling, R.C., Jr. and R.S. Cerveny, 1987: Long-term associations between wind speeds and the heat island of Phoenix, Arizona. *Journal of Climate and Applied Meteorology,* 26, 712-716.

Balling, R.C., Jr. and S.B. Idso, 1989: Historical temperature trends in the United States and the effect of urban population growth. *Journal of Geophysical Research,* 94, 3359-3363.

Balling, R.C., Jr. and S.B. Idso, 1990: Confusing signals in the climatic record. *Atmospheric Environment,* 24A, 1975-1977.

Brazel, A.J., 1987: Urban climatology. *The Encyclopedia of Climatology,* Oliver, J.E. and R.W. Fairbridge, Ed., Van Nostrand Reinhold Co., 889-901.

Brazel, S.W. and R.C. Balling Jr., 1986: Temporal analysis of long-term atmospheric moisture levels in Phoenix, Arizona. *Journal of Climate and Applied Meteorology,* 25, 112-117.

Cayan, D.R. and A.V. Douglas, 1984: Urban influences on surface temperatures in the southwestern United States during recent decades. *Journal of Climate and Applied Meteorology,* 23, 1520-1530.

Charlson, R.J., J. Langner and H. Rodhe, 1990: Sulphate aerosol and climate. *Nature,* 345, 22.

Charlson, R.J., J.E. Lovelace, M.O. Andrea and S.G. Warren, 1987: Oceanic phytoplankton, atmospheric sulfur, cloud albedo and climate. *Nature,* 326, 655-661.

Garstang, M., P.D. Tyson and G.D. Emmitt, 1975: The structure of heat islands. *Reviews of Geophysics and Space Physics,* 13, 139-165.

Gilliland, R.L., 1982: Solar, volcanic, and CO_2 forcing of recent climate changes. *Climate Change,* 4, 111-131.

Goward, S.N., 1981: Thermal behavior of urban landscapes and the urban heat island. *Physical Geography,* 2, 19-33.

Hansen, J. and S. Lebedeff, 1987: Global trends of measured surface air temperature. *Journal of Geophysical Research,* 25, 13345-13372.

Hansen, J.E. and A.A. Lacis, 1990: Sun and dust versus greenhouse gases: An assessment of their relative roles in global climate change. *Nature,* 346, 713-719.

Howard, L., 1833: *The Climate of London.* Harvey and Darton.

Hsu, S.I., 1984: Variation of an urban heat island in Phoenix. *The Professional Geographer,* 36, 196-200.

Jones, P.D., 1988: Hemispheric surface temperature variations: Recent trends and an update to 1987. *Journal of Climate,* 1, 654-660.

Jones, P.D., P.Y. Groisman, M. Coughlan, N. Plummer, W.-C. Wang and T.R. Karl, 1990: Assessment of urbanization effects in time series of surface air temperatures over land. *Nature,* 347, 169-172.

Jones, P.D., P.M. Kelly, C.M. Goodess and T. Karl, 1989: The effect of urban warming on the Northern Hemispheric temperature average. *Journal of Climate,* 1, 285-290.

Jones, P.D., S.C.B. Raper, R.S. Bradley, H.F. Diaz, P.M. Kelly and T.M.L. Wigley, 1986a: Northern hemispheric surface air temperature variations: 1851-1984. *Journal of Climate and Applied Meteorology,* 25, 161-179.

Jones, P.D., S.C.B. Raper and T.M.L. Wigley, 1986b: Southern hemispheric surface air temperature variations: 1851-1984. *Journal of Climate and Applied Meteorology,* 25, 1213-1230.

Jones, P.D., T.M.L. Wigley and P.B. Wright, 1986c: Global temperature variations

between 1861 and 1984. *Nature,* 322, 430-434.
Karl, T.R., H.F. Diaz and G. Kukla, 1988: Urbanization: Its detection and effect in the United States climatic record. *Journal of Climate,* 1, 1099-1123.
Karl, T.R. and P.D. Jones, 1989: Urban bias in area-averaged surface air temperature trends. *Bulletin of the American Meteorological Society,* 70, 265-270.
Karl, T.R. and R.G. Quayle, 1988: Climatic change in fact and theory. Are we collecting the facts? *Climatic Change,* 13, 5-17.
Karl, T.R., J.D. Tarpley, R.G. Quayle, H.F. Diaz, D.A. Robinson and R.S. Bradley, 1989: The recent climate record: What it can and cannot tell us. *Reviews of Geophysics,* 27, 405-430.
Karl, T.R. and C.N. Williams Jr., 1987: An approach to adjusting climatological time series for discontinuous inhomogeneities. *Journal of Climate and Applied Meteorology,* 27, 1744-1763.
Karl, T.R., C.N. Williams Jr., P.J. Young and W.M. Wendland, 1986: A model to estimate the time of observation bias with monthly mean maximum, minimum, and mean temperatures for the United States. *Journal of Climate and Applied Meteorology,* 25, 145-160.
Kelly, P.M. and T.M.L. Wigley, 1990: The influence of solar forcing trends on global mean temperature since 1861. *Nature,* 347, 460-462.
Kukla, G., J. Gavin and T.R. Karl, 1986: Urban warming. *Journal of Climate and Applied Meteorology,* 25, 1265-1270.
Landsberg, H.E. 1981: *The Urban Climate.* Academic Press.
Lee, D.O., 1984: Urban climates. *Progress in Physical Geography,* 8, 1-31.
Marotz, G.A. and J.C. Coiner, 1973: Acquisition and characterization of surface material data for urban climatological studies. *Journal of Applied Meteorology,* 12, 919-923.
Mitchell, J.M., Jr., 1953: On the causes of instrumentally observed secular temperature trends. *Journal of Meteorology,* 10, 244-261.
Mitchell, J.M., Jr., 1958: Effect of changing observation time on mean temperatures. *Bulletin of the American Meteorological Society,* 39, 83-89.
Newell, N.E., R.E. Newell, J. Hsiung and Z. Wu, 1989: Global marine temperature variation and the solar magnetic cycle. *Geophysical Research Letters,* 16, 311-314.
Oke, T.R., 1973. City size and the urban heat island. *Atmospheric Environment,* 7, 769-779.
Oke, T.R., 1979: Review of Urban Climatology, 1973-1976. WMO Tech. Note No. 169, Geneva, Switzerland.
Oke, T.R., 1982: The energetic basis of the urban heat island. *Quarterly Journal of the Royal Meteorological Society,* 108, 1-24.
Quinlan, F.T., T.R. Karl and C.N. Williams Jr., 1987: United States historical climatology network (HCN) serial temperatures and precipitation data. NDP-019, Carbon Dioxide Information Analysis Center, Oak Ridge National Laboratory, Oak Ridge, Tennessee, 32 pp.
Schwartz, S.E., 1988: Are global cloud cover and climate controlled by marine phytoplankton? *Nature,* 336, 441-445.
Vinnikov, P.Y., P.Y. Groisman and K.M. Lugina, 1990: Empirical data on

contemporary global climate changes (temperature and precipitation). *Journal of Climate,* 3, 662-667.

Wigley, T.M.L., 1989: Possible climate change due to SO_2-derived cloud condensation nuclei. *Nature,* 338, 365-367.

Wigley, T.M.L. and P.D. Jones, 1988: Do large-area-average temperature series have an urban warming bias? *Climate Change,* 12, 313-319.

Wigley, T.M.L. and S.C.B. Raper, 1990: Climate change due to solar irradiance changes. *Geophysical Research Letters,* 17, 2169-2172.

Wood, F.B., 1988: Comment: On the need for validation of the Jones *et al.* temperature trends with respect to urban warming. *Climatic Change,* 12, 297-312.

Wu, Z., R.E. Newell and J. Hsiung, 1990: Possible factors controlling global marine temperature variations over the past century. *Journal of Geophysical Research,* 95, 11799-11810.

Global Climate Change: Implications, Challenges and Mitigation Measures. Edited by S.K. Majumdar, L.S. Kalkstein, B. Yarnal, E.W. Miller, and L.M. Rosenfeld. © 1992, The Pennsylvania Academy of Science.

Chapter Thirteen

GENERAL CIRCULATION MODEL STUDIES OF GLOBAL WARMING

ROBERT G. CRANE
Earth System Science Center and
Department of Geography
The Pennsylvania State University
University Park, PA 16802

INTRODUCTION

Predictions of global climate change can take various forms: they may be conceptual in nature, where knowledge of how the climate system operates is used to infer its possible response to change; they may be made by analogy to past climates; or they may be derived from numerical models of the climate system. Conceptual models are useful in formulating research design, but, given the complexity of the climate system, they cannot provide a quantitative assessment of the way in which the system will respond to a particular forcing. Analogies to past climates, on the other hand, can often produce both high resolution and quantitative descriptions of the possible distributions of climate change, providing good historic or paleoclimatic data are available. However, since the factors that gave rise to the past climate are very different to those we are concerned with for the future, there is no guarantee that the climate system will respond in the same way (Crowley, 1990, Mitchell, 1990). The alternative is to use numerical climate models that replicate features of the present climate, and examine how these models respond to change. At one end of the modeling spectrum these are simple zero- or one-dimensional models that calculate the energy balance at a point, or in a column, while at the other extreme are coupled three-dimensional models of oceanic and atmospheric dynamics (although few coupled models are available for CO_2 studies at this time).

Three-dimensional General Circulation Models (GCMs) of the atmosphere solve the primitive equations of energy, mass, and momentum at the Earth's surface, and at various heights in the atmosphere. The models include the effects of atmospheric gases and clouds; surface-atmosphere interactions as a function of surface type, elevations, soils, and vegetation; surface hydrology; and some interactions with the ocean surface. Most models include the effects of snow cover over the land and sea ice on the oceans, and attempts are currently being made to couple atmospheric GCMs to models of the oceanic circulation. The present generation of atmospheric GCMs includes annual and diurnal cycles.

These GCMs reproduce many of the large-scale features of the climate system. Their sophistication is such that sensitivity analyses of the model parameters can provide considerable insight to the workings of the present-day climate system, and they have also been used to analyze past climate conditions (e.g. Kutzbach, 1981; Barron and Washington, 1982). GCMs provide the best means presently available for estimating the probable climatic effects of increasing atmospheric CO_2 concentrations.

There are on the order of 8-10 different GCMs regularly used for climate and climate change analysis. Cess (1990) lists about 20 model experiments, but several are with the same model, or different versions of these models. Of the GCMs that have been used in doubled CO_2 experiments, almost all agree on a $+2.5°C$ to $+4.5°C$ global mean surface air temperature change. Most models also agree on the large-scale distribution of this change, with the warming being greatest at the higher latitudes of the winter hemisphere. Despite the broad agreement between the model results, however, it is recognized that these models do not contain all of the relevant physics. Several model intercomparison studies have demonstrated large regional discrepancies between different model predictions (e.g. Schlesinger and Mitchel, 1987; Grotch et al., 1991). Nevertheless, these models provide the focus for most CO_2 climate change predictions, and GCMs form an integral part of the U.S. Global Change Program (Committee on Earth Sciences, 1989). Consequently, this chapter will provide some background into the development of atmospheric GCMs, outline some of the results from different doubled-CO_2 experiments, and discuss some of the problems and limitations of current modeling efforts. A simple description of climate models and recent results from CO_2 climate change experiments can also be found in Houghton et al. (1990). For those wishing a more detailed description of three-dimensional climate modeling, there is an excellent text by Washington and Parkinson (1986) from which much of the background material and the description of GCM techniques presented here is drawn.

BACKGROUND

GCMs had their beginnings in early attempts at numerical weather prediction, the most notable of which was by Lewis Fry Richardson during World War I. Richardson was an ambulance driver who, in order to forecast the weather, spent part of his off-duty time solving the basic equations of atmospheric motion . . .

using a mechanical calculator! The results were presented in the form of a book *Weather Prediction by Numerical Processes*, in which he attempted to predict the weather for a small area within Europe (Richardson, 1922). Little progress was made following this pioneering work until the development of electronic computers in the 1940s. Interestingly, weather prediction was then one of the first problems addressed by Princeton's Institute for Advanced Studies, using their newly developed computer under the direction of John von Neumann. The weather prediction team was led by Jule Charney, and these early results were presented in Charney *et al.* (1950).

The transition from weather prediction models to climate models began in 1956 when Phillips reproduced many of the features of the large-scale circulation in a long-term simulation that became independent of the initial state (Phillips, 1956). The mid-1950s also saw the beginnings of several new modeling groups, including a group at the U.S. Weather Bureau under the direction of Joseph Smagorinsky. This group continues now as part of the National Oceanographic and Atmospheric Administration (NOAA) at the Princeton University Geophysical Fluid Dynamics Laboratory (GFDL). Independent modeling groups later developed at the Goddard Institute for Space Studies (GISS) and the National Center for Atmospheric Research (NCAR). Numerous other GCM modeling groups exist, but these three are the ones most frequently encountered in the climate change literature, together with the United Kingdom Meteorological Office Model (UKMO), and models from the Max Planck Institute in Germany, the Canadian Climate Center, and the French Weather Service. One point to note here is that not only are there many different GCMs, but there are also numerous versions of each one. In some cases these represent an evolutionary development, while in others—the NCAR Community Climate Model (CCM) in particular—several versions of the model have been developed in parallel by different investigators. As all of these models differ from one another to some extent, it is important to note which model is being referred to in any particular analysis of the GCM results.

GCM development continued in the 1960s with the beginnings of ocean models, and in the 1970s and 1980s with increasingly sophisticated ocean, sea ice, surface hydrology, and biospheric components added to the atmospheric GCMs. During this same time period the models developed to include a seasonal cycle (and later a diurnal cycle), and from using prescribed cloud distributions to include both large-scale and convective cloud prediction schemes. These changes have also been accompanied by steadily increasing spatial resolution: earlier GCMs typically had grid sizes of 8° latitude by 10° longitude, while present GCMs are often run at a resolution of approximately 2.5° x 3.5°, or even greater.

MODEL BASICS

General Circulation Models of the atmosphere solve the equations for the conservation of energy, mass, and momentum; an equation for moisture; and an equation of state that relates pressure, density, and temperature. A set of boundary conditions is established that determines geography (surface type and elevation), and the seasonal cycle of insolation. The appropriate gas composition of the atmosphere for the time-period of interest is determined, although some model experiments allow this to change through time (to simulate the climate response of a continuously changing level of atmospheric CO_2, for example). The equations are initialized in space and integrated forward in time until they become independent of the initial state, when they reach an equilibrium representative of the mean climate for that given set of boundary conditions.

The equations are expressed as a set of spatial and temporal derivatives that can be solved through a variety of numerical differencing techniques. The horizontal grid spacing, and the speed with which signals propagate through the atmosphere, constrains the length of the time-step that can be used, with the time-step decreasing as the horizontal resolution increases (this is important for determining model resolution, as we will see later). These equations can also be expressed as wave functions—based on latitude and longitude—and solved using spectral techniques. In this case, the spatial resolution is determined by the series truncation. Approximate horizontal resolutions for several common truncations are listed in Table 1.

As with the horizontal representations, there are several ways of representing the vertical structure of the atmosphere in these models. Most of the current generation of atmospheric GCMs employ sigma coordinates, where sigma is the ratio of the pressure to the surface pressure. Regardless of the method used to solve the horizontal equations, the models use numerical differencing techniques in the vertical. The number of levels varies from model to model, but most will have a level defining the boundary layer and several layers in the stratosphere. Most models will reserve the greatest vertical resolution for the troposphere.

The objective when developing an atmospheric GCM is to include all of the relevant atmospheric physics and the surface-atmosphere exchange processes necessary to simulate accurately the state of the atmosphere at the model resolution. In order to do this, it is also necessary to account for some processes that operate at spatial and temporal scales smaller than the model is able to resolve directly. These processes are included in the model in the form of empirical parameterizations. In other words, the process is represented by an empirical or statistical relationship derived from observations, rather than a numerical representation of the physics involved. GCMs, therefore, are not only characterized by differences in resolution, but also by the processes that are modeled versus those that are parameterized, and by the level of sophistication of the parameterization schemes used.

It is apparent that any increase in model resolution requires more computer

TABLE 1

Approximate Horizontal Resolutions of Different Spectral Truncation Schemes

Truncation	Latitudes	Longitudes	Gridpoints	Spacing
R15	40	48	1,920	4.5°x7.5°
T21	32	64	2,048	5.6°
T42	64	128	8,192	2.8°
T63	96	192	18,432	1.9°

resources. Similarly, replacing an empirical parameterization with a set of physical equations that has to be solved at each grid-point, or simply increasing the sophistication of any parameterization scheme, will require increased computer resources. To put this into context, it takes about 50 hours on a Cray X-MP to carry out a 15-year simulation with a seasonal cycle GCM. It also takes about 15 model years for an atmospheric GCM to reach a quasi-equilibrium state (and considerably longer if the model is coupled to an ocean circulation model). To run the model for thirty years in a quasi-equilibrium state to examine the model's climate variability, works out to about 150 hours on the Cray for a single climate run. A doubling of the resolution on a grid-point model results in a fourfold increase in the number of necessary calculations. But you must also halve the time-step that can be used—again, doubling the number of computations needed. Furthermore, any appreciable increase in horizontal resolution almost invariable requires an increase in vertical resolution as well. Consequently, modelers are very concerned with questions of model resolution, and with determining the processes and level of treatment necessary to obtain a useful solution. This is not a simple task, partly because often it is not known what is important to include and what is not, and partly because what may satisfy one application may not suit another. GCM design and development, therefore, are processes of constant compromise. Because each modeling group approaches this in a different way, it makes it that much more difficult for an "outsider" to interpret the results of any particular model experiment.

There is not the space available here to detail the similarities and differences of all of the GCMs in use today, but a few general observations can be made. As with the real world, the model climate is driven largely by the radiation budget; all present GCMs simulate the annual solar cycle, and many include the diurnal cycle as well. The shortwave and longwave radiation fluxes are calculated over a range of spectral intervals, which vary from model to model, and absorption and emission are calculated for each band at each level in the model atmosphere. In the absence of clouds, the solar radiation is calculated for a Rayleigh scattering atmosphere as a function of the solar zenith angle, the surface albedo, and ozone and water vapor absorption. The longwave absorption and emission are calculated as a function of wavelength for water vapor, carbon dioxide, and ozone (the primary gases responsible for longwave absorption and emission in the atmosphere).

The largest single factor that modifies the radiation budget, however, is the cloud cover. Clouds are an important feature of both the observed and the GCM climates, but they are also one of the features for which we have little reliable observational data, and for which GCM parameterizations are highly simplistic. The influence of clouds on climate is particularly complicated. Clouds reflect a large portion of the incoming solar energy, which results in a cooling effect on climate. At the same time, they absorb the upward longwave loss from the surface and re-emit longwave radiation both upward and downwards, thus reducing the longwave emission to space. In this capacity, clouds have a warming effect on climate. The net radiative effect of the global cloud cover is difficult to determine from the limited observational data. It would appear from observational and modeling studies that the net effect is a cooling (e.g. Ramanathan *et al.*, 1989), but this varies both regionally and with the height of the cloud in the atmosphere (high clouds tend to warm the system, low clouds to cool it). Consequently, it is very important that the effects of the cloud cover should be accurately simulated by GCMs used for climate change purposes. Cloud cover was a fixed quantity in earlier models, based on the present-day observed distribution, and CO_2 climate change experiments using these models resulted, for example, in a 4.1°C mean global temperature increase for a quadrupling of atmospheric CO_2 (Manabe and Stoufer, 1980). Allowing the cloud cover to change by including a cloud prediction scheme in the same model doubled the sensitivity of the model to a CO_2 increase, producing a 4°C increase for only a doubling of CO_2 (Wetherald and Manabe, 1986).

All current GCMs now predict the cloud cover, but, as noted above, the prediction schemes are still highly simplistic. A description of cloud parameterization schemes and an extensive discussion of the sensitivity of atmospheric GCMs to the cloud parameterizations used is given by Cess et al. (1990). Most models will generate both large-scale (stratiform) clouds, and convective clouds. The prediction schemes vary, but large-scale clouds are usually predicted as some simple function of the relative humidity, and a parameterization for moist convection is used to predict the occurrence of convective clouds. When clouds are predicted within a grid cell, some models allow for fractional coverage of all cloud types within a cell. In other models, however, convective clouds usually have fractional coverage, but stratiform clouds occupy the entire cell. In this case, fractional cloud is obtained by averaging through time. The microphysical properties of the clouds, which determine their radiative characteristics, are usually ignored, with the cloud albedo often being a function of optical depth and the solar zenith angle. Water clouds are usually treated as blackbodies in the infrared, and ice clouds have a specified emissivity. The sensitivity of the model climate to clouds, together with the simplistic treatment of the cloud cover, represent the greatest contribution by the atmospheric portion of the models to the uncertainty of the $2xCO_2$ GCM temperature predictions.

The climate system is also very sensitive to the state of the ocean. Ocean currents account for a large proportion of the latitudinal heat flux; the resulting distribution of sea surface temperatures has a significant impact on interannual climate variability, and the deep ocean circulation affects climate over longer (1000 year) time-scales. Again it is obvious that GCMs have to account for these processes if accurate simulations of the climate are to be made. The observed sea surface temperature distribution can be used for simulations of the present-day climate; for climate change experiments, however, sea surface temperatures have to be predicted. The most common method in use with current GCMs is to include an ocean mixed layer (typically 65 m, although this varies between models and, in some cases, may be predicted by the model). This allows for thermodynamic exchanges at the surface, and some oceanic heat storage. Some models (the GISS Model II, for example) include a parameterized ocean transport based on the present-day pattern. This simplified treatment of the oceans represents another large source of uncertainty in climate change experiments that can only be addressed by coupling a dynamic ocean model to the atmospheric model. Coupling an atmospheric and an oceanic GCM is not simple, but several groups now have coupled models, and preliminary analyses suggest that the more realistic treatment of the oceans slightly reduces the magnitude of the doubled CO_2 warming.

Treatments of land surface processes are equally simplified, with the greatest uncertainty being due to the treatment of surface hydrology. Most models, at this point, use a simple 'bucket hydrology' where precipitation falls on a grid-square, collects in a "bucket" at a rate determined by the soil and vegetation characteristics, accumulates to a certain level (usually about 15 cm) beyond which further precipitation overflows the bucket, resulting in runoff. Evaporation and evapotranspiration occur simultaneously as a function of water availability and surface cover. Much more sophisticated schemes have been introduced recently (e.g., Dickinson, 1983; Dickinson et al., 1986; Sellers et al., 1986; Xue et al., 1991). These attempt a more realistic treatment of soil and plant processes with multiple soil and vegetation layers; canopy processes that include the effects of interception loss and stomatal resistance to transpiration; and more complex treatments of surface albedo, infiltration, and soil moisture fluxes. It is not clear, at present, how important such schemes are for improving the overall performance of the GCM, but they are obviously vital for improving predictions of how the surface hydrology will respond to climate change.

Finally, the surface topography also has an important effect on climate. The distribution of land and water affects surface exchange processes, and the height and location of mountain ranges affect both local processes and the large-scale atmospheric circulation. The treatment of the surface topography is largely a function of the model resolution: the finer the resolution, the more realistic the geography. With relatively coarse spatial resolutions, mountains tend to be lowered and smoothed out (Table 2), which has implications for local climate (particularly precipitation), and also for simulating the dispersion of atmospheric energy. This has been addressed by increasing the model resolution, or by introducing a gravity wave drag coefficient that simulates the effect of a barrier mountain range on the airflow across it. The reasons for including the drag coefficient, however, are more complicated than simply poor topography, and are related to subgrid-scale turbulence and energy cascade.

CO_2 CLIMATE CHANGE EXPERIMENTS

A brief history of CO_2 climate change experiments using GCMs is presented by Schlesinger and Mitchell (1987), who describe how these models evolved through

TABLE 2

Maximum Surface Elevation of Selected Mountain Ranges (m) from Boville (1991)

Truncation	Himalayas	Andes	Rockies	Greenland
T21	4,386	2,008	1,949	1,881
T42	5,695	3,008	2,159	2,877
T63	6,518	3,897	2,349	3,135

the early 1980s. At that point, similar model experiments were performed by three different groups: the Goddard Institute for Space Studies (Hansen *et al.*, 1984), the National Center for Atmospheric Research (Washington and Meehl, 1984; Meehl and Washington, 1985a,b; Bates and Meehl, 1985), and the Geophysical Fluid Dynamics Laboratory (Wetherald and Manabe, 1986; Manabe and Wetherald, 1986). Each of the models was used in a doubled CO_2 experiment, and the global mean surface air temperature changes from a $1xCO_2$ to a $2xCO_2$ climate were found to be 4.2°C, 4.0°C, and 3.5C for the GISS, GFDL, and NCAR models respectively. The results from these models and several others are presented in Table 3, adapted from Houghton *et al.* (1990). The NCAR result should be regarded as an underestimate of the change because the simulations had not reached quasi-equilibrium by the end of the experiment (which could not be continued due to limited computer resources). At that point, the $2xCO_2$-$1xCO_2$ temperature difference was still increasing at about 0.19°C per year. All three models can, therefore, be regarded as showing a similar global-scale temperature response to an instantaneous doubling of atmospheric CO_2.

These three models differ in resolution, numerical approach, and the parameterization schemes used. On the other hand, they represent equivalent levels of treatment, and they are a baseline against which the results from more recent model experiments can be compared. The GISS model uses a finite difference grid with a horizontal spacing of 8° latitude by 10° longitude, while the NCAR and GFDL models are both spectral, with an equivalent 4.5° x 7.5° horizontal resolution (higher resolution versions of all three models have since been developed). All three models include a mixed layer ocean, although the GISS model also includes a parameterization for ocean transport that is based on present-day observations. All of the models predict large-scale and convective clouds, with the basic differences being that the GISS model includes a scheme that provides for fractional cloud coverage within a cell; otherwise, large-scale cloud is produced when relative humidity exceeds 80%, 99%, and 100% for the NCAR, GFDL, and GISS models respectively. The GFDL and NCAR models include moist convective schemes, and use values of 100% and 80% relative humidity for assigning cloud cover, while the parameterization in the GISS model includes terms for shallow, middle-level, and deep convection. The treatment of the radiative calculations is similar in all three models (although there are differences in the values used for the solar constant, the surface albedo, and the surface emissivity), and all three models use similar methods for calculating large-scale condensation and snow cover. The remaining areas in which they differ are in the treatment of sea ice, soil moisture, and surface temperature. Again, the NCAR and GFDL models are similar, using a single layer ice model, with each grid-box either ice free or containing 100% ice cover; a single soil layer for calculating the surface temperature and soil moisture; and a 15 cm 'bucket.' The GISS model, on the other hand, has a two layer ice model, allows for fractional ice coverage, has two soil layers, and has a geographically varying 'bucket' size.

More detailed descriptions of all three models, and a detailed intercomparison of the model results, is given by Schlesinger and Mitchell (1987). Figures 1 and 2 (taken from their paper) show the temperature changes predicted by each of the

TABLE 3

Summary of Results from a Selection of GCM Doubled-CO_2 Experiments after Houghton et al. (1990)

	Group	Investigators	Resolution	Vertical Layers	ΔT (oC)	ΔP (%)	Comments
Fixed, zonally averaged cloud; no ocean heat transport								
1	GFDL	Manabe & Stouffer, 1980	R15	9	2.0	3.5	Based on 4 x CO2 simulation
2		Wetherald and Manabe, 1986, 88	R15	9	3.2	n/a	
Variable cloud: no ocean heat transport								
3	OSU	Schlesinger & Zhao, 1989	4° x 5°	2	2.8	8.0	
4			4° x 5°	2	4.4	11.0	As (3) but with revised clouds
5	NCAR	Washington and Meehl, 1989	R15	9	4.0	8.0	
6	GFDL	Wetherald and Manabe, 1986, 88	R15	9	4.0	9.0	As (2) but with variable clouds
Variable cloud; prescribed oceanic heat transport								
7	AUS	Gordon & Hunt, 1989*	R21	4	4.0	7.0	
8	GISS	Hansen et al., 1984	8° x 10°	9	4.2	11.0	
9			8° x 10°	9	4.8	13	As (8) but with more sea-ice control
10	UKMO	Wilson and Mitchell, 1987	5° x 7.5°	11	5.2	15.0	
11		Mitchell et al., 1989	5° x 7.5°	11	2.7	6.0	As (10) but different cloud scheme
12			5° x 7.5°	11	3.2	8.0	As (10) but different ice scheme
High Resolution; variable cloud, prescribed oceanic heat transport								
13	CCC	Boer et al., 1989*	T32	10	3.5	4.0	
14	UKMO	Mitchell et al., 1989	2.5° x 3.75°	11	3.5	9.0	As (12) with gravity wave drag

GFDL = Geophysical Fluid Dynamics Laboratory
OSU = Oregon State University
NCAR = National Center for Atmospheric Research
AUS = Commonwealth Scientific and Industrial Research Organization (CSIRO), Australia
GISS = Goddard Institute for Space Studies
UKMO = United Kingdom Meteorological Office
*Unpublished: Taken from Houghton et al., 1990

models for a doubling of atmospheric CO_2. Figure 1 is a latitude-time cross section of the zonal mean surface air temperature change ($2xCO_2$ - $1xCO_2$) for each of the models. All three models indicate that the change is primarily in the extra-tropical latitudes, and that the greatest changes are at high latitudes in the winter months. This pattern is apparent in all three models, and plotted in this fashion the results would suggest a large degree of agreement among the predictions of the different models. However, there is much disagreement in the geographic distribution of these changes (Figure 2). Most of the change is again confined to the extra-tropics, with the smallest being over the tropical oceans, and the largest being at high latitudes. Beyond this, there is considerable disagreement over the regional distribution of the global change. In December, January, and February, the GISS model produces its greatest differences centered on the Arctic Ocean, and over the Eastern Canadian Arctic, whereas the GFDL model finds the largest changes to be located along the margins of the Arctic Ocean in the Kara/Barents Sea and the Beaufort/Chukchi Sea. In the NCAR model, on the other hand, the greatest changes are found at high latitudes, but in this case located further south in the North Atlantic and in the Bering Sea. In June, July, and August, the models are more consistent, with the greatest changes taking place between 50°S and 70°S; again, however, the GISS model shows the change occurring further poleward than the other two models.

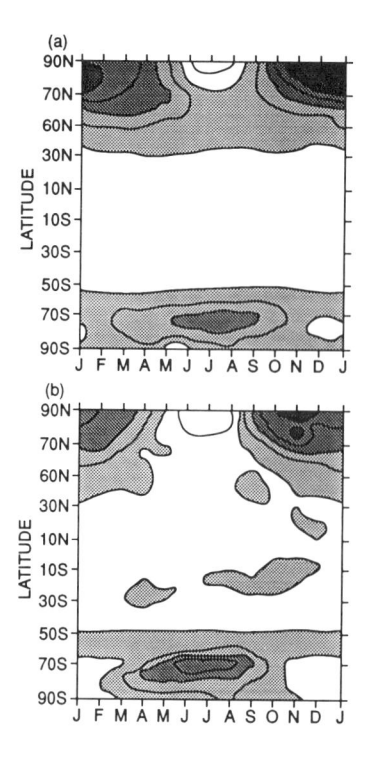

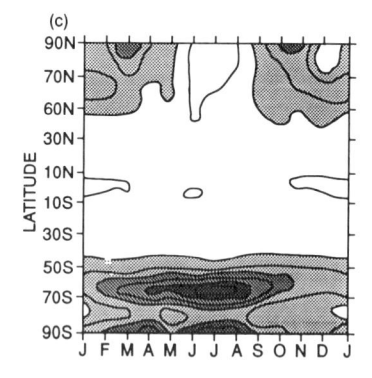

FIGURE 1. Latitude-time cross section of the zonal mean surface air temperature change (°C) ($2xCO_2$ - $1xCO_2$ for (a) the GFDL GCM (Wetherald and Manabe, 1986); (b) the GISS GCM (Hansen et al., 1984) and (c) the NCAR GCM (Washington and Meehl, 1984). Shading is at 4°, 8°, 12° and 16°C. Redrawn from Schlesinger and Mitchell, 1987.

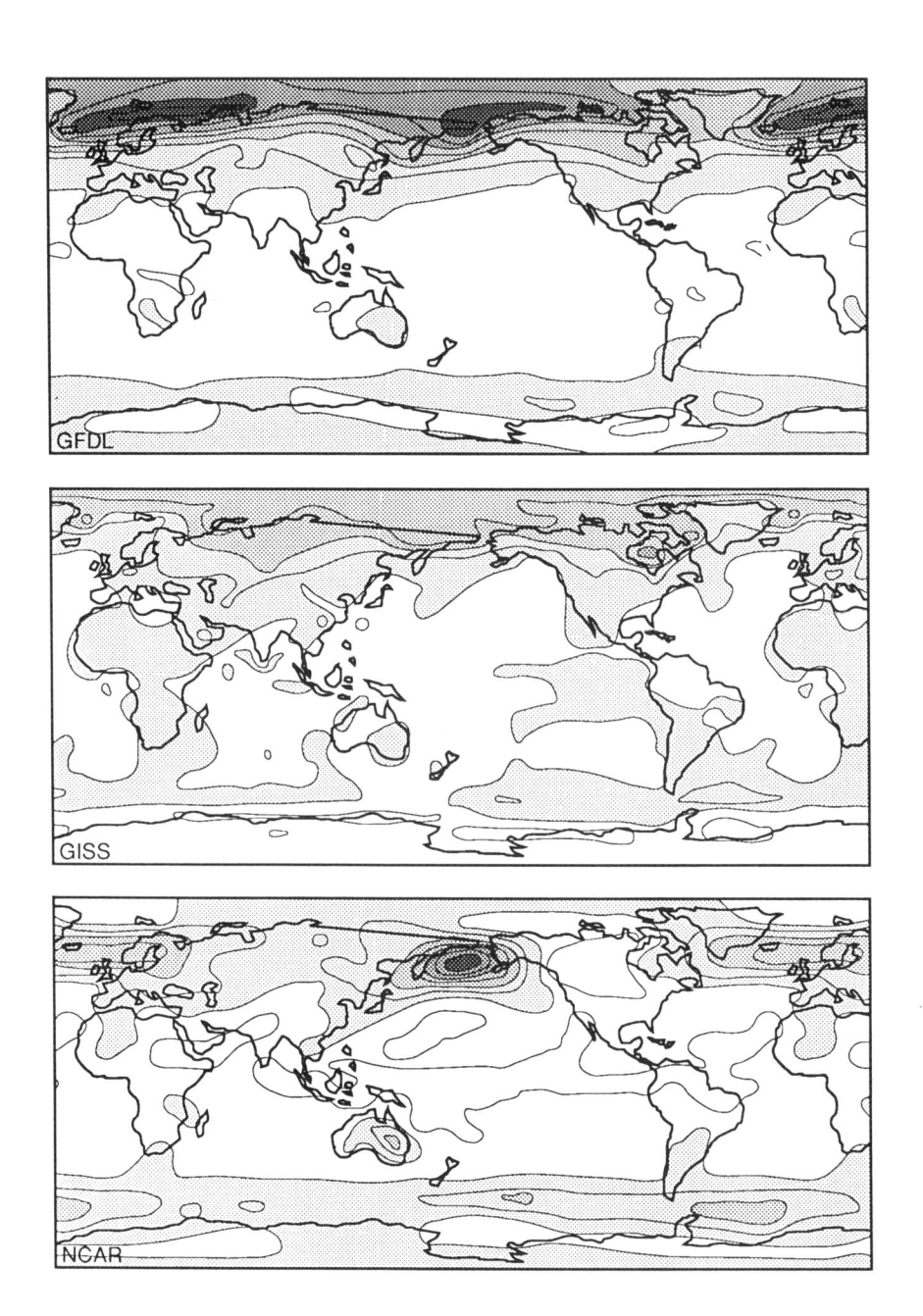

General Circulation Model Studies of Global Warming 199

FIGURE 2. Geographic distribution of $2xCO_2 - 1xCO_2$ temperature change (Dec., Jan., Feb.) for the same models as Figure 1. Shading at 4°, 8°, 12° and 16°C intervals. Redrawn from Schlesinger and Mitchell, 1987.

In each case, the largest changes are occurring in the region in which the models have their greatest difference in sea ice cover between the 1xCO_2 and the 2xCO_2 simulations, emphasizing the importance of ice-temperature feedbacks. Two particular processes are important here, one affecting the albedo and the other the oceanic heat loss. A reduced ice cover results in a lower albedo and thus a higher temperature, and a reduced ice cover in winter also allows greater heat loss from the ocean to the atmosphere, thereby increasing surface air temperatures. The high latitude differences between the GISS model and the GFDL and NCAR models may also be due to the different ways in which they treat the sea ice; the GFDL and NCAR models do not allow for fractional ice coverage and all of the change is confined to the ice margins, unlike the GISS model, which allows fractional coverage of a grid-cell and spreads the change over a larger area. Beyond these spatial differences, there are also differences in magnitude, with maximum changes reaching +18°C in the GFDL model, +16°C in the NCAR model, and +10°C-12°C in the GISS model. Away from the polar regions, the intermodel differences are even greater; for most of the United States the summer and winter differences are about the same in the NCAR model, the summer change is greater than the winter change in the GFDL model, and the temperature change is greater in winter than in summer in the GISS model. Similar degrees of variability are apparent for almost any region chosen.

RECENT DEVELOPMENTS

Several questions arise out of the previous section that warrant further discussion. One is simply the robustness of the ~4°C global mean temperature change predicted by the models for a doubling of CO_2. The direct radiative impact of a doubling of CO_2 would be the equivalent of about a 1°C temperature increase; positive feedback processes in the GCM climate system result in an additional 3°C warming. How reliable that number is depends on the ability of the model to simulate correctly all of the necessary physics, and it may also depend in part on the model resolution. One approach to refining these numbers is, therefore, to try to improve upon the model physics. Much of the work in this area is focused on coupling more sophisticated biosphere and ocean models to the atmospheric GCMs. At this point, most progress has been made with coupling ocean and atmosphere models, and a brief review of developments in this field is presented by Meehl (1990). Coupled atmosphere-ocean circulation models are now being operated by various groups around the world, including China's 2-level atmosphere and 4-level ocean model based on the Oregon State University (OSU) model (Zhang *et al.*, 1989); the UKMO's 2.5°x3.75° 11-level atmosphere and 17-level ocean model (Foreman *et al.*, 1988); and the T21, 16-level atmosphere coupled to a 4°x4° 10-level ocean model at the Max Planck Institute for Meteorology (Cubasch, 1988).

There are some inherent problems with coupled models that are discussed by Meehl (1990) and Washington and Meehl (1989). These problems result in tropical sea surface temperatures that are too low, and high latitude temperatures that are too high. The consequence is to produce sea ice extents that are too small, and these effects combine to reduce the sensitivity of the model to an instantaneous change in CO_2. Washington and Meehl (1989) find a global mean temperature change of about 1.6°C. The problems in the coupled models tend to reduce the sensitivity of the model to a doubling of CO_2 Meehl (1990). There are also problems in the uncoupled (mixed-layer) models, some of which have been discussed here. Whether these problems tend to increase or decrease the model sensitivity to change is not easily determined. Consequently, the real change for an instantaneous doubling of CO_2 may fall somewhere between the 1.6°C and the 4°C predicted by the two types of model, but it is still impossible to say for certain. Two transient response CO_2 experiments have also been carried out in the United States using the GFDL and NCAR coupled ocean-atmosphere models. In a transient response experiment the CO_2 is increased in increments over time, rather than with an instantaneous doubling. Both the NCAR model (Washington and Meehl, 1989) and the GFDL model (Stouffer et al., 1989) show the continents warming more quickly than the oceans, and both models show less warming around Antarctica (which is very different to the results described earlier).

There is also some suggestion that GCM simulations may be sensitive to the horizontal resolution used. Boer and Lazare (1988), using the Canadian Climate Center GCM, found that the simulations change significantly as resolution changes, but that none of the simulations could be regarded as being significantly better than any of the others. Boville (1991), on the other hand, found that, in general, when using the NCAR CCM1 at resolutions from T21 to T63 (a wider range than that used by Boer and Lazare) the simulations improved as the resolution increased. This has given rise to an on-going debate over which approach to take—whether it is better to enhance the model resolution, or improve the model physics. Given the current limitations on computer power it is probably not possible to do both at the same time. One further observation that can be made at this point is that most of the model improvements have tended to reduce the CO_2 temperature response— but not by very much; the tendency of all of the models is still to demonstrate an overall positive feedback that amplifies the original radiative forcing.

A second feature of Figure 2 worth examining more closely is the enhanced response at high latitudes. Virtually all GCM CO_2 experiments show the greatest warming to be at high latitudes, and this has led to some speculation that the polar regions may be one of the first areas to demonstrate a CO_2 induced global warming. Much of the high latitude warming in the GCMs appears to be due to changes in sea ice extent. Most current GCMs, however, have very simplified sea ice models and do a relatively poor job of simulating the present-day arctic climate in general, and the ice distribution in particular. This is due in part to the dependence on ice extent and concentration of ice dynamics (movements of the ice by wind and ocean currents), which is not included in most models. The errors are partly due to inaccurate simulations of polar temperatures, and partly due to the fact that the ice cover, in the North Atlantic in particular, is related to oceanic processes that cannot be simulated without a coupled ocean circulation model. Figures 3 and 4 illustrate the present-day Arctic temperature and pressure fields from the same three models

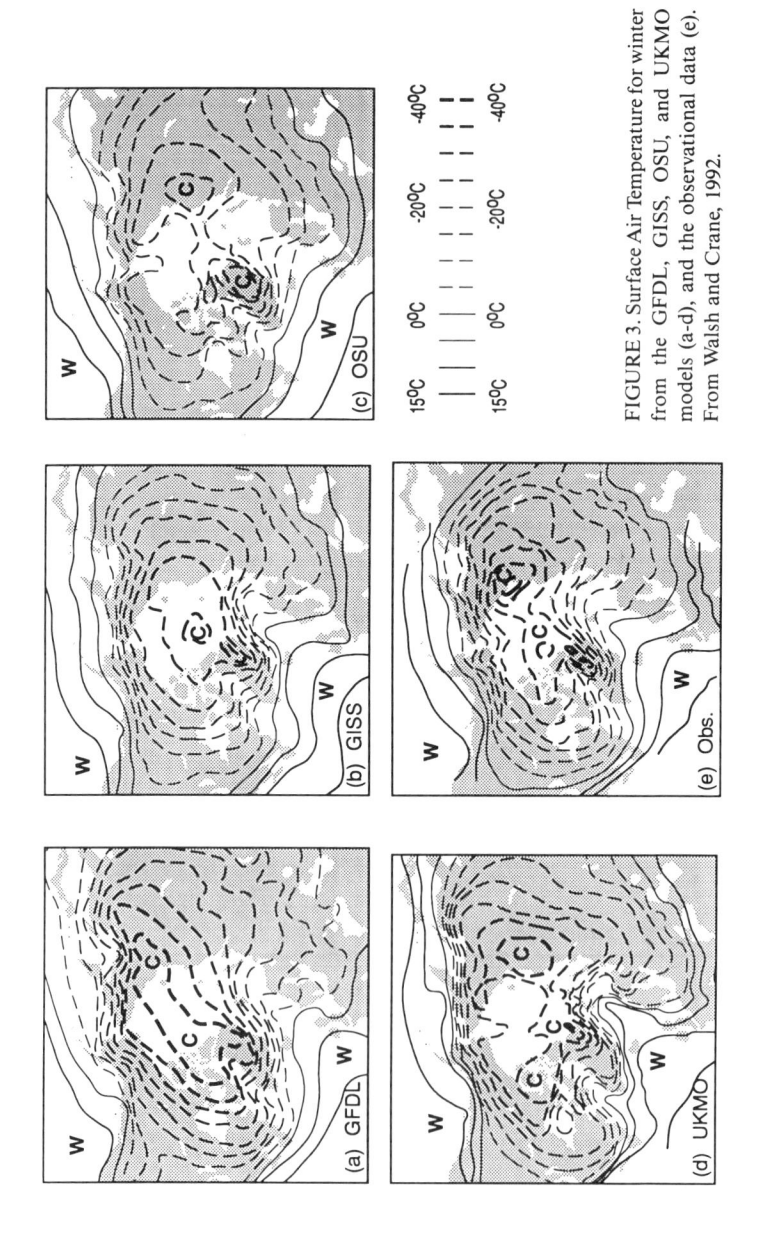

202 Global Climate Change: Implications, Challenges and Mitigation Measures

FIGURE 3. Surface Air Temperature for winter from the GFDL, GISS, OSU, and UKMO models (a-d), and the observational data (e). From Walsh and Crane, 1992.

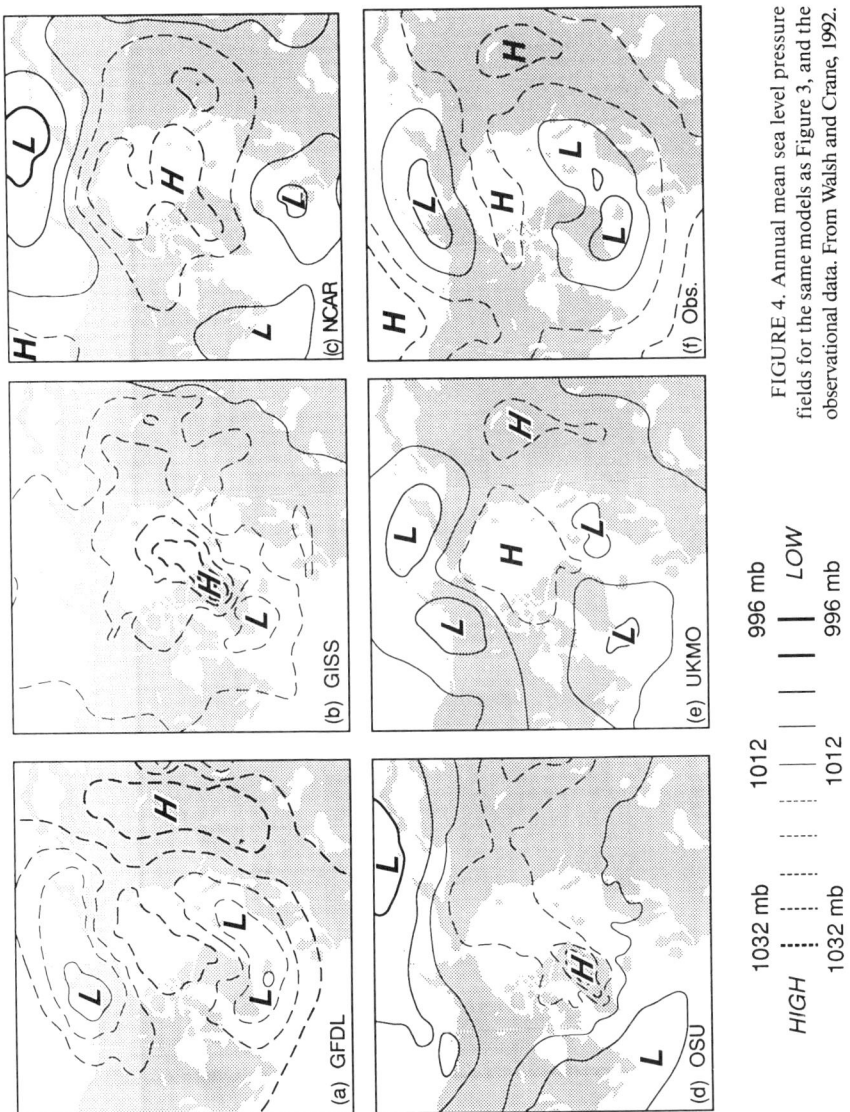

FIGURE 4. Annual mean sea level pressure fields for the same models as Figure 3, and the observational data. From Walsh and Crane, 1992.

described in Section IV, as well as the OSU and the UKMO models. The figures also show the observed fields for comparison purposes. Figure 4 indicates that there are some large areas of disagreement between the modeled and observed temperature distributions. In terms of overall bias, the bias in the UKMO model is very small, the GISS model has a 2°C-4°C cold bias for most of the year, the OSU model has a warm bias that reaches 8°C in November and December, and the GFDL model tends to be 9°C-11°C too cold from October to February (Walsh and Crane, 1992).

As noted above, however, part of the problem is also due to the lack of ice dynamics. Figure 4 indicates that even if the dynamics were included, these models would still have problems because of errors in the sea level pressure field (wind stress due to the pressure gradient is one of the major forcing functions for ice movement). An important quantity here is the pressure difference between north-east Greenland and the north-west tip of Svalbard—this determines the wind stress in a region responsible for most of the ice export from the Arctic Basin. This pressure difference is 4 mb in the annual mean of the observed data set, 5 mb in the GFDL and GISS models, 1 mb in the NCAR model, 3 mb in the OSU model, and 4 mb in the UKMO model (Walsh and Crane, 1992). Given the large degree of uncertainty present in the GCM simulations of the present Arctic climate, one would have to look with some suspicion at the enhanced high latitude changes predicted by most GCMs. Whether such a response is valid cannot be determined until fully coupled dynamical ocean-atmosphere-sea ice models are available, and such models are only just now being developed.

Finally, while the three models depicted in Figure 2 and 3 all agree on the overall magnitude of the global warming, it is apparent from these figures that there is little agreement on the regional distribution of this temperature increase (cf. Grotch and MacCracken, 1991). It is possible that these regional differences will diminish as model resolution increases, and as more sophisticated models that include more realistic treatments of the oceans and of land surface processes are developed. Even so, it will still be difficult to carry out regional analyses at the present GCM grid resolutions. Most current GCMs have a resolution between 4°x5° and 3°x3°. A 3°x3° resolution produces a matrix of about 7x4 grid-points over the Amazon Basin, but only 5 or 6 grid points over the British Isles, and only a couple of gridpoints per state over the United States. Obtaining a resolution sufficient to analyze changes within individual drainage basins does not seem possible anytime in the near future. This problem can be tackled for selected regions by embedding a higher resolution regional (mesoscale) model within the GCM. The GCM is used to establish the boundary conditions for the mesoscale model, which is then used to simulate the climate at a much smaller spatial scale. In one example for the western United States, Giorgi (1990) has driven the 60 km resolution Pennsylvania State University/NCAR mesoscale model (MM4), using the NCAR CCM to set the boundary conditions at each step. For CO_2 climate change predictions, this approach is still limited at present by the uncertainty in the individual GCM grid-point predictions. As improvements continue to be made to the GCMs, however, this type of approach has the potential to obtain the necessary high resolution information, but only for a limited number of areas.

CONCLUSIONS

GCMs have developed very rapidly over the last twenty years, and they will continue to evolve into the foreseeable future. The current generation of atmospheric GCMs are sophisticated models that simulate many of the broad features of the present-day climate system. Due to current limitations in computer power, it would appear that future developments will follow two different paths: one of increased spatial resolution, the other of improved treatments of physical processes. With regard to the latter, one possibility is to improve the coupling between the atmosphere and ocean by integrating the atmospheric model with an ocean GCM. Again, however, it would appear that ocean-atmosphere modelers are also faced with a choice of paths. The discussion in the text indicated that these models have some inherent problems that reduce their sensitivity to climate change. One option here is to accept the inaccuracies in the present-day simulations and discuss the climate change results in the light of the model sensitivity. The other approach is to adjust the present-day results using techniques referred to as 'flux correction' or 'flux adjustment' schemes to bring the model more into line with present-day observed climates, before running the model for the climate change experiment. The problem here is that the adjustment used in the climate change calculations is based on the present-day climate, and may not be representative of the changed climate state.

Although GCMs will continue to improve, at the present time the consensus is an approximate 4°C global mean temperature increase for a doubling of CO_2 in GCMs with a mixed-layer ocean, and a somewhat lower value for fully coupled atmosphere-ocean GCMs (e.g., 1.6°C for the NCAR model). The mixed layer models may be too sensitive to a CO_2 climate change, and the reverse might be true of the fully coupled models. If this is the case, the correct answer may lie somewhere in-between. However, the cloud problem could still drive the models in either direction. One important observation to make is that, while the magnitude of the global warming is still open to debate, almost all GCM results indicate that there will be a net positive feedback in the climate system, which will increase the temperature change beyond the 1°C that can be expected due to CO_2 radiative forcing alone. Finally, while we have a consensus that global warming will occur if atmospheric CO_2 continues to increase, at present we have little agreement as to what the regional consequences of this warming will be.

ACKNOWLEDGEMENTS

I would like to thank Dr. T. Ackerman for his comments on the manuscript. The research on GCMs in polar regions is supported by NSF grant ATM-8913039 to the Pennsylvania State University.

REFERENCES

Barron, E.J. and W.M. Washington. 1984. The role of geographic variables in explaining paleoclimates: Results from Cretaceous climate model sensitivity studies. J. Geophy. Res., 89:1267-1279.

Bates, G.T. and G.A. Meehl. 1985. The effect of CO_2 concentration on the frequency of blocking in a general circulation model coupled to a simple mixed layer ocean model. Mon. Weather Rev., 113:689-701.

Boer, G.J. and M. Lazare, A. Becker and J. Nemec. 1988. Some results concerning the effect of horizontal resolution and gravity wave drag on simulated climate. J. Climate, 1:789-806.

Boville, B.A. 1991. Sensitivity of simulated climate to model resolution. J. Climate, 4:469-485.

Cess, R.D. *et al.* 1990. Intercomparison and interpretation of climate feedback processes in 19 atmospheric general circulation models. J. Geophys. Res., 95:16,601-16,615.

Charney, J.G., R. Fjortoft and J. von Neumann. 1950. Numerical integration of the barotropic vorticity equation. Tellus, 2:237-254.

Committee on Earth Sciences. 1989. Our Changing Planet: The FY 1990 Research Plan: p. 8.

Crowley, T.J. 1990. Are there any satisfactory geologic analogs for a future greenhouse warming? J. Climate, 3:1282-1292.

Cubasch, U. 1988. A global coupled atmosphere-ocean model. Proceedings of the Thirteenth Annual Climate Diagnostics Workshop: 292-297.

Dickinson, R.E. 1983. Land Surface Processes and Climate-Surface Albedos and Energy Balance. Theory of Climate. Advances in Geophysics, 25:305-353.

Dickinson, R.E., A. Henderson-Sellers, P.J. Kennedy, M.F. Wilson. 1986. Biosphere-Atmosphere Transfer Scheme (BATS) for the NCAR Community Climate Model. NCAR Technical Note, NCAR/TN-275+STR.

Foreman, S.J, N.S. Grahame, K. Maskell and D.L. Roberts. 1988. Feedbacks and error mechanisms in a global coupled ocean/atmosphere/sea ice model. Modelling the sensitivity and variations of the ocean-atmosphere system, WCRP-15 (WMO/TD-No. 254):271-279.

Gates, W.L., Y. Han and M.E. Schlesinger. 1985. The global climate simulated by a coupled atmosphere-ocean general circulation model: preliminary results. Coupled Ocean-Atmosphere Models: 131-151.

Giorgi, F. 1990. Simulation of regional climate using a limited area model nested in a general circulation model. J. Climate, 3:941-963.

Grotch, S.L. and M.C. MacCracken. 1991. The use of general circulation models to predict regional climate change. J. Climate, 4:286-303.

Hansen, J., G. Russell, D. Rind, P. Stone, A. Lacis, S. Lebedeff, R. Ruedy and L. Travis. 1983. Efficient three-dimensional global models for climate studies: Models I and II. Mon. Wea. Rev., 111:609-662.

Houghton, J.T., G.J. Jenkins and J.J. Ephraums. (Eds.) 1990. Climate Change: The IPCC Scientific Assessment. Cambridge University Press: 365 pp.

Kutzbach, J.E. 1981. Monsoon climate of the early Holocene: Climatic experiment using the Earth's orbital parameters for 9000 years ago. Science, 214:59-61.

Manabe, S. and R.J. Stouffer. 1980. Sensitivity of a global climate model to an increase of CO_2 concentration in the atmosphere. J. Geophys. Res., 85:5529-5554.

Manabe, S. and R.T. Wetherald. 1986. Reduction in summer soil wetness induced by an increase in atmospheric carbon dioxide. Science, 232:626-628.

Meehl, G.A. 1990. Development of global coupled ocean-atmosphere general circulation models. Climate Dynamics, 5:19-33.

Meehl, G.A. and W.M. Washington. 1985a. Sea surface temperatures computed by a simple ocean mixed layer coupled to an atmospheric GCM. J. Phys. Oceanogr., 15:92-104.

Meehl, G.A. and W.M. Washington. 1985b. Tropical response to increased CO_2 in a GCM with a simple mixed layer ocean: Similarities to an observed Pacific warm event. Mon. Weather. Rev., 114:667-674.

Mitchell, J.F.B. 1990. Greenhouse warming: Is the Mid-Holocene a good analogue. J. Climate, 3:1177-1192.

Mitchell, J.F.B., C.A. Wilson and W.M. Cunnington. 1987. On CO_2 climate sensitivity and model dependence of results. Quart. J. Roy. Met. Soc., 113:293-322.

Mitchell, J.F.B, C.A. Senior and W.J. Ingram. 1989. CO_2 and climate: A missing feedback. Nature, 341:132-134.

Phillips, N.A. 1956. The general circulation of the atmosphere: A numerical experiment. Quart. J. Roy. Meteorol. Soc., 82:123-164.

Ramanathan, V., R.D. Cess, E.F. Harrison, P. Minnis, B.R. Barkstrom, E. Ahmad and D. Hartmann. 1989. Cloud-radiative forcing and climate: Insights from the Earth Radiation Budget Experiment. Science, 243:57-63.

Richardson, L.F. 1992. Weather Prediction by Numerical Process. Cambridge University Press: 236 pp.

Schlesinger, M.E. and Z-C Zhao. 1989. Seasonal climatic changes induced by double CO_2 as simulated by the OSU atmospheric GCM/mixed-layer ocean model. J. Climate, 2:459-495.

Schlesinger, M.E. and J.F.B. Mitchell. 1987. Climate model simulations of the equilibrium climatic response to increased carbon dioxide. Rev. Geophy., 25:760-798.

Sellers, P.J., Y. Mintz, Y.C. Sud and A. Dalcher. 1986. A Simple Biosphere Model (SiB) for use within general circulation models. J. Atmos. Sci., 43:505-531.

Spelman, M.J. and S. Manabe. 1984. Influence of oceanic heat transport upon the sensitivity of a model climate. J. Geophys. Res., 89:571-586.

Stouffer, R.J., S. Manabe and K. Bryan. 1989. Interhemispheric asymmetry in climate response to a gradual increase of atmospheric CO_2. Nature, 342:660-662.

Walsh, J.E. and R.G. Crane. 1992. A comparison of GCM simulations of Arctic climate. Geophys. Res. Lett., 19:29-32.

Washington, W.M. and C.L. Parkinson. 1986. An Introduction to Three-Dimensional Climate Modeling: 422 pp.

Washington, W.M. and G.A. Meehl. 1984. Seasonal cycle experiment on the climate sensitivity due to a doubling of CO_2 with an atmospheric general circulation model coupled to a simple mixed layer ocean model. J. Geophys. Res., 89:9475-9503.

Washington, W.M. and G.A. Meehl. 1989. Climate sensitivity due to increased CO_2: Experiments with a coupled atmosphere and ocean general circulation model. Climate Dyn., 4:1-38.

Wetherald, R.T. and Manabe, S. 1986. An investigation of cloud cover change in response to thermal forcing. Clim. Change, 8:5-23.

Wetherald, R.T. and S. Manabe. 1988. Cloud feedback processes in a general circulation model. J. Atmos. Sci., 45:1397-1415.

Wilson, C.A. and J.F.B. Mitchell. 1987. A doubled CO_2 climate sensitivity experiment with a global climate model including a simple ocean. J. Geophys. Res., 92:13,315-13,343.

Xue, Y. P.J. Sellers, J.L. Kinter and J. Shukla. 1991. A Simplified Biosphere Model for global climate studies. J. Climate, 4:345-364.

Zhang, X.-H., X.-Z. Liang and Q.-C. Zeng. 1989. A numerical world ocean model free from "rigid lid". Abstract Volume, International Conference of Modelling of Global Climate Change and Variability: 47.

Global Climate Change: Implications, Challenges and Mitigation Measures. Edited by S.K. Majumdar, L.S. Kalkstein, B. Yarnal, E.W. Miller, and L.M. Rosenfeld. © 1992, The Pennsylvania Academy of Science.

Chapter Fourteen

CHANGES IN CLIMATE VARIABILITY WITH CLIMATE CHANGE

LINDA O. MEARNS

Interdisciplinary Climate Systems Section
Climate and Global Dynamics Division
National Center for Atmospheric Research[1]
P.O. Box 3000
Boulder, CO 80307-3000

INTRODUCTION

Changed climate variability associated with climate change could significantly affect natural resources. Most resource systems important to society (e.g., agriculture, water resources) experience climate variability through the occurrence of extreme events such as droughts, floods, and heat waves. The frequency of such events are affected by changes in the mean as well as the variance of climate time series.[1-4] However, it has recently been demonstrated[5] that the frequencies of extreme events are more sensitive to changes in the variance than to changes in the mean.

Lack of information on how climate variability may change has limited the completeness of climate change impact studies published so far.[6] It is at present difficult to state unambiguously how climate variability will change with greenhouse gas-induced global warming.[7] However, it is certain that global warming could change the variability of climate and that it *may* have greater impacts on some systems than change in average conditions, depending upon the magnitude of change in the mean climate relative to that of variability.

The goal of this chapter is to articulate what we currently know about how climate

---

[1] The National Center for Atmospheric Research is sponsored by the National Science Foundation.

variability may change under conditions of global warming. Most of our information comes from general circulation model (GCM) experiments, the results of which will form the focus of the chapter. First, the nature and causes of climate variability are described. A brief review of what information is available from the observed climate record is then presented. Then, how well the GCMs reproduce the present variability is analyzed, and finally, analyses of changed variability in perturbed climate experiments are discussed.

THE NATURE AND CAUSES OF CLIMATIC VARIABILITY

Variability is an inherent characteristic of climate[8] and is closely related to the concept of climate change. However, there is no universally accepted distinction made between the terms "climate variability" and "climatic change". Both terms refer to fluctuations in climate from some expected or previously defined mean climate state. Berger[9] makes the distinction that climate change refers to a secular trend which produced a change in the average, whereas variability refers to the oscillations about that mean. Distinctions can only be made relative to the time scales of concern. Climate change as discussed in this book generally refers to a change from the mean global climate conditions we have experienced in roughly the past few centuries. On a longer time scale (i.e., thousands of years), however, this climate "change" would be viewed as an instance of climate variability, i.e., as one of many fluctuations around mean conditions prevailing over several thousand years. For the purpose of this chapter, climate variability is defined as the pattern of fluctuations about some specified mean value (i.e., a time average) of a climatic element.

The causes of climatic variability are largely time scale dependent and may be divided into two major categories: 1) those arising from internal dynamics that produce stochastic (random) fluctuations (and possible chaotic behavior) within the climate system; and 2) those arising through external forcing of the system. Table 1, (after Berger[9]), summarizes different causes of climatic variability on different time scales. On very long time scales (e.g., 100,000 years) astronomical factors account for much variability (orbital parameters in Table 1.) Long-term variations in the shape of the earth's orbit and in the orientation of the Earth's axis affect the seasonal and latitudinal distribution of solar radiation reaching the top of the atmosphere. These astronomical factors are external forcings to the climate system. On much shorter time scales, the earth's rotation and the axial tilt at any one time result in the climatic variations of the diurnal cycle and the seasons. These are the periodic climatic variations most familiar to us.

Variations of climate on a year-to-year basis (interannual variability) can arise from external forcings such as volcanic eruptions or from slowly varying internal processes including, as part of the internal system, interactions between the atmosphere and oceans, soils, and variations in sea ice extent.[10] These interactions can result in shifts in locations of major circulation features or changes in their intensity.[11] The largest effect, presumably, is due to variations in sea surface temperatures, such as occur in ENSO events (El Niño Southern Oscillation).[12-14]

Daily variability of a non-periodic nature largely results from variations in synoptic scale weather processes, such as cyclones and anticyclones and upper atmosphere wind streams which direct the movement of such features[15] (atmosphere autovariation in Table 1). These features interact with local topography to provide location specific variability. (Variations caused by these weather processes are largely stochastic and internal to the climate system). The daily climatic variability of a given location will be a function of the frequency of occurrence of different air masses, which is partially determined by its position with respect to the air mass source regions, to orography and to the mean locations of the circulation features such as cyclones and anticyclones.[16]

In this chapter we are mainly concerned with variations on time scales of several years or less, that is, from interannual to daily variability. These are the time scales which have most often been considered in modeling experiments of greenhouse gas increases.

EMPIRICAL STUDIES OF CHANGING CLIMATE VARIABILITY

One of the methods available for possibly gaining some insight into how climate

TABLE 1

Major Processes Involved in Climate Fluctuations for Different Time Scales

			Earth's history	Quaternary ice ages	History	Instruments
			Years 10^{10} 10^9 10^8	10^7 10^6 10^5	10^4 10^3	100 10 1

NATURAL POTENTIAL CAUSAL MECHANISMS

EXTERNAL — affecting available incoming radiation:
- galactic dust
- Sun's evolution
- solar variability
- orbital parameters

INTERNAL (related to):

geophysical boundary conditions:
- plate tectonics
- epeirogeny, orogeny
- isostasy

net radiation:
- atmospheric evolution
- volcanic activity
- tropospheric dust
- surface cover:
  - vegetal
  - snow
  - sea ice
  - glacier
  - ice sheet

feedbacks:
- atmosphere-cryosphere-lithosphere
- atmosphere-ocean
- atmosphere autovariation

HUMAN ACTIVITIES:
- land use
- traces gases
- aerosols
- heat pollution

Source: Berger (1980).

variability may change in a generally warmer climate is to investigate past relationships in the climate record between mean climate change and changes in variability. However, past research efforts to determine changes in climate variability and/or mean climate conditions and changes in climate variability in the historical record have not resulted in a clear consensus. Below the major research efforts in this area are briefly reviewed.

Van Loon and Williams[17] found significant differences in interannual temperature variability in North America during two different 51 year periods. However, no single connection between trend in temperature and trend in its interannual variability was found. Specifically they assert that their results do not support the postulated association between cold periods and high variability of temperature. Similarly, Chico and Sellars[18] found there were changes over time of interannual winter temperature variability in the U.S. on a decadal basis (an increase from 1900 to 1930; then a decrease from 1930 to 70) but they did not relate these changes to changes in mean temperature conditions. Diaz and Quayle,[19] in a more thorough analysis of the U.S. climate (temperature and precipitation), found no systematic relationship between changes in mean temperature and precipitation and their corresponding variances.

Brinkmann[16] analyzed the relationship between mean temperature and variability in Wisconsin using climate data from three stations. She found no relationship between mean temperature and interannual variability, but that there was a negative correlation between winter mean temperature and the day to day variability, and a corresponding positive relationship for summer conditions. Brinkmann explains these relationships on the basis of Wisconsin's location with respect to general circulation patterns.

Lough et al[20] analyzed the association between mean temperature and precipitation and variability in their creation of climate change scenarios for a warmer world, through the analysis of historical climate data (i.e., the analogue approach). Two periods were selected when Arctic temperatures were particularly warm and cold (1934-53 and 1901-20). They then mapped out the temperature differences between these two periods for the region of Europe. They also mapped out differences in interannual temperature variability (as measured by standard deviations) for the warm minus the cold period. Results indicate that the regions of lower winter temperatures roughly coincides with the region of increased variability, but the coincidence is far from perfect.

Finally, Schuurmans and Coops,[21] in their study of seasonal mean temperatures of Europe found very little relation between interannual variability and mean temperature, except in winter at high latitudes. There the often assumed association of warm periods with low variability and cold periods with high variability held true.

These studies indicate that there have been significant changes in both interannual and day-to-day climate variability in historical times, but that simple or distinct relationships between changes in mean climate conditions and changes in variability have not been established. Moreover, the value of seeking such relationships in the past as a key to the future is potentially limited since the causes of very short term warming or cooling in the past are not known, but in any event, are not caused by increases in greenhouse gases. Even if clear relationships in the historical record had been established, this would not necessarily indicate anything about the future relationships under CO_2-induced climate change, since the causes of the changes would be different. There is reason to believe that different forcings may result in different patterns of changes in climate.[22,23]

The failure of historical climate records to provide an empirically consistent and causally coherent scenario of possible changes in climate variability contributes to the necessity of examining climate variability in climate modeling experiments. General circulation models have limitations, but they have one definitive strength over empirical attempts to analyze future climate change: the modeling experiments are constructed such that the response of the climate system to the true cause of the change (i.e., increased greenhouse gases in the atmosphere) is simulated. Therefore, analyzing the changes in climate variability from such modeling experiments should provide insights into coherent possible pictures of future variability changes of significant climate elements in a CO_2 warmed world. Examination of the analysis of climate variability in climate models is the subject of the next two sections.

VALIDATION OF CLIMATE VARIABILITY IN GCM CONTROL RUNS

In order to determine if the variability statistics of trace gas perturbed experiments are valid, how well the GCMs reproduce the variability of the present day climate must be examined. If there are serious errors in the control run simulations, then it becomes more difficult to give credence to the perturbed climate run results, since the model is obviously not properly simulating the processes giving rise to climate variability. Moreover, Mitchell et al.[24] demonstrated that the response of climate models to perturbations is highly dependent on the model's simulation of the present day climate.

Studies comparing variability statistics of observed time series with variability statistics of general circulation model (GCM)-generated time series of climate variables relevant to climate impacts are not numerous in the atmospheric sciences literature, although studies first appeared in the early 1980s.[25,26] The number of studies has particularly increased in the past several years.

A number of the more recent and relevant works that address comparison of variability are briefly reviewed in this section. Very often, these studies also involve a study of the change in variability under doubled CO_2 conditions. Those perturbed climate results are presented in the next section. The studies presented in this section are largely grouped according to the type or time-scale of variability investigated where possible and, hence, are not necessarily presented chronologically.

Interannual Variability

Chervin[27] investigated interannual climate variability in the NCAR, CCM0(A) (National Center for Atmospheric Research Community Climate Model). He designed a special version of the model to eliminate external variability of interannual time series of climate variables (i.e. fixed sea surface temperatures (SSTs) and fixed sea ice boundaries were used), so that discrepancies between modeled and observed variability would reflect the external component of variability present in the observed data. The variability of mean sea level pressure and 700-mb geopotential height (which roughly corresponds to the height above the surface where the atmospheric pressure equals 700 mb, and is related to large-scale wind patterns) were analyzed in the Northern hemisphere, with particular focus on the United States. Results indicated no significant differences between modeled and observed variabilities of mean sea level pressure over the United States and only limited areas of differences in the variability of 700 mb geopotential height.

Hansen et al.[28] used the Goddard Institute for Space Studies (GISS) model II GCM to simulate the global climate effects of time-dependent variations of atmospheric trace gases and aerosols. Several different scenarios of trace gas increase from 1958 to the present were used. A one hundred year control run was also produced. From this run it was determined that globally the model only slightly underestimates the observed interannual variability. However, the model's variability tends to be larger than observed over land.

Portman et al.[29] concentrated on regional validation of free atmosphere daily and interannual temperature variability (850, 500, and 300 mb heights) in one version of the NCAR CCM1 for several regions in the U.S. Using Monte Carlo techniques, they produced large sample sizes for both observed and model values. In general they found that there was little significant difference between observations and temperature for seasonal interannual temperature variability, but significant differences (i.e., model overestimates) for daily variability in summer. In addition, the differences decreased with increasing altitude.

Recently Houghton et al.[30] analyzed the interannual variability of four climate variables from two 100-year simulations of two versions of a low resolution GCM: one with prescribed external boundary conditions (i.e., SSTs and sea ice); and one with simple ocean and sea-ice models. Variables included surface temperature, surface pressure, area covered by sea ice, and 700 mb temperature. Overall the authors conclude that the models reproduce the spatial and temporal characteristics of climate variability accurately enough to be useful for analysis of sources of interannual variability through feedbacks within the climate system.

The control run of a version of the NCAR CCM1 with a simple mixed layer ocean produced by Oglesby and Saltzman[31] has been examined for its accuracy in reproducing the interannual variability of surface temperature and precipitation for certain regions in the U.S.[32] The model tends to overestimate temperature variability (especially in summer), and errors in the estimation of precipitation variability were positively correlated with the errors in mean precipitation.

Santer and Wigley,[33] using a large battery of statistical methods presented in Wigley and Santer,[34] compared the interannual variability of mean sea level pressure (MSLP) of 4 GCMs (two versions of the Oregon State University (OSU) model, the low resolution GISS model, and the European Center for Medium Range Forecasting (ECMWF) T21 resolution model with prescribed sea surface temperatures) with observations. They found that for January three of the models in general underestimate the variance in the study region (North America/Atlantic/Europe), but all models underestimate it over the United States. In July (shown in Figure 1), the GISS and ECMWF models overestimate the variance over the entire study area, whereas the errors are mixed for the other two models. They also found much greater statistically significant differences in the mean values compared to that of the variances.

Blocking Events

A blocking event is roughly defined as a well developed stationary anticyclone (high pressure center) usually accompanied by surrounding lows, which tends to persist for five to 10 days. These persistent anomalies are a source of climate variability and are associated with persistent anomalous weather events such as droughts, heat waves, and cold snaps.

Bates and Meehl[35] used the CCM to investigate changes in the frequency of blocking events on a global scale under doubled CO_2 conditions. Their version of the CCM included a seasonal cycle, computed hydrology, and a simple mixed layer ocean. The statistics of the 500-mb height field (the height above the surface at which atmospheric pressure equals 500 mb) are examined, as are blocking events defined as persistent positive height anomalies (which indicate high pressure). In comparing modeled output with observed data for a 10-year period, they found that the model does a "reasonable" job of simulating 500-mb height standard deviations, in winter in both hemispheres, but tends to underestimate the variability in summer in both hemispheres. The model generally produces too few extreme blocking events.

A similar investigation of blocking events was performed for the Canadian Climate Centre GCM by Gough and Lin.[36] They found that the model blocks compared favorably with observations regarding spatial distribution and frequency, but that the model tended to underestimate the duration and magnitude of the height anomalies. These deficiencies were attributed to the use of prescribed sea surface temperatures in the model and relatively low model resolution.

Daily Variability

Two studies were conducted on local or regional scales using the U.K. Meteorological Office 5-layer GCM. Reed[37] analyzed observed versus model control run results for one gridpoint in eastern England. He anlayzed the mean and variance of daily temperature for a 3-year integration of model runs. Compared to observations, the model tended to produce temperatures that were too cool and variability that was too high as measured by the standard deviation. For precipitation, the model produced too many rain days but did not successfully simulate extreme rain events of greater than 20 mm/day.

More recently, Wilson and Mitchell[38] examined the modeled distribution of extreme daily climate events over Western Europe, using the same model. Comparisons

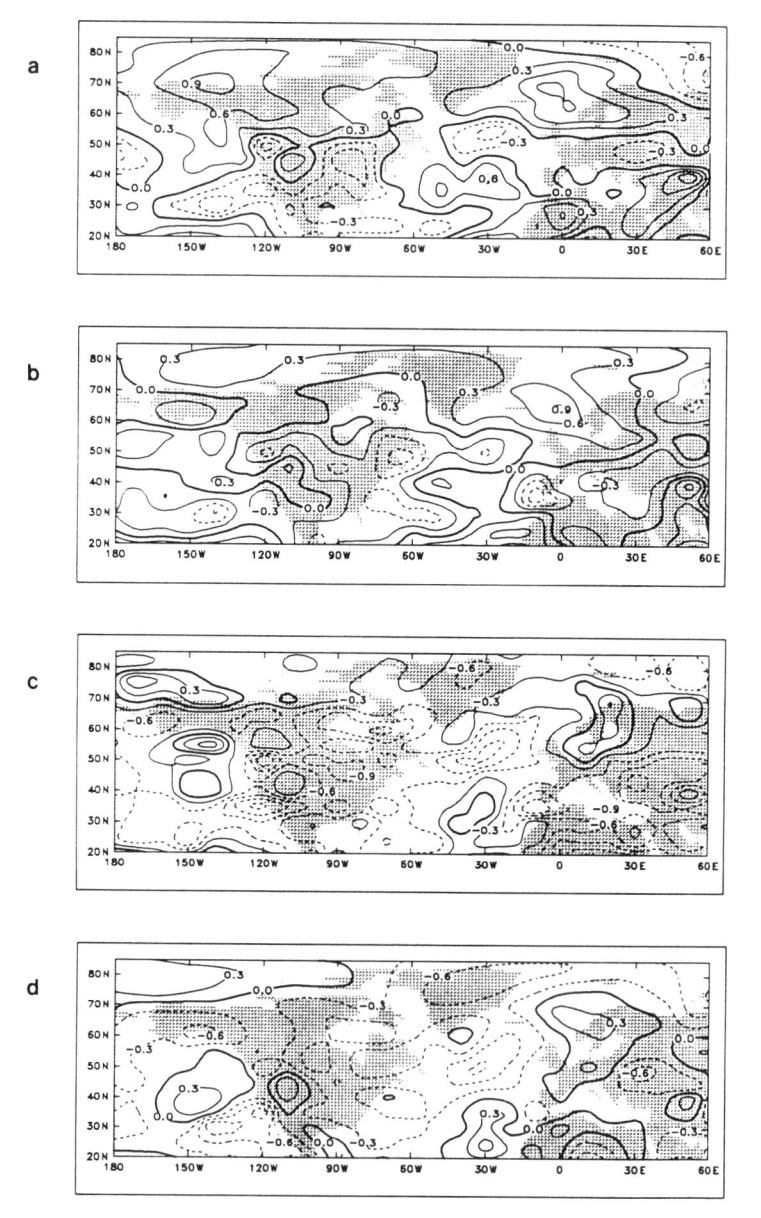

FIGURE 1. Variance ratios for July mean sea level pressure, UKMO observed (1971-80) divided by (a) OSU AGCM, (b) OSU CGCM, (c) GISS AGCM, and (d) ECMWF T21 model. AGCM indicates an atmospheric GCM and CGCM indicates a coupled ocean-atmosphere GCM. The isopleths show the logarithm of the variance ratio in order to identify unusually high or low ratios. Negative isopleths denote areas where the model variance is greater than observed. Source: Sanger and Wigley (1990).

were made of minimum surface temperatures and precipitation. Again, the model produced temperatures that were too cold, and hence, extreme minimum temperatures were overestimated. This problem was most pronounced in grid boxes away from the coasts. The model also produces too much precipitation in general, but does not successfully reproduce observed highest daily totals. The number of rain days is overestimated.

Rind et al.,[39] in a regional investigation of four areas in the U.S. found that in the GISS model interannual variability of temperature tended to agree with observations in most months, but was overestimated in the summer. Regarding precipitation, absolute interannual variability was in general overestimated. On a daily basis, temperature variability was usually overestimated as was daily precipitation. However, the relative variability of precipitation was generally in good agreement with observations.

In a study parallel to that of Rind et al.,[39] Mearns et al.[40] found the interannual variability of temperature in the Chervin[27] version of the NCAR CCM to be underestimated and the variability of precipitation to be overestimated. On a daily basis, several different versions of the CCM were examined. Most versions overestimated daily temperature variability (except one version with a more sophisticated surface package, which in general accurately reproduced the variability). Most versions also overestimated the daily variability of precipitation.

Meehl and Washington[41] found in two different versions of the CCM0 (one with an altered snow-sea-ice albedo parameterization) that daily variability of temperature tended to be overestimated, particularly in northern latitudes.

The possible connection between extreme daily sea surface temperature events and coral reef bleaching motivated Mearns[42] to examine daily sea surface temperatures in several versions of the NCAR CCM with simple mixed layer oceans, and one with the addition of horizontal ocean heat transport. She found that in several subtropical regions the daily variability of sea surface temperatures was greatly underestimated in all versions and that in one version the seasonal cycle dampened out too rapidly toward the equator. These results are consistent with those of Meehl and Washington,[43] who found daily standard deviations of SSTs in their mixed layer model to be underestimated by 50%. Figure 2 displays the observed and modeled daily January standard deviations from their work. It is unlikely that daily variability of SSTs would be much better represented in more complex ocean models, since the resolution and modeling of the mixed layer is still generally crude.[44]

The studies reviewed above indicate some important shortcomings of GCMs with regard to their ability to faithfully reproduce observed variability statistics. For example, most of the models overestimate daily temperature variability over land. More research is needed to further determine the sensitivity of the models to changes in physics, resolution, and so forth, with regard to the determination of variability. Studying the higher moments (e.g. variance) of climate variable statistics, and carefully verifying the models' ability to reproduce observed variability on regional scales, are the necessary prerequisites to rigorously analyzing possible changes in these statistics under doubled CO_2 conditions.

218 Global Climate Change: Implications, Challenges and Mitigation Measures

CHANGES IN VARIABILITY IN PERTURBED CLIMATE MODEL RUNS

The studies described below are good representatives of the most recent work analyzing changes in variability in climate models. However, none of these results should be taken as definitive predictions of changes in climate variability. The errors in control runs described above should be kept in mind when considering these results.

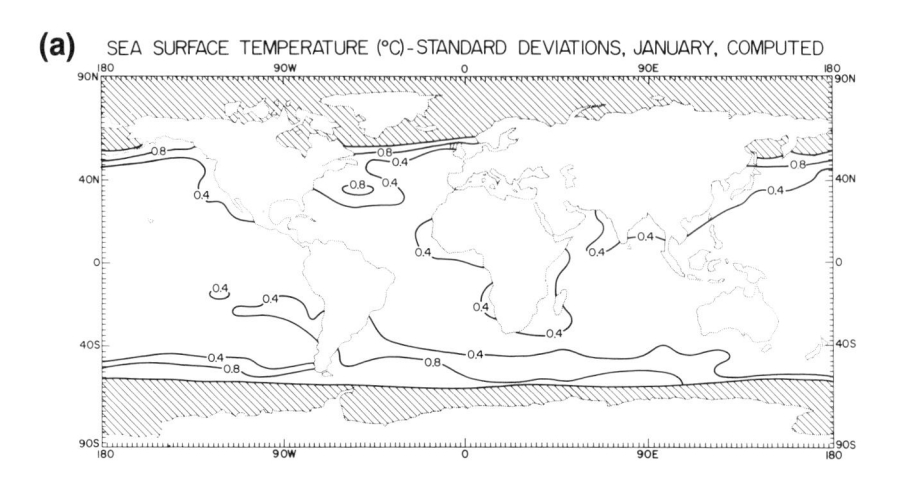

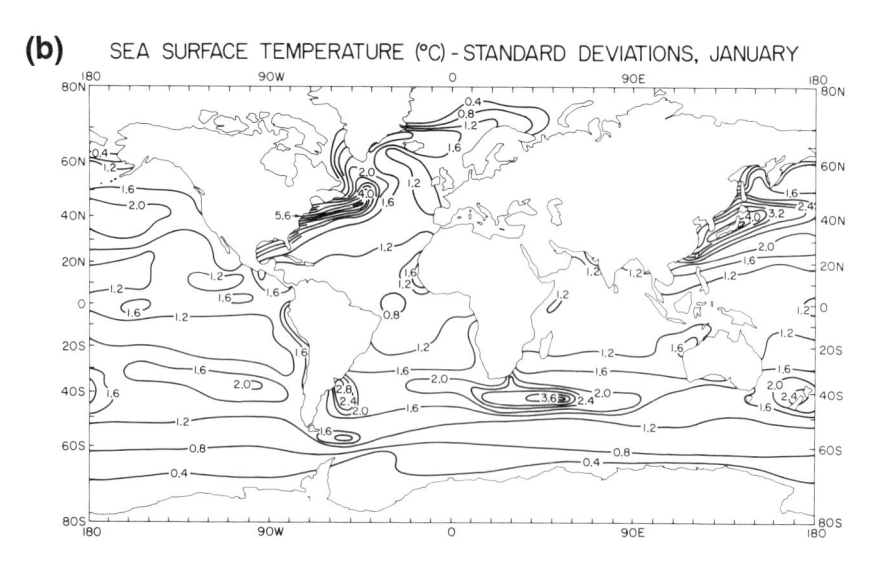

FIGURE 2. Standard deviations of daily SSTs from long-term monthly means (°C): (a) January, computed; (b) January, observed. Source: Meehl and Washington (1985).

Interannual variability

Rind et al.[39] found a general decrease in interannual temperature variability and an increase in precipitation variability in their regional analysis of the U.S.

Rind[45] provides some further evidence from the GISS model in a series of experiments in which past climates were simulated (e.g., the last ice age, 18,000 years B.P.; the Younger Dryas, 10,000 years B.P.; and the warm Cretaceous, 65 million years B.P.), as well as further analysis of a doubled CO_2 experiment, that there is in general a decrease in interannual temperature variability under warmer average conditions and an increase under colder average conditions. Figure 3 portrays the results for the doubled CO_2 experiment for January. The general decrease in variability is widespread, although there are small areas of increase. Results for precipitation for these experiments were mixed.

In the doubled CO_2 runs of Oglesby and Saltzman,[31] Mearns,[46] in a global analysis of January and July, found mixed results for changes in interannual variability of temperature and largely increases in the variability of precipitation (correlating with areas of increased mean precipitation).

In a review of changed interannual variability of temperature of five of the models evaluated in the Intergovernmental Panel on Climate Change Report[7] Mitchell et al.[47] found no meaningful patterns apart from reductions in the vicinity of sea-ice margins in winter. Similar results were found for two of the models with regard to daily temperature variability.

Giorgi et al.,[48] in the first completed perturbed climate experiment using a mesoscale model nested within a GCM, provide a brief analysis of changes in interannual variability of temperature and precipitation for Europe. They found in

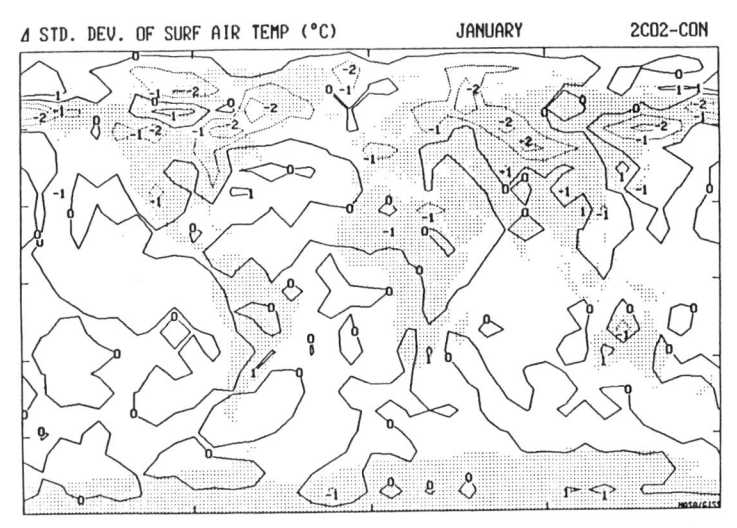

FIGURE 3. Change in the interannual standard deviation of surface air temperature during January between the doubled CO_2 experiment and its current climate control. Values are calculated from 10-year simulations each. Source: Rind (1991).

general decreases in temperature variability and increases in precipitation variability for January, April, July, and October. However, the sample size is very small (five years of control and doubled CO_2 runs), and no statistical tests were performed.

Blocking Events

Under doubled CO_2 conditions, Bates and Meehl[35] found that standard deviations of 500-mb height and blocking activity mainly decreased in all seasons (i.e., the variability of blocking events decreased). More specifically there was a small decrease in blocking frequency in the North Pacific region and the Icelandic area, and a substantial decrease in events over northern Siberia. These results are portrayed in Figure 4. Events were generally also reduced in the Southern Hemisphere.

Daily Variability

Wilson and Mitchell[38] examined changes under quadrupled CO_2 conditions and found that daily variability of temperature most often decreased. However, when the authors compared the changes in daily temperature variability using the rigorous procedures of Katz[49] as well as using the more standard F-test, both tests indicated a significant decrease only in winter temperature variability, whereas results for the other seasons were mixed or not significant.

Rind et al.[39] found results which paralleled those for interannual variability. However, the decreases in daily temperature variability were for the most part statistically insignificant. Results for daily precipitation variability indicated a general increase in variability which tended to be positively correlated with the change in mean precipitation.

Mearns et al.,[40] found more mixed results (compared to Rind et al.[39]) in analyzing changes in daily temperature and precipitation variability in the NCAR CCM comparing the control and the doubled CO_2 runs in several regions of the U.S. Very few of the temperature changes were statistically significant and both increases and decreases were found. There was more agreement in changes in daily precipitation variability, which in most cases, increased.

DISCUSSION AND CONCLUSIONS

The studies described above show some consistency in their results. Although there is some indication from these studies that temperature variability (both daily and interannual) may decrease, and precipitation variability may generally increase as the climate warms, no firm conclusion can be drawn at this point. It still must be stated that the direction of change of climate variability in a future warmer climate remains uncertain, and that it is quite possible that different regions could experience different directions of change, especially when considering precipitation.

Rind[45] provides a reasonable dynamical argument regarding the expected decrease of temperature variability based on the decreased equator to pole gradient of temperature, which would reduce the temperature contrast of air masses and reduce

Δ BLOCKING EVENTS
2×CO₂ MINUS CONTROL
DJF

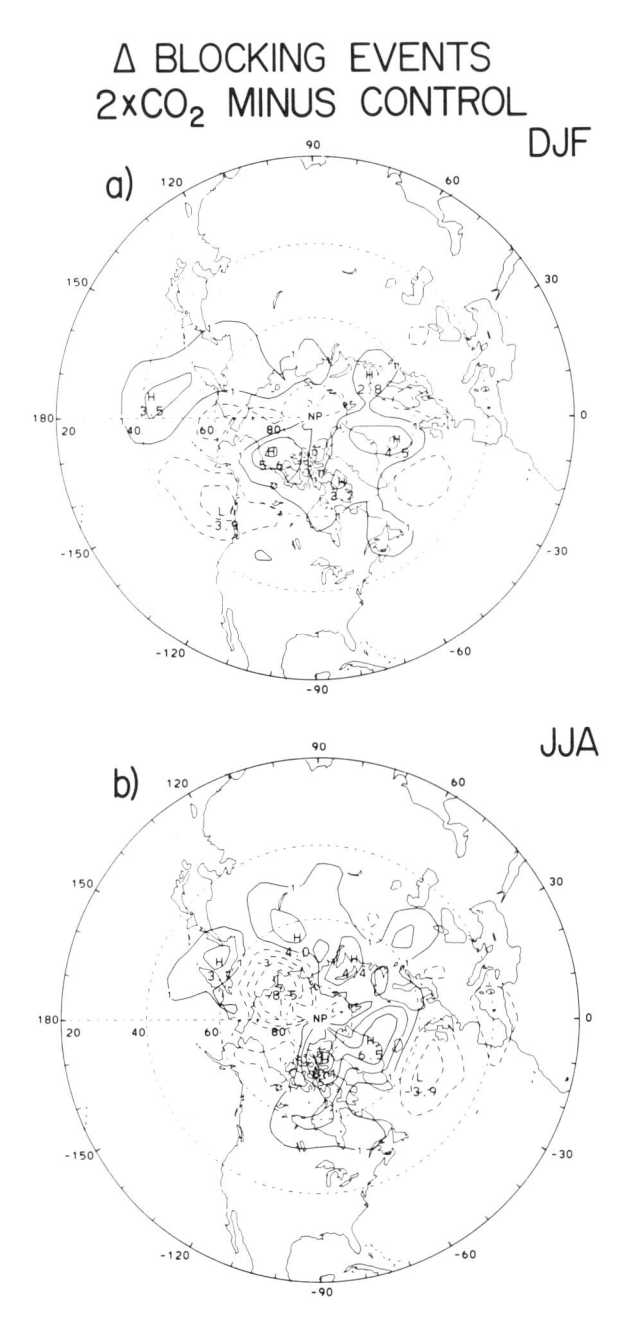

JJA

FIGURE 4. Differences between number of Northern Hemisphere blocking events per 15 years in the NCAR CCM, 2 x CO_2 minus control in (a) winter and (b) summer. The contour interval is two events. Negative isopleths indicate a decrease in blocking frequency. Source: Bates and Meehl (1986).

the intensity of extratropical storms. However, shifts in general circulation patterns, even though some aspects may be weakened, could result in shifts in temperature variability in either direction for a given region. There is far from universal agreement among modelers and climate dynamicists concerning the expected direction of change of climate variability.

It should also be noted that the studies described in this chapter underline the importance of viewing climate change results of the models in the context of how well they reproduce the present climate. Model deficiencies can be expected to limit the reliability of climate change results, and faith in quantitative results are probably unwarranted. A major model deficiency is the inability to resolve subgrid scale atmospheric phenomena which contribute to climate variability, such as fronts and intense cyclones (hurricanes); and important variations in atmosphere-ocean coupling such as ENSO (El Niño Southern Oscillation) events. (More sophisticated GCMs incorporating complete ocean models now do produce ENSO type events,[13,50] and should be more successful at reproducing interannual variability). However, model results do give crude estimates as to the importance of some physical processes responsible for variability and what must be done to improve them (e.g., the importance of surface/atmosphere interactions in producing daily temperature variability).

There is a need for further testing to determine how the models' deficiencies in reproducing present-day climate affects predictions for a CO_2-warmed future climate.

The research reported above clearly indicates that research of changes in climate variability with climate change is in its infancy. Future research needs include further investigation of variability in presently existing GCM control and perturbed runs. Results summarized here represent only an initial effort at looking at variability in GCMs. Other time scales of variability should also be examined such as 7 to 10 day scales, which correspond to the life time of many frontal storms, etc. In addition, the analytical and statistical techniques used must be carefully examined, and only appropriate ones employed. Katz[51] points out some problems with some of the statistical techniques used so far.

Furthermore, the model deficiencies mentioned above must be rectified so that future models will be better able to reproduce variability on different time scales. For example, a number of experiments concerning improved spatial resolution[52-54] have demonstrated the importance of resolution in properly reproducing climate. Other improvements underway include the implementation of improved surface packages (vegetation/atmosphere interactions),[55-57] better cloud parameterizations,[58,59] and coupling of atmospheric GCMs with full ocean GCMs.[44,60,61] It can be anticipated that significantly more will be known from modeling experiments about possible changes in climatic variability in the next several years.

ACKNOWLEDGMENTS

This work was partially supported by a grant from the U.S. Environmental Protection Agency.

REFERENCES

1. Mearns, L.O., R.W. Katz, and S.H. Schneider. 1984. Extreme high-temperature events: Changes in their probabilities with changes in mean temperature. *Journal of Climate and Applied Meteorology* 23:1601-1613.
2. Parry, M.L. and T.R. Carter. 1985. The effect of climatic variations on agricultural risk. *Climatic Change* 7:95-110.
3. Wigley, T.M.L. 1985. Impact of extreme events. *Nature* 316:106-107.
4. Wigley, T.M.L. 1988. The effect of changing climate on the frequency of absolute extreme events. *Climate Monitor* 17:44-55.
5. Katz, R.W. and B.G. Brown. 1992. Extreme events in a changing climate: Variability is more important than averages. *Climatic Change* (in press).
6. Smith, J.B. and D.A. Tirpak (Eds.). 1989. Report to Congress on the Potential Effects of Global Climate Change on the U.S. EPA-220-05-89-050, EPA, Washington, D.C.
7. Houghton, J.T., G.T. Jenkins and J.J. Ephraums (Eds.). 1990. *Climate Change: The IPCC Scientific Assessment,* Report prepared for the IPCC by Working Group I. Cambridge: University Press, 365 pp.
8. Gibbs, W.J., J.V. Maher and M.J. Coughlan. 1975. Climatic variability and extremes. In: Pittock, A.B. *et al.* (Eds.) *Climatic Change and Variability.* Cambridge U. Press, Cambridge, U.K. pp. 135-150.
9. Berger, A. 1980. Spectrum of climate variations and possible causes. In: Berger (Ed.) *Climatic Variations and Variability: Facts and Theories.* D. Reidel, Dordrecht, Holland. pp. 411-432.
10. Walsh, J.E. and C.M. Johnson. 1979. Interannual atmospheric variability and associated fluctuations in Arctic sea ice extent. *J. of Geophys. Res.* 84(C):6915-6928.
11. Pittock, A.B. 1975. Patterns of variability in relation to the general circulation. In Pittock, A.B. *et al.* (Eds.) *Climatic Change and Variability*, Cambridge U. Press, Cambridge, U.K. pp. 167-178.
12. Pitcher, E.J., M.L. Blackmon, M.L. Gates, G.T. Bates, and S. Munoz. 1988. The effect of north Pacific sea surface temperature anomalies on the January climate of a general circulation model. *J. Atmos. Sci.* 45:173-188.
13. Meehl, G.A. 1990. Seasonal cylcle forcing of El Niño-Southern Oscillation in a global coupled ocean-atmosphere GCM. *J. of Climate* 3:72-98.
14. Gordon, H.B. and B.G. Hunt. 1991. Droughts, floods, and sea-surface temperature anomalies: A modeling approach. *Int. J. of Climatology* 11:347-365.
15. Mitchell, J.M. 1976. An overview of climatic variability and its causal mechanisms. *Quaternary Research* 6:481-493.
16. Brinkmann, W.A.R. 1983. Variability of temperature in Wisconsin. *Mon. Wea. Rev.* 111:172-180.
17. van Loon, H. and J. Williams. 1978. The association between mean temperature and interannual variability. *Mon. Wea. Rev.* 106:1012-1017.

18. Chico, T. and W.D. Sellers. 1979. Interannual temperature variability in the United States since 1896. *Climatic Change* 2:139-147
19. Diaz, H.F. and R.G. Quayle. 1980. The climate of the United States since 1895: Spatial and temporal changes. *Mon. WEa. Rev.* 108:249-266.
20. Lough, J.M., T.M.L. Wigley and J.P. Palutikof. 1983. Climate and climate impact scenarios for Europe in a warmer world. *J. of Clim. and Appl. Meteorol.* 22:1673-1684.
21. Schuurmans, C.J.E. and A.J. Coops. 1984. Seasonal mean temperatures in Europe and their interannual variability. *Mon. Wea. Rev.* 112:1218-1225.
22. Crowley, T.J. 1990. Are there any satisfactory geologic analogs for a future greenhouse warming? *J. Clim.* 3:1282-1292.
23. MacCracken, M.C. and J. Kutzbach. 1991. Comparing and contrasting Holocene and Eemian warm periods with greenhouse-gas-induced warming. In: Schlesinger, M., (Ed.), *Greenhouse-Gas-Induced Climatic Change: A Critical Appraisal of Simulations and Observations.* Elsevier: New York. pp. 17-34.
24. Mitchell, J.F.B., C.A. Wilson and W.M. Cunnington. 1987. On CO_2 climate sensitivity and model dependence of results. *Quart. J. Roy. Meteorol. Soc.* 113:293-322.
25. Manabe, S. and D.G. Hahn. 1981. Simulation of atmospheric variability. *Mon. Wea. Rev.* 109:2260-2286.
26. Chervin, R.M. 1981. On the comparison of observed GCM simulated climate ensembles. *J. of Atmos. Sci.* 38:885-901.
27. Chervin, R.M. 1986. Interannual variability and seasonal predictability. *J. of Atmos. Sci.* 43:233-251.
28. Hansen, J., I. Fung, A. Lacis, S. Lebedeff, D. Rind, R. Ruedy, G. Russell, P. Stone. 1988. Global climate changes as forecast by the Goddard Institute for Space Studies three-dimensional model. *J. of Geophys. Res.* 93(D8):9341-9364.
29. Portman, D.A., W.C. Wang and T.R. Karl. 1990. A comparison of general circulation model and observed regional climates: daily and seasonal variability. In: *Proceedings of the Fourteenth Annual Climate Diagnostics Workshop.* U.S. Dept. of Commerce, National Oceanic and Atmospheric Administration: Washington, D.C. pp. 282-288.
30. Houghton, D.D., R.G. Gallimore and L.M. Keller. 1991. Stability and variability in a coupled ocean-atmosphere climate model: Results of 100-year simulations. *J. of Clim.* 4:557-577.
31. Oglesby, R.J. and B. Saltzman. 1990. Sensitivity of the equilibrium surface temperature of a GCM to systematic changes in atmospheric carbon dioxide. *Geophy. Res. Letters.* 17:1089-1092.
32. Mearns, L.O. 1990. (Personal Communication).
33. Santer, B.D. and T.M.L. Wigley. 1990. Regional validation of means, variances, and spatial patterns in general circulation model control runs. *J. of Geophys. Res.* 95(D):829-850.
34. Wigley, T.M.L and B.D. Santer. 1990. Statistical comparison of spatial fields in model validation, perturbation, and predictability experiments.
35. Bates, G.T. and G.A. Meehl. 1986. The Effect of CO_2 concentration on the

frequency of blocking in a general circulation model coupled to a simple mixed layer ocean model. *Mon. Wea. Rev.* 114:687-701.
36. Gough, W.A. and C.A. Lin. 1987. Distribution of blocks in data from the Canadian Climate Centre general circulation model. *Clim. Bull.* 21:3-15.
37. Reed, D.N. 1986. Simulation of time series of temperature and precipitation over eastern England by an atmospheric general circulation model. *J. of Clim.* 6:233-257.
38. Wilson, C.A. and J.F.B. Mitchell. 1987. Simulated climate and CO_2-induced climate change over Western Europe. *Climatic Change* 10:11-42.
39. Rind, D., R. Goldberg and R. Ruedy. 1989. Change in climate variability in the 21st century. *Climatic Change* 14:5-38.
40. Mearns, L.O., S.H. Schneider, S.L. Thompson an L.R. McDaniel. 1990. Analysis of climate variability in general circulation models: Comparison with observations and changes in variability in 2XCO_2 experiments. *J. Geophys. Res.* 95:20,469-20, 490.
41. Meehl, G.A. and W.M. Washington. 1990. CO_2 climate sensitivity and snow-sea-ice albedo parameterization in an atmospheric GCM coupled to a mixed-layer ocean model. *Climatic Change* 16:283-306.
42. Mearns, L.O. 1991a. Changes in sea-surface temperature extremes in the tropics under global warming. Paper presented at the Workshop on Coral Bleaching, Coral Reef Ecosystems, and Global Climate Change, held in Miami Florida, June 18-21, 1991. pp. 15.
43. Meehl, G.A. and W.M. Washington. 1985. Seas surface temperatures computed by a simple mixed layer coupled to an atmospheric GCM. *J. of Phys. Oceanography* 15:92-103.
44. Washington, M. and G.A. Meehl. 1989. Climate sensitivity due to increased CO_2: Experiments with a coupled atmosphere and ocean general circulation model. *Climate Dynamics* 4:1-38.
45. Rind, D. 1991. Climate variability and climate change. In: Schlesinger, M., (Ed.), *Greenhouse-Gas-Induced Climatic Change: A Critical Appraisal of Simulations and Observations.* Elsevier: New York. pp. 69-78.
46. Mearns, L.O. 1991b. Changing climate variability with greenhouse warming and its possible impacts. Paper presented at the First Nordic Inter-disciplinary Conference on the Greenhouse Effect. Copenhagen, 16-18 September 1991. (Proceedings in preparation).
47. Mitchell, J.F.B., S. Manabe, V. Meleshko, T. Tokioka. 1990. Equilibrium climate change—and its implications for the future. In Schlesinger, M., (Ed.), *Greenhouse-Gas-Induced Climatic Change: A Critical Appraisal of Simulations and Observations.* Elsevier: New York. pp. 17-34.
48. Giorgi, F., M.R. Marinucci and G. Visconti. 1992. A 2XCO_2 climate change scenario over Europe generated using a limited area model nested in a general circulation model. II. Climate change scenario. *J. Geophys. Res.* (in press).
49. Katz, R.W. 1988. Statistical procedures for making inferences about changes in climate variability. *J. Clim.*, 1:1057-1058.
50. Meehl, G.A. 1991. El Niño-Southern Oscillation and CO_2 climate change.

submitted to *J. of Clim.*

51. Katz, R.W. 1992. The role of statistics in the evaluation of general circulation models. *Climate Research* (in press).
52. Boer, G.J. and M. Lazare. 1988. Some results concerning the effect of horizontal resolution and gravity wave drag on simulated climate. *J. Clim.* 1:789-806.
53. Kiehl, J.T. and D.L. Williamson. 1991. Dependence of cloud amount on horizontal resolution in the National Center for Atmospheric Research Community Climate Model. *J. Geophys. Res.* 96(D): 10955-10980.
54. Boville, B.A. 1991. Sensitivity of simulated climate to model resolution. *J. Clim.* 4:469-485.
55. Sellers, P.J., Y. Mintz, Y.C. Sud and A. Dalcher. 1986. A simple biosphere model (SiB) for use within general circulation models. *J. Atmos. Sci.* 43:505-531.
56. Dickinson, R.E. and A. Henderson-Sellers. 1988. Modelling tropical deforestation: a study of GCM land-surface parameterizations. *Quart. J. of Roy. Met. Soc.* 114:439-462.
57. Verseghy, D.A. 1991. Class — A Canadian land surface scheme for GCMs. I. Soil model. *Int. J. Climatol.* 11:111-134.
58. Mitchell, J.F.B., C.A. Senior and W.J. Ingram. 1989. CO_2 and climate: a missing feedback. *Nature* 341:132-134.
59. Slingo, A. and J.M. Slingo. 1991. Response of the National Center for Atmospheric Research Community Climate Model to improvements in the representation of clouds. *J. Geophys. Res.* 96(D):15,341-15,357.
60. Stouffer, R.J., S. Manabe and K. Bryan. 1989. Interhemispheric asymmetry in climate response to a gradual increase of atmospheric CO_2. *Nature* 342:660-662.
61. Cubasch, U. 1991. Preliminary assessment of the performance of a global coupled atmosphere-ocean model. In: Schlesinger, M., (Ed.), *Greenhouse-Gas-Induced Climatic Change: A Critical Appraisal of Simulations and Observations.* Elsevier: New York. pp. 137-150.

Global Climate Change: Implications, Challenges and Mitigation Measures. Edited by S.K. Majumdar, L.S. Kalkstein, B. Yarnal, E.W. Miller, and L.M. Rosenfeld. © 1992, The Pennsylvania Academy of Science.

Chapter Fifteen

ZONAL COMPARISONS OF GLOBAL CIRCULATION MODEL: Temperature and Precipitation Data with Historical Climate for North America and Eurasia

ANTHONY J. BRAZEL[1]
and
ROBERT A. MULLER[2]

[1]Office of Climatology
Department of Geography
Arizona State University
Tempe, Arizona 85287-1508
and
[2]Southern Region Climate Center
Department of Geography and Anthropology
Louisiana State University
Baton Rouge, Louisiana 70803

INTRODUCTION

Currently, in the field of Climatology much emphasis is being placed on the evaluation of Global Circulation Models (GCMs) and their applicability in order to project global changes of the climate system, particularly in relation to effects of present and projected changes of atmospheric CO_2 levels and of other greenhouse gases (e.g., Bolin, *et al.* 1986). As part of this evaluation process, and as an overall procedure in the scientific method, models should be verified for their validity in representing an observable part of the climate system, before much faith is put in long-term projected results of CO_2 doubling simulations (Grotch and MacCracken, 1991). The predominant procedures for verification as suggested by Schneider (1990) are: (1) checking overall model-simulation skill against the real climate for today (e.g., Grotch and MacCracken, 1991), (2) testing individual physical subcomponents of the model in isolation (e.g., Gutowski, Gutzler, and Wang, 1991), (3) comparing daily variability of model-generated and observed grid-point statistics (e.g., Rind, Goldberg, and Ruedy, 1989), and (4) verifying the model's ability to simulate the very different climates of the ancient earth or even of other planets (COHMAP, 1988).

This chapter intercompares four GCMs in common use today with the so-called RAND climate dataset (Schutz and Gates, 1971) for selected latitudinal transects across North America and Eurasia. The four model runs are by the Geophysical-Fluid Dynamics Laboratory (GFDL-Manabe and Wetherald, 1987), Goddard Institute for Space Studies (GISS-Hansen, et al. 1984), Oregon State University (OSU-Schlesinger and Zhao, 1989), and the United Kingdom Meteorological Office (UKMO-Mitchell, 1988). They are listed in Table 1, together with selected model characteristics, such as date of model run, resolution, atmospheric layers modeled, treatment of diurnal processes, and general global projections for doubled CO_2. This intercomparison arose from an EPA-funded project (Kalkstein, 1991), in which the earth's land surface areas (outside of the polar regions) were assessed.

The two continents present similiar challenges in the GCM modeling process. Models must treat topography, land surface conditions, surface hydrology, and continental border geography distinct from oceanic conditions. Manabe and Broccoli (1990) have simulated global climate with and without orography in order to investigate the role of mountains in maintaining extensive arid climates in middle latitudes of the Northern Hemisphere. Ruddiman and Kutzbach (1991) recently analyzed the influence of slow uplift mountain-making processes and the resultant modeled paleoclimatic change effects occurring on the two continents, using a version of the NCAR CCM model. They present paleoclimatic evidence for the verification of modeled projections of the mountain and plateau uplift effects.

TABLE 1
Characteristics of Four GCM Models*

Model	Date	Grid Resolution	Layers	Diurnal	2xCO_2 Temp	Precip
GFDL	1984-85	4.44 x 7.50°	9	No	+4.0°C	+8.7%
GISS	1982	7.83 x 10.00°	9	Yes	+4.2°C	+11.0%
OSU	1984-85	4.00 x 5.00°	2	No	+2.8°C	+7.8%
UKMO	1986	5.00 x 7.50°	11	Yes	+5.2°C	+15.0%

*After Jenne, R.L. (1988).
Date = date of model run.
Grid resolution = latitude x longitude.
Layers = number of layers in atmosphere modeled.
Diurnal = whether diurnal processes treated in models.

These two studies simulated our thinking for a contemporary intercomparison and verification of GCMs for the two continents—especially how the models compare with climate data for areas of complex topography and land surface variation. Thus, in the case of North America, how models perform in the western versus eastern sectors of the continent is a critical comparison of the topographic orientation of mountain ranges, and of the contrasts in southerly latitudes between western deserts and eastern humid environments. In the case of Eurasia, the variations from high latitudes (relatively uniform surface conditions) to the southern periphery of the study area (Himalayan region) is deemed critical, as these comparisons will highlight topographic variation influences and the model-RAND comparisons for the vast deserts of the Middle East and China. Therefore, our analysis expresses the geographical commonalities and differences between North America and Eurasia in relation to model agreement with contemporary climate data.

METHODS OF ANALYSIS

Grid point output of temperature and precipitation data for the four models (1 x CO_2 equilibrium model runs) and RAND climate were obtained from the National Center for Atmospheric Research (D. Joseph) under a program initiated by the Environmental Protection Agency's Office of Policy Analysis (Kalkstein, 1991). These data were obtained in digital, PC compatible form, with accompanying software for tabulating these data into sets of parameter outputs arranged in geographic latitude/longitude matrices. No spatial transformation or interpolation programs were applied to frame outputs into common grid cells.

The raw temperature and precipitation data for each grid cell are to the nearest hundredths of a °C and mmd^{-1}. We have relaxed our comparisons of model and RAND zonal means by rounding data to whole °C values and tenths of mmd^{-1}. However, in the text we mention specified raw data values from time to time. Furthermore, precision in the comparison may be relaxed somewhat, since model versus RAND comparisons are also compounded by the fact that temperature simulations in each model do not necessarily represent exact "shelter height" conditions of the surface observations (Grotch, 1988).

Longitudinal transects in North America and Eurasia, spanning land areas across 58°, 50°, 42°, and 34°N latitude, are analyzed. The focus is at two levels of detail or scale. The first level of generalization is zonal comparisons of January and July mean temperatures, December-January-February (DJF) and June-July-August (JJA) mean daily precipitation rates for the land areas at the specified latitudes indicated above (similar to analyses of Grotch, 1988 and Grotch and MacCracken, 1991).

The second comparison is a graphical presentation of individual grid values of temperature and precipitation, from the four models and the RAND data, for selected times and places. The purpose of the graphics is to provide insights into longitudinal variations of model differences from RAND. This latter method expresses, in a qualitative manner, regional variations of model differences from present climate on the continents—variations that may not be well expressed by the zonal mean and which may result from topography and inhomogeneity of land surface conditions.

The inferences made from the mean zonal tabular information are presented with the caveat that at these four latitudes, models do not directly conform to RAND spacing and data frequency. However, the spacing differences of the 34°, 42°, 50°, and 58°N latitude grids of RAND from model grid center points are at their minimum, and the "overlapping" of grids is at a maximum for these latitude transects. In fact, the RAND grid and the OSU model grid systems are centered along these exact four latitudes (and are at identical resolutions). The other model center points are quite close, but are different from RAND and OSU. Thus, in the final analysis, the RAND/OSU comparison is deemed more accurate relative to the grid mismatch problem.

Table 2 lists the exact grid center points chosen for the comparisons with RAND data and the differences in latitude between RAND center points and model points. Differences range from 0.0° to 2.5° of latitude. The GISS grid cells, of course, represent larger regions than the other models, since resolution is largest—7.8° x 10°. In descending order, UKMO, GFDL, and OSU represent smaller areas per grid. Thus, the results of the zonal mean comparisons of RAND and models, and the raw graphical output of grid transects, are not directly comparable and may contain a certain degree of bias. In descending order, the largest to smallest biases due to the grid latitude mismatches exists for UKMO (at 50°), GFDL (at 58°), UKMO (at 34°), GISS (at 34°), and GFDL (at 50°). All these deviations of RAND latitude centers from model grid point centers are >1.0° of latitude. For the rest of the comparisons, these deviations are <1.0°. Distortions in areal representation certainly may also exist due to resolution differences among models. No corrections are made for these differences in this paper. Methods of grid transformations to a common grid reference system were thought to present similar biases to those mentioned above and may compound the difficulties in attempting to understand the RAND-model intercomparisons (Kalkstein, 1991).

Zonal means were areally weighted similar to the procedure used by Grotch and MacCracken (1991). Model minus RAND values were multiplied by the cosine of the latitude (retained in Table 4), summed over the four latitudes, averaged, and

TABLE 2
Grid Center Latitudes (in °) of Models Used to Represent Given Latitudes of Rand Climate

Model	RAND	GFDL	GISS	OSU	UKMO
58°N	58	59.99 (1.99)	58.70 (0.70)	58	57.5 (−0.5)
50°N	50	51.11 (1.11)	50.87 (0.87)	50	52.5 (2.5)
42°N	42	42.22 (0.22)	43.04 (1.03)	42	42.5 (0.5)
34°N	34	33.33 (−.67)	35.22 (1.22)	34	32.5 (−1.5)

N.B.: Differences from RAND values shown in ( )

divided by the average cosine of the four latitudes to express the overall differences over the range of latitudes studied (see Table 5).

The RAND dataset has been compared to a more recent dataset after Oort (1983) by Grotch and MacCracken (1991). Their comparison indicates an average absolute temperature difference of <0.7°C. For precipitation, Grotch and MacCracken (1991) used the Jaeger (1976) dataset to characterize seasonal values. We use the older Schutz and Gates (i.e., RAND) data.

The reader is cautioned on the degree of confidence to be placed on the use of gridded forms of observed point data to characterize the climate of various regions of the earth. Much research is still needed to determine how to better represent the real climate of the earth and how to compare these data with simulated gridded datasets (e.g., Karl, et al. 1990). Several biases may be imbedded in point and regionally averaged statistics and thus in gridded datasets: (a) urban effects, (b) temporal variations in quality control procedures of governmental weather services, (c) differing sensor and shelter deployment through time, (d) changes in observing schedules over time, (e) local horizontal and elevational relocations of stations,

TABLE 3

Zonal Means* for North America (NA) and Eurasia (EA) Land Areas — Rand Climate and Models

	North America				Eurasia				
Model:	RAND	GFDL	GISS	OSU	UKMO	RAND	GFDL	GISS	OSU	UKMO
Lat:										
January Temp:
58°N	−22	−22	−20	−13	−18	−18	−23	−18	−18	−21
50°N	−14	−14	−7	−8	−11	−13	−17	−12	−12	−18
42°N	−4	−8	−4	−4	−4	−4	−11	−2	−5	−6
34°N	6	2	3	6	7	−5	−18	−9	−10	−16
July Temp:
58°N	13	14	10	14	13	17	18	18	19	17
50°N	17	16	16	19	19	19	22	23	21	18
42°N	23	21	17	21	20	22	21	22	20	22
34°N	27	24	22	24	28	19	17	14	15	17
DJF Precip:
(mm/day x 10)
58°N	12	24	14	13	13	11	19	13	10	10
50°N	19	29	27	18	22	12	16	15	14	11
42°N	18	37	29	22	29	12	17	20	19	17
34°N	28	31	24	21	16	10	22	22	15	12
JJA Precip:
(mm/day x 10)
58°N	17	23	27	16	26	20	26	22	11	27
50°N	21	27	31	16	36	19	18	27	13	31
42°N	22	30	31	20	35	12	25	25	17	27
34°N	25	27	27	15	27	32	37	44	22	52

*Means are unweighted by latitude.

N.B.: Temperatures rounded to nearest whole °; precipitation rounded to tenths of a mm. Model and RAND raw data output are typically stated to nearest 1/100ths °C and mm/d^{-1}.

(f) the numbers of stations representing regional grid averages, and (g) low numbers of stations to represent remote and/or mountainous regions (Jones, *et al.* 1990; Karl, *et al.* 1989).

We make the assumption that the RAND dataset represents observations against which modeled data can be verified. However, some caution should be exercised in interpretating differences of modeled output from the RAND dataset. This caution is particularly important for mountainous locations (e.g., Gates, 1985; Brazel and Marcus, 1991). A larger departure pattern of modeled output from RAND values, that might be evident for data deficient regions of the continents, should not necessarily infer a proportionally larger "error" in models. Also, since spatial variations in precipitation are large and cause difficulty in areal averaging processes applied to real data, apparent "errors" of the models for this parameter may appear larger than for temperature. Thus, it is expected that modeled temperatures would be relatively closer to RAND data than modeled precipitation, but not solely because of model construct errors for this parameter. On the other hand, it also holds that precipitation is a more difficult parameter to model in a GCM (e.g., Hansen *et al.* 1983).

TABLE 4

North America (NA) and Eurasia (EA)
July-January Temperature Range (°C) Rand Climate and Models

Model:	RAND	GFDL (GFDL-RAND)	GISS (GISS-RAND)	OSU (OSU-RAND)	UKMO (UKMO-RAND)
Latitude:		North America			
58°N	36	37 (1)	29 (−7)	26 (−10)	31 (−5)
50°	31	30 (−1)	23 (−8)	28 (−3)	31 (0)
42°N	27	28 (1)	21 (−6)	24 (−3)	24 (−3)
34°	21	27 (6)	19 (−2)	19 (−2)	21 (0)
		Eurasia			
58°N	35	41 (6)	35 (0)	37 (2)	38 (3)
50°N	32	39 (7)	35 (3)	33 (1)	36 (4)
42°N	26	32 (6)	24 (−2)	25 (−1)	29 (3)
34°N	24	35 (11)	23 (−1)	25 (1)	29 (5)

N.B.: Temperatures are rounded to nearest whole°.

RESULTS OF MEAN ZONAL COMPARISONS

Tables 3-6 list the comparisons between the four models and RAND, arranged by latitude sectors across the Eurasian and North American continental land areas. Table 3 shows the zonal mean temperature and precipitation values for the models and RAND for each latitude transect; Table 4, the model area weighted July-January mean temperature values; Table 5, the 34° to 58°N latitude areal weighted values of model-RAND differences; and Table 6, the % departure [i.e., (model-RAND) x 100%/RAND] for seasonal precipitation.

TABLE 5
Zonal Areal Weighted Mean Differences, Models - Rand for Land Areas
of North America (NA) and Eurasia (EA)

	North America				Eurasia			
	GF-R	GI-R	OS-R	UK-R	GF-R	GI-R	OS-R	UK-R
Jan. Temp:								
58°N	0	1	5	2	-2	0	0	-1
50°N	0	4	3	2	-2	0	1	3
42°N	-3	0	0	0	-5	2	-1	-2
34°N	-3	-3	-1	1	-10	-3	-4	-9
Domain	-2.3	1.1	2.9	1.5	-7.1	-0.1	-1.5	-5.4
July Temp:								
58°N	1	-2	0	-1	1	0	1	0
50°N	-1	-1	1	1	2	2	1	-1
42°N	-2	-4	-2	-2	-1	-1	-2	0
34°N	-3	-4	-2	0	-1	-4	-3	-5
Domain	-1.6	-4.1	-1.0	-0.4	0.1	-0.7	-1.1	-2.0
DJF Precip:								
(mm/day x 10)								
58°N	6	1	1	1	4	2	-1	0
50°N	6	5	-1	2	2	2	1	1
42°N	2	2	1	1	14	8	3	8
34°N	3	-4	-6	-10	11	10	4	2
Domain	11	4	-1	0	8	7	4	2
JJA Precip:								
(mm/day x 10)								
58°N	3	5	-1	5	3	1	-5	10
50°N	6	7	-3	10	0	5	-4	8
42°N	14	6	-1	9	10	10	4	11
34°N	3	2	-8	2	4	7	-8	17
Domain	10	7	-5	9	6	10	-4	15

GF-R = GFDL-RAND
GI-R = GISS-RAND
OS-R = OSU-RAND
UK-R = UKMO-RAND

ZONAL TEMPERATURES

Table 3 provides opportunities for visual comparisons of zonal temperature and precipitation in North America and Eurasia according to RAND and the 4 GCMs. For example, RAND suggests that the zonal temperature transects at higher latitudes in January and July are slightly colder for North America than Eurasia, but warmer for North America at subtropical latitudes. These patterns are thought to be related to the distribution and orientation of mountain barriers and seas and oceans. Comparisons of model outputs are also interesting. For example, the UKMO zonal transects for January are warmer than RAND for North America, but colder than RAND for Eurasia. In July, GISS is cooler than RAND for North America, but warmer or about the same as RAND for all but the southernmost transect in Eurasia. The precipitation patterns are even more complex, but perhaps even more interesting.

A major consideration is how the models perform over the annual cycle (Schneider, 1990). We have chosen to illustrate this by analyzing the July-January temperature range for GCM models in comparison to RAND (Table 4). Two aspects are to be noted: (a) model temperature range changes with latitude, and (b) the differences in the temperature ranges from the RAND dataset. In the RAND dataset, the temperature ranges become smaller from 58° to 34°N latitude. For North America, the range changes by $-0.6°C/°$lat. from north to south. For Eurasia, the rate is $-0.46°C/°$lat. All models show a similar tendency for a decline in the July-January temperature range with decreasing latitude. Rates vary among models

TABLE 6

Percentage Departures of Models from Rand for Zonal Mean Areally Weighted Precipitation for North America (NA) and Eurasia (EA)

Model:	North America				Eurasia			
	GFDL	GISS	OSU	UKMO	GFDL	GISS	OSU	UKMO
Latitude:								
DJF Precip:								
(mm/day x 10)								
58°N	55	12	9	5	40	9	−5	−2
50°N	33	26	−5	10	19	14	8	−7
42°N	77	42	14	41	34	51	42	29
34°	10	−13	−22	−34	107	103	42	20
JJA Precip:								
(mm/day x 10)								
58°N	18	30	−3	28	16	6	−25	18
50°N	31	33	−14	47	−2	29	−19	43
42°N	64	29	−6	43	82	84	35	94
34°N	12	6	−33	7	14	32	−24	54

N.B.: Percentages calculated as:

$$\% \text{ Departure} = \frac{(\text{model precip} - \text{RAND precip}) \times 100\%}{\text{RAND precip}}$$

from −0.31 to −0.45°C/°lat. for North America, and from -0.26 to −0.52°C/°lat. for Eurasia. In the North American case, model temperature ranges do not drop as fast as RAND values with decreasing latitude. For Eurasia, model temperature range drops with decreasing latitude at least bracket the RAND values.

The differences between model temperature ranges and RAND are also shown in Table 4. For individual comparisons, there are large discrepancies. In other cases, there are close matches. No one model appears free of some large discrepancies for both the continents. The pattern of differences of model-RAND July-January temperature ranges is a mixed one and is as follows: (a) for GFDL, model-RAND values are mostly positive (larger for Eurasia); (b) for GISS, they are mostly negative (more so for North America); (c) for OSU, they are mostly negative (more so for North America); and (d) for UKMO, they are negative in North America, positive (larger) in Eurasia.

Table 5 shows the areal weighted zonal mean values of model-RAND by latitude, and the continental (domain) mean zonal model-RAND values. For temperatures in January, zonal mean differences between models and RAND range from −3.18°C to 5.07°C (note here we quote raw values to specified accuracies in the datasets) for North America and −9.84°C to 1.52°C for Eurasia. For July, these ranges are −4.25°C to 1.39°C for North America, and −4.71°C to 2.25°C for Eurasia.

Individual temperature departures by model are listed in Table 5 for the respective months. For North America, there is a tendency for GFDL, GISS, and OSU simulated temperatures to display considerable underestimation (i.e., mostly >2°C) of temperatures of 34°N latitude. UKMO temperature simulations appear best for this latitude. In winter, GISS, OSU, and UKMO considerably overestimate temperatures in the higher latitudes of the domain studied (by >1°C in some cases). Thus, no one model "out-performs" all others for all months and locations.

For Eurasia, in the southern latitudes of the domain studied, large departures of models from RAND are evident, particularly for the month of January (departures range from −3°C to −10°C from RAND). There is also a tendency for similar underestimation of temperatures in summer, but the magnitude of the underestimation is reduced in the models. For January, at the higher latitudes, two of the models are very close to RAND (GISS and OSU), whereas the other two (GFDL and UKMO) underestimate temperatures by −1 to −3°C. For July, comparisons are generally close for the higher latitudes. Again, there is no clear-cut overall "winner" among models in comparison to RAND.

ZONAL PRECIPITATION

Referring to Table 5, the magnitude of precipitation departures for winter and summer between models and RAND are shown to be quite comparable. For North America, departures for models range from −0.95 to 1.41 mmd^{-1} in DJF; −0.83 to 1.41 mmd^{-1} for JJA. These numbers appear smaller in the mmd^{-1} unit, relative to the temperatures cited above, and can be misleading as a departure statistic. Therefore, we have converted precipitation departures of models-RAND to relative departures (in %) from the expressed mean values of RAND. Table 6 shows the % departure (model-RAND divided by the RAND mmd⁻¹) by model, by continent, and by season. Departure values are surprisingly low for certain models and times of year (e.g., GFDL's -2.1% in JJA, along 50°N lat. in Eurasia and OSU's -2.9% in JJA, along 58°N lat. in North America). However, there are also very large % departures, particularly along latitude zones which include mountains and/or arid conditions (e.g., GFDL's +107% in DJF, along 34°N lat. in Eurasia; GISS's +103% in DJF, along 34°N lat. in Eurasia; and UKMO's °94% in JJA, along 42°N lat. in Eurasia). Very large departure percentages are evident for mountainous and arid zones of Eurasia in comparison to the generally lower departures evident for North America. Given that climate change estimates for 2 x CO_2 states are for a ca. 7-15% increase in precipitation, the % departures in the model-RAND statistics are, indeed, quite large. However, these differences are difficult to intepret as entirely model error, as suggested above.

INTRA-ZONAL VARIATIONS

Details of continental land area zonal variations for the four models and RAND data are provided in Kalkstein (1991) and are not fully shown here. We have simply selected the 34°N latitude transects across the land areas of respective continents to demonstrate the variability that exists within the zonal mean data, especially for mountainous and arid regions. For Eurasia, the transect data illustrated in Figure 1 span the range of ca. 70°E to 150°E longitude. The transect starts in the highlands of Pakistan, bridges the Great Himalayan range into Tibet, and crosses China to the East China Sea.

For North America (Figure 2), the 34°N transect starts at 130°W longitude and terminates at 70°W. This transect starts over the Pacific Ocean, makes landfall at ca. 120°W, moves across rugged topography and desert conditions, over the continental divide between 105° and 110°W, and gradually into the humid southeastern United States to the east coast (75°W). Elevational diferences used in all models are shown for each continental transect in Figures 1 and 2. Overall, there are much higher elevations encountered in the Eurasian 34°N transect region (three-fold increase) compared to that for the United States.

The mean monthly January and July temperatures are shown in Figures 1 and 2 for the two continental transects. Trends of the modeled data across the transects follow the RAND data trends, for the most part, but the absolute temperature values show wide disparities in relation to RAND values. Brazel and Marcus (1991) discuss the heating effect of the Tibetan Plateau evident in RAND data, which appears underestimated by the models. RAND longitudinal gradients of temperature east of 100°E (rapid elevational drop eastward off the Tibetan plateau) do not compare very favorably with modeled data, particularly for January. In the U.S. transect, some of the models depart significantly from RAND, particularly for mountain and desert sectors of the transect.

Precipitation data (Figures 1 and 2) for Eurasia along the 34°N latitude transect also show very wide departures of models from RAND trends in both summer

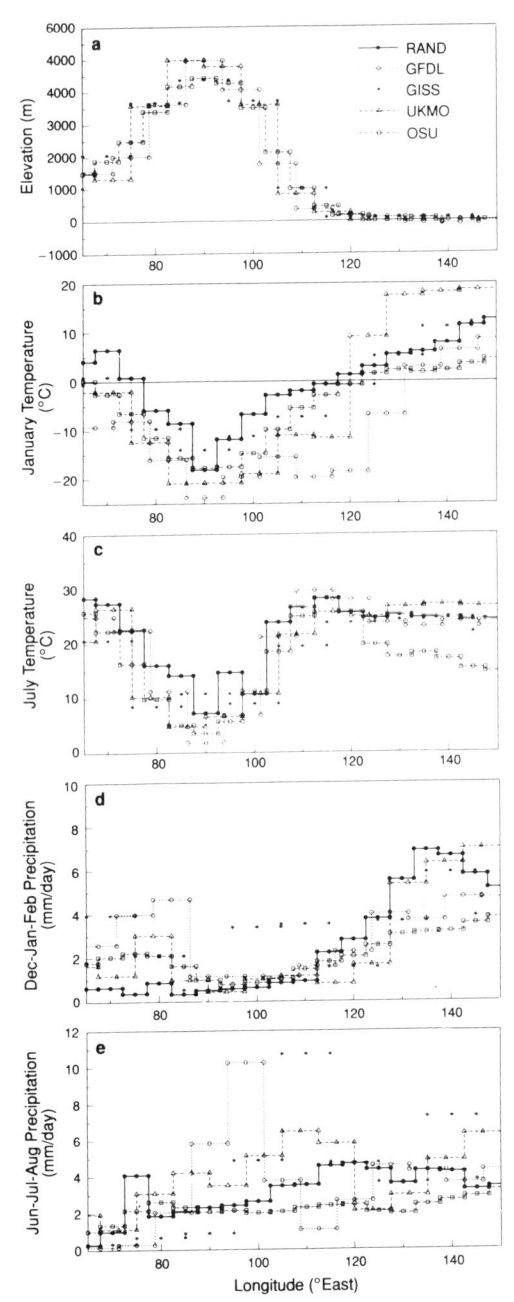

FIGURE 1. Eurasian longitudinal grid cell transect data along 34°N of RAND and models for: (a) elevations specified in models, (b) January mean temperature, (c) July mean temperature, (d) DJF precipitation, and (e) JJA precipitation.

238 Global Climate Change: Implications, Challenges and Mitigation Measures

FIGURE 2. North American longitudinal grid cell transect data along 34°N of RAND and models for: (a) elevations specified in models, (b) January mean temperature, (c) July mean temperature, (d) DJF precipitation, and (e) JJA precipitation.

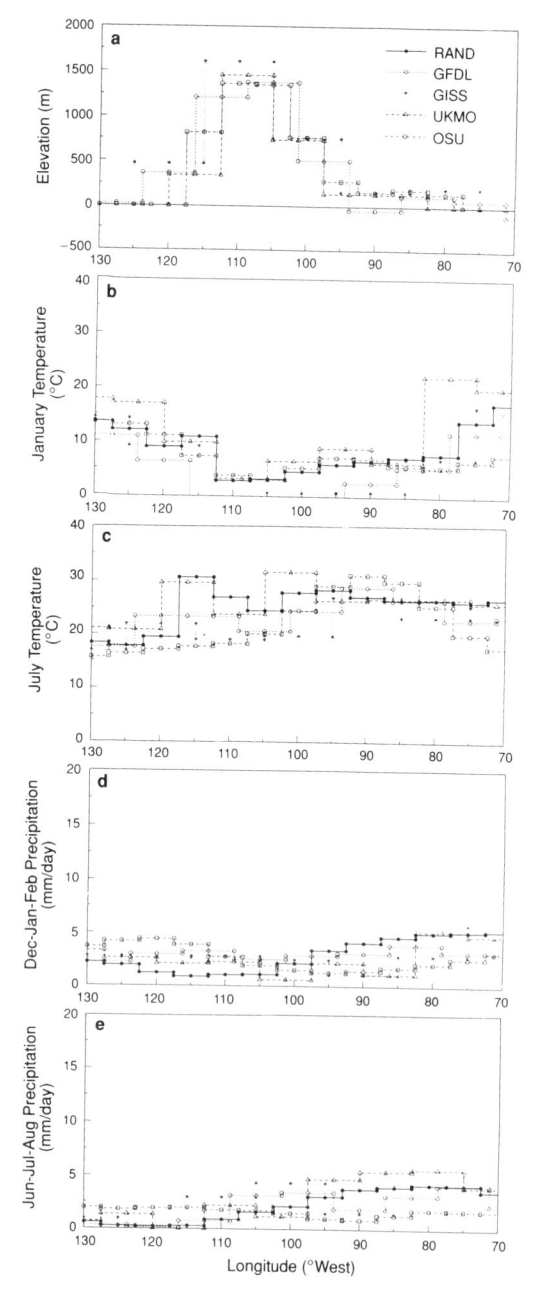

and winter, illustrating possible modeling problems with the enormous topographic variations and the specification of the monsoon regime. For the U.S., in most cases regional west-east gradients of precipitation are in the same direction as the RAND trends, especially for summertime. However, for winter, there appears to be a longitudinal out-of-phase placement of the precipitation field, especially for the OSU model—higher precipitation in the west, less in the east. The graphical transect data tend to further illustrate and substantiate possible source regions that account for much of the significant zonal mean departures of modeled data from RAND climate.

CONCLUSIONS

The comparisons on a zonal mean and regional basis illustrate possible modeling problems with topography, specification of climate at smaller scales, and treatment of the boundary layer. These have certainly been previously suggested by the modeling community as possible shortcomings in models (e.g., Dickinson, 1989; Giorgi and Bates, 1989; and Giorgi, 1990). For more complex environments, representation of "real" climate in gridded formats compounds the process of disentangling modeling problems from those related to geographically varying earth surface observations. Better spatial representation of earth's surface climate remains a fundamental research goal (e.g., Legates and Willmott, 1989, 1990), so that the disentanglement processes of model error detection from better representations of real-world climate can eventually be accomplished. This conclusion leads us to make a plea for: (a) the establishment of specialized regional networks dedicated to monitoring present climate, and (b) further development of procedures to increase areal accuracy of present networks that are monitoring earth's mountainous regions. In the long run, policy studies will only be helpful when we can understand true model errors that may be imbedded in projections of future climate.

ACKNOWLEDGEMENTS

The authors express sincere appreciation to D. Joseph (NCAR), L. Kalkstein, J. Smith, and D. Tirpak (EPA) for making this study possible; and to our colleagues, J. Grymes (Louisiana State University), P. Robinson (Univ. of North Carolina), and W. Wendland (Illinois Water Survey), who worked together with us on an EPA funded project to evaluate the land areas of the globe. We also thank P. Robinson for the graphics we use in this paper and B. Trapido for further cartography on these graphs.

REFERENCES

Bolin, B., B.R. Doos, J. Jager and R.A. Warrick, (eds.) 1986. The Greenhouse effect, Climate Change, and ecosystems, John Wiley & Sons, Inc., Chicester, Scope 29, 541 pp.

Brazel, A.J. and M.G. Marcus. 1991. July temperatures in Kashmir, India: comparisons of observations and general circulation model simulations, *Mountain Research and Development*, Vol. 11, No. 2, 75-86.

COHMAP members (33 authors). 1988. Cimatic changes of the last 18,000 years: observations and model simulations, *Science*, Vol. 241, 1043-1052.

Dickinson, R.E., R.M. Errico, F. Giorgi and G.T. Bates. 1989. A regional climate model for the western United States, *Climatic Change*, Vol. 15, 383-422.

Gates, W.L. 1985. The use of general circulation models in the analysis of the ecosystem impacts of climatic change, *Climatic Change*, Vol. 7, 267-284.

Giorgi, F. 1990. Simulation of regional climate using a limited area model nested in a general circulation model. *Journal of Climate*, Vol. 3, 941-963.

Giorgi, F. and G.T. Bates. 1989. The climatological skill of a regional model over complex terrain, *Monthly Weather Review*, Vol. 117, 2325-2347.

Grotch, S.L. 1988. Regional Intercomparisons of General Circulation Model Predictions and Historical Climate Data, Report Prepared for the U.S. Department of Energy, Office of Energy Research, Office of Basic Energy Sciences, Carbon Dioxide Research Division, Contract No. W-74505-ENG-48, NITS, Springfield, Va., 291 pp.

Grotch, S.L. and M.C. MacCracken. 1991. The use of general circulation models to predict regional climatic change, *Journal of Climate*, Vol. 4, 286-303.

Gutowski, W.J., D.S. Gutzler, and W-C Wang. 1991. Surface energy balances of three general circulation models: implications for simulating regional climate change, *Journal of Climate*, Vol. 4, 121-134.

Hansen, J., G. Russell, D. Rind, P. Stone, A. Lacis, S. Lebedeff, R. Ruedy and L. Travis. 1983. Efficient three-dimensional global models for climate studies: models I and II, *Monthly Weather Review*, Vol. 111, No. 4, 609-662.

Hansen, J., A. Lacis, D. Rind, G. Russell, P. Stone, I. Fung, R. Ruedy and J. Lerner. 1984. Climate sensitivity: analysis of feedback mechanisms, In, Climate Processes and Climate Sensitivity (eds. J. Hansen and T. Takashasi), Maurice Ewing Series, No. 5, American Geophys. Union, 130-163.

Jenne, R.L. 1988. Data From Climate Models; The CO_2 Warming, Draft Copy, NCAR, 26 August (Rev. 6 March, 1989), 33 pp.

Jones, P.D., P. Ya. Groisman, M. Coughlan, N. Plummer, W-C Wang, and T.R. Karl. 1990. Assessment of urbanization effects in time series of surface air temperature over land, *Nature*, Vol. 347, 169-172.

Kalkstein, L.S. (ed.). 1991. Global Comparisons of Selected GCM Control Runs and Observed Climate Data, Report Prepared for United States Environmental Protection Agency, Office of Policy, Planning, and Evaluation, Climatic Change Division, EPA Contract No. 68-W8-0113, 251 pp.

Karl, T.R., J.D. Tarpley, R.G. Quayle, H.F. Diaz, D. Robinson and R.S. Bradley. 1989. The recent climate record: what it can and cannot tell us, *Reviews of Geophysics*, Vol. 27, No. 3, 405-430.

Karl, T.R., W-C Wang, M.E. Schlesinger, R.W. Knight and D. Portman. 1990. A method of relating general circulation model simulated climate to the observed local climate. Part I: seasonal statistics, *Journal of Climate*, Vol. 3, 1053-1079.

Legates, D.R. and C.J. Willmott. 1989. Mean seasonal and spatial variability in global surface air temperature. *Theoretical and Applied Climatology,* Vol. 41, 11-21.

Legates, D.R. and C.J. Willmott. 1990. Mean Seasonal and Spatial Variability in Gauge-Corrected, Global Precipitation, *International Journal of Climatology,* Vol. 10, 111-127.

Manabe, S. and A.J. Broccoli. 1990. Mountains and arid climates of middle latitudes, *Science,* Vol. 247, 192-195.

Manabe, S. and R.T. Wetherald. 1987. Large scale changes in soil wetness induced by an increase in carbon dioxide, *Journal of Atmospheric Sciences,* Vol. 44, 1211-1235.

Mitchell, J.F.B. 1988. Local effects of greenhouse gases, *Nature,* Vol. 332, 399-400.

Oort, A.H. 1983. Global Atmospheric Circulation Statistics, 1958-1973, NOAA Prof. Paper 14, U.S. Govt. Printing Office, Washington, DC.

Rind, D., R. Goldberg, R. and R. Ruedy. 1989. Changes in climate variability in the 21st century, *Climatic Change,* Vol. 14, 5-37.

Ruddiman, W.F. and J.E. Kutzbach. 1991. Plateau uplift and climatic change. *Scientific American,* March, 66-75.

Schlesinger, M.E. and Z.C. Zhao. 1989. Seasonal climatic changes induced by doubled CO_2 as simulated by the OSU atmospheric GCM/mixed-layer ocean model, *Journal of Climate,* Vol. 2, 459-495.

Schneider, S.H. 1990. The global warming debate heats up: an analysis and perspective, *Bulletin of the American Meteorological Society,* Vol. 71, No. 9, 1292-1304.

Schutz, C. and W.L. Gates. 1971 and 1972. Global Climatic Data for Surface, 800 mb, 400 mb, January, July, Reports R-915-ARPA and R-1029-ARPA, Rand Corporation, Santa Monica, CA [NTIS - AD-760283, Springfield, Va.].

Global Climate Change: Implications, Challenges and Mitigation Measures. Edited by S.K. Majumdar, L.S. Kalkstein, B. Yarnal, E.W. Miller, and L.M. Rosenfeld. © 1992, The Pennsylvania Academy of Science.

Chapter Sixteen

SATELLITE MONITORING OF KUWAIT'S EFFLUENT:
Smoke from the Oil Fires

SANJAY S. LIMAYE
Space Science and Engineering Center
University of Wisconsin-Madison
1225 West Dayton Street
Madison, Wisconsin 53706

Summary

The deliberate ignition of over seven hundred oil wells in Kuwait during the six week long Persian Gulf conflict during 15 January - 28 February 1991 represents an unprecedented event in human history. Between March and July, an estimated 4 million barrels of oil went up in smoke every day in an area of less than 300 km^2. The smoke produced from these fires was detectable in weather satellite observations as far north as the Caspian Sea, as far east as Pakistan and as far West as Libya, covering an area of typically between one half and one million square kilometers. Despite the extensive spread of the smoke and the amount of crude oil and natural gas burnt, the impact on the global climate appears minimal for several reasons. The daily amount of crude being burnt represented only about 2% of the world's daily consumption of crude oil, and the resulting smoke rarely reached the stable layers of the stratosphere, remaining below 3-5 km on most days. Extraordinary efforts by as many as 28 firefighting teams had capped all of the destroyed wells by the first week of November 1991, ending this tragic assault on the environment.

INTRODUCTION

The recent Persian Gulf conflict (January 15 - February 28, 1991) resulted in the deliberate destruction of hundreds of active and inactive oil wells in Kuwait. Most of these wells burnt uncontrolled, while many others gushed crude and created lakes. This event is unique in terms of any number of statistics—number of wells on fire at any one time (732 out of 749 destroyed), amount of crude oil and natural gas being burnt (nearly 4 million barrels/day of crude[1]), amount of smoke produced, number of cities affected by the smoke (major population areas in the Persian Gulf cities of Kuwait, Dhahran, Bahrain, and Riyadh), amount of energy expended (nearly $12 billion worth of crude went up in smoke over the period that the fires burned, using up about 600 million barrels of crude, equivalent to about three months global consumption at the present rates), persistence of the fires, duration and area covered by the smoke (in May and June Bahrain reported smoke on nearly 60% of the days), resources allocated to the control of these fires (nearly 10,000 workers from 34 nations assisted firefighters from USA, Canada, Germany and China) as well as the number of scientists expected to study the results (as many as five international teams of scientists undertook aircraft studies of the smoke plumes between March and July of 1991). Despite such impressive numbers, the impact on the global climate is not expected to be devastating, or even significant because the bulk of the smoke did not rise into the stable layers of the upper atmosphere ("stratosphere" where the temperature increases with height), and the area where the smoke was thick enough to block sunlight remained a small fraction of the earth's surface area.[2,3]

Occasionally, smoke was detected in the skies of the Basra, and more infrequently Baghdad, Shiraz and even Cairo, and was also seen in satellite images as far to the north as the Caspian Sea with some reports of "black rain" coming from as far to the east as Pakistan. However, pollution was not restricted to the atmosphere alone. Surely the local landscape has been charred near the wells, covered with many lakes of unburned crude surrounding the damaged wells, and blackened by the smoke fallout over a wide area of the Arabian peninsula. An estimated 4-6 million barrels of crude also found their way into the Persian Gulf from damaged tankers, oil terminals and trenches filled with crude along some of the Kuwait shoreline. Much of this crude was carried by the winds and gulf circulation southwards for nearly 300km. A fair fraction was recovered by a concerted effort. The recovered crude in the form of crude cakes is contained within pits in the desert surface.

Within the immediate vicinity of the oil fields the air pollution from the smoke has undoubtedly affected the health of many thousands. However, the air quality further away from the burning oil fields was not significantly worse than that encountered in some of the major metropolitan areas of the world. The final tally of the environmental effects of the destruction of the oil wells is not yet at hand. This paper presents a summary of the preliminary assessment of the observations of the airborne smoke from the oil-well fires.

PROLOGUE

The Persian Gulf conflict represents the first military conflict which has had scientific interest from the very beginning, even before the start of the actual hostilities in January 1991. Further, the capability to monitor the environment was at hand. The possible environmental impact arising out of the conflict by deliberate destruction of Kuwait's oil was noted by King Hussein of Jordan in November 1990, and a meeting was held in London to assess the climatological impacts.[4] Compared to earlier major conflicts, progress in remote sensing technology for weather, natural resources, geographic information as well as intelligence applications has culminated in a pool of many satellites as well as instrumented research aircraft capable of providing data on the environmental effects. Advances in communications allowed data to be acquired in near real time even though the conflict was thousands of miles away, halfway around the world. For example, not only were the observations of the Persian Gulf region from weather satellites received in near real time at several locations around the world, capabilities existed to *analyze* the data as it was received. These factors allowed the environmental impacts to be monitored at least partially in real time from remote observations from civilian or commercial spacecraft as the conflict developed, progressed, ended, and afterwards. Thus, shortly after the Iraqi forces marched into Kuwait on August 2, 1990, some of the fires set by the skirmishes were seen in the images acquired from the European remote sensing satellite SPOT during August 1990. A significant amount of crude oil was also discharged into the Persian Gulf during the conflict, affecting the coastal regions along west shore of the Persian Gulf. The oil spill was seen in the satellite observations acquired during late January and February, 1991.[5] About 6 million barrels are estimated to have found their way into the gulf waters, much less than the nearly 13 million barrels estimated to have entered the gulf during the earlier conflict between Iran and Iraq. For comparison, it is estimated that nearly 250,000 barrels of oil polluted the gulf waters each year prior to the conflict.

Beyond the scientific curiosity and the technological capability to observe the environment, a greater awareness of the environment was also responsible for the interest in pursuing the conflict by not only scientists but also governmental and non-governmental agencies as well as the media[6] for a variety of reasons. The motivations might include the health and safety of the military and civilian personnel, of the local population, and for damage reparations from the aggressor. Significant among the environmental concerns regarding a military conflict were those that arose out of the work on nuclear winter scenario investigated by Crutzen and Birks, Turco *et al.*,[7] and others. Briefly, the concern was that if the oil fields in Kuwait were set afire, then the resulting smoke could affect the global climate if it rose into the stratosphere. Important in the suspected rise of the smoke into the upper layers of the atmosphere was the probable ability of the smoke to "self loft." The hot black smoke could absorb sunlight during the day, thereby increasing the buoyancy of the smoke causing it to rise even higher. If the smoke reached the stratosphere, then it would remain there for a long time as there would be no effective means whereby the smoke particles could be removed from the atmosphere, causing it to spread over large areas of the world, blocking out sunlight, thereby reducing average temperature at the surface. Associated with this is the enhancement of atmospheric carbon dioxide and other optically active gases that contribute to the warming of the atmosphere by their ability to absorb infrared radiation—the greenhouse effect.

In view of the recent efforts to control the rate of greenhouse warming as well as the effects of previous environmental disasters such as oil spills and volcanic eruptions, the scientific observations were sought with more than a casual interest. Efforts by many international parties have now resulted in a fairly large inventory of observations that are now being scrutinized. Preliminary indications are that many of the concerns expressed before the conflict were reasonable and some have indeed been realized, but only to a limited degree. The smoke did not rise in sufficient amount to the stratosphere because the heat output from the fires was diluted by the distance between the wells, and because self lofting was inefficient, the major impact of the fires is restricted to Kuwait. It is interesting to note that Mt. Pinatubo volcano in Phillipines erupted a few times in June 1991, injecting a substantial amount of volcanic debris and gases into the upper stratosphere. It is likely that the global impact of this short lived event will be much larger than that of the total smoke generated by over 500 million barrels of crude over seven to eight months.

The "ground truth" observations from at least a few key surface measurements, however, remained elusive during the conflict and even beyond for security and logistical reasons. To acquire these, organized efforts were necessary and undertaken. Thus the National Oceanic and Atmospheric Administration set up sixteen towers in the vicinity of the Kuwaiti oilfields to monitor the near surface atmosphere. Various international agencies also made efforts to sample the air quality in the region as well as to monitor the soil. Such observations are still being gathered and are expected to be analyzed in the near future.[8]

ENVIRONMENTAL OBSERVATIONS

Whatever was known about the smoke produced from oil well fires was learnt from only a few experiments during which small oil fires were lit and observations acquired. The smoke and other combustion products produced depend on the type of crude as well as the efficiency of the combustion. It was expected that the smoke from the oil fires would be different from that produced by the forest fires. For example, it was known that oil fire smoke would be "blacker" than the smoke from forest fires, and that the bulk of the particles would be smaller than one micron in diameter. However, the particle size distribution of the smoke changes rapidly with age due to settling as well as coagulation, efficiency of combustion and scavenging, and no real life observations of smoke oil fires were available until the conflict. Thus it was natural that efforts were mounted to study the smoke extensively that included satellites, instrumented aircraft and conventional surface sampling. Satellite observations are able to provide a good view of the geographic spread of the smoke while the in-situ aircraft and surface observations tell more about the properties of the smoke. They are described below.

246 Global Climate Change: Implications, Challenges and Mitigation Measures

Observations of the Smoke from Space

Smoke can be readily seen in satellite images acquired in reflected sunlight at visible wavelength. First reports of the Kuwait oil wells on fire came from the military pilots in the last week of January and were widely quoted in the media. Prior to that the fires resulting from the aerial bombardment in Iraq were seen in satellite observations.[9] The smoke plume from the oil fires in Kuwait was visible in many different satellite observations.[5] Because the infrared radiation emitted by a hot body increase much more rapidly with temperature in the 3-4 micron wavelength range compared to the 10-15 micron wavelength range, the short infrared observations are a particularly useful means of detecting fires, although they are smaller than the spatial resolution of the satellite sensors. This property has been used to detect forest fires and gas flares[10], and has been useful in detecting and monitoring the oil well fires from the weather satellites.

The satellite data generally fall into two categories: continuous low spatial resolution observations from weather satellites in both geostationary and polar orbits and relatively higher resolution satellites in polar orbits which provide data less frequently. Continuous coverage from METEOSAT satellite is particularly useful for the study of smoke dispersal because relatively thin smoke can be detected in time lapse which would otherwise be hard to detect in the higher spatial resolution polar orbiting satellite data such as from DMSP or NOAA 10, 11 and 12[11]. Table 1 lists the satellites from which data were acquired that show the smoke from the oil fires in Kuwait.

Figure 1 shows the Persian Gulf region as observed from the NOAA-10 satellite on February 23, 1991, a week before the conflict. Taken in the 3.7 micron channel, this image shows the hot spots (black areas) corresponding to the oil fields on fire.

TABLE 1
Satellites that observed the oil-well fires

SATELLITE	Orbit	Frequency	Spectral Bands and resolution
METEOSAT	Geo	30 min	Vis, IR, 5 km
INSAT	Geo	3 hr	Vis, IR, 8 km
NOAA-10*	Polar	12 hr	2 Vis and 2 IR 1.1km
NOAA-11	Polar	12 hr	2 Vis, 3 IR, 1.1km
NOAA-12**	Polar	12 hr	2 Vis, 3 IR, 1.1km
DMSP	Polar	12 hr	5 vis, IR, 500 m
SPOT	Polar	2-3 weeks	3 visible, 10 m
LANDSAT	Polar	2-3 weeks	
ERS-1	Polar	2-3 weeks	radar, vis
IRS-1	Polar	2-3 weeks	3 visible, 100 m
IRS-2	Polar	2-3 weeks	3 visible, 100 m
STS	Low	Sporadic	Hand held astronaut photography: 70 mm still, 16mm cine, 6x6 inch film, video

*Till September 11, 1991
**Replaced by NOAA-12, September 12, 1991

The wells on fire were in two fields to the north, and four to the south and west of Kuwait city. The white plume extending to the southwest represents smoke that appears cooler than the land in this infrared image and probably indicates the presence of some larger chains of coagulated unburned crude and soot particles several microns long. Table 2 lists the distribution of these wells in the different fields. Pre-invasion production of crude by Kuwait was about 1.5 million barrels, extracted from about 300 wells. Figure 2 shows the same area as observed on November 15, 1991. The fires have been put out and this visible channel image from the NOAA-11 satellite shows the outlined areas surrounding the fires where the desert surface has been considerably darkened due to the settling of smoke particles and unburned crude.

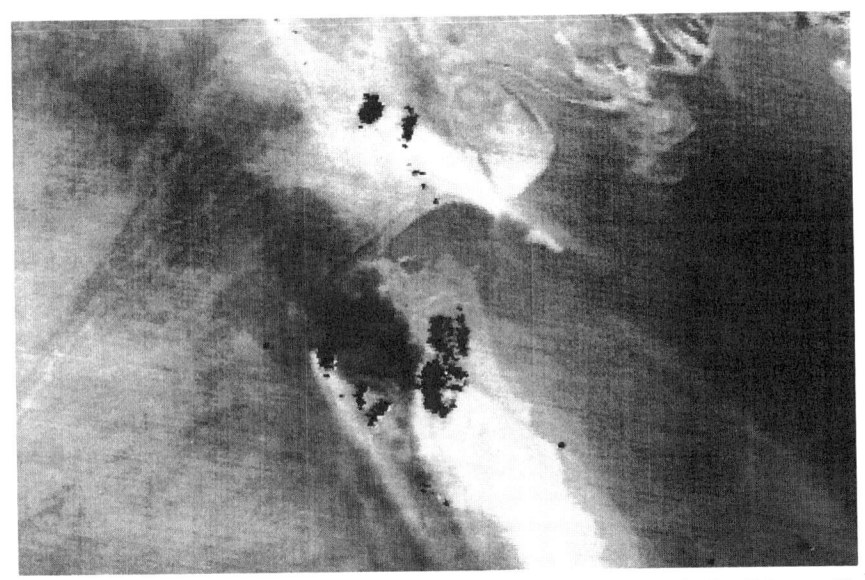

FIGURE 1. Kuwait as seen in the NOAA-10 early morning observations acquired on February 23, 1991 in AVHRR infrared (3.7 micron) observations. The dark areas in the infrared image indicate the burning oil fields.

TABLE 2: *Kuwait Oil Fields*

Oil Field	# of Wells Burning	Crude Burning rate Barrels/day	Gas burnt million m^3
Raudhatain	50	200,000	2.828
Rumaila	75	300,000	4.242
Bahrah	25	100,000	1.414
Mina Gish	50	200,000	2.828
Umm Gudair	50	200,000	2.828
Burgan	350	1,400,000	19.796
Wafra	25	8,750	0.124
Total	625	2,408,750	34.060

*Satellite observations indicated a total of 749 fires. A final official tally is not yet available.

248 Global Climate Change: Implications, Challenges and Mitigation Measures

The smoke generally travelled for several hundred km to the southeast in fairly narrow plumes of 50-100 km width. Its narrow width, which was somewhat of a surprise, indicate that the smoke did not mix much horizontally with the ambient environment and also that the winds were very steady. Indeed, climatological data indicates that the persistence of the winds in the gulf region is very high. This also explains why the desert surface that is markedly darkened by the fallout is elongated along a direction that is essentially the same as the summer winds. Figure 3 shows a mercator projection view obtained from the METEOSAT satellite on June 4, 1991. A raging sandstorm is visible in Saudi Arabia (bright streaks), a frequent occurrence in the summer months in this region. Whether the settling of the smoke and crude particles on the desert surface has a noticeable effect on the frequency and intensity of dust storms in the vicinity of Kuwait is of interest as is also the time required for natural resurfacing of the damaged areas seen in Figure 2 by blowing sand.

The spatial resolution of the polar orbiting weather satellite observations is too low (1.1 km at nadir at best) to show the individual oil wells. The fireballs from the wells on fire are however visible individually in the high resolution images acquired from the SPOT satellite.[12] As the wells in a given field are typically only about a

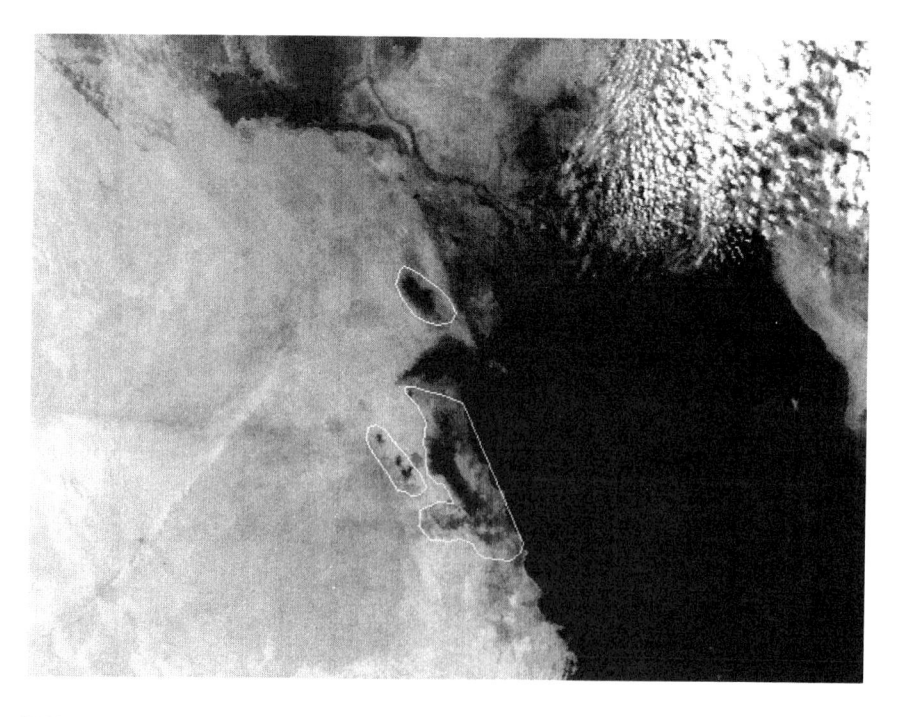

FIGURE 2. Kuwait as seen in the NOAA-11 local noon observations acquired on November 15, 1991 in visible light. The smoke is gone, but the soot fallout has left the underlying ground darkened (outlined areas). The steadiness of the winds in this region has clearly left relatively unaffected areas between the oil fields. The infrared image shows the hot-spots from the normal oil field operations.

kilometer apart, the individual smoke plumes rapidly merged into a larger one covering the entire field, and as the wind carried such merged plumes, they frequently merged with the plumes from the other fields, forming a giant "super plume" that was visible for over a thousand kilometers during the summer (Figure 3).

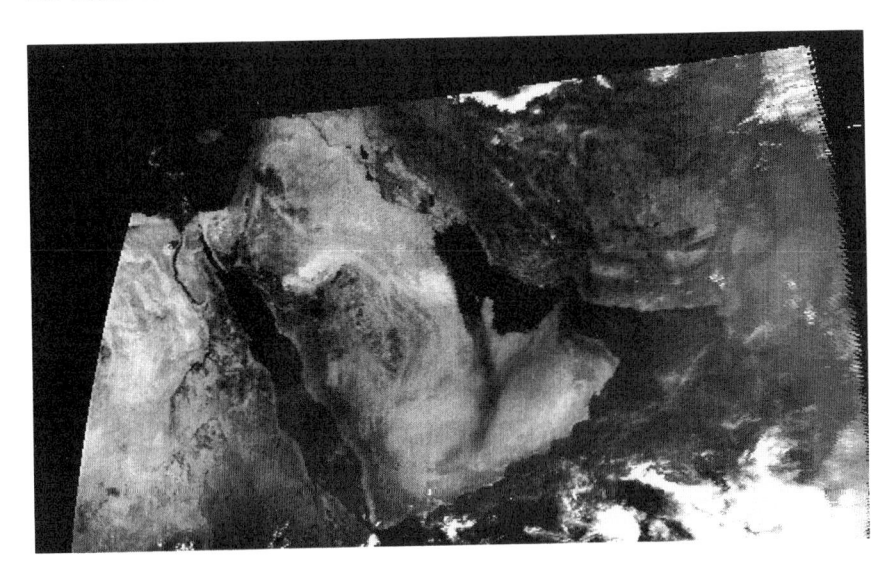

FIGURE 3. Sections of a pair of METEOSAT visible images acquired on 4 June 1991 (a) and 5 June 1991 (b) showing the plume position (dark narrow band from Kuwait along the Persian Gulf that breaks up into two parts, one going southwest and the other northwest) and its change over a day.

The amount of crude and natural gas being burnt was not measurable directly. As the Kuwaiti oil wells under natural pressure (as much as 14,000 lb/sq. inch), the daily estimated production of 1.5 million barrels/day is a lower limit for the amount of crude that burned initially. The rate was estimated to be 3.9 ± 1.6 Mbarrels/day indirectly from measurements of sulfur flux[13], and 4 ± 2 Mbarrels/day from the total heat production due to combustion of the crude and natural gas from weather satellite observations. Over the period that the wells burnt, over 600 million barrels of crude is estimated to have burned.

The vertical penetration of the smoke from the Kuwait oil fires is not easy to determine precisely, if it is even possible at all. It is hoped that simultaneous satellite observations (e.g., METEOSAT and combinations of the geosynchronous and polar satellites) at various times will allow stereo composites which lead to better estimates of the vertical level of the smoke, as has been done in the past for estimating cloud heights or topography. Comparison of the plume shape with concurrent pressure surface topography provides a rough estimate of the vertical level of the smoke. Typically, even during the warm summer afternoons, the smoke plume shape usually matched the shape of the 700 hPa or the 850 hPa pressure surfaces, corresponding to a level of roughly between 3000 and 1500 meters above mean sea level. The aircraft measurements confirmed that this was typical and that the top of the smoke layer was indeed at around the 3 km level[13]. However, these observations also indicated that the smoke was layered. Vertical temperature soundings showed temperature inversions at the corresponding levels. Frequently the layering of the smoke (that perhaps was evident due to lofting of the smoke) was evident from the different directions in which different parts of the smoke drifted. Continuous METEOSAT images acquired on 22 May 1991 (Figure 4) illustrate such a case when

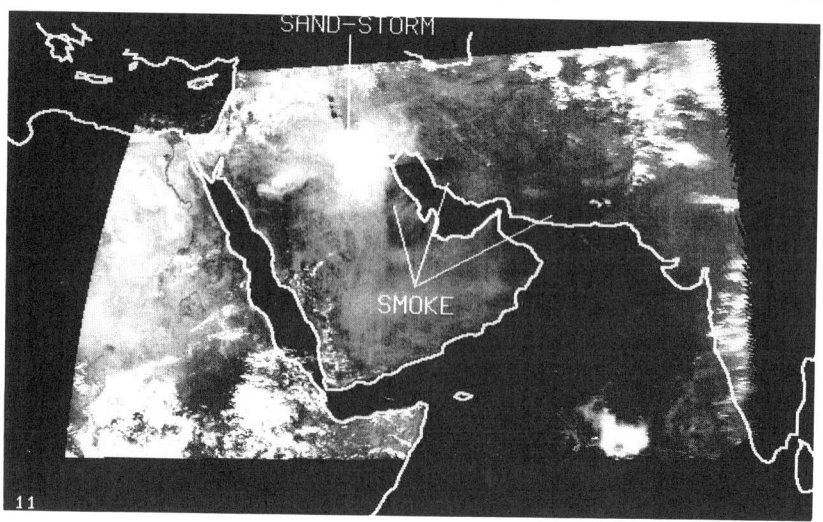

FIGURE 4. A METEOSAT visible image section showing the Persian Gulf on 22 May 1991. The plume is layered, one part travelling along the gulf shore, the other rising to a higher level and travelling across the gulf waters. A sandstorm is visible including some mixing of the sand and dust.

some smoke penetrated to higher levels and was carried eastwards across the Persian Gulf over Iran, while a substantial fraction remained lower and drifted to the south.

The Kuwait oil fires were observed by the space shuttle astronauts during April, June and July, 1991. Besides being able to see the glow due to the fires at night from the 180 nm orbit, the astronauts got a unique view of the smoke drift and its spread.[14] These observations also confirmed that the bulk of the smoke stayed close to the surface, below the 3-5 km level. Further, by virtue of being able to view the sunlight, they were able to photograph the lakes of crude oil from space in sun-glint.

The interplay between the effects of the smoke on the local atmosphere and the wind conditions over the region lead occasionally to disproportionately large accumulations of smoke hundreds of kilometers away from Kuwait. Thus, skies in Kuwait city, which is surrounded by the oil fields on three sides, were more often clear of smoke than skies over the other larger gulf cities such as Dhahran and Manama (Bahrain) or even in Riyadh, located in central Saudi Arabia! During April, May and June, the daily maximum, minimum and average temperatures were lower by an average of 3 to 5°F in Bahrain compared to the last 30 years, and most of this drop is attributable to the continual presence of smoke over Bahrain blown by the winds nearly 400 kilometers away from the burning oil fields. On some days the smoke pall is thick enough to reduce the visibility at the ground to less than 3 kilometers![1] On such days one can actually smell the smoke and the reduction in the sunlight is so much that often sunspots can be seen easily. In the last few months there have been increased reports (50% in Sharjah, UAE) of respiratory problems in cities affected by the smoke![15]

For predicting the spread of the smoke over the larger region, the key requirements are a knowledge of the low level winds and the heat intensity. Both of these however are not very easy to obtain in the gulf region. The number of weather stations is small compared to many other parts of the world, and no observations from Iraq have been available for some time. Four months after the end of the conflict surface weather observations from Kuwait were still spotty, and upper air observations non-existent. The continuous images of the region from the operational weather satellites thus provided a convenient, and often the only means, of monitoring the spread of the smoke.

Numerical models were also used to predict the spread of the smoke[16] as well as to assess the climatological impact.[17] These models were generally good at predicting the plume position of fresh smoke, but assessment of the prediction of the level of the smoke layer has not yet been done. The climatological impact assessment reflected the sense that the impacts were likely to be restricted to the gulf region, and that the concerns about the Indian monsoon were not valid.

The satellite observations show that the area covered by the smoke was as large as 100 to 400 thousand km^2, which compared to the surface area of the earth (511 million km^2), is insignificant. Thus the total reduction of the sunlight falling on the surface of the earth is also insignificant and well within normal variability (mostly due to local cloudiness variation and solar output variation). Further, the smoke did not rise very quickly into the stratosphere, but instead remained below 6-7 kilometers within about 2000 kilometers away from the oil fields, or, for at least four to six or seven days since the smoke was created. Thus the smoke has a greater chance of being scavenged out of the atmosphere in a matter of weeks thousands of kilometers away from Kuwait. Although such scavenging occurred during February and March in the gulf region (resulting in "black" rain), since April clouds have been nearly absent in the Persian Gulf region in the vicinity of the smoke. While the larger particles (mostly unburned crude) fall out of the atmosphere, a fraction of the smaller smoke particles probably remained in the atmosphere and it is not unlikely that some by now have probably travelled around the world. Scientists have reported abnormally high amounts of soot from observations in Hawaii and Wyoming, but whether it came from the Kuwait oil fires is not yet known. In any case, these particles are in small enough numbers to be of any concern regarding weather or climate. Reportedly the eruptions of the Pinatubo volcano in the Philippines may have a more noticeable effect on the global temperatures than the oil well fires in Kuwait despite the fact that many burned for nearly eight months. Only half of the 749 fires were controlled during the first six months after the end of the conflict, while the remaining half were brought under control in about two months.[1]

The primary effect of the smoke on the atmosphere is the reduction of sunlight. The fact that the smoke was nearly transparent to infrared radiation, allowing the heat from the surface to escape, was indicated by the nighttime infrared observations.[3] Thus, during the day the smoke absorbs the sunlight causing the underlying surface to remain dark, and hence cool. At night the radiation emitted by the surface largely escapes through the smoke, thereby cooling the surface further. That this process created noticeable surface inversions of temperature is evident in the preliminary analysis of hourly observations from Bahrain. During the months of May-July, Bahrain recorded smoke near the surface or at elevated levels nearly 30% of the days. The monthly average temperatures however were no more than 2.5°C lower than the preceding 30 year average.[11]

Aircraft Studies of the Smoke

Several sampling studies of the smoke from the oil fires in Kuwait were undertaken from aircraft. The first was undertaken by the UK Meteorological Office in March. Subsequently Goethe University, National Center for Atmospheric Research (NCAR), University of Washington, Seattle, US Department of Energy, and National Aeronautics and Space Administration also conducted many hours of research aircraft flights studying the smoke plumes during May - August 1991. These data are still being analyzed. Preliminary analysis confirms the expectations that the environmental effects are restricted to the local region.

Gulf Data Archive

The effort to analyze the data acquired during the course of the fires is just beginning. In view of the scope of the environmental impacts—desert, gulf waters and the atmosphere, the international scope, and the uniqueness of the oil well fires and the importance of the data collected during the aircraft flights and from the vicinity of the wells is being gathered together (Baumgardner). Gulf Region Atmospheric Measurement Program (GRAMP) is a WMO-sponsored effort to archive the data relevant to the environmental and health aspects of the smoke from the Kuwait oil-well fires.[8] The archive will be maintained at the National Center for Atmospheric Research (NCAR) located in Boulder, Colorado, USA. It is expected that the conventional meteorological observations (surface and upper air), surface air pollution data on particulates and trace species, some satellite observations, and NMC analyzes will be entered into the archive. NOAA has set up a network of 15 towers near the oil fields, and these data are also expected to be entered into the archive. In addition, data from the research aircraft flights made by the UK Meteorological Office (March 22-31, 1991), as well as by the German group is also expected to be entered into the archive. It is expected that as the analysis proceeds, much more will be known about the effects of the fires not only on the environment but also on the population affected directly. The economic effects, however, are much harder to tally.

SUMMARY

A conclusion to the assessment of the destruction of the environment and the impact of the smoke on the regional weather and climate awaits systematic analysis. It appears that the warnings about impending disaster were valid in spirit if not in extent. It is clear that the intensity of the oil-well fires was not sufficient to propel the resulting smoke high into the atmosphere. However, had the oil wells been located much closer than the nearly 0.8-1.6 km spacing between them, it is quite possible that the resulting impacts would not have been so small. The rising plumes could then have combined and would have been diluted less by the mixing of ambient air, and risen higher. Calculations have shown that for a heat flux of about 9×10^4 Wm^{-2}, the smoke injection height would have been about 9 km or higher.[9] The controlling factors in the smoke injection height, other than the heat flux, are the environmental moisture abundance and the latent heat release from the moisture created by the fires themselves. This suggests that if the wells were not as far apart as they were in Kuwait, the smoke could have risen higher into the atmosphere and that more significant effects on the climate could have occurred. Thus, fortunately, the oil-well fires posed less of a hazard than they could have otherwise posed, for a quirk of nature. Since the Kuwaiti oil wells are harvested under natural pressure, placing the wells too close actually reduces the well-head pressure resulting in a lower production. Thus there is no benefit to a closer network of the wells.

BIBLIOGRAPHY

1. Wald, M.L., *The New York Times,* 7 November 1991, Amid ceremony and ingenuity, Kuwait's oil-well fires are declared out.

2. Small, R.D., *Nature,* 350, 11-12 (1991).
3. Limaye, S.S., V.E. Suomi, C. Velden, and G. Tripoli, 1991a, Satellite Monitoring of smoke from oil fires in Kuwait, *Science,* 251, 1536-1539.
4. Aldous, P., 1991, Oil-well climate catastrophe?, *Nature,* 349, 96.
5. Williams, R.S., J. Heckman, and J. Schneeberger, 1991, Environmental consequences of the 1990-1991 Persian Gulf War-a guide to remote sensing datasets of Kuwait and environs, National Geographic Society, Committee for Research and Exploration, Washington D.C.
6. Turco, R.P., O.B. Toon, T.P. Ackerman, J.B. Pollack, and C. Sagan, 1990, *Science,* 247, 166.

   Crutzen, P.J., and J.W. Birks, 1982, The atmosphere after a nuclear war: twilight at noon, *Ambio,* 11, 114-125.
7. Warner, Sir Frederick, 1991, The environmental consequences of the Gulf war, Environment, 33, 6-9, and 25-26.

   Sample Media reports:

   J. Horgan, 1991, The muddled cleanup in the Persian Gulf, *Scientific American,* October 1991 issue, p. 106-107, and The danger from Kuwait's air pollution, p. 30.

   Booth, W., 1991, War's oil spill still sullies gulf shore. Washington Post, 8 April 1991, p. A1 and A13.

   E. Schmitt, *The New York Times,* 3 March 1990, p. 13.

   R. Tyson and S.V. Meddis, *USA Today,* 15 to 17 March 1991.

   Anonymous, *Space News,* April 1, 1991, p.8.
8. Baumgardner, D., and R. Friesen, 1991; Management of Data Collected in Gulf Region Atmospheric Measurement Program (GRAMP), NCAR Technical Note TN-363, Proceedings of a WMO workshop held 22-24 July 1991 at NCAR, Boulder, Colorado.
9. Matson, M., and J. Fishman, 1991: Fires and smoke during desert storm - a satellite perspective. Paper presented at 1991 AGU Spring Meeting, Conference abstracts, p. 64.
10. Matson, M., and J. Dozier, 1981, Identification of subresolution high temperature sources using a thermal IR sensor. *Photogrammetric Engineering and Remote Sensing,* 47, 1311-1318.
11. Limaye, S.S., P.M. Fry, S. Ackerman, M. Isa, H. Ali, G. Ali, A. Wright, and A. Rangno, 1992, Satellite monitoring of smoke from the Kuwait oil fires. *J. Geophys. Res.,* in press.
12. Canby, T.Y., and S. McCurry, National Geographic, 180, No. 2, August 1991, 2-35.
13. Johnson, D.W., C.G. Kilsby, D.S. Mckenna, R.W. Saunders, G.J. Jenkins, F.B. Smith, and J.S. Foot, 1991, Airborne observations of the physical and chemical characteristics of the Kuwait oil smoke plume, *Nature,* in press.

Limaye, S.S., 1992, Monitoring the oil-well fires from weather satellites. *In preparation.*

Hobbs, P.V., and L.F. Radke, 1992, The United States Interagency airborne study of the smoke from the Kuwait oil fires, *Science,* (256), 987-991.

14. Lulla, K., and M. Helfert, 1991, Smoke palls induced by Kuwaiti oilfield fires mapped from space shuttle imagery, *Geocarto International,* 6, 71.
15. Duraid Al Baik, N. Srinivasan, and R. Owais, 1991, Special report of the environment, (Smoky nightmare, Has rise in air pollution increased health hazards?, Recurring maladies, and Doctors differ on causes) Gulf News, Saturday, June 8, 1991, Bahrain, p. 3.
16. Browning, K.A., R.J. Allam, S.P. Ballard, R.T.H. Barnes, D.A. Bennetts, R.J. Maryon, P.J. Mason, D. McKenna, J.F.B. Mitchell, C.A. Senior, A. Slingo, and F.B. Smith, Environmental effects from burning oil wells in Kuwait, *Nature,* 251, 363-366 (1991).
17. Bakan, S., A. Chlond, U. Cusbach, J. Feichter, H. Graf J. Grassl, K. Hasselmann, I. Kirchner, M. Latif, E. Roeckner, R. Sausen, U. Schlese, D. Schriever, I. Schult, U. Schumann, F. Sielmann, and W. Welke, Climate response to smoke from burning oil wells in Kuwait, *Nature,* 351, 367-371 (1991).
18. See Chapter 4 of *Environmental Consequence of Nuclear War, SCOPE 28,* Volume I: Physical and Atmospheric Effects, A.B. Pittock, T.P. Ackerman, P.J. Crutzen, M.C. MacCracken, C.S. Shapiro, and R.P. Turco, Eds., J. Wiley and Sons, New York.

Global Climate Change: Implications, Challenges and Mitigation Measures. Edited by S.K. Majumdar, L.S. Kalkstein, B. Yarnal, E.W. Miller, and L.M. Rosenfeld. © 1992, The Pennsylvania Academy of Science.

Chapter Seventeen

SEA LEVEL RISE:
Implications and Responses

STEPHEN P. LEATHERMAN
Laboratory for Coastal Research &
Department of Geography
1113 LeFrak Hall
University of Maryland
College Park, MD 20742

INTRODUCTION

Throughout geologic history sea level has fluctuated greatly. During the last Ice Age (approximately 18,000 years ago), sea level was as much as 100 meters below present levels. The earth at this time was about five degrees Celsius colder than today. During warm interglacial periods, sea level has been at times several meters higher than present. Because of the historic relationship between climate and sea level position, it is expected that anthropogenic (human-induced) global warming could cause a significant rise in sea level. Warmer temperatures will cause expansion of near-surface ocean waters and melt land-based ice, notably mid-latitude glaciers and the Greenland ice pack, raising sea levels.

Climate can effect sea level position by heating and thereby expanding (or conversely cooling and contracting) surface sea water. This process can occur over relatively short periods of time. At present the majority of mid-latitude mountain glaciers are still retreating. Although most of the glaciers have melted since the last Ice Age, there is still enough water in polar glaciers to raise sea level by more than 70 meters. Over longer periods of time, significant rises in sea level could be caused by disintegration of the West Antarctic ice sheet, which is marine-based and subject to temperature increases.

In the last century, surface temperatures have shown a gradual increase based on National Weather Service data, and tide gauges have recorded about a 30 centimeter rise along the U.S. Atlantic coast (National Research Council, 1987). Some of this relative rise in sea level (relative to the land surface) can be explained by the natural compaction and subsidence of unconsolidated coastal sediments. However, part of this rise (about 18 cm) can be attributed to thermal expansion of surface ocean waters and glacier recession (Douglas, 1991), resulting from the observed warming of 0.5°C during the last century. Figure 1 shows the strong correlation between global temperature and sea-level rise.

FUTURE SEA-LEVEL RISE

Concern about a possible acceleration in the rate of sea-level rise stems from measurements showing that concentrations of carbon dioxide and other "greenhouse" gases produced by human activities are increasing in the atmosphere. Because these gases absorb (trap) long-wave radiation (heat) in the atmosphere, it is generally expected that the earth will warm substantially in the future. The National Academy of Sciences (1983, 1985) has convened two panels to review all the evidence and concluded that warming will take place. More recently, the Intergovernmental Panel on Climate Change (IPCC, 1990) has issued similar predictions.

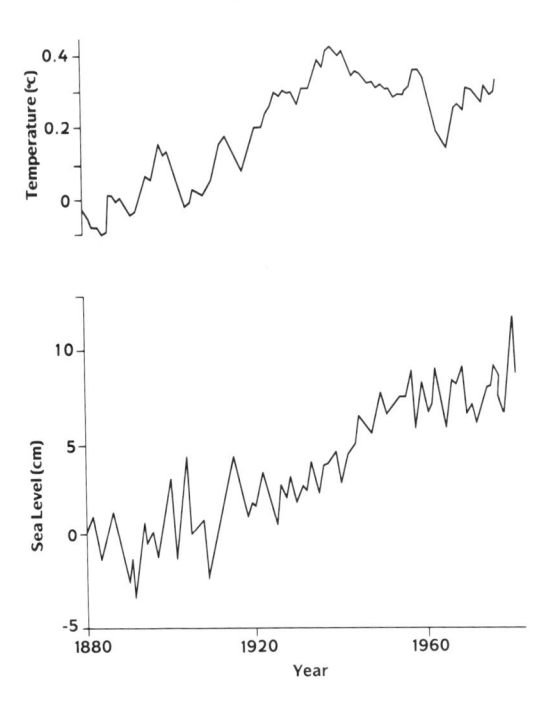

FIGURE 1. Global temperatures and sea level rise in the last century.

There is no doubt that the concentration of greenhouse gases is increasing and will do so in the foreseeable future. However, considerable uncertainty exists regarding the amount of warming; it is generally agreed that a doubling of the greenhouse gases will raise the earth's average surface temperature by about 1°C if nothing else has changed. It appears that most of the climatic factors will amplify the direct effects, but some negative feedbacks (such as increased cloud cover to offset part of the warming) cannot be ruled out. Nevertheless, two panels of the National Academy of Sciences (NAS) and the IPCC have concluded that a doubling of greenhouse gases will eventually induce a warming between 1.5° and 4.5°C.

Based on current trends, Revelle (NRC, 1983) estimated that sea level could rise by 70 cm (30 cm due to thermal expansion, the balance attributed to glacial melting) by the next century. There have been a range of estimates made by NAS, EPA, and various scientific investigators. The most recent and authoritative estimates of future sea-level rise are by IPCC (1990). They estimated a rise by the year 2100 in the range 0.31 m to 1.10 m, with the most likely rise being 0.66 m (Figure 2). Even if the greenhouse forcing is totally arrested, global sea levels are predicted to continue rising as a commitment has been made by the huge quantity of radiative gases already pumped into the atmosphere.

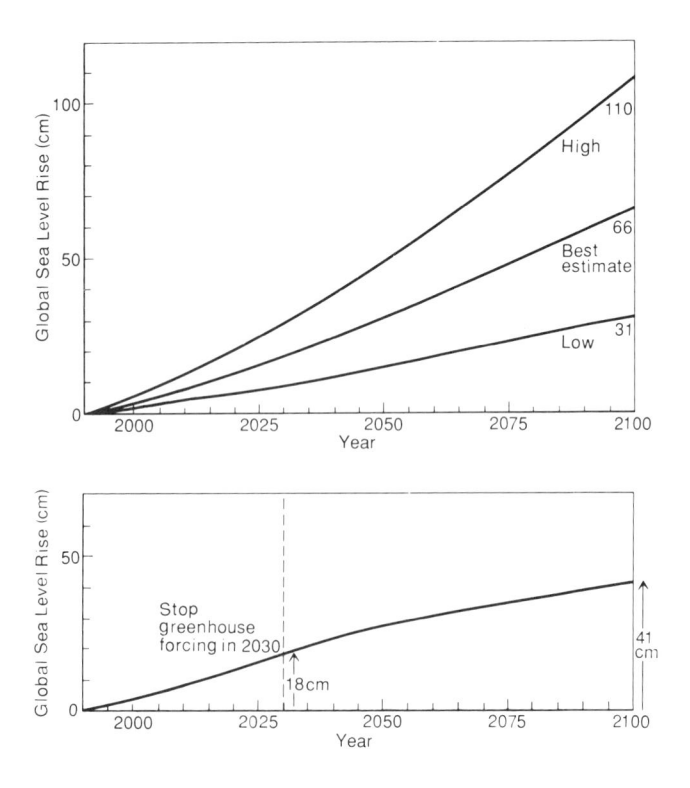

FIGURE 2. Global sea level rise scenarios (IPCC, 1990).

EFFECTS OF SEA-LEVEL RISE

The principal effects of sea-level rise are increased flooding and wave-induced erosion. Salt-water intrusion can also be a problem in some areas, particularly affecting surface waters.

Flooding and Submergence

A rise in sea level represents a raising of water base level. Therefore, storm waves and surges can reach higher and further inland. This can result in accelerated beach erosion as explained later, and major flooding will occur more often. For example, "100-year" storm can occur in the future on a 15-20 year averaged basis by virtue of higher base levels when considering frequency-magnitude relationships of coastal flooding (Leatherman, 1983).

The most significant impact of higher sea levels will be the submergence of coastal wetlands. Intertidal salt marshes can adapt only to relatively moderate rates of sea level rise; rapid increases in sea level can literally drown these wetlands, converting them to shallow bodies of open water. It is worth noting that the present U.S. tidal marshes postdate the previous maximum rate of sea level rise during the Holocene (since the last Ice Age), when sea level rose approximately one meter per century (1 cm/year; Gehrels and Leatherman, 1989).

Much of the Louisiana coastal zone is experiencing a rapid *relative* rise in sea level (up to 1 cm/year) largely due to human-induced subsidence. Without adequate supplies of sediment to raise the elevation of the marsh surface, these wetland plants become water-logged and eventually die. Presently, Louisiana is losing 100 km^2 of wetlands annually, and entire parishes (counties) will be underwater within the next 50 to 100 years (NRC, 1987). During periods of only gradual sea-level rise, plant-generated (organic) sediment and inorganic materials from rivers, uplands, and the sea could maintain the marsh surface plain relative to sea level positions. However, the Louisiana marshes dramatically illustrate the problem of rapid water level changes, and can serve as useful analogs of what will happen elsewhere along the world's coastline with accelerated sea-level rise.

A one-meter rise in sea level could drown most of the coastal wetlands without necessarily creating new salt marshes inland. Even in natural areas, marshes will often contract because of the sloping nature of the land above the marsh plain (Figure 3). Where marshes are backed by urbanized areas, such as along much of the Long Island, New York coast for example, these habitats will be squeezed out of existence with future sea-level rise.

Coastal Erosion

Sea level is one of the principal determinants of shoreline position. There are several reasons why sea-level rise would induce beach erosion or accelerate on-going shore retreat: (1) waves can get closer to shore before dissipating their energy by breaking, (2) deeper water decreases wave refraction and thus increases the capacity for longshore transport, and (3) with a higher water level, the wave and current erosion processes are acting further up the beach profile, causing a readjustment of that profile.

Most sandy shorelines are presently eroding on a worldwide basis (Bird, 1985). Historical records indicate the prevalence of shore retreat during at least the past century. The National Shoreline Study by the U.S. Army Corps of Engineers (1971) was the first overall appraisal of shore erosion problems in the continental United States. This study showed that 43 percent of the shoreline is undergoing significant erosion, excluding Alaska. In fact, this report indicates that most all of the U.S. ocean shoreline is undergoing erosion (excluding hard-rock coasts). Accretion is restricted to coastal areas where locally excess sediment is supplied by river sources or where the land is being elevated by tectonic (earthquake or glacial rebound) activity.

There are several different approaches that can be used to project new shorelines. Slope is the controlling variable, such that gently-sloping shores will undergo a much broader area of inundation for a given sea-level rise compared to steep-sloped areas. This is the preferred methodology to apply to immobile (rocky or armored) coasts or for sheltered coasts, such as small bays and estuaries.

The other approaches that have been employed to date are largely based on the erosional potential of sea-level rise: (1) extrapolation of historical trend, (2) Bruun Rule, (3) sediment budget analysis, and (4) the dynamic equilibrium model. These methodologies, including applications and limitations, are discussed in more detail elsewhere (Leatherman, 1990). Case studies along the U.S. coast yielded a comparable range of rates of shore retreat as predicted by the different approaches. A severe limitation to our forecasting future erosion rates is lack of good quantitative data on historical rates of shore retreat for much of the world's coastline.

For open ocean sandy beaches, at least a doubling and perhaps a five-fold increase in erosion rates can be forecast, depending upon the realized rates of accelerted sea-level rise. Many urbanized beaches are already critically narrow and continue to

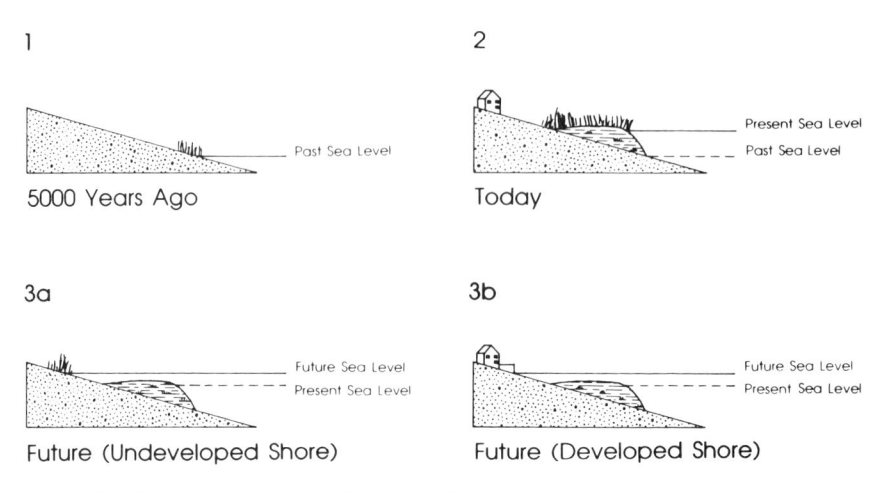

FIGURE 3. Salt marsh evolution with sea level rise.

erode. As a result accelerated sea-level rise will exacerbate an already serious problem (Leatherman, 1986). Presently many U.S. recreational beaches are being nourished (e.g., Miami Beach, Florida, 1980s, $65 million) or will be renourished in the near future (e.g., Ocean City, Maryland). More beaches will have to be replenished in the future and more often in order to maintain their recreational quality and provide storm protection for the landward-flanking coastal development (Leatherman, 1989).

RESPONSES

A dominant and growing proportion of the world's population, facilities and development are located on coasts in sensitive balance with local sea levels. Even the lower estimates of accelerated sea-level rise will place many of the world's coastal cities in jeopardy in the coming decades. The low-lying coastal fringe is subject to catastrophic events of flooding. For example, the 1970 cyclonic flooding that killed over 300,000 people in Bangladesh will become a more frequent occurrence in the future as sea levels rise and the third world population continues to explode.

An accelerated rise in relative sea level will force people who live on the coasts to make a number of important decisions. The choice of a response strategy will depend upon several factors, including environmental, economic, and social factors (NRC, 1987).

There are essentially three possible responses:(1) fortify the shore, (2) retreat from the coast, and (3) nourish the beach. Armoring the shore with such coastal engineering structures as seawalls, breakwaters, and bulkheads is very expensive, and often compromises the resources (loss of a recreational beach). Planned retreat is the best option for sparsely developed areas. Beach nourishment is the most attractive alternative where highly urbanized areas must be protected and high value is placed on maintaining a recreational beach. Selection of the appropriate response will be site-specific on the basis of existing conditions. The costs and benefits of stabilizations vs. retreat must be carefully considered as the cost of either case is likely to be quite high.

SUMMARY AND CONCLUSIONS

1. There appears to be a strong co-relationship between earth warming and sea-level rise. During the past century, sea level has risen about one foot (30 cm) for which about one-half can be attributed to global causes, concurrent with warming of the earth's surface by 0.5°C. The chief uncertainty lies in how rapidly the polar glaciers may melt.
2. Several groups have projected sea-level rise, notably the National Academy of Sciences (NAS) and the Environmental Protection Agency (EPA). Two panels of NAS have been convened and concluded that sea level will increase by 70 cm by 2100 (1983 Revelle report) with best estimates by the IPCC (1990) of a global sea-level rise of 0.66 m by the year 2100. For planning and engineering purposes, government officials and decisionmakers should utilize the one meter bench-

mark, because this estimate is well within the probable range of change.
3. Wetlands will be much affected by accelerated sea-level rise, resulting in significant losses. Wetlands can shift inland, but their area will drastically shrink due to the sloping nature of the flanking mainland. Where urbanized, wetlands will be essentially squeezed out of existence. Clearly, some areas will lose more marsh than others, depending upon topographic conditions and anthropogenic controls. People will be unwilling to abandon urbanized areas to allow for wetlands invasion concurrent with sea-level rise so that the continuing urban sprawl along bay and estuarine shorelines will have severe long-term implications.
4. Sea-level rise will promote increased coastal erosion. Already approximately 70% of the world's sandy coastlines are eroding, and accelerated sea-level rise will only exacerbate this critical problem. Rates of shore erosion will probably at least double and may increase five-fold based on the realized rate of water level changes. As a rule of thumb, a 30 cm rise in sea level will result in 30 + meters of erosion along the U.S. Atlantic coast. This means that most recreational beaches would be lost since so many are already critically narrow. Artificial nourishment is being used to restore some U.S. beaches, but the costs are high. Accelerated sea-level rise will increase the quantity and frequency of beach restoration projects.
5. It is certain that these potential problems will only worsen in the near future. Within the next 40-50 years, sea level will probably rise by about 30 cm, resulting in major impacts to coastal environments. Rather than triggering dramatic change, sea-level rise will promote gradual erosion and invariably increase the vulnerability of human development as well as culminate in significant losses of wetlands. These impacts are perhaps more insidious than the short-lived, dramatic storm-induced damages to coastal areas.
6. There are three general responses to accelerated sea-level rise: fortify, retreat, or nourish. The proper response will be based on site-specific information on environmental and socioeconomic conditions on a community or coastal sector basis. In any case, the long-term costs are likely to be high along the world's low-lying and sedimentary (erodible) coasts.

ACKNOWLEDGEMENTS

This review was supported under the aegis of a grant from the W. Alton Jones Foundation.

REFERENCES

Bird, E.C.F. 1985. Coastline changes: A Global Review, Chickester Wiley Interscience p. 219.
Douglas, B.C. 1991. Global sea-level rise. Journal of Geophysical Research 96:6981-6992.
Gehrels, R. and S.P. Leatherman. 1989. Sea level rise: animator and terminator of coastal marshes. Vance Bibliography, Monticello, Illinois, 39 pp.
Intergovernmental Panel on Climate Change. 1990. Scientific assessment of climate change. Report prepared for the Intergovernmental Panel on Climate Change (IPCC) by Working Group I.
Leatherman, S.P. 1983. Historical and projected shoreline changes, Proceedings of Coastal Zone 83, ASCE, San Diego, CA p. 2902-2910.
Leatherman, S.P. 1986. Coastal geomorphic impacts of sea-level rise on coasts of South America Proceedings of United Nations Environmental Programme Conference, Washington, D.C. p. 73-82.
Leatherman, S.P. 1989. Beach response strategies to accelerated sea level rise Proceedings of the Second North American Conference on Preparing for Climate Change, The Climate Institute, Washington, D.C. p. 353-358.
Leatherman, S.P. 1990. Modelling shore response to sea-level rise on sedimentary coasts Progress in Physical Geography 14:447-464.
National Research Council. 1983. Changing Climate, National Academy of Science Press, Washington, D.C. p.496.
National Research Council. 1985. Glaciers, Ice Sheets, and Sea Level, National Academy of Science Press, Washington, D.C.
National Research Council. 1987. Responding to changes in sea level: Engineering implications. National Academy of Science Press, Washington, D.C. p.148.
U.S. Army Corps of Engineers. 1971. National Shoreline Inventory, Washington, D.C., several volumes.

Global Climate Change: Implications, Challenges and Mitigation Measures. Edited by S.K. Majumdar, L.S. Kalkstein, B. Yarnal, E.W. Miller, and L.M. Rosenfeld. © 1992, The Pennsylvania Academy of Science.

Chapter Eighteen

USING A GIS TO ANALYZE THE EFFECTS OF SEA LEVEL RISE

GARY OSTROFF[1] and SUSAN TUCKER
Hunter College
Department of Geography
New York, NY 10021

INTRODUCTION

The mean sea level as measured at Sandy Hook, N.J., has been rising for the last fifty years at an approximate rate of 0.014 feet per year and is anticipated to continue to rise at the same rate for the indefinite future (Lyles *et al.*, 1988). This increase in sea level is attributed to general geophysical factors acting on the eastern seaboard that are causing the coastal region to sink and are not related to any increase in the volume of sea water in the Atlantic Ocean. Despite the lack of contemporary evidence for sea level rise induced by global warming, concern over this possibility is growing, encouraged by the debate over the inevitability or avoidability of the greenhouse effect and the consequences to be expected from it.

    The goal of this project was to perform an initial analysis of the effect of sea level rise on the portion of the Long Island coast in the vicinity of Mecox Bay using a GIS and a SPOT image. A general methodology for performing such analyses was to be developed, feasibility of economic valuations were to be determined, and technical problems were to be noted so that future efforts in this area can be more refined. In addition, the project would yield insight into the general usefulness of the method and the limitations imposed by specific site conditions. As a final result of the analysis, an outline of economic, social, and topographical effects of sea level rise on the area would be determined.

Present address: Coastal Environmental Services, Inc., 2 Research Way, Princeton, NJ 08540.

It was concluded that a GIS is an effective tool for studying the effect of sea level rise on selected areas, and that it is feasible to arrive at an assessment of projected effects. The technical problems associated with the task, in particular the registration of the SPOT image to the relevant map coordinates, were more difficult than anticipated, and the limits of the system's resolution, the accuracy of the land use classification, and terrain digitizing, all required that adjustments to the process of analysis be made. Many of these were site specific and can yield only general insights to the analysis process.

Finally, the effect of sea level rise on the study area would be marked, although not as severe as anticipated. The reason for this is that the local topography is not as low lying as had been believed.

SEA LEVEL RISE

Sea level rise due to coastal subsidence has been observed at many locations worldwide, including several, such as the area near the Los Angeles Harbor, where human extraction of groundwater or oil is the cause. At this time, any effects of the projected global warming are masked by the relative rise in sea level that is of geophysical origin (National Academy of Science, 1987).

If the more dire predictions of warming come to pass, and if the projected sea level rise that is associated with that phenomenon does occur, many urban areas, agricultural regions, and recreational areas would be flooded. The economic and social dislocations resulting from this would be extreme, although the gradualness of their realization might mitigate their effect. In order to plan appropriate responses to sea level rise, whatever its origin, data on the effect of rise scenarios must be gathered so that affected areas can be tabulated. Planners will need information on road networks, sewer outfalls, croplands, and residential areas, among other data sets, that would be flooded or isolated by rising waters so that rational decisions on whether to abandon or protect the properties may be made. In addition, such projections can be a useful tool for planning future development to avoid the impact of projected sea level rises.

Calculation of the effect of sea level rise can easily be performed using the "sunken valley" approach. A given increment of rise is postulated and topographical contour maps are consulted to determine the location of the new coastline. Naturally, those areas with steeply sloped coastal regions will suffer the least flooding, while low lying, flat areas, such as south Florida or Bangladesh, will suffer a large inland dislocation of the coastline. The effect of such a water level rise is indicated by the scum lines, in place of contour map lines, left on the banks of a reservoir with a falling water level.

The sunken valley approach, however, does not account for the migration of the coastline that is caused by erosion. This is an especially important phenomenon in this project's study area where the local coastal dynamic of the barrier island and sheltered beach is dominated by beach creation and erosion. Ignoring coastline migration due to erosion here would underestimate the loss of land to the rising sea.

Determining the degree of erosion for a given unit of sea level rise is a difficult undertaking not amenable to the formulation of clear-cut general rules since coastal dynamics vary widely for each type of coastal morphology and for specific coasts. Various researchers have developed formulae that express the equilibrium profile of beach regimes and predict coastal recession associated with sea level rise. For this project, since no detailed data on the parameters involved was available, and because the study is on a regional scale, use of such relationships was deemed unnecessary and perhaps productive of a false impression of precision.

Instead, a historical approach was used based on data and methodologies described in the Responding to Changes in Sea Level (National Academy of Science, 1987). The average yearly recession rate for Maine-New York barrier islands of 0.3 meters was taken as a base erosion rate (May 1983, cited in National Academy Science, 1987). Cumulative erosion figure over the last fifty years was projected backwards (50 x 0.3m = 15m), and this was divided by the concurrent rise in sea level at Sandy Hook, N.J. of 0.7 feet. This yielded a working ratio of approximately 22 meters of beach recession for every foot of sea level rise.

This technique has been successfully applied to low lying, easily erodible terrain from Texas to Maryland (Leatherman 1984, cited in National Academy of Science, 1987). The approach with the GIS was to employ the sunken valley approach for a first estimate of the coastline change to be modified by the addition of a buffer representing the additional recession due to erosion. The degree of accuracy afforded is adequate since the GIS used in this project has a cell resolution of 30 meters per side rendering further precision impossible. Naturally, the size of the added buffer was subject to significant rounding off, which was performed in the upward direction to produce worst case effects.

METHOD OF INVESTIGATION

The first step in the investigation was to choose a suitable study area. Long Island was chosen because of its topical interest, its relevant topography, i.e. barrier islands sheltering relatively low lying terrain, and easy availability of satellite images. The immediate study area was limited to a small portion of the island so that all capture of elevational data would be limited to one USGS quadrangle map. The Mecox Bay area was chosen because its topography closely matched those criteria listed above and it exhibited a variety of land uses in close proximity.

The relevant portion of a SPOT image of the eastern half of Long Island was extracted by examination and entered into its own file. The resulting image was then geo-registered to the UTM coordinate system of the USGS quadrangle by selecting ground control points on the map, locating them in the image, and assigning the image points the associated UTM coordinates. The ERDAS GIS software rubbersheeted the image to conform to the map geometry. Although locating the control points in the image was difficult, and in some sense highly subjective, the resulting error was small, less than one image cell, because the overall rectified area was relatively small.

There was no readily available digitized terrain data for the study area so manual entry of digitized data was required. This introduced two sources of error for the overlay of terrain data onto the satellite image that was to follow: entry of the elevational data is itself a source of error, particularly when it is performed by novices; the set-up of the USGS map on the digitizer pad entails some error that may result in misalignment in subsequent overlays. Data entry errors were particularly noticeable along the coast where exact alignment of the overlays was desired. In addition, the interpolation of captured data to produce an elevational grid produced some oddities, such as above sea level projections located many meters off the beach in the open sea, which were safely ignored since they were of no relevance to the goals of this study.

The original SPOT image was converted to a GIS file through the unsupervised classification module. Land use was classified into the following categories: urban/residential, cropland, wetlands, water bodies, and undeveloped. The accuracy of this classification process directly affected, of course, the validity of final statistics gathered from the study, and it presented some severe difficulties since the urban/residential areas studied are very low density and not easily distinguished from all other categories. The barrier island also presented difficulties since its spectral signature resembled that of uncultivated cropland with which it was initially classed. Moreover, the USGS quadrangle used for digitizing has not been updated for thirty years and provided little assistance in checking on the results of the unsupervised classification.

Prior to digitizing elevations, the existing shoreline elevation had to be determined. The USGS map indicates that the mapped shoreline represents the mean high water line and that the range between high and low water is 2.6 feet. Mean sea level was taken to be the zero datum about which the level oscillates from -1.3 to $+1.3$ feet, and the coastline was digitized as a high water elevation 1.0 feet, rather than 1.3 feet, since no decimal entries are allowed by the software. In the end, the analysis was determining the impact of a given increment of increase of the high water elevation on the location of the new shoreline, under the assumption that the range of high and low water levels would remain the same.

The overlay procedure followed was to extract a slice from the elevational data, from one foot above sea level to two feet above for a one foot rise in level for instance, and add an appropriate buffer to represent recession due to erosion. The result is a polygon that includes the new area to be lost to a one foot sea level rise, and it was recoded with value zero while the surrounding area was recoded as fifteen, an arbitrary high value. This file was overlaid onto the land use file with a maximum dominance output so that the submerged areas, with land use values between one and ten, showed up, while the unflooded areas were masked. This method allows for easy extraction of statistics with the LISTIT function and provides visual information on what land features, natural and man-made, would be lost to flooding.

Additional adjustments and recoding on the basis of in-depth knowledge of local coastal dymanics would result in more accurate overlays for this area. For example, the barrier islands, fragile environments subject to severe erosion by surges, were not lost to a one or two foot sea rise when the facts are that they most likely would be eroded by such a development. This level of detail was not included in this study. In general, the analysis assumes that the dymanics of the system would remain constant in the face of drastically altered morphology, a greatly simplifying assumption.

A primary source of error in the analysis was the imperfect registration of images due to difficulties already discussed. The shorelines of the elevational data file and the land use file were not absolutely in alignment, but were shifted in an east-west direction relative to one another. The result of this was that land on the western side of certain fingers of water that extend inland was not submerged in our flooding scenarios, particularly where the shore had a relatively sharp slope, such as along Hayground Cove. The effect of this error is mitigated for the area-wide analysis because flooding of land to the east was increased, resulting in some rough canceling of the error in the aggregate analysis.

One anticipated source of site specific error did not occur, that is, the pseudo-inundation of inland low lying areas that would, in fact, not be effected by shoreline changes. The study area comprises the outwash area of a glacial moraine and has a continuous slope without significant inland depressions. If this error had occurred, it would have had to be removed through examination and recoding.

The creation of new wetland areas as part of the modification of the coast under a new sea level regime was not accounted for in this analysis. Isolating areas likely to be transformed into wetlands would involve the description of criteria for such sites at another level of analysis.

FINDINGS

The purpose of this exercise was to develop a method which would enable one to study effects on land use due to a rise in sea level. Utilizing this tool could enable federal, state and local planners to predict the extent of damage attributable to shoreline flooding. The creation of different scenarios would be helpful for formulating coastal management policies by assessing the potential damage and evaluating appropriate means of mitigation and their costs, long before the event takes place.

The study area includes a 9 kilometer stretch of shoreline extending from the village of Flying Point, eastward to Sagaponack. The study area consists of 10,754.91 hectares, of which 2,612.52 hectares are water: Table 1 lists the area of each land use category and its percentage of the overall study area. The shoreline area was not included in the analysis since any area lost would be recreated elsewhere. Areas that are currently inland water bodies were also deemed irrelevant to this analysis.

TABLE 1
Land Use Types in Study Area

Land Use	Land Area Size (ha.)	Percentage of Study Area
Undeveloped	909.18	8.42
Cropland	6683.04	62.28
Wetlands	200.61	1.86
Urban/Residential	310.86	2.88
Shoreline	38.7	0.36
Water Bodies	2612.52	24.20

The image from which land use data was classified and extracted was a recent SPOT image of eastern Long Islnad. Cropland clearly makes up the largest percent of the study area and urban/residential land is widely distributed; the largest cluster lying to the north is Bridgehampton.

As a means of assessing the potential land use categories damaged due to a rise in the sea level, six different scenarios were developed. The first five represent an increase of one foot intervals, beginning at one foot; the sixth represents a ten foot rise. Within each scenario, the ERDAS system generated a list of the number of pixels for each land use affected by the rise in sea level. The number of hectares was calculated by multiplying the number of pixels by their given size, 0.09 hectares per pixel. Graph 1 depicts the effect of sea level rise in each scenario while Table 2 breaks down the flooded land parcels by land use type.

The results were not as dramatic as anticipated; the topographical relief was much greater than what the analysts had attributed to this type of coastal area. Photographs of each scenario result were produced and comparing these images, it was clear that even the worst case scenario is not a dramatic inundation, particularly since the study area contains a great deal of water already.

The logarithm of the area of flooded land for each land use category was plotted on a semi-log grid against the sea level rise increment (Graph 2) and a straight line fit of the points was attempted. The slope intercept equation of the resulting line was transformed into an exponential equation and plotted on a regular grid, resulting in Graph 3. Clearly, the rate at which cropland is flooded grows significantly

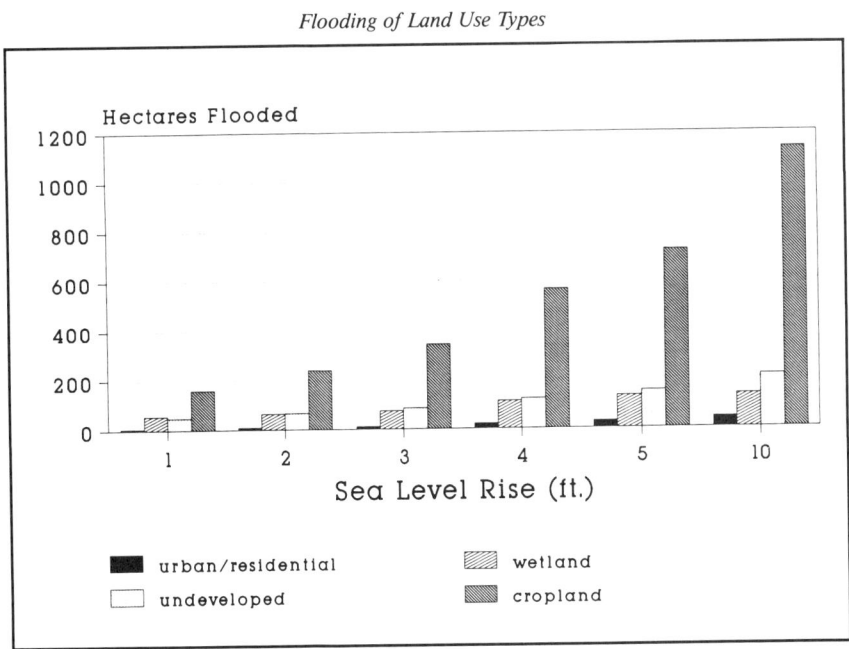

GRAPH 1

Flooding of Land Use Types

as the level of rise increases, while the other land use categories follow a more nearly linear progression. This is presumably because the cropland in the study area was concentrated away from the shoreline. A second attempt to derive an equation from the semi-log plot took account of the apparent break point near a five foot level rise. Two lines were fitted: one for a one to five foot rise; one for a five to ten foot rise, and the results were plotted in a similar fashion in Graph 4. It shows that after a five foot rise, even the inundation of cropland follows a nearly linear progression.

TABLE 2
Areas Flooded by Sea Level Rise (ha.)

	1 ft. rise	2 ft. rise	3 ft. rise	4 ft. rise	5 ft. rise	10 ft. rise
Urban/Res.	6.57	8.46	11.61	19.80	27.36	42.66
Wetlands	55.89	63.54	74.16	111.24	129.87	134.91
Undevel.	48.06	66.24	85.14	121.05	150.57	213.21
Cropland	160.02	238.77	342.27	563.94	721.62	1132.92

GRAPH 2
Semi-Log Plot

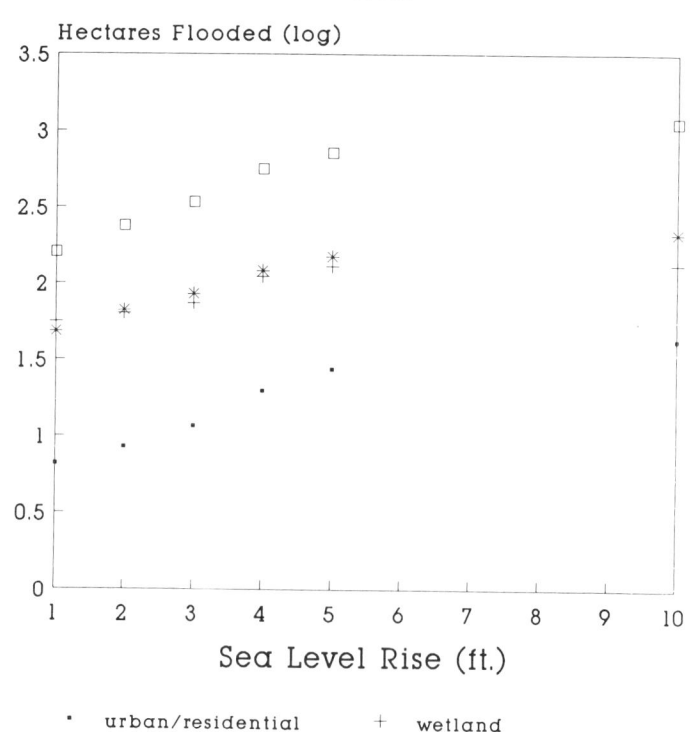

It also accurately reflects the fact that wetlands begin with a higher flooding rate than undeveloped land, a plausible result for a coastal area with a sheltered bay.

The graphs developed from the analysis constitute a predicative tool applicable to coastal systems similar to the project area. Long Island coastal communities from Montauk Point at the easternmost tip of the island to the Queens-Nassau County boundary are situated in areas that are similarly protected by barrier islands, and the upland geomorphology of Long Island is dominated by its axial moraine. Thus, the utility of these graphs for predicting the effects of sea level rise on the Atlantic coast of the island is limited only by the relative mix of land uses as compared to this study area.

In the study area, under the one foot rise scenario, a total of 6.58 hectares of urban/residential property would be affected. Telephone interviews to local realtors indicated that residential properties along the shoreline are valued as high as $1,000,000 per acre. Converting that into a per hectare value would result in total land value losses of $16,252,000. Real estate values decrease as one moves inland. Prices per acre drop dramatically to a range between $50,000 to $350,000 per acre. Therefore, under the worst case flood scenario, a total of 42.66 hectares of

GRAPH 3

Flooding Rates I

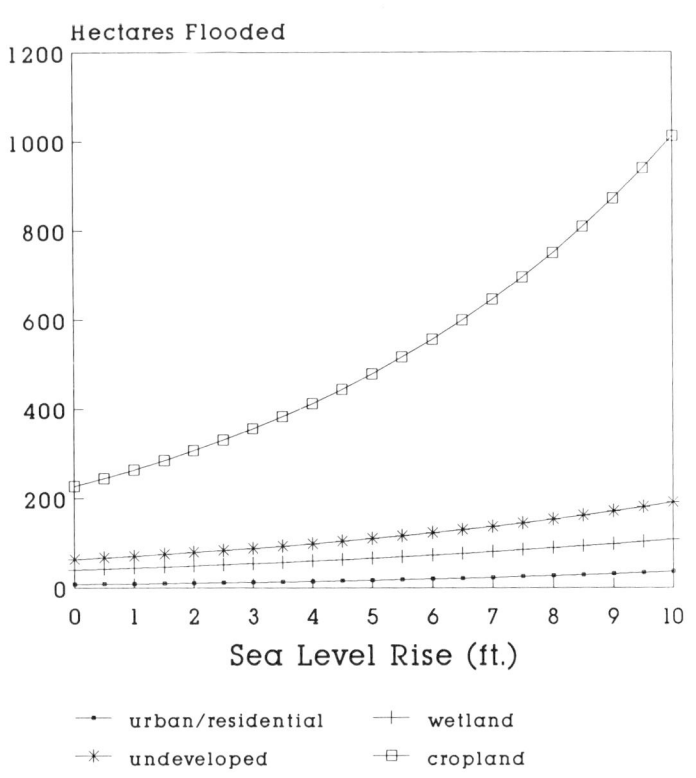

urban/residential properties would be destroyed at a cost of at least $20,708,000. Assessing specific values to undeveloped land uses was not appropriate in this study since it is not known what zoning applies to the land parcels. If more detailed information were available, a cost valuation could be determined. Discussions with realtors indicated that most of the property in our study area is now zoned for residential and local commercial uses.

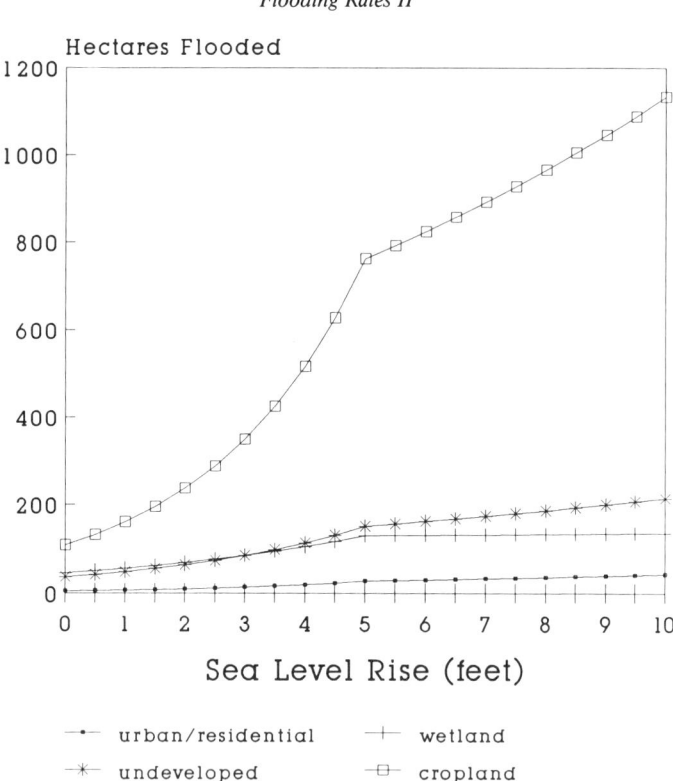

GRAPH 4

Flooding Rates II

- urban/residential
- wetland
- undeveloped
- cropland

LAND USE MANAGEMENT TECHNIQUES

Using a GIS to analyze the effects of sea level rise could be very beneficial for coastal management purposes. There are numerous planning issues which can be attributed to any significant rise in sea levels, such as the following:

1. stabilization and protection of coastal shore
2. saltwater intrusion into fresh water supplies

3. protection of limited shoreline recreation facilities
4. economic impact on local residential and business communities
5. destruction of wetlands

Each of these issues represents a significant potential environmental and economical impact on the local and regional area. Utilizing this GIS model could assist coastal management planners in identifying the numerous alternatives available, based on the extent of the flooding, and provide raw data to work out a cost valuation for the alternatives thought to be the most appropriate.

Coastal management policy is very complex since private and public interests must be weighted against one another. Controversies often arise due to the differences in approaches which can be taken. Generally there are two main approaches associated with coastal management: using engineering techniques, which include beach nourishment, dune improvement, stabilization of vegetation, sea wall construction, and use of hydraulic fill; using land use management techniques which rely on regulatory solutions such as coastal zoning, specific building codes, flood plain management, delineation of flood hazardous areas, and public acquisition of beach front property. The controversy arises in trying to choose between the technical solutions to control the flooding, or allowing it to take its course. Having the proper tools to predict the potential damage and to apply different solutions to the model before the damage occurs can greatly assist those who are developing the coastal management policies.

The effect of flooding on wetlands would be of particular interest in Long Island since the local aquifers which currently suffer from salt front intrusion problems are the region's sole potable source. A rising sea level would increase the head on the salt water intrusion and could exacerbate this problem by pushing the salt front further inland to well points. This represents a planning problem where a GIS system could be of significant aid to local officials with its ability to develop groundwater potential coverages and to identify sites where recharge might be a cost-effective operation.

This GIS exercise presents one way in which imagery data can assist planners, engineers, and policy makers in making more intelligent decisions. The model could be elaborated to include the effect of building sea walls around the more densely populated areas and to analyze the resulting flooding effects, although sea walls themselves are also controversial in terms of their long term effects on beach erosion. The results of this particular exercise were not as important as the creation of more detailed models. Integration of more detailed raw data is crucial, but the exercise tells us what kind of information can be derived from such a model.

Global Climate Change: Implications, Challenges and Mitigation Measures. Edited by S.K. Majumdar, L.S. Kalkstein, B. Yarnal, E.W. Miller, and L.M. Rosenfeld. © 1992, The Pennsylvania Academy of Science.

Chapter Nineteen

FORESTS AND GLOBAL CLIMATE CHANGE

DANIEL B. BOTKIN, ROBERT A. NISBET and LLOYD G. SIMPSON

Departments of Biological Sciences and
Environmental Studies Program
University of California
Santa Barbara, CA 93106

INTRODUCTION

Forests have always been important to civilization, but there has also always been a duality in human attitudes toward forests. On the one hand, early civilizations depended on timber for construction materials and fuel, and therefore forests were a source of necessary resources. On the other hand, the dark interior of forests were viewed as threatening and dangerous, and were the inspiration for many myths. Thus both economic pressures and fear of forests led to clearing of vast areas. Cleared lands were seen as civilized; forested lands as wild and threatening. But the clearing of forests destroyed the very resources on which early civilizations depended. Plato recognized the undesirable results when he wrote that the hills of Attica were "skeletons of their former selves."

Perlin has suggested that an important motivation in exploration of new lands was the need to find new sources of timber to compensate for those exhausted; this story has been repeated over and over during the history of civilization.[1] Today, we seem to be in the process of carrying forest clearing to its logical conclusion; except for areas of the boreal forests and tropical forests, most forests of the world have been cut at least once, and large areas that were original forests have been cleared for other uses or have not regenerated. Boreal forests are subject to intense utilization; for example, plans for commercial Canadian boreal forests, which cover 1,923,950 km^2, lead to harvest of the entire area once a century.[2] The rapid deforestation of the tropics is of course well publicized.

With the concern that human activities may be leading to a rapid climate change, a new awareness of the global importance of forests has awakened. It is now recognized that forests can affect climate at a local, regional, and global level. Not only might rapid climate change have great effects on the availability of forest resources, but also changes in forests may have important effects on climate. Such global roles of forests provide a new set of reasons to believe that continued deforestation is not a wise course of action, and that it is important to develop means to project possible effects of rapid climate change on large forested areas.

HOW FORESTS MIGHT AFFECT CLIMATE

Forests can affect climate through four mechanisms: surface reflection of electromagnetic radiation; surface roughness which affects windspeed and the transfer of energy from the atmosphere to the land; evaporation of water; and rates of transfer of other greenhouse gases between the atmosphere and the land, especially carbon dioxide and methane. The net global effect of all of these mechanisms could be a positive or negative feedback between climate change and forests. For example, if global warming leads to world-wide deforestation, this would lead to massive decomposition of organic matter which would in turn provide an additional source of carbon dioxide to the atmosphere, further accelerating global warming. On the other hand, if the major effect of global warming on forests were to increase evapotranspiration, and the evaporated water primarily condensed into clouds, this would cool the surface, reducing the rate of global warming. Determining the net direction of all effects on a global scale requires that we know the status of forests world-wide in regard to the four mechanisms, which means that we must know such things as the total carbon storage with some degree of accuracy. We must also understand in greater detail than we do at present the factors governing the rates of processes by which forests affect climate and climate affects forests. With this knowledge, we must also develop means to forecast effects of climate change on forests. In this chapter we will consider the current state of our abilities in comparison to these needs.

METHODS TO PROJECT EFFECTS OF RAPID CLIMATE CHANGE ON FORESTS

Computer Simulation of Forest Growth

During the past four years, we have been involved in an attempt to project the possible effects of global warming on forests, under support from the Office of Policy,

Planning, and Analysis of Environmental Protection Agency, the Pew Charitable Trusts, and the USDA Forest Service. In this work, we have taken the projections of changes in temperature and precipitation from general circulation models and used these as input to a computer model of forest growth. In our work, we are using the most recent version of the JABOWA model of forest growth, originally developed in 1970 by Botkin, Janak, and Wallis.[3,4] Descendants of the JABOWA model (FORET, LINKAGES) have been used for the same purpose.[5,6,7,8]

For those not familiar with that model, a brief description follows; a complete description of the most recent version can be found in Botkin, 1992.[9] The forest model simulates the process of forest growth by calculating the regeneration of each species, in terms of the number of new stems added to a plot each year, and the growth and mortality of individual trees with annual time steps on small 10 x 10m plots. The annual growth in height and diameter of each tree is a result of life history characteristics for each species and their response to the simultaneous environmental limitations of light, air temperature, soil moisture, soil nutrient conditions, and stand crowding. In our newest version of JABOWA, JABOWA-II, a complete water balance is calculated for each month which depends on the monthly rainfall and temperature, soil depth, soil moisture holding capacity (which in turn is a function of the average soil particle size), and depth to the water table. The modeled responses of growth to each of these factors is based on the long history of research on the physiology of vegetation in general and woody plants in particular.[10,11,12,13]

Reproduction is limited by the same environmental constraints that affect tree growth. Mortality is a stochastic function of the maximum possible age of each species, and the risk of mortality increases for trees that are growing poorly. There has been considerable experience in the use of this model during the past two decades, and it is well established that the model is realistic in its projections.

Pollen Analysis and Past Climates

Another, independent approach to the same problem involves correlations between depositions of pollen and reconstructions of past climate. These correlations are strong, and from them it is possible to project future steady-state distributions of vegetation.[14] The results of these two methods, computer simulation and pollen analysis, are consistent, in that each method projects great alterations in the present distributions of forest tree species and similar changes in the small locations where comparision have been made. Each method provides unique information. Pollen analysis provides information about the rate of migration of species across the landscape and large-scale maps of possible redistribution of tree species under a future steady-state between climate and vegetation. An important limiting assumption of this approach is that the future relationship between rainfall and temperature will resemble past relationships.

Forest models provide information about changes in existing stands, such as rates of die-back, during the transition from present to future climatic conditions, but as presently formulated do not provide projections about rates of seed migration across the landscape. These models also allow investigation of the response of forests to novel climates, with different combinations of precipitation and temperature. Together, the two methods provide considerable insight into the possible responses of forests to rapid climate change. Because of limitations of space, in this chapter, we will focus on the results from computer simulation.

In our work we emphasized projections of a transition in temperature and precipitation from 1980 to 2070 produced by the NASA GISS general circulation model, mainly using "transient-A", which assumes a "business as usual" per capita production of carbon dioxide in the future.[15] We have applied this transient to modify a local weather record, as we have explained elsewhere.[16]

RESULTS

Here we will focus on the results of our own work, but we will summarize some work of others using related models, referred to earlier. What generalizations can we reach, based on computer simulation? Have we resolved the issues sufficiently so that scientists can move on to other issues? Or is there more work to be done?

Our research shows that projected effects of global warming are severe in some cases, as shown in Figure 1. This figure shows the projected change in above ground biomass for a mature, white birch dominated, boreal forest site in the Superior National Forest of Minnesota, which includes the million-acre Boundary Waters

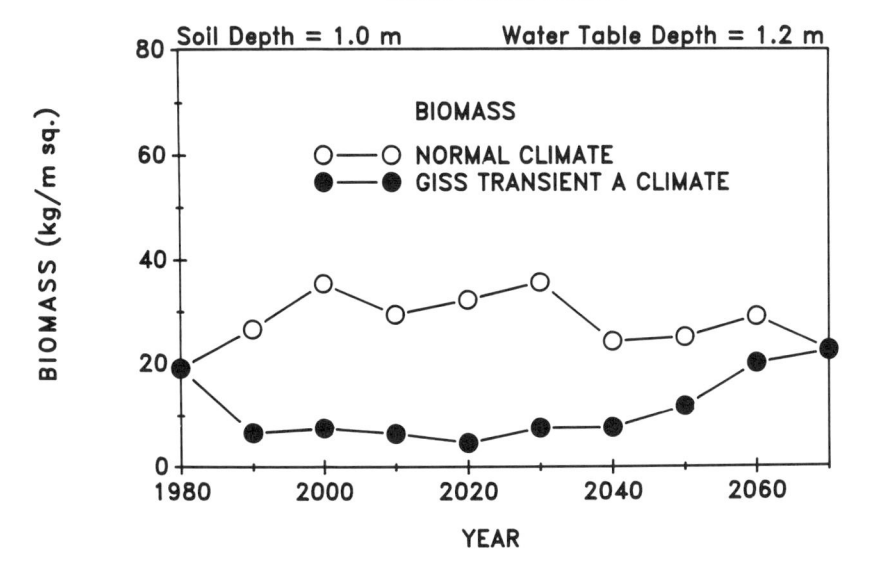

FIGURE 1. Above ground biomass for a white birch site near The Boundary Waters Canoe Area, Minnesota under normal and global warming conditions.

Canoe Area, a legally designated wilderness. Thus the results are of local interest for commercial timber production, conservation, recreation, and other uses of forest ecosystems.

Since carbon is a fairly constant percentage of biomass (about 45% by dry weight), the carbon storage would decline at the same rate as shown in Figure 1, if these projections come to pass. This suggests that the carbon content of mature white birch-dominated stand in the southern part of the North American boreal forest would undergo a decline from 19.08 kg/m^2 in 1980 to 4.63 by the year 2020, a decline in storage to 23% of its present amount! If this rapid loss in carbon storage in live trees occurred over a large region, it would create a positive feedback, further increasing the rate of global warming. It is therefore useful to ask how much carbon might such a change add to the atmosphere?

Extrapolations from One Study, Using our Biomass Estimates, to a Larger Scale

To extrapolate the results from this single sample area to the entire boreal forest, we need two kinds of information: (1) the total carbon stored in that forest; and (2) whether the forest is sufficiently homogenous in current state and in response to climate change, that one can extrapolate from a single sample to the entire forest.

In other work, we have conducted the first study to provide a statistically valid estimate of carbon storage for any large vegetated region of the Earth. Based on direct, field samples, we have estimated that the carbon content of the boreal forests of North America to contain 9.7 ± 2 gigatons.[17] *If all the North American boreal forest declined at the same rate as shown in our figure for a white birch dominated site in the southern part of this forest, then 76.8% of the carbon in live, above-ground biomass would be released in 40 years.* This would be an average release from North American boreal forests of 0.17 gigatons per year.

Extrapolating these carbon storage estimates, we have estimated that the total carbon content of the boreal forests of the world would be approximately 27 gigatons. If the boreal forest declined by 76.8% in 40 years, the averge rate of carbon release would be 0.52 gigatons/year (gt/y). Since the estimated total production of carbon from burning of fossil fuels is about 5 gt/y, the release from the boreal forests alone would represent ten percent of the current fossil fuel contribution, and would be a significant additional source, and would likely lead to a significant increase in global warming.

Can We Accept an Extrapolation from One-site to the Entire Boreal Forest?

But can we expect all forests to respond as this single site? This seems unlikely, but readers involved in policy-making know that, given political pressures and budget limitations, there is a strong pressure to use an absolute minimum number of sites, and extrapolate from these to a global projection. A desire for newness and novelty reinforces this pressure.

Fortunately, with support from EPA, the U.S. Forest Service, and the Pew Charitable Trusts, we and others have been able to consider a number of additional sites. In this chapter we will focus on the results of our studies, but those by others reinforce the general conclusions.

For example, one of the first projections of the effects of global warming on forests, using a combination of output from a general circulation model projection and a model derived from JABOWA was done by Solomon (1986). His results, for six sites in the boreal forest of North America, are consistent with those for the birch stand mentioned here. Forest stands would undergo a die-back in southern areas, with some areas becoming treeless before invasion by southern deciduous species.[18] In this earlier application, the climate transition from present conditions to future global warming climate was simulated as a linear increase between two steady-state projections of a general circulation model: a 1980 steady-state climate, and a "twice-CO_2," steady-state climate, assumed to be reached by year 2070.

However, projections vary with species and site conditions and with geographic location.

Another boreal forest site we have been able to consider is a Siberian larch dominated site near Ust'-Nera in Siberia. For these sites, field measurements were obtained for the forest conditions in 1990, and projections were made from these initial conditions. These results are especially interesting because the general circulation models indicate that the greatest changes in temperature will occur at mid and high latitudes. Therefore, one would expect a comparatively large change in temperature at a high latitude Siberian forest. One would expect the change to be at least as great as that for northern Minnesota, which is at a lower latitude.

Contrary to that expectation, projections of the JABOWA-II forest model indicates that, if the projections of the NASA GISS transient-A are correct, *nothing* will happen to these forests (Figure 2). This is in striking contrast to our projections for a white birch site in the Superior National Forest. Why is this? When we obtained this result, we were also puzzled, so we graphed the projected change in temperature for this area from the GISS model, which is shown in Figure 3.

True, this graph shows that there are changes in temperature, but almost all the warming takes place in the middle of winter, when the temperature warms from −40° to −35°. This has *no* effect on tree growth in the model, and would likely have little effect on tree growth in reality (Low winter temperatures can be a problem to trees at the extreme southern limit of their range, where they are already under great stress and where warming and cooling periods can affect tree physiology and forest structure. These are not the conditions for the trees in the Ust'-Nera forests). During the growing season, little temperature change occurs. Thus, temperature changes during the season that does not affect tree growth, but does not change during the seasons when trees are actively growing.

The comparison of results for two sites far apart suggests that there wil be important regional differences in the response of forests to global warming. What about local variations, such as might occur from stands with different species or growing on different soils? To gain insight into those possibilities, we have considered several additional cases, two of which we will review here. These are projections for two sites that are near to each other near Antigo, Wisconsin, but have different soil conditions. These sites are in a transition between the boreal forest to the north

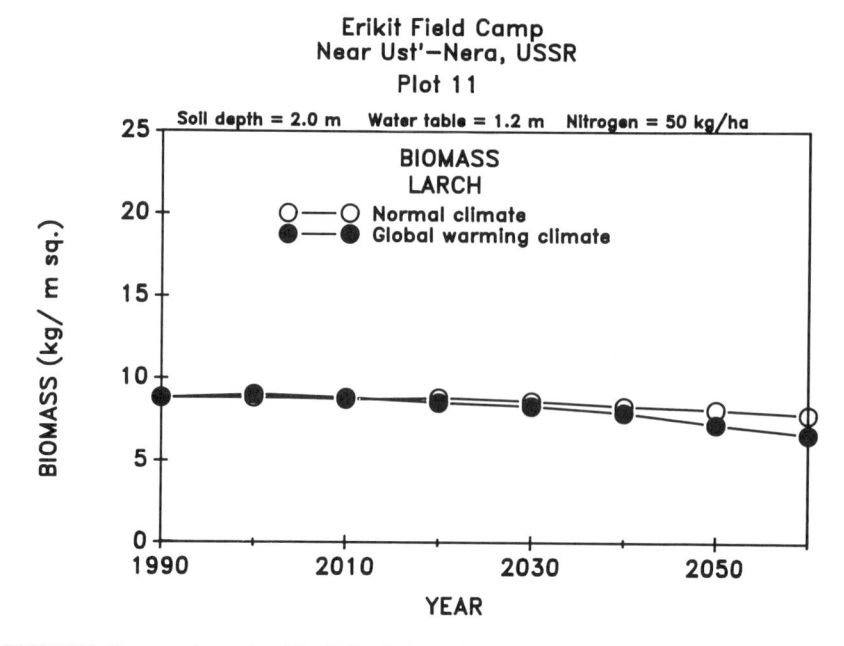

FIGURE 2. Biomass change in a Siberial larch site at the Erikit Field Camp near Ust'-Nera, Siberia.

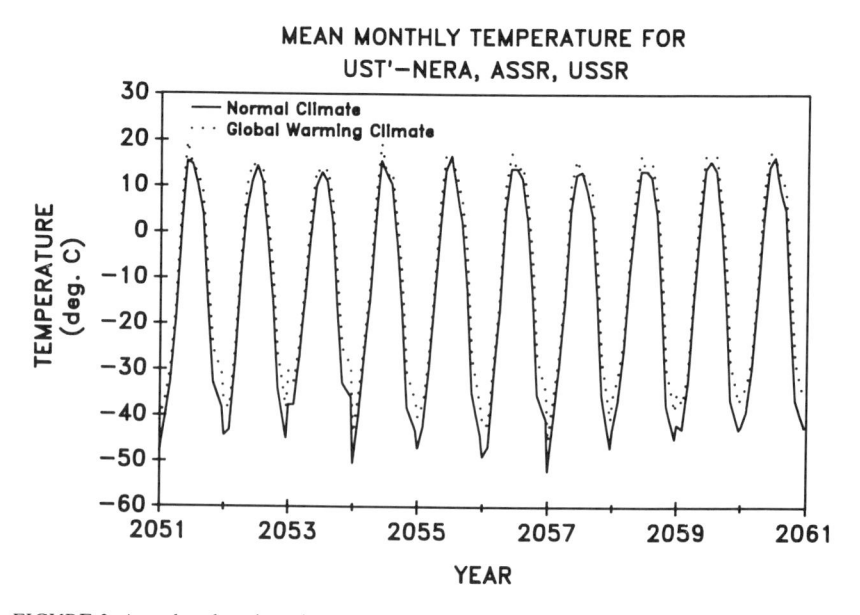

FIGURE 3. Actual and projected temperature between 2051 and 2061 for Ust'-Nera, Siberia.

and northern hardwoods, to the south.

In the first stand, there is little change (Figure 4), while the second shows an *increase* in biomass (Figure 5). The second Antigo site is *wetter* and *too wet under present climatic conditions* for optimum tree growth. Global warming dries the soil and improves the site conditions. Furthermore, this site is at the northern end of the range of sugar maple, and warming also benefits that species. These figures show that there will be soil and species specific differences in responses. The first Antigo site has adequate moisture, and the drying effects of global warming are insufficient to produce a major change.

We have made projections for a number of other sites, including jack and white pine stands in central Michigan, oak-dominated hardwood forests in southern Michigan, a variety of sites in the Superior National Forests, and selection of sites at different elevations in Alaska. The majority of these sites show a rapid decline in existing forests. Some, like jack pine stands in central Michigan, undergo an even more rapid decline than the birch stands in Figure 1.

An analysis using another model derived from JABOWA, which includes an additional module, a soil component and also soil-forest nitrogen cycling and feedbacks, was caried out by Pastor and Post in 1988.[9] They also found that effects vary with soil conditions. Productivity and biomass increased on sites with adequate moisture and nitrogen availability, and decreased on drier sites, particularly those with lower

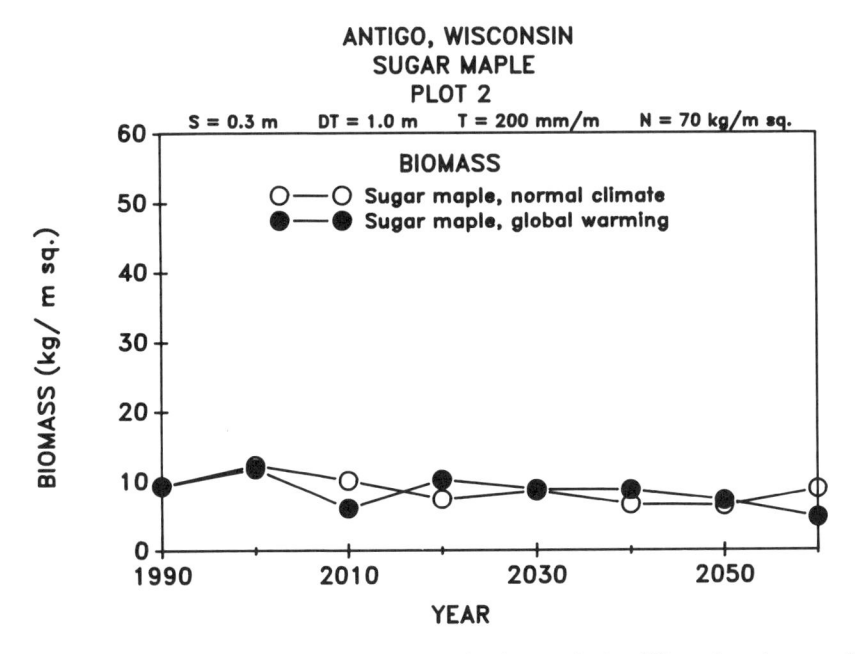

FIGURE 4. Growth of sugar maple on a moderately dry site near Antigo, Wisconsin under normal and global warming conditions.

nitrogen availability. Like the study by Solomon (1986), Pastor and Post (1988) projected the climatic transition as a linear interpolation between 1980 steady-state and 2070 twice-CO_2 steady-state climatic projections.

If we have to base policy simply on projections available now, then we would have to say that most sites in the boreal forest will undergo an intitial, large decline in carbon storage, and that our best estimate is that the boreal forests will provide a large source of carbon dioxide to the atmosphere, further accelerating global warming. However, these results warn us that effects will depend on what species are present, where the site is located in relation to the range of each species, and on the local soil conditions, which in turn will be influenced by bedrock and topography. Thus these results suggest that an assessment that could be used to project the net change in carbon storage would require simulation for a statistically valid sample of sites across the boreal forest.

Possible Mitigating Factors that Might Reduce the Response of Forests to Climate Change.

In contradiction to the results given to this point, some scientists believe that global warming will have little net impact on forests, because increased CO_2 concentrations will augment tree growth and balance growth reductions imposed by global warming.[20] The effect of CO_2-fertilization on plant growth has been known

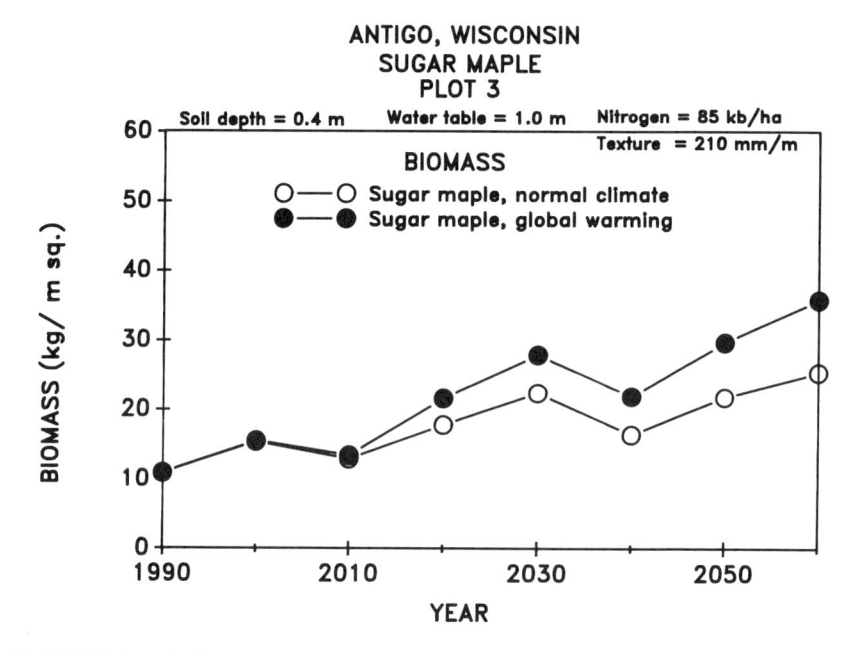

FIGURE 5. Growth of sugar maple on a moderately wet site near Antigo, Wisconsin under normal and global warming conditions.

since an experiment by Priestly, but has been studied primarily under laboratory conditions where all other factors were non-limiting.[21] Few data exist for tree species, and these are restricted to seedlings and saplings, most under uniform laboratory conditions,[22] or with variation of one environmental factor in addition to CO_2.[23] No experiments have examined responses of mature trees under field conditions to a combined increase in CO_2-concentration, warming of air and changes in soil moisture, three changes that would occur in global warming. In a much earlier paper, several of us showed that a forest under a constant climate (with no global warming) and subject only to the fertilization effects of an increase in CO_2 concentration would undergo much smaller increases in growth and biomass than one would project from the response of a single seedling in a greenhouse grown under non-limiting environmental conditions. The fertilization effect would increase the amount of leaves, and thus increase the shading within the forest. Competition among trees of different species for light would compensate for the increase in carbon dioxide concentration.[24]

More recently, we have approached the same question, to which we have added projections of global warming. However, because of the lack of the necessary data for each species, we calculated CO_2-fertilization effects from published studies of seedlings and saplings of four North American tree species, each representing a set of species with similar ecological and successional characteristics:[25,26]

$$Y = -0.050 + 0.0042X - 0.000002X^2 \quad (Populus\ deltoides); \tag{1}$$
$$Y = 0.840 + 0.0005X - 0.0000002X^2 \quad (Platanus\ occidentalis) \tag{2}$$
$$Y = 0.457 + 0.0018X - 0.0000008X^2 \quad (Pinus\ banksiana); \tag{3}$$
$$Y = 0.625 + 0.0012X\ \ \ \ 0.0000005X^2 \quad (Picea\ glauca); \tag{4}$$

where X = CO_2 concentration (μl/L) and Y is the ratio between control (340 PPM CO_2) and treatment CO_2 concentration.[27] Early successional broadleafs are represented by *Populus deltoides*; late successional by *Platanus occidentalis*; early successional conifers by *Pinus banksiana*; late successional by *Picea glauca*.

Estimation of increases in CO_2 concentrations follows the GISS projections:

$$X = 340 + 1.2373t + 0.0205t^2 - 0.00004678t^3 \tag{5}$$

where X is the CO_2 concentration at year t after 1980.[28] The forest model first calculates growth for each tree each year given the light, temperature, soil moisture, and nutrient conditions. Then growth of each tree is increased by the CO_2-fertilization factor.

Projections were made for sites near to the birch site of Figure 1, in the southern part of the Superior National Forest, near Virginia, MN, characterized by boreal forests, but near to the transition to northern hardwood forests. On an old-age stand on relatively deep, moist soil, balsam fir is typically dominant. Under "normal" (1950-1980) weather conditions, the forest model projects that the abundances of the major species vary slightly over time, but fir remains dominant. Under global warming, balsam fir declines rapidly, reaching a significantly lower abundance by 1990, and is replaced by sugar maple, which reaches a significantly larger value than under the normal conditions by year 2010 (Figure 6). Projections with CO_2-fertilization are indistinguishable from projections without CO_2-fertilization.

This site is near the southern limit of balsam fir where effects of warming and drying of soils might be most pronounced. A second site was located to the north, at the optimum temperature regime for balsam fir. Again CO_2-fertilization leads to no change from projected effects of a global warming climate without the fertilization effect (Figure 7). Similar results were found for an early successional jack pine forest subject to frequent fires near Grayling, Michigan (not shown).

Two factors contribute to this result. First, even under constant climate, competition among trees of different species and sizes counteracts the fertilization effect. As CO_2 fertilizes growth, trees increase in size, but their leaf weight also increases. Under the canopy trees the forest becomes shadier, slowing growth and counteracting the fertilization effect. Second, CO_2 response functions give a small growth increase especially in the first decades when warming and drying of soils is less severe; warming and increases in evapotranspiration are so great that the small CO_2-fertilization has little relative effect. The model warns us that believing the CO_2-fertilization will compensate for global warming is like believing that fertilizing corn in a drought year will make up for lack of rain.

Limitations of the Projections of Carbon Dioxide Fertilization:

The uncertainty in the application of one of four growth response curves to all 34 species may generate a bias in growth estimates by the model. This bias can be greatly reduced by future research on the effects of increased CO_2 on growth of each of the 34 species; variables such as light intensity and water-use efficiency, as well as temperature and water availability should be analyzed as CO_2 is increased in the experiments.

Additional uncertainties involve the validity of current projections from general circulation models as input to characterize a time-course of climate for the forest growth model. GCM models simulate temperature and precipitation averages for 75,000 square mile areas on the Earth, while the forest model simulates changes in forest composition on 10m x 10m plots. Modification of local weather records with large regional climate averages may mask local effects of rivers, lakes, oceans and mountains. Higher-resolution GCM models are needed in the future for better linkage with the forest growth model.

In spite of these potential limitations, the consistency of the results for several sites, ranging from the southern boundary of a major species to optimum temperature conditions for that species, suggest that the projections are robust. In addition, these results are consistent with earlier applications of the forest model to study the effects of CO_2-fertilization on forests under a constant climate.[29] From this we conclude that the direct fertilization effect of CO_2 (even at levels of 1.5X and 2X normal growth increments) is not likely to compensate for the effects on current forests of changes in air temperature and soil moisture resulting from an increase in the atmospheric concentration of greenhouse gases.

These results also illustrate the necessity for changes in forest planning to cope

Forests and Global Climate Change 285

**BWCA
EFFECT OF CO2 FERTILIZATION ON
BALSAM FIR**

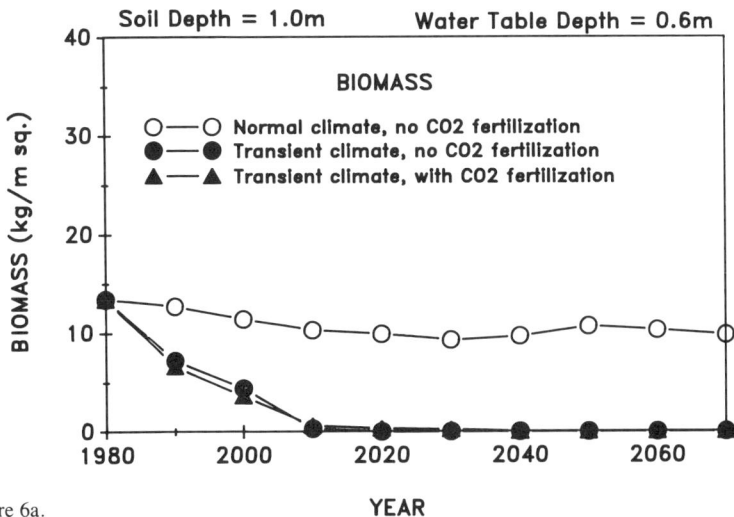

Soil Depth = 1.0m   Water Table Depth = 0.6m

BIOMASS
○──○ Normal climate, no CO2 fertilization
●──● Transient climate, no CO2 fertilization
▲──▲ Transient climate, with CO2 fertilization

Figure 6a.

**BWCA
EFFECT OF CO2 FERTILIZATION ON
SUGAR MAPLE**

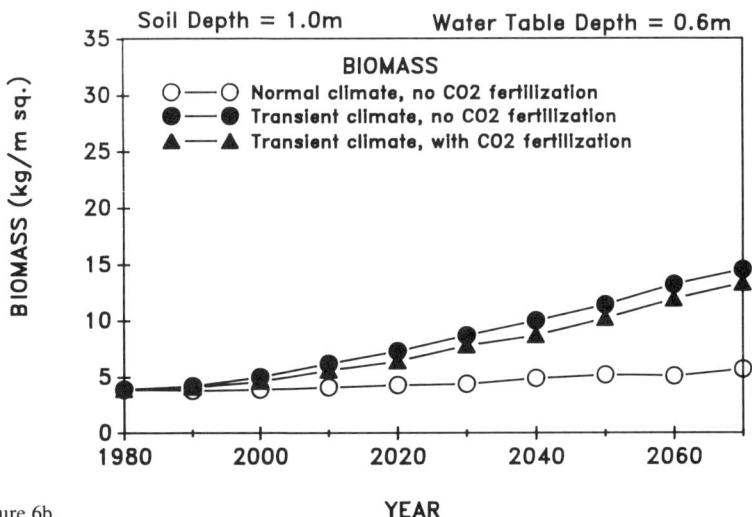

Soil Depth = 1.0m   Water Table Depth = 0.6m

BIOMASS
○──○ Normal climate, no CO2 fertilization
●──● Transient climate, no CO2 fertilization
▲──▲ Transient climate, with CO2 fertilization

Figure 6b.

FIGURE 6. Effect of CO_2 fertilization on forest biomass growth near Virginia, Minnesota under normal and global warming conditions. (A) Balsam fir (B) Sugar maple.

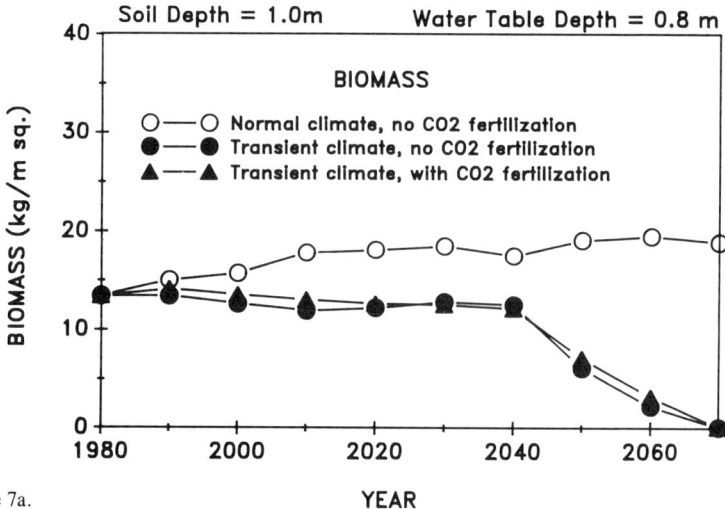

Figure 7a.

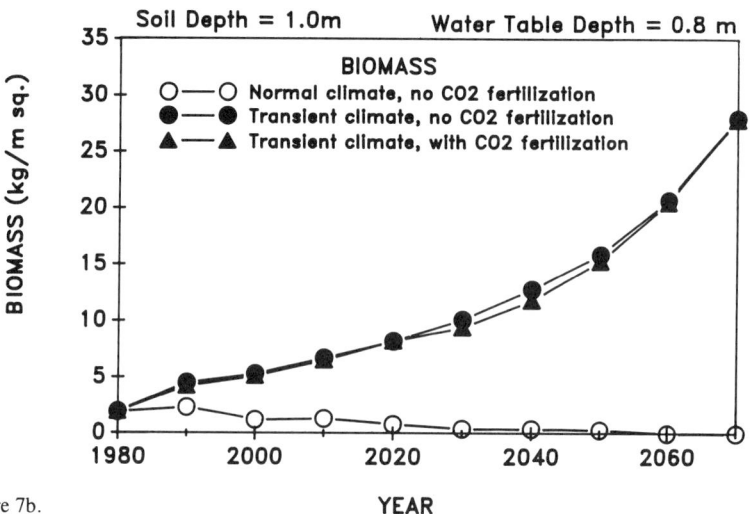

Figure 7b.

FIGURE 7. Effect of CO_2 fertilization on forest biomass growth on a northern site optimal for balsam fir growth; (A) Balsam fir (B) Sugar maple.

with the probability of large-scale forest change during the next 90 years. The focus of forest management policy, which are of significant ecological, economic, and recreational value, should begin to shift away from current tree species in many areas (which may not grow there at all in the future) to those that will grow in these areas.

SUGGESTIONS FOR A NEW PROGRAM: AN EARLY WARMING SYSTEM FOR GLOBAL CLIMATE CHANGE

Our results suggest that a program should be initiated to obtain projections for a statistically reliable sample of sites, representative of the boreal forests. From our biomass studies, we have demonstrated that we know how to set up a statistical sampling scheme that gives a representative estimate with an approximately 20% error (the 95% confidence interval is within 20% of the mean). We suggest using this method to select a network of forest sites. At each site, record necessary initial conditions. Project the effects of global warming for all these sites. The result will provide the best available projection that we can do today for the large scale response of boreal forests to global warming. Let me emphasize that these results demonstrate the implications of our present understanding of global warming and forest dynamics. Rather than saying that the projections would be "true" we would say that these projections represent the implications of what we know today.

We believe that this is an important project, useful as one of the first steps in linking climate dynamics and the global carbon cycle. Doubtless the reader can think of other viable approaches to investigating the linkages between the global carbon cycle and climate dynamics. These, we believe, will be an important next step in our attempt to understand how human activities may influence the environment at a global scale.

CONCLUSIONS

If global warming occurs as projected by general circulation models, then many forested areas of mid and high latitudes will undergo surprisingly rapid and severe changes. On many sites existing trees will die and there will be a net decline in carbon storage until species adapted to warmer and drier conditions migrate into the area. However, the response of a specific stand of trees depends on the species present, the location of the site in relation to the entire geographic range (and therefore physiological capacities) of each species, and local soil conditions. Sites that are too wet for maximum tree growth and not too near the maximum thermal limits of the species present might undergo an increase in growth. Other stands in areas which experience warming primarily in the winter, will undergo little change.

Because forests can affect climate through reflection of electromagnetic radiation, surface roughness, evaporation of water, and rates of transfer of CO_2 and methane, wide scale changes in forests could have major effects on climate. The net global effect of all forest responses might lead to either a positive or negative feedback.

While we have some of the tools necessary to assess the likely direction of these effects, we need to develop a new world-wide program, which we have referred to as "an early warming system for global climate change." In this program, projections of the effects of global warming would be conducted for a statistically valid sample of stands throughout the major forest types. Only with this kind of information can the overall effect of forests on climate be projected.

LITERATURE CITED

1. Perlin, J. 1989. *A Forest Journey: The Role of Wood in the Development of Civilization*, W.W. Norton, New York.
2. Bickerstaff, A., W.L. Wallace and F. Evert. 1981. Growth of forests in Canada Part 2: A quantitative description of the land base and mean annual increment. Petawawa National Forest Institute, *Canadian Forest Service Information Report* PI-X-1.
3. Botkin, D.B., J.F. Janak and J.R. Wallis. 1970. A simulator for northeastern forest growth: a contribution of the Hubbard Brook Ecosystem Study and IBM Research. *IBM Research Report* 3140. Yorktown Heights, NY. 21 pp.
4. Botkin, D.B., J.F. Janak and J.R. Wallis. 1972. Rationale, limitations and assumptions of a northeast forest growth simulator. *IBM J. of Research and Development* 16:101-116.
5. Solomon, A.M., M.L. Tharp, D.C. West, G.E. Taylor, J.W. Webb and J.L. Trimble. 1984. Response of unmanaged forests to CO_2-induced climate change: available information, initial tests, and data requirements. *U.S. Dept. Energy, Technical Report.* 009, 93 pp.
6. Solomon, A.M. 1986. Transient response of forests to CO_2-induced climatic change: simulation modeling experiments in eastern North America, *Oecologia* 68:567-580.
7. Solomon, A.M. and D.C. West. 1987. pp. 189-217 in W.E. Shands and J.S. Hoffman (ed). *The Greenhouse Effect, Climate Change, and U.S. Forests*. The Conservation Foundation, Washington, D.C. They use climate projections from J.F.B. Mitchell, *O.J.R. Meteorl. Soc.* 109:113 (1983).
8. Pastor, J. and W.M. Post. 1988. Response of northern forests to CO_2-induced climate change, *Nature* 334:55-58.
9. Botkin, D.B. 1992. *Forest Dynamics: An Ecological Model*, Oxford University Press (in press; publication expected September, 1992).
10. Kozlowski, T.T. 1968. *Water Deficits and Plant Growth,* Vol. 1 (1968), Vol. 2. 1970, Vol. 3. 1972. Academic Press, N.Y.
11. Kozlowski, T.T. 1982. Water supply and tree growth, Part I Water Deficits. *Forestry Abstracts*, Commonwealth Forestry Bureau 43:57-95.
12. Kozlowski, T.T. 1982. Water supply and tree growth, Part II Flooding, *Forestry Abstracts,* Commonwealth Forestry Bureau 43:145-162.
13. Kozlowski, T.T., P.J. Kramer and S.G. Pallardy. 1991. *The Physiological Ecology of Woody Plants*. Academic Press, San Diego.
14. Zabinski, C. and M.B. Davis. 1989. Hard Times Ahead for Great Lakes Forests:

A Climate Threshold Model Predicts Responses to CO_2-Induced Climate Change, Chapter 5, pp. 5-1 to 5-19 In: J.B. Smith and D.A. Tirpak (ed.) *The Potential Effects of Global Climate Change on the United States: Appendix D-Forests* Office of Policy, Planning and Evaluation, U.S. Environmental Protection Agency, Washington, D.C. EPA-203-05-89-054.
15. Hansen, J., I. Fung, A. Lacis, D. Rind, S. Lebedeff, R. Ruedy and G. Russell. 1988. Global climate changes as forecast by Goddard Institute for Space Studies Three-Dimensional Model. *J. Geophysical Research* 93:9341-9364.
16. Botkin, D.B., R.A. Nisbet and T.E. Reynales. 1989. Effects of Climate Change on Forests of the Great Lake States, pp. 2-1 to 2-31 in J.B. Smith and D.A. Tirpak (ed.) *The Potential Effects of Global Climate Change on the United States*. U.S. Environmental Protection Agency, Washington, D.C. EPA-203-05-89-054.
17. Botkin, D.B. and L. Simpson. 1990. The First Statistically Valid Estimate of Biomass for a Large Region, *Biogeochemistry* 9:161-174.
18. Solomon. 1986. *Loc cit.*.
19. Pastor, J. and W.M. Post. 1988. *Loc cit.*
20. Wittwer, S.H. 1979. Future technological advances in agriculture and their impact on the regulatory environment. *Bioscience* 29:603-610.
21. See for example: J.E. Hardh. 1966. CO_2 enrichment in raising young vegetable plants. *Acts. Hort.* 4:126-128; Krizek, D.T., R.H. Zimmerman, H.H. Klueter, and W.A. Bailey. 1969. Accelerated growth of birch and crabapple seedlings. *Plant Physiol.* (Suppl.) 55:15.
22. See for example: Funsch, R.W., R.H. Mattson and G.R. Mowry. 1970. Carbon dioxide-supplemented atmosphere increases growth of *Pinus strobus* seedlings. *Forest Science* 16:459-460; Jurik, T.W., J.A. Weber, and D.M. Gates. 1984. Short-term effects of CO_2 on gas exchange of leaves of bigtooth aspen (*Populus grandidentata*) in the field. *Plant Physiol.* 75:1022-1029; Tolley, L.C. and B.R. Strain. 1985. Effects of CO_2 enrichment and water stress on gas exchange of seedlings grown under different irradiance levels. *Oecologia* 65:166-172; Sionit, N., B.R. Strain, and H.H. Helmers. 1985. Long-term atmospheric enrichment affects growth and development of *Liquidambar styraciflua* and *Pinus taeda* seedlings. *Can. J. Forestry Res.* 15:468-471; Rogers, H.H., J.F. Thomas, and G.E. Bingham. 1983. Response of Agronomic and Forest Species to Elevated atmospheric carbon dioxide. *Science* 220 (4595):428-429; Leverenz, J.W. and D.J. Lev. 1987. Effects of carbon dioxide induced climate changes on the natural ranges of six major commercial tree species in the western United States, pp. 123-156. In: *The greenhouse effect, climate change, and U.S. forests*, Eds. W.H Shands and J.S. Hoffman. The Conservation Foundation, Wash., D.C.
23. Tolley, L.C. and B.R. Strain. 1985. Effects of CO_2 enrichment and water stress on gas exchange of seedlings grown under different irradiance levels. *Oecologia* 65:166-172.
24. Botkin, D.B., J.F. Janak and J.R. Wallis. 1973. Estimating the effects of carbon fertilization on forest composition by ecosystem simulation, pp. 328-344, In: G.M. Woodwell and E.V. Pecan, eds., *Carbon and the Biosphere*, Brookhaven

National Laboratory Symposium No. 24, Technical Information Center, U.S.A.E.C., Oak Ridge, TN.
25. Kimball, B.A. Carbon dioxide and agricultural yield: an assemblage and analysis of 770 prior observations. *Water Conservation Laboratory Rep. 14* (USDA Water Conservation Laboratory, Phoenix, AZ, 1983).
26. CO_2 response curves for *Populus deltoides* and *Platanus occidentalis* were calculated from Carlson, R.W. and F.A. Bazzaz. 1980. The effects of elevated carbon dioxide on growth, photosynthesis, transpiration, and water use efficiency in plants, pp. 609-612 In: J. Singh and A. Deepak (ed.) *Proc. Symp. on Environmental and Climate Impact of Coal Utilization,* (Academic Press, N.Y.; CO_2 response curves for *Pinus banksiana* and *Picea glauca*) were calculated from Yeatman, C.S. 1970. Enriched air increased growth of conifer seedlings. *Forest Chron.* 46:229-230.
27. Keeling, C.D., R.B. Bacastow T.P. Whorf. 1982. Measurement of the concentration of carbon dioxide at Mauna Loa Observatory, Hawaii. pp. 377-385. In: *Carbon Dioxide Review 1982*, W.C. Clark, Ed. (Oxford Univ. Press.)
28. Hansen, *et al.* 1988, who give ΔT_0, the rate of change of annual mean global surface air temperature, between 1960 and 2050 as a function of x, the CO_2 concentration:

$\Delta T_0 (x) = f(x) - f(x_0)$
where,
$f(x) = \ln(1 + 1.2x + 0.005x^2 + 1.4 \times 10^{-6}x^3)$,

x is the projected CO_2 concentration, and $x_0 = 340 \, \mu l/1$ (estimated CO_2 concentration in 1980).
From their Figure 2, we determined CO_2 concentration as a function of temperature:

$Y = 340 + 1.2373x + 0.0205x^2 - 0.00004678x^3$

where Y is the CO_2 concentration at year x expressed as years after 1980. Hansen, J. I. Fung, A. Lacis, D. Rind, J. Lebedeff, R. Ruedy, G. Russell, and P. Stone. 1988. Global climate changes as forecases by the Goddard Institute of Space Studies three-dimensional model. J. Geophys. Res. 93:9341-9364.
29. D.B. Botkin, J.F. Janak and J.R. Wallis, in *Carbon and the Biosphere*, G.M. Woodwell and E.V. Pecan, Eds. (Brookhaven National Laboratory Symposium No. 24, Technical Information Center, U.S.A.E.C., Oak Ridge, TN, 1973) pp. 328-344.

Global Climate Change: Implications, Challenges and Mitigation Measures. Edited by S.K. Majumdar, L.S. Kalkstein, B. Yarnal, E.W. Miller, and L.M. Rosenfeld. © 1992, The Pennsylvania Academy of Science.

Chapter Twenty

ECOLOGICAL EFFECTS OF RAPID CLIMATE CHANGE*

DEXTER HINCKLEY[1] and GERALDINE TIERNEY[2]
Office of Policy Analysis, Climate Change Division
U.S. EPA, PM 221
Washington, DC 20460

INTRODUCTION

Over the eons, the Earth's climate has gone through many changes. There was a 150-million-year period in the Mesozoic when the world was ice-free, with an average temperature 10°C higher than now. Then, there have been ice ages; the first beginning perhaps 2.3 billion years ago and the most recent ending only 10,000 years ago! Some forms of life, such as bacteria, have survived all these fluctuations, while others have been driven to extinction. Humankind is recent on the scene and can claim collective experience with only a rather short sequence of glacial and interglacial periods.

The paleontological record shows the continued interplay between life and climate. Species go extinct, only to be replaced by newly-evolved species. Patterns of distribution change; plants, animals and humans retreat towards the equator as the glaciers advance, then advance poleward and up the mountain valleys as the glaciers retreat. What is striking about changes in these rhythms and patterns is their leisurely pace. The replacement of species by others occupying comparable niches may take millions of years. The northward migration of trees following a glacial retreat may take thousands of years.[2] Life can live with relatively slow climate changes; it moves, adapts and evolves. In fact, the diversity of life may be greatest

---

*This chapter does not present policies of the U.S. Environmental Protection Agency or any other agency of the U.S. government.
[2]Bruce Co.

where disturbance is frequent—but not excessive.[3] If the disturbance is sudden and drastic—a comet impact, for example—the consequences include massive extinctions.[4]

How does anthropogenic global warming compare with these natural phenomena? Will the projected increases over the next 100 years of 1.5 to 4.5°C globally, (twice as much in the high latitudes), fall within the range of variation humans and other species can tolerate? What are the likely responses of different ecosystems to temperature increases, changing pattern of precipitation, and other climatic shifts? Will the anthropogenic climate change be stimulating—or devastating?

RATES OF CHANGE

The rapid rate of anthropogenic global and regional warming may be the most important factor determining the extent and severity of ecological effects.[5] If that rate is much faster than natural rates of adaptation or migration, many species will go extinct and many communities of plants and animals will be greatly modified. Fortunately, paleoecologists have been able to learn enough from past changes to help modelers develop plausible scenarios of future changes.

Various combinations of geological, atmospheric and ecological indicators have been used to reconstruct past climates and estimate the rates at which these climates have changed. While these studies, especially those using pollen deposits, have been extremely meticulous, we should never forget that we have no direct record of past climates and must rely on surrogates. We should be most confident of our interpretations if multiple indicators point independently to the same climatic conditions.

Focusing on one parameter of climate, the average annual temperature, the table below shows some of the problems associated with picking a time-frame for estimating the rate of climate change:

The first three represent the maximum rates estimated for transition periods in the temperate north (Europe and North America).**

TABLE 1
Past and Future Rates of Temperature Change

Time-period	Warming, °C	Rate, °C/century
Last glacial (15,000 BP) to Allerod (11,500 BP)	10	0.3
Younger Dryas (10,500 BP) to climatic optimum (7000 BP)	7	0.2
5000 BP to 2500 BP	4	0.2
Last 10,000 years	5	0.05
Last 100 years	0.5	0.5
Next 100 years	2.5	2.5
Next 100 years (high latitudes)	5.0	5.0

**While these are maximum rates sustained over decades or centuries, there was a short period (c. 20 years) near the end of the last glaciation when the North Atlantic region apparently had a 5°C warming, presumably related to the renewal of a warm current coming up from the tropical Atlantic.

The climate modelers predict that global warming will be 1.5 to 4.5°C in the next century, and we have every reason to believe that it will keep rising for some time thereafter (see "The Case for Global Warming" in this volume).

PROBLEMS WITH ADAPTATION

Confronted with severe climatic stress, an organism really has but two options— "adapt" or "move". Adaptation to climatic stress has genetically defined limits. Leaf stomata can close only so much and, for an animal, time spent in the shade is limited by the time required to forage and seek mates.

Mechanisms for adaptation typically come into play when an organism is exposed to environmental conditions near the limits of its tolerance. These extreme conditions, such as heat waves or periods of drought, are episodic but not unprecedented. The ancestors of the organism were exposed to stresses similar in severity and duration. Through natural selection, genomes with adaptive mechanisms evolved. Now, however, these well-adapted species may be exposed to events more extreme in frequency, length, or extent than any in their evolutionary history. They simply may not be programmed to cope.

Animals have patterns of adaptive behavior helping them survive extreme heat by "timing of daily activity," "local and long distance movements," and "utilization of favorable microenvironments."[6] If these instinctive tactics do not suffice, terrestrial vertebrate animals must rely on "physiological heat defence" with the activities of panting or sweating imposing on such animals the "task of reconciling the antagonistic requirements of thermoregulation and maintenance of water balance."[7]

Terrestrial insects, such as *Drosophila* spp., may not be able to adapt to simultaneous increases in temperature and desiccation. Each type of stress makes different demands on their defenses and their genomes represent a compromise of adaptive characteristics partially responding to warming and partially to drying. If these stresses are superimposed on habitat destruction, extinctions are likely.[8] Aquatic organisms, such as fish, may not have to contend with desiccation but have limited ability to adapt to seasonal warming. If regional warming forces ambient water temperatures above a certain threshold, the fish no longer grow; if the temperatures go still higher, the fish die.[9] Some of the early indicators of climate change described later in this chapter are freshwater or marine organisms pushed beyond their thermal limits.

PROBLEMS WITH DISPERSAL

All organisms have some potential for dispersal. If this dispersal involves a round-trip, it is called "migration." Even forests can be said to "migrate," moving towards the equator as the glaciers advance, then back towards the poles as the glaciers retreat.[10] Many species in the temperate zone have been doing this successfully for several hundred thousand years. However, their dispersal mechanisms may not be able to cope with climatic changes far more rapid than those under which these species evolved.

For example, a 3°C warming in the next 100 years would be equivalent of a northward shift of 300 to 450 km in one century. Those trees, such as elm and maple, which produce wind-borne seeds have historically been able to advance no more than 100 km/century. Those with large nuts, such as oak, chestnut, and walnut, migrate at significantly lower rates, from 20 to 50 km/km century.[11]

The dispersal of oaks may be especially dependent on the caching behavior of blue jays[12] since mammals rarely transport acorns much beyond the crown area of individual trees.[13] There is the interesting possibility that the recently extinct passenger pigeon, *Ectopistes migratorius*, was an important contributor to the postglacial migration of nut-bearing trees. If this is true, such trees may be severly handicapped in the race north and may be excluded by species already established.

Barriers, natural and man-made, will be another problem constraining the dispersal of trees and many other species.[14] These barriers include the Great Lakes, east-west mountain ranges, farm lands, highways, and urban areas. One of the greatest challenges facing conservation biology in the decades to come is the development of strategies for helping species circumvent barriers during a period of rapid climate change. Conservation corridors or stepping stones have been suggested but it is not clear how these would help many species, especially those that require large areas of closed forest.

POSSIBLE ECOSYSTEM EFFECTS OF RAPID CLIMATE CHANGE

Much of the literature on the potential responses climate change emphasizes impacts on boreal and temperate forests (see "Effect of Global Warming on Forests; a Case Study for the Great Lakes" in this volume.) There has also been some interesting speculation on both beneficial and adverse effects on fisheries.[15] However, knowledge of probable responses by many types of ecosystems is fragmentary at best.

TERRESTRIAL ECOSYSTEMS:

Polar/Tundra

As the IPCC Working Group II indicated, permafrost covers 20-25% of the landmass in the Northern Hemisphere and could experience "significant degradation."[16] They did not discuss specific ecological aspects of this rapid degradation, nor did they cover the ecological consequences of retreating and fragmenting sea ice.

Historic and prehistoric information on the responses of arctic birds and mammals to climatic fluctuations in Greenland has been used to relate the distribution and abundance of many species to the status of drift-ice.[17] Grazers, such as reindeer and muskoxen are adversely affected by snow and ice covering forage during the winter. Thus global warming may put some grazers at greater risk while it may allow other species to expand their range northward.

Alpine/Montane

Montane and alpine communities of plants and animals may be vulnerable even to slow rates of global warming. A 3°C warming, over any period of time, will force zones of vegetation, and associated animals, 500 m up the mountains. As their habitats shrink, they will be more susceptible to extinction.[18]

Boreal and Temperate Forests

As we indicated, there have been numerous assessments of potential climatic effects on these ecosystems, partly because they are economically important and partly because the paleoecological record of their responses to glacial advances and retreats is quite useful. Direct effects are expected to result from warming and drought[19] and indirect effects from outbreaks of pests and pathogens[20] and increased incidence of forest fires.[21]

Tropical Forests

The literature on climate impacts affecting tropical forests is sparse. There are some indications that projected warmer and wetter conditions will lead to expansions of rainforests into zones now occupied by vegetation adapted to dry seasons. However, that will only be possible in areas where there is little pressure to convert forests into cropland or pasture, and it assumes that tropical rainforest species have the necessary adaptations for rapid dispersal. Furthermore, there is concern that changes in the timing and intensity of precipitation will upset delicate linkages between highly co-evolved tropical species.[22] Also, there is the disturbing possibility that short droughts will increase in some tropical rainforest zones, leading to forest fires such as those that destroyed huge areas in Kalimantan (Borneo) during the El Niño period of 1982-83.[23]

Grasslands and Deserts

Grass species have the ability to colonize suitable habitats rapidly enough to keep up with anticipated rates of warming and drying, but desert plants do not have adaptations for such rapid dispersal. It seems likely that the grasslands will be subject to more frequent outbreaks of pests.[24]

AQUATIC ECOSYSTEMS:

Lakes and Streams

Schindler and his colleagues reported a 2°C warming over the past two decades in northwestern Ontario. As they stated, their "observations on a variety of ecological processes provide a preview of how climatic change may affect boreal lakes and catchments in the next century."[25] Consequences included lower water renewal, increased concentrations of chemicals, and reduced summer habitats for late trout and opossum shrimp. If the lake trout die out, they might be replaced by warm-water species, such as bass, but these may be too contaminated to eat.

Freshwater Wetlands

Recent declines in the populations of several duck species in North America can be attributed to reduction in the Prairie Pothole wetlands where they breed in Canada and the U.S.[26] If the droughts of the 1980's continue through the 1990's, exacerbated by climate change, this trend will continue. Twelve of the amphibian populations in the northeast U.S. that have declined are at the edge of their range, suggesting that their wetland habitats may be adversely affected by climate change, although acid deposition, increased ultraviolet-B (UV-B), and other factors may be involved.[27] Since the amphibian populations had been stable for many decades, some threshold may have been crossed and the species may now be jeopardized by anticipated rates of climate change throughout their entire ranges.

Coastal Wetlands

Coastal wetlands, and their inhabitants, may not have to contend with drought but they will have to cope with sea level rise. Their future is determined by the rates at which they can migrate inland, barriers to that migration, and their continued ability to accumulate sediments. The IPCC Impacts Working Group assumed a sea level rise of 30 to 50 cm by 2050 and 100 cm by 2100, with consequent loss of salt, brackish and fresh marshes, as well as mangrove and other swamps.[28] EPA estimated that half the U.S. coastal wetlands would be inundated in a 1-meter sea level rise by 2100.[29] The IPCC Impacts Working Group anticipated a temporary benefit for fisheries from nutrient release during the early part of inundation but concluded that the overall impact on fisheries and wildlife is likely to be negative by 2050.[30]

Seagrass die-offs related to warm water may have mixed consequences. Rasmussen felt that the destruction of the eelgrass opened the way for a new infauna, richer both in species and individuals.[31] Other authors have suggested that the decline of eelgrass in the 1930s had adverse effects on the Atlantic Brant, *Branta bernicla hrota*, and the Bay Scallop, *Argopecten irradians*.[32] It even led to the extinction of the eelgrass limpet, *Lottia alveus*, the only marine mollusc known to have become extinct in historic times.[33]

Mangrove ecosystems line, and protect, 25% of tropical coastlines. Paleoecological studies in the Caribbean and the Pacific indicate that those surrounding low tropical islands can keep pace with a sea-level rise of 8 to 9 cm/100 years and are under stress at rates between 9 to 12 cm/100 years.[34] Since the IPCC Impacts working group used scenarios with a sea-level rise of 30 to 50 cm by 2050 and 100 cm by 2100, projected rates may far exceed those mangroves can tolerate.[35] Even the conservative scenario used by the IPCC Scientific Assessment working group postulated a 65 cm rise by 2100.[36]

Coral Reefs

If there is a general warming of the tropical seas, the prospects for the coral reefs are grim. Water temperatures above 32°C cause coral polyps to lose their symbiotic microalgae (zooxanthellae). Without their symbionts, the coral polyps can no longer build the reef framework and keep ahead of predation and erosion, let alone cope with rising sea-levels. The entire reef ecosystem will degrade.

There also will be adverse effects on commercial, traditional and recreational fisheries. Coral reefs protected in underwater parks from physical damage or fishing pressure have been shown to provide breeding stocks of many valuable species, including groupers and spiny lobsters.[37]

EARLY INDICATORS OF CLIMATIC STRESS

There are a number of phenomena that can be considered as early indicators of climate change. These are typically associated with stress caused by warming. If the following examples have been triggered by a 0.5°C global warming over the past 100 years, or by extreme warmth during the past decade, they can be taken as harbingers of future climatic stress and ecosystem response.

Coral Bleaching

During the severe 1982-83 El Niño event, water temperatures gradually warmed to 32°C in the Eastern Pacific, causing coral polyps to lose their symbiotic microalgae (zooxanthellae).[38] The exact mechanism is not clear but, without their microalgae, the coral polyps cease to grow and they will die if the symbionts are not restored. This phenomenon had been observed in 1979-80 and was very widespread during the 1986-88 period.[39] A recent report suggest that the 1982-83 warming led to "one probable extinction and one range reduction of eastern Pacific reef-building hydrocoral (*Millipora*) species."[40]

Red Tides

Blooms of toxic dinoflagellates, the "red-tides," have been recorded in the Book of Exodus and the writings of Homer. Like coral bleaching, the incidence of red tides appears to be increasing world-wide. This is of concern because red-tides cause potentially lethal human illness, called "Paralytic Shellfish Poisoning (PSP)," when shellfish contaminated by the dinoflagellates are consumed. Mass mortality of dolphins and other marine organisms also may be linked to their consumption of contaminated fish.

Many factors have been implicated as possible causes of these toxic tides. One likely suspect is "creeping eutrophication," the gradual increase in off-shore nutrient loading from a number of sources, one of which may be related to climate change.[41] Bakun showed that upwellings caused by alongshore wind stress have increased, during the 1955-1985 period, off the coasts of California, Portugal, Morocco and Peru but did not discuss the possible connection with toxic tides.[42] Others have suggested that upwellings containing the elements required by dinoflagellates could trigger the blooms.[43]

Brown Tides

The "brown tides" plaguing Atlantic coastal embayments since 1985 do not seem related to upwelling, but they do appear simultaneously in a number of locations, suggesting that there is a climatic connection. These diatom blooms have contaminated blue mussels harvested off Prince Edward Island, causing a new type of poisoning called Amnesic Shellfish Poisoning (ASP). Drought conditions, high salinity, and water warmed to 20-25°C may trigger the blooms.[44]

Wasting Disease of Eelgrass

There is a large body of literature on periodic die-offs of eelgrass, *Zostera marina*, and the effects of such destruction on substrates and fauna. Rasmussen eliminated a number of contradictory or localized explanations for the die-off in the 1930s, concluding that a combination of mild winters and unusually warm summers was lethal to eelgrass.[45] Recent observations suggest that the wasting disease started to increase again during the 1980s on both sides of the Atlantic.[46] This appears to be correlated with the period of maximum water temperatures and decreasing light levels in late summer and early fall.

Changes in Populations of Marine Fish

The dynamics of a marine fish population are directly dependent on parameters affected by climate change. Changes in sea surface temperature affect migration, growth, spawning, and survival.[47] During the first half of the century, there were incidental sightings of many fish species far north of their accustomed ranges. Some species of importance in commercial fishing also moved north. These included cod, herring and tilefish. Such changes were correlated with a half degree increase in air temperature in the Northern Hemisphere, implying that the threshold for fishery effects could be less than 1°C, "although it is not clear that temperature alone was responsible for the observed changes."[48]

The northward expansion of important commercial fisheries would seem to be all "good news." However, there is a major risk to such populations. For example, the tilefish appeared in large numbers off New England in 1879. Three years later it experienced massive mortality, with over a billion dead fish floating on the Atlantic. Apparently, northern gales had cooled the water below their lower thermal limit.[49]

Changes in the Environment of Canadian Lakes

The recent report from Canada by Schindler and his colleagues has some very disturbing implications.[50] Air and lake temperatures have increased by 2°C during the last 20 years in northwestern Ontario, although it is not known whether this is a regional variation or a symptom of global warming. However, the warming has been a factor in the observed increase in forest fires which, in turn, have led to many changes in the watersheds of lakes and the lakes themselves.

CONCLUSIONS

1. Organisms have evolved in changing environments, developing various mechanisms of adaptation and migration to survive.
2. Projected rates of warming may be far more rapid than past rates of change.
3. Genetically determined limits on adaptation, and physical barriers to dispersal, may hamper the ability of many species to cope with rapid climate change.
4. Species extinctions and ecosystem degradation are highly probable.
5. Some ecological effects, generally related to warming, have already occurred and may be early ecological indicators of climate change.
6. Efforts to slow the rate of climate change will help reduce the extent and severity of adverse ecological effects.

REFERENCES

1. Levenson, Thomas. 1989. *Ice Time: Climate, Science, and Life on Earth.* Harper & Row, New York. pp. 19-25.
2. Davis, Margaret B. 1983. Holocene Vegetational History of the Eastern United States, In: H.E. Wright, Jr. (Ed.) *Late-Quaternary Environments of the United States*, Vol. 2. U. of Minnesota Press, Minneapolis.
3. Colinvaux, Paul A. 1989. The Past and Future Amazon. *Scientific American.* May, pp. 102-108.
4. Kerr, Richard A. 1991. Dinosaurs and Friends Snuffed Out? *Science:* 251:160-162.
5. Smith, Joel B. and Dennis A. Tirpak (Eds.). 1989. *The Potential Effects of Global Climate Change on the United States.* p. 153.
6. Dawson, William R. 1991. Animals: Overview of Physiological Responses to Climatic Variables. In: R.L. Peters (Ed.) *Consequences of Greenhouse Effect for Biological Diversity.* Yale University Press, New Haven, CT.
7. Dawson, W.R. 1991. loc cit.
8. Parsons, Peter A. 1989. Conservation and Global Warming: A Problem in Biological Adaptation to Stress. *Ambio* 18(6):322.
9. Stefan, Heinz. 1991. Personal communication.
10. Huntley, Brian and Thompson Webb, III. 1989. Migration: Species' Response to Climatic Variations Caused by Changes in the Earth's Orbit. *Journal of Biogeography* 16:5-19.
11. Shugart, H.H. *et al.* 1986. CO_2, Climatic Change and Forest Ecosystems, In: B. Bolin *et al.* (Eds.) *The Greenhouse Effect, Climatic Change, and Ecosystems* SCOPE 29, John Wiley & Sons.
12. Johnson, W. Carter and Thompson Webb, III. 1989. The role of blue jays (*Cyanocitta cristata* L.) in the postglacial dispersal of fagaceous trees in eastern North America. *Journal of Biogeography* 16:561-571.
13. Aizen, Marcelo A. and William A. Patterson, III. 1990. Acorn size and geographical range in the North American oaks (*Quercus* L.). *Journal of Biogeography* 17:327-332.

14. Peters, Robert L. 1989. Effects of Global Warming on Biological Diversity, In: *The Challenge of Global Warming,* Island Press, Washington, DC, pp. 82-95.
15. Magnuson, John J. *et al.* 1989. Potential Response of Great Lakes Fishes and Their Habitat to Global Climate Warming, pp. 2-1 to 2-42. In: J.B. Smith and D.A. Tirpak (Eds.). *The Potential Effects of Global Climate Change on the United States: Appendix E - Aquatic Resources.* Washington, DC.
16. IPCC Working Group II. 1990. *Intergovernmental Panel on Climate Change - Policymakers' Summary of the Potential Impacts of Climate Change* p. 39.
17. Vibe, Christian. 1967. *Arctic Animals in Relation to Climatic Fluctuations.* C.A. Reitzels Forlag, Copenhagen, p. 227.
18. Peters, Robert L. and Joan D.S. Darling. 1985. Potential Effects of Greenhouse Warming on Natural Communities. *BioScience* 35:707-717.
19. Bonam, G.B., H.H. Shugart, and D.L. Urban. 1990. The Sensitivity of Some High-latitude Boreal Forests to Climatic Parameters. *Climatic Change* 16:9-29.
20. Zabinski, Catherine and Margaret B. Davis. 1989. Hard Times Ahead for Great Lakes Forests: a Climate Threshold Model Predicts Responses to CO_2-Induced Climate Change, pp. 5-1 to 5-19. In: J.B. Smith and D.A. Tirpak (Eds.) *The Potential Effects of Global Climate Change on the United States, Appendix D, Forests.* Washington, DC.
21. Fosberg, M.A., J.G. Goldammer, D. Rind and C. Price. 1990. Global Change: Effects on Forest Ecosystems and Wildfire Severity. *Global Change 21, Ecological Studies.*
22. Hartshorn, Gary. 1991. Possible Effects of Global Warming on the Biological Diversity in Tropical Forests. In: R.L. Peters (Ed.) *Consequences of the Greenhouse Effect for Biological Diversity.* Yale University Press, New Haven, CT.
23. Leighton, M. and N. Wirawan. 1986. Catastrophic Drought and Fire in Borneo Tropical Rain Forest Associated with the 1982-1983 El Niño Southern Oscillation Event, pp. 75-102. In: G.T. Prance (Ed.) *Tropical Rain Forests and the World Atmosphere*, Westview Press, Boulder, CO.
24. Mabbutt, J.A. 1989. Impacts of Carbon Dioxide Warming on Climate and Man in the Semi-arid Tropics. *Climatic Change* 15:191-221.
25. Schindler, David W. *et al.* 1990. Effects of Climatic Warming on Lakes of the Central Boreal Forest. *Science* 250:967-970.
26. LeBlanc, Alice, Daniel J. Dudek and Luiz Fernando Allegretti. 1991. *Disappearing Ducks: the Effect of Climate Change on North Dakota's Waterfowl.* Environmental Defense Fund, New York, NY, pp. 36.
27. Wyman, Richard L. 1990. What's Happeninbg to the Amphibians? *International Conservation News* 4(4):350-352.
28. IPCC II. 1990. op cit. p. 33.
29. Smith, Joel B. and Dennis A. Tirpak (eds.) 1989. *The Potential Effects of Global Climate Change on the United States: Appendix B - Sea Level Rise.*
30. IPCC II. 1990. op cit. p. 35.
31. Rasmussen, E. 1977. The Wasting Disease of Eelgrass (*Zostera marina*) and Its Effects on Environmental Factors and Fauna. In: C.P. McRoya and C. Helfferich (Eds.) *Seagrass Ecosystems: a Scientific Perspective.* Marcel Dekker.

32. Thayer, G.W., W.J. Kenworthy and M.S. Fonseca. 1984. *The Ecology of Eelgrass Meadows: a Community Profile.* FWS/OBS-84/02.
33. Gould, Stephen Jay. 1991. On the Loss of a Limpet. *Natural History* 6/91:22-27.
34. Ellison, Joanna C. and David R. Stoddart. 1991. Mangrove Ecosystem Collapse During Predicted Sea-Level Rise: Holocene Analogues and Implications. *Journal of Coastal Research* 7(1):151-165.
35. IPCC II. 1990. op cit. p.8.
36. IPCC Working Group I. 1990. J.T. Houghton, G.J. Jenkins and J.J. Ephraums (Eds.) *Climate Change, the IPCC Scientific Assessment*, Cambridge University Press, Cambridge, UK, p. xxx.
37. Holloway, Marguerite. 1991. Hol. Chan. *Scientific American* 264(5):32.
38. Glynn, P.W. 1984. Widespread Coral Mortality and the 1982-83 El Niño Warming Event. *Environmental Conservation* 11(2):133-146.
39. Bunkley-Williams, L. and E.H. Williams Jr. 1990. Global Assault on Coral Reefs. *Natural History* 4/90:47-54.
40. Glynn, P.W. and W.H. de Weerdt. 1991. Elimination of Two Reef-Building Hydrocorals Following the 1982-83 El Niño Warming Event. *Science* 253:69-71.
41. Cherfas, J. 1990. The Fringe of the Ocean - Under Siege from the Land. *Science* 248:163-165.
42. Bakun, Andrew. 1990. Global Climate Change and Intensification of Coastal Ocean Upwelling. *Science* 247:198-201.
43. Blasco, D. 1975. Red Tides in the Upwelling Regions; and Prakash, A. 1975. Dinoflagellate Blooms - an Overview. In: V.R. LoCicero (Ed.). *Proceedings of the First International Conference on Toxic Dinoflagellate Blooms.* Massachusetts Science and Technology Foundation.
44. Beltrami, E.J. 1989. Brown Tide Dynamics as a Catastrophic Model. In: E.M. Cosper *et al.* (Eds.). *Novel Phytoplankton Blooms; Causes and Impacts of Recurrent Brown Tides and Other Unusual Blooms.* Springer-Verlag.
45. Rasmussen, E. 1977. op cit.
46. Short, F.T., B.W. Ibelings and C. Den Hartog. 1988. Comparison of a Current Eelgrass Disease to the Wasting Disease in the 1930s. *Aquatic Botany* 30:295-304.
47. Frye, Richard. 1983. Climatic Change and Fisheries Management. *Natural Resources Journal* 23:77-96.
48. Sibley, T.H. and R.M. Strickland. 1985. Environmental Factors Relevant to Fisheries In: *Characterization of Information Requirements for Studies of CO_2 Effects.* U.S. DoE/ER-0236.
49. Freeman, B.L. and S.C. Turner. 1977. *Biological and Fisheries Data on Tilefish, Lopholatilus chamaeleonticeps, Goode and Bean.* Sandy Hook Laboratory, Tech. Series Report No. 5; p. 10.
50. Schindler, D.W. *et al.* 1990. op cit.

Global Climate Change: Implications, Challenges and Mitigation Measures. Edited by S.K. Majumdar, L.S. Kalkstein, B. Yarnal, E.W. Miller, and L.M. Rosenfeld. © 1992, The Pennsylvania Academy of Science.

Chapter Twenty-One

GENERAL CIRCULATION MODEL ESTIMATES OF REGIONAL PRECIPITATION

DAVID R. LEGATES[1]
and
GREGORY J. McCABE, JR.[2]

[1]Department of Geography
College of Geosciences
University of Oklahoma
Norman, Oklahoma 73019
and
[2]U.S Geological Survey
Denver Federal Center
Mailstop 412
Denver, Colorado 80225

INTRODUCTION

Atmospheric concentrations of carbon dioxide (CO_2) are expected to double that of pre-industrial levels sometime in the next century. This is presently a cause of concern since this increase could alter both temporal and spatial precipitation patterns (Bolin, 1986). In turn, these changes in regional precipitation could have an adverse effect on agricultural production, water supply and demand, and the frequencies of droughts and floods. Presently, general circulation models (GCMs) can provide physically-based prognostications of the regional precipitation changes that may be induced. Although GCMs differ in their level of physical detail and mathematical complexity, they are able to reproduce the large-scale features associated with precipitation. Estimates of climate on regional scales (10^4 to 10^6 km^2), however, vary markedly among different GCMs.

Modeling precipitation at the spatial scales employed by GCMs is a rather difficult task for several reasons. Horizontal resolution is seldom finer than 100km over much of the earth's surface and most models are usually limited to no more than fifteen atmospheric layers. Precipitation and cloud-forming processes, however, occur at scales that are much smaller than these model resolutions and present knowledge of these processes is rather limited. In addition, GCM simulations of precipitation are often adversely affected by numerical instabilities that can arise due to a wide range of derived moisture amounts (Washington and Parkinson, 1986) as well as by difficulties in modeling other components of the hydrologic cycle, including evapotranspiration and soil moisture.

In this chapter, an evaluation of GCM simulations of the present climate and their prognostications of doubled-CO_2 conditions is presented. This evaluation includes model representations of observed precipitation, model-simulated changes in precipitation under doubled CO_2 conditions, and an assessment of the uncertainty in GCM estimates of regional precipitation patterns. Model simulations of the present climate and the impact of doubling CO_2 on precipitation have been extensively investigated (*cf.* Manabe and Wetherald, 1975; Manabe and Stouffer, 1980; Manabe and Wetherald, 1980; Watts, 1980; Manabe *et al.* 1981; Mitchell, 1983; Schlesinger, 1984; Washington and Meehl, 1984; Manabe and Wetherald, 1987; Schlesinger and Mitchell, 1987; Wilson and Mitchell, 1987; Meehl and Washington, 1988; Rind, 1988; Lockwood, 1989; Mitchell, 1989; Schlesinger and Zhao, 1989; Gates *et al.*, 1990; Mearns *et al.*, 1991).

REGIONAL PRECIPITATION

Knowledge of present-day regional precipitation patterns is limited for several reasons. First, reliable estimates of oceanic precipitation are scant because surface observations are limited and biased (Quayle, 1974) and satellite estimation techniques are limited (Arkin and Ardanuy, 1989). This lack of data is of concern since oceans cover nearly three-quarters of the earth's surface and approximately 81% of the Southern Hemisphere. Second, terrestrial station time-series are subject to discontinuities resulting from changes in station location, instrumentation, recording practices, and siting characteristics (Eischeid *et al.*, 1991). Furthermore, precipitation measurements are undercatches of the actual precipitation due to the effects of the wind, wetting on the interior walls of the gage, evaporation from the gage, and the blowing and drifting of snow.

A global precipitation climatology that attempts to address some of these concerns, particularly with respect to gage measurement errors, has been compiled (Legates, 1987; Legates and Willmott, 1990). This database contains 24,635 independent terrestrial station records (corrected for errors in gage measurement) and oceanic estimates from the most reliable indirect technique available. A complete discussion of this climatology including the bias-correction procedure, accuracy of the fields, and comparison with other climatologies is given by Legates (1987). Although these precipitation data have been adjusted for gage measurement biases, some uncertainties remain, especially over the oceans.

To illustrate the ability of GCMs to simulate present-day precipitation patterns and model prognostications of doubled-CO_2 conditions, two widely respected GCMs will be examined—the Goddard Institute for Space Studies (GISS) and the Geophysical Fluid Dynamics Laboratory (GFDL) models. The GISS model is a grid-point GCM with a horizontal resolution of 4° of latitude by 5° of longitude and includes a 65m deep mixed layer ocean and nine vertical atmospheric levels (Hansen et al. 1988). Both seasonal and diurnal solar cycles are simulated and snow cover and depth, cloud cover, and sea ice extent are computed. By contrast, the GFDL model is spectrally-based with a horizontal resolution of approximately 2.25° of latitude by 3.75° of longitude (R30 truncation) and nine unevenly-spaced vertical levels (Manabe and Broccoli, 1990). Prescribed, albeit seasonally-varying, distributions of cloud cover, sea surface temperature, and sea ice also are included. Solar insolation varies seasonally although mean daily insolation is used instead of a daily solar cycle.

Precipitation Under Current Climate Conditions

Observed precipitation is greatest along the Intertropical Convergence Zone (ITCZ). Since the ITCZ migrates seasonally, the zone of maximum precipitation follows the ITCZ reaching its most northerly position in August and its most southernly position in February (Figure 1). Consequently, precipitation patterns

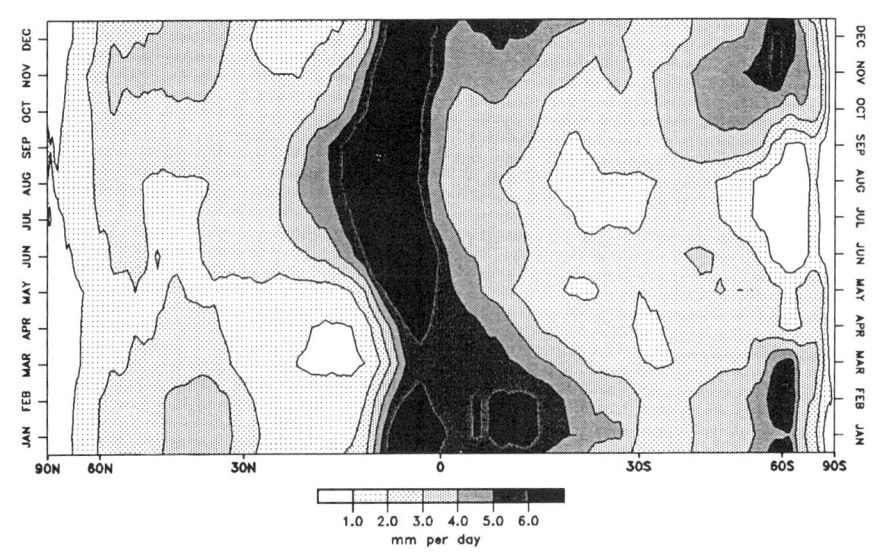

FIGURE 1. Seasonal variations in zonally-averaged observed precipitation. Observed data are taken from Legates (1987) and Legates and Willmott (1990).

in the Northern Hemisphere tropics generally are opposite of those in the Southern Hemisphere. In mid-latitudes of the northern hemisphere, precipitation is greatest in the winter months and a minimum in the summer. This pattern occurs due to the summer expansion of the subtropical high pressure which results in decreased precipitation in the mid-latitudes. In both hemispheres, precipitation is lowest near the poles.

Both GCMs exhibit areas of low precipitation (less than 2.0mm day^{-1}) between 20°S and 30°S that are larger in both spatial and temporal extent than observed (Figure 2). Both models also reproduce northern hemisphere precipitation better than precipitation in the southern hemisphere. For the GISS model, precipitation rates are about the same magnitude as the observed although the seasonal migration of the ITCZ is not reproduced. By contrast, the GFDL GCM exhibits a seasonal migration of the ITCZ that closely follows the observed but considerably underestimates the magnitude of precipitation. Decreased summer precipitation in northern hemisphere mid-latitudes also is reproduced by both models as is the decrease in precipitation toward the poles.

Changes in Precipitation Under a Doubling of CO$_2$

Zonally-averaged prognostications of the temporal changes in precipitation caused by a doubling of CO_2 are presented (Figure 3). Although the GISS model exhibits larger changes than those simulated by the GFDL model, both models agree on increases of nearly 10% per month (Legates and McCabe, 1991) which are commensurate with, although slightly larger than, evaluations from other models (*e.g.,* Mitchell, 1983; Schlesinger, 1984; Lockwood, 1989; Schlesinger and Zhao, 1989).

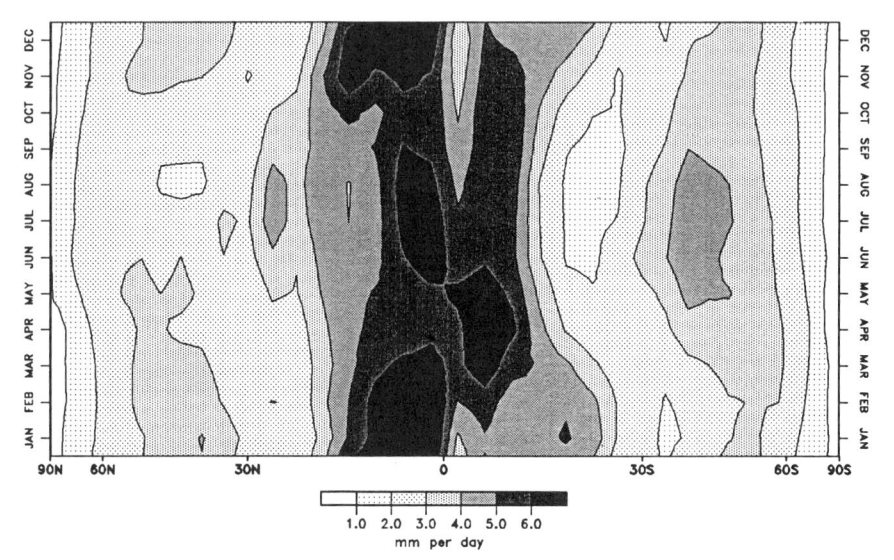

FIGURE 2a. Seasonal variations in zonally-averaged precipitation for the present-day climate as simulated by the Goddard Institute for Space Studies (GISS) model.

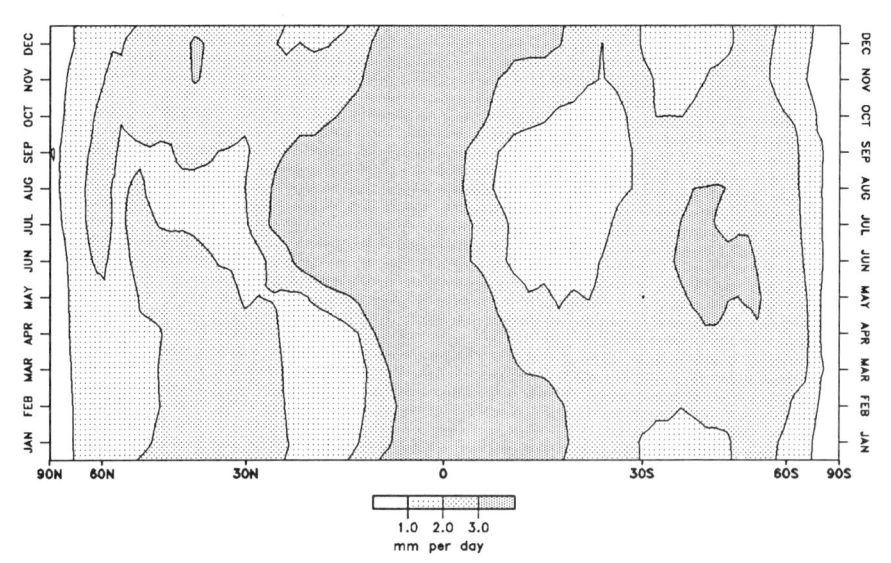

FIGURE 2b. Seasonal variations in zonally-averaged precipitation for the present-day climate as simulated by the Geophysical Fluid Dynamics Laboratory (GFDL) model.

Both models also simulate the greatest increases along the ITCZ where the highest precipitation rates presently are observed. Considerable differences between the two models exist, however, in the subtropics (between 10°N and 25°N) and mid-latitudes of the northern hemisphere. In addition, the GFDL model exhibits widespread dryness in summer while only a small decrease is observed in the GISS simulation. This could be of great concern because it coincides with the period of increased water demand and could adversely impact agriculture. Both models, however, indicate modest increases in precipitation north of 55°N.

In the southern hemisphere, the GISS model indicates increased precipitation south of the ITCZ to about 25°S for the entire year with decreased in most of the mid-latitudes (Figure 3). The GFDL model, by contrast, indicates decreased precipitation between 20°S and 30°S in the low-sun season (June through September) and between the equator and 20°S in January, May and August. Both models agree, however, on increases poleward of 50°S, particularly during the winter months.

Spatial patterns of precipitation changes under a doubling of carbon dioxide are presented here for January and July using only the GISS model for brevity. Simulated increases in January precipitation are not uniformly distributed along the equator but are greatest over the eastern equatorial Pacific, the Indian Ocean, and equatorial southeast Asia (Figure 4a). North of the ITCZ, large areas of decreased precipitation are simulated for Central and northern South America, and over the western equatorial Pacific from about 165°W to 120°E and extending northward into eastern China. Considerable increases in precipitation are indicated for much of Southeast Asia.

Precipitation changes in January are simulated to be small over most of Northern Africa, Asia, and Europe. Areas with precipitation increases greater than 1mm per day are evident over eastern France, north of the Black Sea, and in central

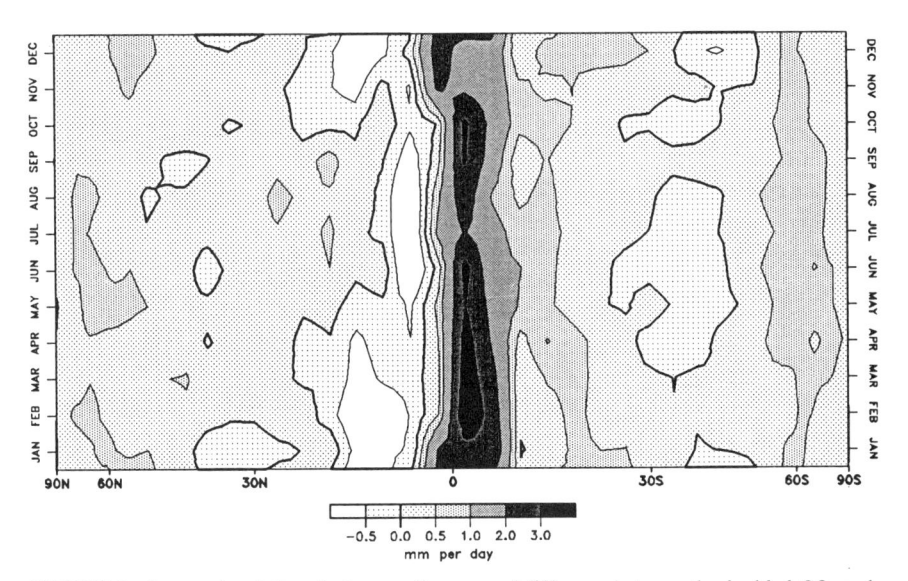

FIGURE 3a. Seasonal variations in the zonally-averaged difference between the doubled-CO_2 and present-day precipitation simulations of the Goddard Institute for Space Studies (GISS) model.

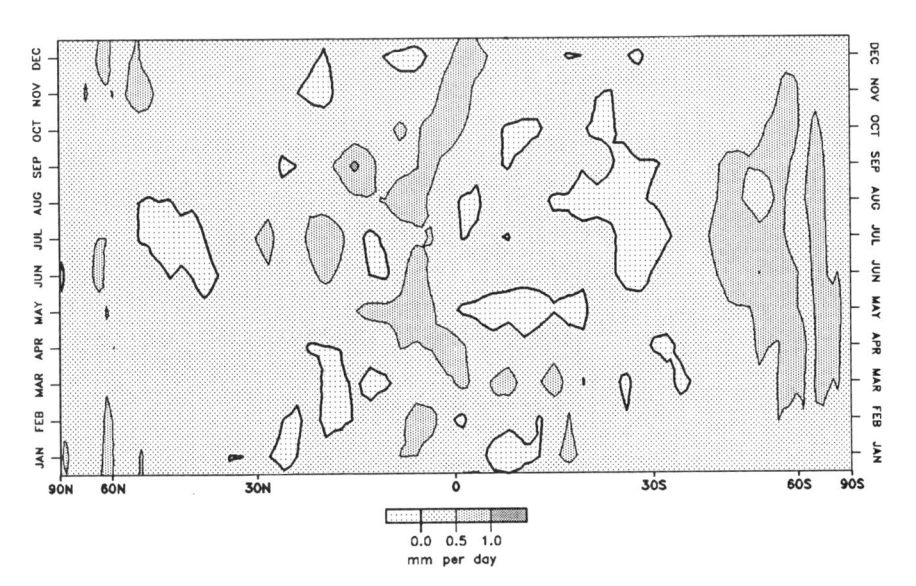

FIGURE 3b. Seasonal variations in the zonally-averaged difference between the doubled-CO_2 and present-day precipitation simulations of the Geophysical Fluid Dynamics Laboratory (GFDL) model.

Russia. In North America, however, increases greater than 2mm per day are simulated over the desert southwest of the United States, southern Greenland and the central North Atlantic, and the Caribbean while more modest increases are found over eastern Alaska and from Arkansas northeastward into southern Ontario. Decreases in precipitation are found over the western Atlantic from 30°N to 40°N, the Pacific Northwest of the United States, and in the North Pacific Ocean just east of the dateline. In the Southern Hemisphere, large areas of precipitation increases are simulated over the central Pacific between the ITCZ and 25°S, Argentina, southwestern Africa, western Australia, and over the Southern Ocean south of New Zealand. Decreases are simulated only over western Australia, western Africa near Angola, and in a small area in the equatorial South Pacific Ocean.

As in January, the July simulation of the GISS model has the largest precipitation increases in July over the Indian Ocean, the western equatorial Pacific, northern South America, and across portions of equatorial Africa (Figure 4b). Decreased precipitation is found just north of the ITCZ over the western and central equatorial Pacific and most of the equatorial Atlantic including Surinam. In the Southern Hemisphere, large areas of increased and decreased precipitation are found over most of the Pacific Ocean north of 45°S while smaller increases are scattered across parts of the Southern Ocean. Overall, only small changes in regional precipitation patterns are simulated for land areas.

Large areas of precipitation increases in the Northern Hemisphere are simulated over eastern India and Bangladesh, north-central China, southeastern parts of the Middle East, western Russia and Finland, the Iberian Peninsula and parts of northern Africa, the southeastern United States, and much of northwestern North America. Mid-latitude precipitation decreases, however, are greatest over southeast China, northwestern China and Taiwan, and, in particular, over much of the Midwest of the United States east of the Missouri River.

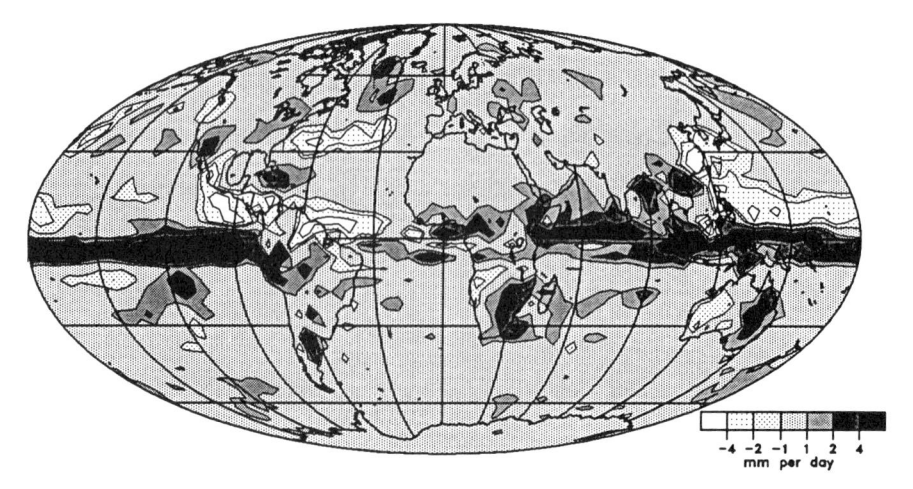

FIGURE 4a. The simulated doubled-CO_2 minus the present-day precipitation for January using the Goddard Institute for Space Studies (GISS) model.

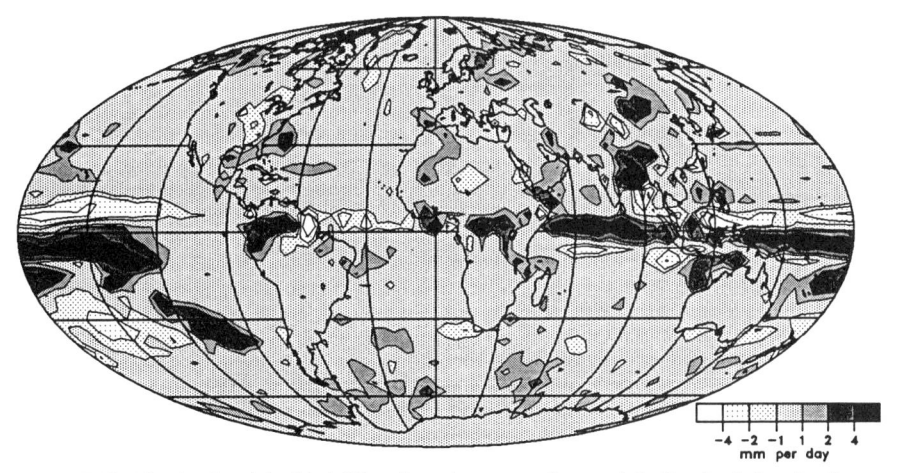

FIGURE 4b. The simulated doubled-CO_2 minus the present-day precipitation for July using the Goddard Institute for Space Studies (GISS) model.

RELIABILITY OF MODEL ESTIMATES

An important concern with regard to the uncertainty of GCM estimates is how the estimated precipitation changes from present-day to doubled-CO_2 conditions compare to the discrepancies between observed precipitation and the model simulations of current conditions. Legates and McCabe (1991) indicate that these discrepancies are larger than the predicted changes in precipitation, especially in the Southern Hemisphere. For January and July, the difference between the observed and present-day model simulations is from two to six times greater than the simulated changes in precipitation due to a doubling of CO_2. These differences are illustrated for the GISS model (Figure 5) which shows that they can extend over large regions and are not simply dependent on latitude. A comparison with the model-simulated changes in precipitation caused by a doubling of CO_2 (Figure 4) illustrates that the discrepancies between the climatology and the present-day model simulation are usually much larger. Differences are greatest along equator due to the inability of the GISS model to reproduce the seasonal migration of the ITCZ.

Legates and Willmott (1992) also compared the Legates (1987) climatology with the January and July simulations of the current climate from four GCMs including coarser resolution versions of the GFDL and GISS models. Evaluations included comparisons by 10° latitudinal bands and spatial distributions. Their results indicated that considerable differences exist between the model simulations and the observed climatology, and even between the four model simulations. Similar results were obtained by Gates et al. (1990) who concluded that differences between station observations, satellite-derived estimates, and indirect measurements from ship observations are usually less than the differences between climatological estimates and the model simulations. Although "true" precipitation may differ from the climatological representation, the inaccuracy and variability of GCM precipitation estimates for current climate conditions introduces a large degree of uncertainty into estimates of regional precipitation changes, even for latitudinal bands as large as 10°.

In a comparison of four GCMs, Grotch and MacCracken (1991) found that GCM estimates of CO_2-induced changes in regional precipitation were more in agreement for winter than summer and that the consensus for large-area averages was better than for subcontinental scales. They concluded a possible cause for the differences among the GCM simulations "is almost certainly related to the limitations in the quality of model simulations of the present climate" (p. 302).

The variability among GCM simulations of precipitation for current climate conditions is related not only to difficulties in modelling precipitation, but also to differences in model parameterization of global atmospheric and oceanic circulation,

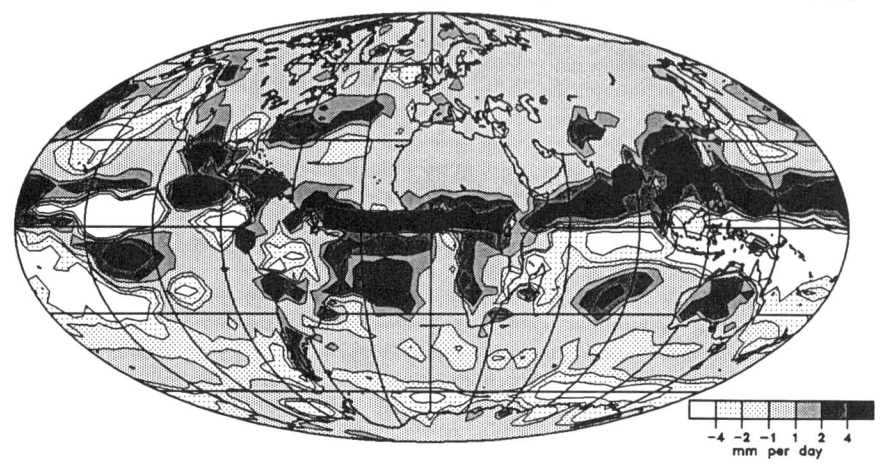

FIGURE 5a. The present-day precipitation simulated by the Goddard Institute for Space Studies (GISS) model minus the Legates (1987) precipitation climatology for January.

FIGURE 5b. The present-day precipitation simulated by the Goddard Institute for Space Studies (GISS) model minus the Legates (1987) precipitation climatology for July.

the hydrologic cycle, surface topography, and spatial (both horizontal and vertical) resolution. For example, the magnitude of decreased summer in the Northern Hemisphere mid-latitudes has been shown to depend on the model representation of the hydrologic cycle. Many studies have examined the simulated Northern Hemisphere "summer dryness" because it has serious consequences for the agricultural Midwest of the United States. Suggested reasons for the simulated decrease in summer precipitation include 1) soil moisture decreases caused primarily by increased evapotranspiration, 2) decreased snowmelt due to decreases in snowfall and accumulation, 3) an earlier initiation of the snowmelt season, and 4) shifts in mid-latitude rainbelts — all induced by the simulated increase in surface air temperature (Manabe et al., 1981; Manabe and Wetherald, 1987).

Mitchell and Warrilow (1987) reexamined this "summer dryness" phenomenon and concluded "the magnitude of the simulated summer drying is dependent on the physical attributes of the soil, and in certain regions can be reduced or even reversed by an alternative, but equally plausible, treatment of run-off over frozen ground" (p. 238). They argue that many GCMs assume that all precipitation and snowmelt directly increase the moisture content of the soil until the soil becomes saturated. After the soil reaches field capacity, additional moisture becomes runoff. In reality, however, snowmelt may occur over frozen ground and runoff may occur so quickly that little water infiltrates into the soil even though the soil is not saturated (Mitchell and Warrilow, 1987). When this process is incorporated into the hydrologic cycle of a GCM, soil moisture simulated under current climatic conditions is lower in the spring than when this process is not included. These differences in soil moisture storage affect evapotranspiration and ultimately precipitation. This example illustrates that GCM precipitation simulations are dependent not only on parameterizations of precipitation-forming processes, but also are highly dependent on model representations of other components of the hydrologic cycle.

Horizontal resolution also can significantly affect model representation of global circulation and, consequently, precipitation. Boer and Lazare (1988) and Boville (1991) demonstrated that model simulations of tropospheric circulation improve dramatically as horizontal resolution is increased. In particular, improvements in the Southern Hemispheric circulation are greater than those in the Northern Hemisphere. Accuracy of precipitation simulations is dependent on the model representation of vertical momentum and the horizontal transport of atmospheric moisture. Consequently, the horizontal resolution of the model can significantly influence regional precipitation for both present-day and doubled-CO_2 scenarios. Changes in vertical resolution, however, seem to have little effect on the simulation of global circulation (Boville, 1991).

CONCLUSION

General circulation models are able to reproduce the general latitudinal patterns of present-day precipitation. Overall, Northern Hemisphere simulations match the observed data more closely than those of the Southern Hemisphere although the accuracy of the observed climatology may be somewhat less in the Southern Hemisphere. While the models correctly simulate the highest precipitation rates along the ITCZ, its seasonal variation is not always reproduced (*e.g.*, the GISS GCM).

The general consensus among most climate model simulations of doubled CO_2 is that global precipitation will increase in all months due primarily to increases in surface air temperatures in the lower troposphere. Most GCMs also indicate that precipitation associated with the ITCZ will increase more than in any other area—by more than 2mm per day in the GISS simulation and by more than 1mm per day in the GFDL model. Both models also indicate that 1) precipitation will increase in all months in the high latitudes of both hemispheres, 2) summer precipitation in mid-latitudes of the Northern Hemisphere will decrease, and 3) winter precipitation between near 30°S will decrease.

The GFDL and GISS models disagree, however, in the Southern Hemisphere mid-latitudes where an increase in precipitation is simulated by the GFDL model but widespread decreases exist in the GISS GCM. In addition, the GISS model simulates considerable decreases in the low-latitudes of the Northern Hemisphere while the GFDL simulation indicates an increase. These discrepancies most likely are due to differences in model parameterizations of precipitation, topography, hydrology, and other processes and serve to underscore the difficulties in GCM precipitation modeling.

Comparisons of observed precipitation with model-simulations of present conditions indicate that some significant differences exist. In particular, the GISS simulation does not exhibit the observed seasonal migration of the ITCZ while the overall precipitation rates simulated by the GFDL model are much less than the observed. Latitudinal patterns such as increased precipitation over the ITCZ and in the upper mid-latitudes and decreased precipitation in the subtropics and high-latitudes are well-simulated by both GCMs. Model simulations of changes in precipitation due to a doubling of CO_2 are two to six times smaller than the errors in GCM simulations of current climate conditions, however.

Thus, GCMs can be used to simulate large-scale precipitation patterns and, consequently, provide generalized prognostications of large-scale changes in precipitation caused by a doubling of atmospheric carbon dioxide. Using GCMs to forecasting precipitation changes on smaller spatial scales can be very misleading, however, due to the difficulties in simulating precipitation and our limited knowledge of global precipitation.

ACKNOWLEDGEMENTS

The authors would like to thank A.J. Broccoli and R.T. Wetherald for discussions regarding the GFDL GCM and D. Rind for data and discussions of the GISS GCM. They also would like to thank T. DeLiberty for reading an earlier draft of the manuscript.

REFERENCES

Arkin, P.A. and P.E. Ardanuy. 1989. Estimating climatic-scale precipitation from space: A review. *Journal of Climate*, 2:1229-1238.

Boer, G.J. and M. Lazare. 1988. Some results concerning the effect of horizontal resolution and gravity-wave drag on simulated climate. *Journal of Climate,* 1:789-806.

Bolin, B. 1986. How much CO_2 will remain in the atmosphere?, in *The Greenhouse Effect, Climate Change, and the Ecosystem,* B Bolin, B.R. Doos, J. Jaeger, and R. Warrick, eds., John Wiley & Sons, New York, 93-155.

Boville, B.A. 1991. Sensitivity of simulated climate to model resolution. *Journal of Climate,* 4:469-485.

Eischeid, J.K., H.F. Diaz, R.S. Bradley and P.D. Jones. 1991. *A Comprehensive Precipitation Data Set for Global Land Areas.* Department of Energy, DOE/ER-69017T-H1, 81 pp.

Gates, W.L., P.R. Rowntree and Q.-C. Zeng. 1990. Validation of climate models, in *Climatic Change: The IPCC Scientific Assessment,* J.T. Houghton, G.J. Jenkins, and J.J. Ephraums, eds., Cambridge University Press, Cambridge, 93-130.

Grotch, S.L. and M.C. MacCracken. 1991. The use of general circulation models to predict regional climatic change. *Journal of Climate,* 4:286-303.

Hansen, J., I. Fung, A. Lacis, D. Rind, S. Lebedeff, R. Ruedy and G. Russell. 1988. Global climate changes as forecast by Goddard Institute for Space Studies three-dimensional model. *Journal of Geophysical Research,* 93:9341-9364.

Legates, D.R. 1987. A climatology of global precipitation. *Publications in Climatology,* 40(1), 86 pp.

Legates, D.R. and C.J. Willmott. 1990. Mean seasonal and spatial variability in gauge-corrected, global precipitation. *International Journal of Climatology,* 10:111-127.

Legates, D.R., and G.J. McCabe, Jr., 1991: Reliability of precipitation estimates for doubled-CO_2 scenarios simulated with two general circulation models. *Proceedings, Special Session on Hydrometeorology, Eighth Conference on Agricultural and Forest Meteorology,* American Meteorological Society, Salt Lake City, Utah, forthcoming.

Legates, D.R., and C.J. Willmott, 1992: A comparison of GCM-simulated and observed mean January and July precipitation. *Global and Planetary Change,* forthcoming.

Lockwood, J.G., 1989: Hydrometeorological changes due to increasing atmospheric CO_2 and associated trace gases. *Progress in Physical Geography,* 13:115-127.

Manabe, S., and A.J. Broccoli, 1990: Mountains and arid climates of middle latitudes. *Science,* 247:192-195.

Manabe, S., and R.J. Stouffer, 1980: Sensitivity of a global climate model to an increase of CO_2 concentration in the atmosphere. *Journal of Geophysical Research,* 85:5529-5554.

Manabe, S., and R.T. Wetherald, 1975: The effects of doubling of the CO_2 concentration on the climate of a general circulation model. *Journal of the Atmospheric Sciences,* 32:3-15.

Manabe, S., and R.T. Wetherald, 1980: On the distribution of climate change resulting from an increase in CO_2 content of the atmosphere. *Journal of the Atmospheric Sciences,* 37:99-118.

Manabe, S., and R.T. Wetherald, 1987: Large-scale changes of soil wetness induced by an increase in atmospheric carbon dioxide. *Journal of the Atmospheric Sciences,* 44:1211-1235.

Manabe, S., R.T. Wetherald, and R.J. Stouffer, 1981: Summer dryness due to an increase of atmospheric CO_2 concentration. *Climate Change,* 3:347-386.

Mearns, L.O., S.H. Schneider, S.L. Thompson, and L.R. McDaniel, 1991: Analysis of climate variability in general circulation models: Comparisons with observations and changes in variability in 2xCO_2 experiments. *Journal of Geophysical Research,* 95:20469-20490.

Meehl, G.A., and W.M. Washington, 1988: A comparison of soil-moisture in two global climate models. *Journal of the Atmospheric Sciences,* 45:1476-1492.

Mitchell, J.F.B., 1983: The seasonal response of a general circulation model to changes in CO_2 and sea temperatures. *Quarterly Journal of the Royal Meteorological Society,* 109:113-152.

Mitchell, J.F.B., 1989: The "greenhouse" effect and climate change. *Reviews of Geophysics,* 27:115-139.

Mitchell, J.F.B., and D.A. Warrilow, 1987: Summer dryness in northern mid-latitudes due to increased CO_2. *Nature,* 330:238-240.

Quayle, R.G., 1974: A climatic comparison of ocean weather stations and transient ship records. *Marine Weather Log,* 18:307-311.

Rind, D., 1988: The doubled CO_2 climate and the sensitivity of the modelled hydrologic cycle. *Journal of Geophysical Research,* 93:5385-5412.

Schlesinger, M.E., 1984: Climate model simulations of CO_2-induced climatic change. *Advances in Geophysics,* 26:141-235.

Schlesinger, M.E., and J.F.B. Mitchell, 1987: Climate model simulations of the equilibrium climatic response to increased carbon dioxide. *Reviews of Geophysics,* 25:760-798.

Schlesinger, M.E., and Z. Zhao, 1989: Seasonal climatic changes induced by doubled CO_2 as simulated by the OSU atmospheric GCM/mixed-layer ocean model. *Journal of Climate,* 2:459-495.

Washington, W.M., and G.A. Meehl, 1984: Seasonal cycle experiment on the climate sensitivity due to a doubling of CO_2 with an atmospheric general circulation model coupled to a simple mixed-layer ocean model. *Journal of Geophysical Research,* 89:9475-9503.

Washington, W.M., and C.L. Parkinson, 1986: *An Introduction to Three-Dimensional Climate Modeling,* University Science Books, Mill Valley, California, 422pp.

Watts, R.G., 1980; Climate model and CO_2-induced climatic changes. *Climatic Change,* 2:387-408.

Wilson, C.A., and J.F.B. Mitchell, 1987: A doubled CO_2 climate sensitivity experiments with a global climate model including a simple ocean. *Journal of Geophysical Research,* 92:13315-13343.

Global Climate Change: Implications, Challenges and Mitigation Measures. Edited by S.K. Majumdar, L.S. Kalkstein, B. Yarnal, E.W. Miller, and L.M. Rosenfeld. © 1992, The Pennsylvania Academy of Science.

Chapter Twenty-Two

FEDERALISM:
A Regional Approach to Global Environmental Problems

JAMES L. HUFFMAN
Northwestern School of Law
Lewis and Clark College
Portland, Oregon 92219

INTRODUCTION

Global warming poses seemingly intractable problems for the leaders of the world's nations. They are faced with scientific disagreement on the extent of climate change and on the likely impacts of the rise in average global temperatures. Robert M. White, President of the United States National Academy of Engineering states: "We find ourselves in a classic dilemma of policy formulation: possibly severe but unknown levels of risk of undesirable consequences of climatic change in the face of great uncertainty about causes, costs, and consequences" (BNA 1989, p. 230). The difficulties associated with this uncertainty are exacerbated by the transboundary nature of the problem and the realities of an international political system rooted in the fundamental concept of national sovereignty.

The thesis of this paper is that climate change and other global environmental problems can be effectively addressed at the regional level through the use of federal governmental arrangements. International efforts to cope with transboundary environmental problems have seldom been effective because of the inability of international institutions to respond to the many competing interests which motivate the actions of nation states. Federal structures bridge national boundaries while preserving many of the essential and valuable attributes of sovereignty.

Because the problem of climate change is predicted to have global consequences, the tendency is to pursue global solutions. The earth's atmosphere is a commons which nations, like individuals, will exploit to extinction in the absence of adequate institutional arrangements. In his classic commentary on the "tragedy of the commons,"Garrett Hardin prescribes "mutual coercion mutually agreed upon" (1968, p.). In a small community, perhaps even in a large nation, this prescription may be practicable. But at the level of global environmental regulation the prospects are dim.

International environmental regulation is made difficult, writes Lars Björkbom, by "the lack of *one* unifying government" (1988, p. 132). But is world government in some form the only answer? Surely not. World government will be difficult to achieve, as the efforts of the League of Nations and the United Nations evidence. More importantly, world government is clearly undesirable given the many objectives which compete with environmental stability on the agenda of human values.

Economic productivity, fairness in the distribution of world wealth, national identity, and human freedom all vie with ecological stability for the attention of those who would design the social institutions of tomorrow. Although human existence is clearly dependent upon a viable world ecology, few will accept the sacrifice of human dignity and freedom to the preservation of that ecology. In pursuit of a solution to the potentially serious problems of climate change and environmental degradation, we must not forget the lessons which our ancestors learned, and which millions of the earth's population continue to learn, about the evils which can result from well-intentioned, not to mention ill-motivated, centralization of governmental power.

Federalism is a concept that deserves attention in this search for effective and acceptable governmental arrangements. It is a form of government which, if well executed, can permit the benefits of centralization while preserving the advantages of localism. It is a form of government with which the world has considerable experience. Sometimes is has performed well and sometimes poorly. There are lessons to be learned from those experiences. This paper explores some of those lessons and suggests that regional federal structures can contribute importantly to the solution of global environmental problems.

EXISTING APPROACHES TO CLIMATE CHANGE

Multilateral Action

There is widespread and active international effort to cope with the prospect of global warming. The dominant approach reflects the model for international cooperation which was conceived after World War I in the League of Nations, and which was reborn after World War II in the United Nations. It is a model which seeks to build international consensus through diplomatic negotiation, and which relies upon the good will of nations for compliance with agreed standards of performance.

Although climate change has only recently taken center stage in international diplomacy, the environment has attracted growing concern since the Stockholm Conference of 1971. In 1975 a total of 34 nations from east and west Europe signed the Helsinki accord which included relatively strong statements on environmental protection. These nations and the European Community agreed to the Convention on Long-Range Transboundary Air Pollution in 1979, an agreement with direct implications for global warming. Also of direct relevance to global warming are the 1985 Vienna Convention on the Protection of the Ozone Layer and the 1987 Montreal Protocol on Substances that Deplete the Ozone Layer.

International activity on climate change has been fast and furious over the past three years. The Intergovernmental Panel and Climate Change (IPCC) was established by UNEP and the World Meteorological Organization in 1988. The IPCC issued an interim report for the Second World Climate Conference held in late 1990. In 1988 the United Nations General Assembly adopted Resolution 43/53 on Protection of Global Climate for Present and Future Generations of Mankind. A March 11, 1989 Declaration of the Hague, signed by 24 countries, called for "new institutional authority . . . responsible for combating any further global warming . . ." (BNA 1989, p. 215). Later that year, nine additional nations joined the original Hague signatories while a UNEP meeting called for a "new or revamped U.N. authority" to deal with global warming (BNA 1989, p. 287).

Although there is a significant disagreement about the extent of the problem and its likely impacts, there seems to be a general agreement on the need for a single global solution. The 68 nations which convened in Noordwijk, The Netherlands, last November could not agree to a deadline for the stabilization of global carbon dioxide emissions, but agreed that the ultimate objective should be an international convention with "the adherence of the largest possible number . . . of countries" (BNA 1989, p. 627). Mostafa Tolba, Executive Director of UNEP, reflected the widespread agreement in the international community when he told a December, 1989 meeting in Cairo: "In a greenhouse world . . ., [a]ctions must be undertaken on a multilateral level guided by equity and need. And it must be done under the framework of a global convention and mandatory protocols" (BNA 1990, p. 6).

The combination of this rush of diplomatic activity and the possibly urgent nature of the problem suggest that an international solution may be at hand, but there is little reason for optimism. At the heart of the international effort is UNEP, an organization which Ludwik Teclaff has cited as an example of the "relative primitiveness of [international] environmental protection" (1974, p. 33). Although Teclaff wrote those words nearly two decades ago, little has changed in terms of UNEP's real authority. Only last year the president of the U.S. National Academy of Engineering urged that policy makers recognize "[t]he absence of institutions on the international scene in a position to formulate and implement necessary policies" (BN 1989, p. 230).

State Sovereignty

The reasons for the failure to achieve significant international environmental successes are obvious to everyone, even to optimists. The world is divided into an ever-growing number of nation states, and international law, such as it is, is firmly rooted in the associated doctrine of national sovereignty. "The fundamental norm retarding the enactment of an international law is sovereignty. Absolute state sovereignty — or even current doctrines of relative state sovereignty — place the State above the rules of international law" (Gormley 1976, p. 34-35). With not unusual candor, British Prime Minister Thatcher gave full recognition to the significance of the principle of sovereignty in describing the global warming efforts of the IPCC and parent organizations UNEP and WMO as a "sort of good conduct guide for all nations" (BNA 1989, p. 581).

State sovereignty is a firmly established principle of modern international law. Although its consequences are often lamented in the environmental area, no one can doubt that it is the dominant factor in international environmental diplomacy. Beginning with Resolution 1803 of December 14, 1962, the United Nations General Assembly adopted a series of resolutions confirming the "[p]ermanent sovereignty over natural resources" of all nations. This principle has subsequently been included in several other United Nations covenants. The problem faced by the 1971 Stockholm Conference was "to link the concept of sovereignty ... with regard to control over resources, with the need perceived by an increasing number of environmentally concerned countries for protection of general environmental interests" (Springer 1983, p. 20). The resulting Principle 21 was "a triumph of diplomatic compromise" (Springer 1988, p. 50). That compromise has remained essentially unchanged over two decades as evidenced by paragraph 6 of the 1989 Noordwijk Declaration: "The Conference recognizes the principle of the sovereign right of States to manage their natural resources independently. The Conference also reaffirms that global environmental problems have to be approached through international co-operation" (BNA 1989, p. 624). The principle of absolute control over national resources, an outgrowth of the exploitation of the colonial era, has been applied not just to traditional raw materials but also to the use of the atmosphere to the extent that many perceive they have what Armin Rosencranz calls "pollution prerogatives." (1983, p. 198)

Although Stockholm Principle 21 recognizes that state sovereignty is limited by national boundaries, this responsibility is difficult to enforce in a world of sovereign nations. The *Trail Smelter Arbitration* is often cited as evidence that the responsible side of the principle can be made to work, but the principle of sovereignty makes it unncessary, and therefore unlikely, for states to submit to international adjudication. When Johan Lammers, legal advisor to The Netherlands Ministry of Foreign Affairs, says that "[s]tates must be aware that their sovereignty in what they allow within their jurisdiction is limited," (BNA 1989, p. 109) he is speaking about an international system he would like to see. When Armin Rosencranz says that ". . . neither principles nor maxims are of much consequence in the case of transboundary air pollution, [because] [n]ations rarely relinquish jurisdiction over cases of pollution emanating from their territory, and even more rarely admit liability for such pollution," (1983, p. 197) he is describing the world in which we live.

The fact of state sovereignty means that international agreements on global warming will most often be aspirational rather than enforceable. Perhaps "nature will continue to challenge our conventional definition of sovereignty" and will someday force us to redefine "our political borders which defy reality," (Carroll 1988, p. 277) but for the present we should seek solutions which accommodate the political realities of the world we live in.

Unilateral Action

Aside from standing in the path of international solutions, state sovereignty makes unilateral state actions ineffective. Of course unilateral actions by some nations, namely those responsible for a disproportionate share of carbon dioxide production or of deforestation, will have greater impacts than actions by other nations. But nations will seldom have the necessary incentives to act unilaterally. Global warming will be only one of many items on every nation's political agenda. Even if the issue of global warming looms large in the politics of a particular country, the fact that unilateral action imposes domestic costs for shared international benefits will make the prevention of global warming a hard sell in domestic politics. Even if individual nations are willing to act, "the net effect of stringent unilateral regulations may be simply to shift a pollution problem abroad" (Springer 1983, p. 19).

A basic characteristic of international law is that it has direct application only to states, while national law has applications to private persons (Brierly 1963, p.1). This distinction has fundamental implications for the resolution of transnational problems like global warming. Even if international law was meaningfully enforceable, its environmental mandates would not apply directly to many of the world's polluters. Except in the case of nationalized industries, effective enforcement of international environmental standards will depend upon the willingness and ability of national governments to regulate through their domestic laws. Of equal importance is the disconnection between individuals and the establishment of international environmental regulations. The interests of nations in a world of sovereign states will not necessarily lead to the same environmental prescriptions as will the interests of individuals if they could somehow be translated into international standards.

Regional Action

Bilateral and regional efforts to deal with environmental problems have been more successful than either international or unilateral endeavors. "[A]s far as post-Stockholm collective accord is concerned," notes Jane Schneider, "regional measures are far in advance of those requiring more extensive collaboration" (1979, p. 37). An illustration of the potential for regional success is in the control of marine pollution. As of 1988 there were "13 regions in the world where a regional intergovernmental agreement on cooperation in combating marine pollution incidents is either in effect or underdevelopment" (Edwards 1988, p. 229).

In part regional successes have resulted from the obvious fact that the involvement of few nations makes it easier to agree. But regions have other advantages. "[M]uch greater progress can be achieved at the regional level between groups of homogenous States, possessing a common interest and even common traditions and heritage" (Gormley 1976 p. 225). A recognition of the importance and foundation of national interests leads to the "obvious conclusion . . . that, consonant with the principle of economy, environmental problems should be dealt with at the lowest possible level of inclusivity" (Schneider 1979, p. 13).

Although regional efforts can be more successful as a matter of pragmatic politics, they will not eliminate the jurisdictional boundaries which stand as obstacles to global solutions. In the case of global warming, however, regionalism may have some prospect for success because of the particular nature of the problem. The principle producers of greenhouse gases are the industrialized countries of Europe and North America. Agreements among these nations can have significant impacts on the prospects for global warming. In addition, the prospects for global agreement among several regional federations is far better than in a world of nearly two hundred sovereign states.

The European Community

Regional agreement will eliminate neither the practical nor the legal constraints of sovereignty. But regionalism is not limited to the traditional forms of international diplomacy. The European Community is a case in point. "The accomplishments of the EEC in environmental matters are at least partly attributable to special features which distinguish the EEC from other regional groupings" (Nanda & Moore 1983, p. 112). If these special features are not unique to Europe, perhaps the European Community can serve as a model for other regions of the world.

Nigel Haigh who is not optimistic about the EEC as a model, says, "[T]here is no other arrangement between sovereign nation states like the European Community anywhere else in the world and the Community does not fit any of the established models for collaboration that do exist." Although Haigh goes on to insist that "the European Community is not a federal nation state," he recognizes the fact that the European Community is unique because "[n]ational governments and parliaments have . . . transferred sovereignty in certain prescribed fields to the EC institutions" (1989, p. 617), thus suggesting that federalism is indeed the model which the European Community represents. Not only does the EEC have some "legislative competence . . . [To] adopt legislation in the field of the environment, which is binding on its member states," it also has "the competence . . . to confer rights directly on individuals which are justiciable in the members' courts" (Nanda & Moore 1983, p. 112).

Although the 1957 Treaty of Rome, which created the European Community, did not provide an authority over environmental matters, Article 130R-S of the Single European Act which took effect in 1987 provides that "environmental protection requirements shall be a component of the Community's other policies." Pursuant to this authority the Community has adopted legislation relating to water pollution, waste, air pollution, chemical marketing and use, wildlife protection, noise pollution, and environmental impact assessment. This legislation is now "so extensive that is it impossible to understand fully the national legislation of any EC Member State without an understanding of the framework of EC law within which the national legislation has to fit." (Haigh 1898, p. 619)

Of course the sovereignty of member states continues to play an important role in European Community politics. Most of the environmental legislation is in the form of directives which are non-binding as to the method of achieving prescribed ends. Stanley Clinton Davis, former EC Environmental Affairs Commissioner, has argued that some 200 pending proceedings against member countries are evidence that "no Community can exist unless there is a rule of law" (BNA 1989, p. 579). In response to such criticisms, the EC is on the verge of establishing a European Environment Agency (BNA 1989, p. 287). In addition, the EC's Economic and Social Committee has recommended that all EC environmental laws be drafted as regulations rather than directives to make those laws immediately applicable in all the member states (BNA 1989, p. 323). The important point is that the European Community is talking seriously about what environmental laws should be immediately binding on all member states. Meanwhile the United Nations Environmental Program was engaged in debate over where to conduct its planning for the 1992 World Conference on Environment and Development (BNA 1990, p. 73). The difference is in no small part a product of the advantages offered by the federal nature of the European Community.

FEDERALISM AND GLOBAL WARMING

Although federalism is not a panacea for the solution of the world's environmental problems, both theory and experience suggest that it has several distinct advantages over existing efforts at bilateral and multilateral cooperation.

Effective Transnational Government

Those familiar with the early history of American federalism will not be surprised at the ineffectiveness of most international environmental initiatives. Like the United States under the Articles of Confederation, existing international organizations are at best what John Marshall described as a league as distinct from the confederation which was created by the 1787 Constitution (1819, p. 404). The essential difference is that members of a league retain their full sovereignty while members of a confederation relinquish some of their sovereignty to a federal government. Under the Articles of Confederation the United States government was dependent on the voluntary cooperation of the states. As a consequence, the United States was powerless to deal with the problems facing the new nation. The parallel to the experience of international efforts at environmental management is not coincidental. "International organizations are *constituent* organizations. In order to perform any useful work at all, and indeed to remain in existence, they must seek a consensus on the part of their constituents" (Perry 1983, p. 34).

Federal structures can create limited central authority sufficient to provide the "compulsion" necessary to "[l]arge-scale collective action" (Schneider 1979, p. 108). A federal system helps overcome "the paradox . . . that international cooperation is impossible without national concurrence, but mere concurrence as a formality is insufficient to insure that effective cooperation will occur," by supplying the "something more . . . [that] is needed to attain the objectives of international environmental cooperation" (Caldwell 1988, p. 16). A well designed federal system can provide a central government with sufficient resources and authority to solve transboundary problems without sacrificing the sovereign claims of the member states.

Accommodating the Sovereign State

The elimination of the doctrine of state sovereignty is neither desirable nor possible. It is not desirable because claims of national sovereignty, or independence in the parlance of revolutionaries since 1776, have been essential to the struggle for human freedom. The American Revolution was rooted in claims of a right of self-governance; a right to exercise sovereignty over the geographical territory occupied by the revolutionists. The ongoing revolutions in eastern Europe are rooted in similar claims of the right of a people to exercise sovereignty over their own lives. Over two centuries ago James Madison urged upon his fellow delegates to the Philadelphia convention the importance to human freedom of competing sovereignties. "England and France have succeeded to the pre-eminence & to the enmity. To this principle we owe perhaps our liberty. A coalition between those powers would have been fatal to us" (Farrand 1966, p. 448).

The truth of Madison's claim that competing sovereignties are essential to human liberty is a product of the seemingly "endlessly propulsive tendency [of power] to expand itself beyond legitimate boundaries" (Bailyn, p. 56). "[T]he essence of the American federal system . . . [is] the division of power along a vertical axis by removing some of it from the central originating point, the states, and shifting some of it up and some of it down, the axis" (McDonald 1979, p. 312). The key to the success of this structural approach for controlling power is the division of power, which explains why, until recently, the Soviet federal system has been a failure in terms of human freedom.

The centralized Soviet federal system has also been a failure in environmental terms, while less centralized federal systems have achieved significant environmental successes. In dealing with global environmental problems, the tendency, as we have seen, is to prefer centralization of authority and to accept state sovereignty as an unfortunate but inevitable obstacle. It is counterintuitive to suggest that divided power might better solve a common problem. But divided power leaves room for experimentation and innovation. Environmental mistakes are readily perceived in the face of environmental successes. The identification and the resolution of environmental problems is far less likely when everyone is obliged to follow a centrally prescribed course.

The perpetuation of state sovereignty is inevitable and thus a federalist approach to the global environment has the distinct advantage of embracing that sovereignty while overcoming its constraints on enforceable transboundary regulation. Neither the rulers nor the ruled are inclined to sacrifice their national identities to counter even severe threats to the global environment. But just as individuals are generally willing to relinquish some liberty to a government over which they have some measure of control, so nations may be willing to give up some sovereignty to a central government of limited powers, if that central government will help to resolve the shared problems of the federating states. Historically, from the formation of the United States to the founding of the European Economic Community, the shared problems of security and commerce have led to confederation. Today the shared problem of environmental risk can provide the necessary incentives for the states to give up some, but not all, of their sovereignty.

Bridging the Borders Which Nature Does Not Recognize

The political demarcations, which give meaning to the concept of transboundary environmental problems, are more or less deeply rooted in history and culture. They cannot be ignored, nor can they be overcome by international edict. But some of them can be more readily bridged than others. "The challenge," writes Allen Springer, "is to create an effective international legal framework in which both . . . [economic development and environmental protection] can be pursued cooperatively by states whose priorities differ" (1983, p. 24), meaning that challenge is more likely if cooperation is first achieved among states whose priorities are similar. The European Community is in no small part the product of *"philosophical bases of a consensus,"* (Gormley 1976, p. 35) (emphasis in Original) evidencing that "much greater progress can be achieved at the regional level between groups of homogenous states, possessing a common interest and even common traditions and heritage" (Gormley 1976, p. 225).

The seeds of potential federalist structures exist in most parts of the world. Indeed these seeds have occasionally germinated in places like Central America, South America and Africa, only to wither from a lack of political will and necessity. But the fact that people have talked about confederation in the past suggests that they might talk about it in the future. If environmental problems threaten national agendas, federation will be an attractive alternative to the sacrifice of national sovereignty and identity. More probably it will be the combination of environmental and economic pressures which can lead nations to seek and accept confederation with familiar neighbors.

Of course if regional federations will not eliminate political borders, internal and external boundaries will persist. The obstacles posed by internal boundaries will depend upon the nature of the particular federation. The European Community and the pre-glasnost Soviet Union represent a range of possibilities. External boundaries will be unaffected by regional federations, but the bridging of these borders will be facilitated by the simple reduction of the number of world players on issues over which federations have authority. UNEP's Register of International Treaties testifies to the participation of the European Community as a single entity acting on behalf of all of its members. The transaction costs of international diplomacy can only be reduced when nations are able to act in concert through their federations.

Redistributing the Costs of a Stable Global Environment

In the politics of international environmental diplomacy, perhaps the most persistent source of disagreement has been the disparity of wealth and economic development among the world's nations. "Without adequate financial and technical assistance," says Jan Schneider, "many states may remain incapable of applying some international environmental law" (1979, p. 96). Many states will also remain unwilling. The IPCC Subgroup on Agriculture, Forestry and Other Human Activities has concluded that "[n]o agreement on [tropical] forest [cutting] and global climate change will be reached without commitments by developed countries on greenhouse gas emissions . . ." (BNA 1990, p. 52). This statement reflects a fundamental split between the industrialized and non-industrialized countries. The less developed nations assert that since the beginning of the colonial era, " a few greedy nations" have stripped the world of its resources and polluted the environment (Fouéré 1988, p. 31).

While not necessarily accepting avarice as the reason, the signatories to the 1989 Noordwijk Declaration agreed that "[i]ndustrialized countries . . . have specific responsibilities . . . [to] support, financially and otherwise, the action by countries to which the protection of the atmosphere and adjustment to climate change would prove to be an excessive burden . . ." (BNA 1989, p. 624). UNEP's executive director, Tolba, has asserted that "some mechanism, . . . almost certainly a global fund, must be created from which developing nations can draw, thus ensuring that their development aspirations are in no way compromised by their commitment to environmental safety" (BNA 1989, p. 335).

But Tolba and others who support the concept of an international fund to redistribute the costs of world environmental protection understand well the practical problems of implementing such a plan. After UNEP approved a budget projection of a relatively modest $100 million in contributions for 1992, Tolba underscored that achieving the objective would require a 35% increase in voluntary donations (BNA 1989, p. 279). No government has ever succeeded on the basis of voluntary financial contributions. The power to tax is both the power to destroy and the power to create. Sovereign states will find mandatory financial contributions more acceptable in a federal structure which promises to limit the central authorities power to collect revenues. As a form of regional organization, federations "may have a potentially valuable function in organizing cooperative plans of action for carbon dioxide control strategies, and for compensating member countries for damages resulting from climate change" (Schware & Kellogg 1983, p. 83). Existing federations, including the European Community, serve to illustrate the wealth redistribution capacities of federations.

Economic Productivity, Markets and Liberty

It is not only the less developed countries which are interested in economic growth. Industrialized nations also seek to maintain and improve the standard of living of their citizens. Although some projections of the effects of global warming would lead to the conclusion that life style changes are inevitable, few nations are willing to accept this prospect as evidenced by the widesperad endorsement of the concept of sustainable development. Paragraph 9 of the Noordwijk Declarations states: "While striving to preserve the global environment, it is important to work at the same time to ensure stable development of the world economy, in line with the concept of 'sustainable development' " (BNA 1989, p. 625).

Sustainable economic development is dependent upon institutional arrangements which provide incentives for both economic productivity and environmental protection. While it is widely accepted, (even in the historically socialist nations of eastern Europe), that private markets lead to productivity, much national environmental legislation has been based on the assumption that markets are a primary contributor to environmental degradation. This supposed dichotomy has contributed to the belief among many environmentalists that economic development and environmental protection are incompatible objectives—that sustainable development is a concept contrived by industrialized nations to mollify the less developed countries.

There is, however, strong evidence and widespread support for the view that markets are important to both economic productivity and environmental protection. The theory behind this point of view is that market mechanisms can create incentives for innovative pollution control, while command and control regulation creates disincentives for such innovation. Proposals for marketable pollution rights have gained increasing support in some nations, and would meld well with proposals like that of a Dutch commissioned study for a world carbon budget of 150 billion tons for the period 1985 to 2100 (BNA 12:584). Such a budget could be allocated to the world's nations on the basis of wealth distribution and development factors, and a market in carbon rights could then function much as do markets in goods and services.

The effectiveness of markets, whether for pollution rights or for goods and services, depends in part upon the willingness of nations to refrain from imposing restraints on trade. Federal structures have proven to be excellent mechanisms for facilitating free trade. A primary motivation for the formation of the United States federal system was the elimination of interstate barriers to trade. The European Economic Community was similarly created to be a common market, an objective which is set to be accomplished in 1992. Other efforts at federation, such as the Latin American Free Trade Association and the Central American Common Market, have sought to overcome the economic consequences of restraints on trade. These latter efforts have foundered, but the prospects for achieving both environmental and economic objectives through federation may yet revive old and inspire new visions of regional cooperation.

United States President George Bush told the IPCC in February of 1990 that: "Strong economies allow nations to fulfill the obligations of environmental stewardship. Where there is poverty, the competition for resources gets much tougher. Stewardship suffers" (BNA 13:43). Surely the record of environmental protection around the world confirms the truth of the President's assertion. It is the wealthy nations of the world which have led the way. Market mechanisms, said the U.S. President, have been essential to strong economies. Markets can also provide incentives for enviornmental stewardship and for the development of technologies necessary to that stewardship. Federations can contribute significantly to the success of such markets.

Finally, well designed federal governments have proven to be effective in protecting the rights and liberties of individuals. The vertical division of powers between federal and state governments, like the horizontal separation of powers among executive, legislative and judicial branches of government, serves to constrain the abuse of power. Emergencies, both real and imaginary, have long served to justify the centralization of powers, often at the expense of human freedom and uncounted human lives. Global warming and other environmental problems may well present such emergencies, but no environmental prospect should lead us to forget the enormous price which our predecessors have paid for freedom.

CONCLUSION

Global warming, as British Environment Secretary Chris Patten has said, "is among the more difficult, complicated, and challenging issues ever tackled by the world community" (BNA 1990, p. 282). Its solution will require creative thinking about science, technology, and social organization. Thomas Kuhn once explained the impact of human nature on the structure of scientific revolutions (1962). Those impacts are only multiplied in the structure of social revolutions. Indeed the constraints upon fundamental change in human social organization are so powerful that revolution, in the sense of totally new institutional arrangements, has seldom if ever succeeded. Fundamental institutional change has always occurred slowly, because institutions must fit the people they govern. Even the American revolution was rooted in a desire to restore the colonists' rights as Englishmen, not to create a totally new society.

Thus those who seek the solution to global environmental problems, exclusively in international law and international organizations, are ignoring a fundamental fact of human history. It has been urged that "time does not allow us the luxury of trying to create someting new, . . . [that] [w]e have institutions enough already" (BNA 1989, p. 281). But our international institutions have failed in over four decades to impact significantly on the deeply rooted tradition of state sovereignty. Nor will internationalization result inevitably from the pressures of environmental degradation. People around the world feel as strongly about other goals, including national identity and individual liberty, as they do about the protection of the global environment. Regional federations can facilitate the pursuit of these diverse goals. Federalism may be an old idea whose time has come.

BIBLIOGRAPHY

Bailyn, B. 1967. *The ideological origins of the American Revolution.* Cambridge: Harvard Univ. Press.
Björkbom, L. 1988. Resolution of environmental problems: the use of diplomacy. In *International environmental diplomacy,* ed. J.E. Carroll, pp. 123-137. New York: Cambridge Univ. Press.
Brierly, J.L. 1963. *The law of nations.* New York: Oxford Univ. Press.
Bureau of National Affairs. 1989. *International environment reporter,* vol. 12. Washington, D.C.: Bureau of National Affairs.
Bureau of National Affairs. 1990. *International environment reporter,* vol. 13. Washington, D.C.: Bureau of National Affairs.
Caldwell, L.K. 1988. Beyond environmental diplomacy: the changing institutional structure of international cooperation. In *International Environmental Diplomacy,* ed. J.E. Carroll, pp. 13-27. New York: Cambridge Univ. Press.
Carroll, J.E. 1988. Conclusion. In *International environmental diplomacy,* ed. J.E. Carroll, pp. 272-275. New York: Cambridge Univ. Press.
Edwards, E. 1988. Review of the status of implementation and devlopment of regional arrangements on cooperation in combating marine pollution. In *International environmental diplomacy,* ed. J.E. Carroll, pp. 229-272. New York: Cambridge Univ. Press.
Farrand, M. (ed.) 1966. *The records of the federal convention of 1787.* New Haven: Yale Univ. Press.
Fouéré, E. 1988. Emerging trends in international environmental agreements. In *International environmental diplomacy,* ed. J.E. Carroll, pp. 29-44. New York: Cambridge Univ. Press.
Gormley, W.P. 1976. *Human rights and environment: the need for international cooperation.* Leyden: A.W. Sijhoff.
Haigh, N. 1989. The environmental policy of the European Community and 1992. *International environment reporter* 12:617-623.
Hardin, G. 1968. The tragedy of the commons. *Science* 162:1243.
Kuhn, T..S. 1962. *The structure of scientific revolutions.* Chicago: Univ. of Chicago Press.
Marshall, J. 1819. McCulloch v. Maryland. *United States reports* 4:400-437.
McDonald, F. 1965. *E pluribu unum: the formation of the American republic, 1776-1790.* Boston: Houghton Mifflin.
Nanda, V.P. and P.T. Moore. 1983. Global management of the environment: regional and multilateral initiatives. In *World climate change: the role of international law and institutions,* ed. V.P. Nanda, pp. 33-45. Boulder, Colorado: Westview Press.

Perry, J.S. 1983. International organizations and climate change. In *World climate change: the role of international law and institutions,* ed. V.P. Nanda, pp. 33-45. Boulder, Colorado: Westview Press.

Rosencranz, A. 1983. The international law and politics of acid rain. In *World climate change: the role of international law and institutions,* ed. V.P. Nanda, pp. 33-45. Boulder, Colorado: Westview Press.

Schneider, J. 1979. *World public order of the environment: towards an international ecological law and organization.* Toronto: Univ. of Toronto Press.

Schware, R. & Kellogg, W.W. 1983. International strategies and institutions for coping with climate change. In *World climate change: the role of international law and institutions,* ed. V.P. Nanda, pp. 33-45. Boulder, Colorado: Westview Press.

Springer, A.L. 1983. *The international law of pollution: protecting the global environment in a world of sovereign states.* Westport, Connecticut: Quorum Books.

Springer, A.L. 1988. United States environmental policy and international law: Stockholm Principle 21 revisited. In *International environmental diplomacy*, ed. J.E. Carroll, pp. 44-65. New York: Cambridge Univ. Press.

Teclaff, L.A. 1974. International law and the protection of the oceans from pollution. In *International environmental law,* ed. L.A. Teclaff & A.E. Utton, pp. 104-139. New York: Praeger Publishers.

Global Climate Change: Implications, Challenges and Mitigation Measures. Edited by S.K. Majumdar, L.S. Kalkstein, B. Yarnal, E.W. Miller, and L.M. Rosenfeld. © 1992, The Pennsylvania Academy of Science.

Chapter Twenty-Three

THE NEED FOR HIGHER RESOLUTION IN THE ECONOMIC ANALYSIS OF GLOBAL WARMING:
An Assessment and Expanded Research Agenda

ROBERT AYRES*
and
ADAM ROSE**[1]

*Department of Economics
Institut Européen d'Administration des Affaires
77305 Fountinbleau Cedex, France
and
**Department of Mineral Economics
The Pennsylvania State University
University Park, PA 16802

INTRODUCTION

A major focus of scientific research on global warming is the class of general circulation models (GCMs). These are meteorological simulation models used to predict possible climate change over the Earth's surface. The major criticisms leveled

---

[1]The authors are, respectively, Sandoz Professor of Environment and Management, INSEAD; and Professor of Mineral Economics, The Pennsylvania State University. The authors wish to thank Joel Darmstadter, Faye Duchin, Diana Liverman, and Katherine McClain for their valuale comments.

at these models today are: (1) they do not adequately characterize the physics of cloud formation and precipitation, (2) they do not generally include the feedback from ocean currents, and (3) they are too geographically coarse—the best of them is only able to break-up the Earth's surface into grids of 500 square kilometers (see, e.g., Schneider and Rosenberg, 1989). This last limitation means an averaging of more extreme effects in important sub-areas. Most scientists have acknowledged the need for further geographic refinement of GCMs, i.e. a higher degree of resolution. (In fact, requests for increased funding for geophysical and climate research are largely based on the need for better models and improved data). Improvements would enable scientists to make more precise predictions about the effects of global warming, especially at the regional meso scale. This would enable scientific results to be applied much more directly to national and regional political entities, the major decision-making units for global warming policy.

An analogous situation exists with regard to the socioeconomic analysis of global warming. The focus is almost always on nations as the primary, if not only, unit of analysis, even in cases where nations are large in size and/or are composed of socio-economically and environmentally diverse regions. Most studies employ aggregate estimates of benefits and aggregate estimates of costs (damages), thus averaging out impacts across subnational areas, income groups, and racial/ethnic minorities. Simulations of the economics of global warming avoidance, or mitigation policies, typically examine effects on the macroeconomy as a whole. Any sector specific attention is devoted mainly to those emitting greenhouse gases, such as agriculture and electric utilities, as well as sectors judged to be "climate sensitive." Moreover, quantitative analysis is often confined to direct effects (e.g., of fuel combustion) with little or no attention to second-order interactions with other sectors. And finally, nearly all studies focus on a limited number of effects (including spillovers), often citing the lack of data or conceptual difficulties.

The purpose of this paper is to demonstrate the need for a higher level of resolution in the economic analysis of global warming and to provide examples of how this might be accomplished. We contend that there is no need to wait until research is completed on more aggregative themes. In fact, there are many reasons to work concurrently on the distributive effects, as well as to expect benefits from the cross-fertilization of ideas between the two tracks.

In the following section we evaluate the degree of resolution of economic analyses performed by and research agendas proposed by noted experts in the field. We then proceed to discuss each of the four areas where a finer level of analysis is needed: (1) technological and sectoral, (2) natural receptor, (3) regional, and (4) socio-economic. In each case we discuss why this higher degree of resolution is important, including modeling considerations and policy implications. We also suggest ways in which future research agendas can be expanded to accommodate our basic theme.

CURRENT RESEARCH AGENDAS ON THE ECONOMICS OF GLOBAL WARMING

Economists believe they have a great deal to contribute to the analysis of global warming. Economic activity—both production and consumption—is the major cause of the release of greenhouse gases, as well as the destruction of forests that might otherwise be able to absorb a large fraction of the CO_2. Economists are often called upon to value the potential damages from global warming, as well as the costs of avoidance, mitigation and adaptation. Finally, economists are quick to point out that, just as current institutional arrangements provide an incentive to release CO_2, CFCs, etc., these institutions can be altered and policy instruments devised to provide incentives for pollution control.

In order to assess the potential role of economics in the study of global warming, a number of expert panels, workshops, and conferences have been organized over the past two years. They have focused on the questions of what we know and what we do not know about the economics of global warming. The derivative of this effort is to analyze what we need to know, or to establish a research agenda for the future.

A good example of such an effort is the recent New Haven Workshop, held in May 1990, and sponsored by the National Oceanographic and Atmospheric Administration, the National Science Foundation, and National Aeronautics and Space Administration. The report of this Workshop, released in January 1991, indicates that its focus was on more basic areas of research (NOAA, 1991). It stated research priorities for the economic study of global climate change as follows:

1. Provide preliminary estimates of costs and benefits. The report however, does not mention the need to identify and characterize a broader range of receptors. We are concerned that, in the rush to provide preliminary estimates, unmeasurables—such as the impact of climate warming on species disappearnce—will be omitted from the discussion. Furthermore, potential benefits of avoidance or mitigation are also likely to be understated as compared with the benefits of adaptation.
2. Establish data collection priorities and begin gathering more complete data. Ironically, many of these collection efforts would likely come from the "bottom up", i.e., from local sources. Thus, we suggest that consideration be given to tabulating these data at the regional level for important analyses of the effects on geographic sub-areas and their role in the policy-making process.
3. Evaluate and revise existing economic models and other forecasting tools. To date, these are typically macro-oriented models (see, e.g., Manne and Richels, 1990; Nordhaus, 1990). They lack the sectoral detail that some alternative modeling frameworks provide (see, e.g., Ayres, 1991; Jorgenson, 1990; Duchin, 1991; Rose and Chen, 1991)).
4. Estimate the value of natural and man-made resources that may be affected. There is even some mention of formally including these into a revised system of national accounts. Suggestions such as this have been offered for more than 20 years[2] without ever gaining much momentum until recent work by the World Resources Institute (Repetto, 1987, 1991), the World Bank (Ahmad

---

[2] See Tobin and Nordhaus (1972); Peskin (1972, 1980, 1989); Ayres (1978); Hueting (1980).

et al., 1990) and the UN Statistical Office (Bartelmus *et al.*, 1989). It is time for these accounting frameworks to be integrated more fully into research on global warming.
5. Examine alternative policy instruments in a global context. Most of the existing evaluations of these instruments are based on considerations of economic efficiency or intergenerational equity. Although there is brief mention of static equity considerations in the report, this consideration often winds up being a step-child of economic inquiry. More work is needed on modeling distributional impacts and evaluating their implications for policy.

The New Haven Workshop report is representative of the emphases in all major economic research assessments known to the authors. These include the Workshop on Economic/Energy/Environment Modeling for Climate Policy Analysis (a joint U.S. - Japanese effort) held in Washington in October 1990; the Energy Modeling Forum, (EMF 12) Global Warming: Greenhouse Gases and Energy-Sector Adjustments held at Stanford University in September, 1990; and the Conference on Global Warming held at MIT in July 1989.[3] Several special issues of leading economic periodicals have had similar orientations, e.g., the *Energy Journal* in 1990 and 1991 and *Energy Policy* in 1990 and 1991.

SECTORAL AND PROCESS DISAGGREGATION

Anthropogenic greenhouse gases are produced by a few specific economic activities, primarily fuel consumption and agriculture. To summarize briefly, the most important greenhouse gas overall is carbon dioxide (CO_2), which accounts for roughly half of the total effect. In turn, carbon dioxide is produced by two processes: (i) the combustion of carbonaceous fuels in proportion to their carbon content, and (ii) the decay of organic matter accompanying deforestation. The relative contribution of these two is somewhat uncertain, but appears to be roughly in the ratio 3:1. In partial compensation, there is probably some CO_2-induced enhancement of photosynthesis resulting in some reabsorption. Also, it appears that about half of the excess CO_2 emitted since the mid-19th century has been absorbed by the oceans.

The other three major greenhouse gases are methane (CH_4), nitrous oxide (N_2O) and some of the chlorofluorocarbons, notably CFC11 and CFC12. Methane and nitrous oxide are both mainly associated with agriculture. Non-agricultrual anthropogenic sources of methane include coal mining, gas flaring, leakage from natural gas pipelines, and swamps. Nitrous oxide is produced by the breakdown of synthetic nitrogen fertilizers, as well as by several natural decay processes. CFCs are purely artificial chemicals not found in nature. They were first synthesized in the

---

[3]The joint U.S.-Japan Conference proceedings are available in Wood and Kaya (1991): a volume based on the EMF Workshop is currently in press; and the MIT Conference papers are published in Tester, Wood and Ferrari (1990). The reader is also referred to the *Economic Report of the President* for the Bush Administration's implicit research agenda, which is also similar to the above.

early 1930's as refrigerants, and continue to be used for that purpose, as well as some other applications such as foam blowing and cleaning electronic components. CFCs are extremely effective absorbers of infrared radiation and account for about 15% of the total warming problem. They have also been the fastest growing component until recently.

There are a number of tradeoffs to be considered in evaluating policy options with respect to greenhouse gas emissions. For instance, a policy commonly recommended is to substitute natural gas for coal and oil. Hence, for a given energy demand, it appears obvious that, *ceteris paribus*, gas is the preferable fuel. However, this conclusion is simplistic, because methane is also a greenhouse gas and 20 times more potent in terms of radiative forcing, molecule for molecule, than carbon dioxide. Increased use of natural gas as a fuel could result in increased leakage, with counterproductive results. On the other hand, to complicate things further, decreased use of coal would also decrease the methane released by coal mining. The point here is that a fairly detailed sectoral (and process) analysis is necessary to come to grips with these tradeoffs.

Another tradeoff is between reduced use of fuel for space heating and increased need of electric power for refrigeration and air-conditioning. Since neither heating nor air-conditioning are needed year-round, these tradeoffs are very location specific. Heating requirements are normally proportional to the number of days in which temperature drops below a certain point, while air-conditioning requirements are related to the number of days the temperature climbs above a certain level. These measurements tend to be non-linear functions of average annual temperature differentials. For instance, a 2 degree rise in average year-round temperature could increase the number of air conditioning days (in which maximum temperatures exceed 90°F) by a factor of 2 or 3. On the other hand, it would decrease the number of heating days, but by a much smaller factor. The carbon equivalents of heating days and air conditioning days are not the same either. Perhaps most important, increased use of air-conditioners would probably result in increased release of CFCs, though not necessarily in proportion. Again, the real impact of global warming depends on technological and regional specifics.

One last example is worthy of attention in this context. Macroeconomic analysis using aggregative models often tend to assess "costs" of greenhouse gas abatement by the following set of approximations:

a. Costs of abatement are computed in terms of costs of reducing CO_2-equivalents.

b. CO_2 equivalents are assumed to be proportional to energy use by fuel type.

c. Costs of energy use reduction are assumed to be equivalent to "lost GNP", where GNP is assumed to be a function of energy input (i.e., energy is taken to be a factor of production), and reduced inputs automatically result in reduced outputs.

d. The mathematical relationship between energy consumed and economic output is assumed to be given by a set of "elasticities" (i.e., ratios of differentials), which are imputed by statistical analysis of time series data. The tenuous nature of some of the underlying assumptions, however, is rarely acknowledged and is not obvious.

All of the heroic assumptions noted above would be forgivable as "the best that can be done under the circumstances" if only those who make them were less insistent on the superiority of the approach. The problem is that disaggregated approaches lead to very different results. In particular, the macroeconomic approach sketched above necessarily concludes that CO_2 reduction (much of it due to energy conservation) must be costly, to some degree (see, e.g., Manne and Richels, 1990; Hogan and Jorgenson, 1991). The reason for this result is that it is built into the production function model. In fact, it is a basic axiom of macroeconomics that the economy is in equilibrium in this regard, having already selected the least-cost the economy is in equilibrium in this regard, having already selected the least-cost technological options. Moreover, such models do not, and cannot, reflect either the endogenous nature of technological change or the distortions and disequilibria due to institutional barriers or wrong prices.

There is very strong empirical evidence that the economy has not selected the least-cost options (see, e.g., Ayres, 1990). Disaggregative engineering-type analysis often in the form of "process" models, points in the opposite direction from aggregative analysis. It suggests strongly that significant energy savings could be accompanied by significant monetary savings, although this would vary considerably across processes or sectors (see, e.g., Ayres, 1991; Mills *et al.*, 1991).[4] This conclusion, if supported by further research, has obvious and important policy implications.

Another reason to pursue work on process models is that they provide a stronger foundation for the various stages of subsequent analysis.[5] They are grounded in engineering realities, but with room to grow on engineering possibilities. They allow for the explicit inclusion of behavioral considerations found to be so important in bounding previous policy initiatives to change energy use patterns. Process models are also an excellent informational framework for technology transfer, the basis of a major global warming strategy that will help LDC's leap frog the profligate technologies used by industrial nations during this century.

A solid foundation is also necessary to carry the weight of models that will be needed to adequately address the issue. A global problem requires a global solution (see, e.g., Grubb, 1989) and ultimately a world economic modeling system will be needed (see, e.g., Duchin, 1991). Ideally, the natural progression of models—from process to industry to region to nation to world — will build on such foundations.

An Expanded Research Agenda — Sectoral and Process Aggregation

1. Improve the estimation of the effect on climate change on greenhouse gas emissions from individual production processes.

---

[4] See, especially, work by Repetto.

[5] For an excellent survey of the relative merits of "top-down" vs. "bottom-up" approaches, the reader is referred to Darmstadter (1991). Some notable examples of "bottom-up" (process-oriented) studies do exist (see Chandler, 1990; Schipper, 1991), but few have addressed the situation in the U.S.

2. Integrate differences in warming potential among various GHGs into policy analysis.
3. Make more explicit (and more realistic) the nature of basic assumptions regarding energy elasticities.
4. Examine the potential of disequilibrium models, i.e., those that do not implicitly assume energy is being used efficiently.
5. Give more attention to successful efforts and promising candidates for energy conservation and examine the potential for transferring these successes to other processes.

DISAGGREGATION BY RECEPTOR SYSTEM

A primary receptor of the greenhouse effect will be vegetation. As a consequence of overall climate warming, all GCMs predict regional variability in the temporal and spatial distribution of temperature, precipitation, evapotranspiration, clouds, and air currents, all important determinants of vegetation type and biome. The non-linear character of the climate system makes it likely that this regional variability is even more pronounced than current GCMs reveal.

Computations carried out to date, comparing equilibrium for the so-called $2xCO_2$ condition (doubling of the atmospheric carbon-dioxide concentration) with control runs for current climate, show very non-uniform responses. The global mean temperature (GMT) is expected to rise between 3°C and 5.5°C for the $2xCO_2$ case (Schneider, 1989), but changes in the equatorial regions will be considerably smaller, while changes in the polar regions will be greater. It is also suggested by some experts that there will be increased seasonal variability.

Global precipitation is likely to increase by 7-10% (moderate confidence), while regional changes are projected to range from -20% to +20% (low confidence). Higher temperatures (high confidence) will probably increase evapotranspiration by 5-10% on a global average. Soil moisture is controlled by precipitation, evapotranspiration and run-off, but regional changes are projected (medium confidence) to be in the range of plus or minus 50%. Globally, surface run-off would probably increase, but, again, on a regional scale changes of -50% to +50% are expected. This follows from changes in evapotranspiration (which is strongly influenced by temperature) and precipitation (Schneider and Rosenberg, 1989).

Simulation studies on arid and semi-arid river basins in the United States suggest that relatively small changes in temperature and precipitation can have multiplier effects on run-off. It appears that run-off will increase in winter in high latitudes, and decrease in summer in mid and low latitudes. These changes in patterns would affect the probabilities of seasonal flooding, not to mention the availability of water for irrigation and hydro-electric power generation.

Sensitivity of vegetation to climate has been investigated on a regional basis by (Emmanuel *et al.,* 1985 and Leemans, 1990) using the so called Holdridge classification of "life zones." This classification is based on three climatic variables, viz. *bio-temperature* (a function of the average temperature and length of the growing season, in degree-days), *mean annual precipitation,* and *potential evapotranspiration* (a measure of humidity). Since the broad scale distribution of terrestrial ecosystems is determined to a large part by the regional climate, one can relate the character of natural vegetation to climatic variables such as average annual temperature and precipitation.

A world map of current Holdrige life zones (grid resolution of 0.5°x0.5° latitude and longitude intervals) has been generated by Leemans (on the basis of interpolated data from about 8 thousand meteorological stations). It has been compared to a corresponding map computed for a climate expected to result from doubling the atmospheric CO_2 concentration (2xCO_2). The differences are dramatic. Statistical analysis indicates that in response to a 2xCO_2 climate, 48% of the global terrestrial surface would change its Holdridge classification, and consequently its vegetation type, with sizable variation among regions.[6]

Interpretation of these results is difficult because we have no experience with such rapid and broad scale changes in environmental conditions. The question that arises is *not* whether nature can adapt to changing climate in general, since it has done so many times in the past several billion years. The question is: how fast can adaptation occur in a natural, unmanaged ecosystem, in response to global warming 10 to 60 times faster than any known change in the historical record (Schneider and Rosenberg, 1989) and at what cost? It has been calculated that the necessary migration rate of tree species to accommodate to a 2xCO_2 warming in 100 years has to be ten times higher than rates calculated for North America over the past 10 thousand years (Sedjo and Solomon, 1989). The actual climate changes, being a non-linear phenomenon, could conceivably be even faster.

Barriers, such as agricultural areas and urban areas, will tend to retard natural migration rates. Hence, in the absence of "massive reforestation programs," rapid disappearance of some tree species from some areas can be anticipated (Sedjo and Solomon, 1989; Solomon and West, 1985). As a result, a few tree species that have small, wind-dispersed seeds, easily transported over large distances, might dominate future landscapes. Non-linear response to changes in temperature and precipitation variability, as well as non-uniform biological response to increased CO_2 concentration, will cause shifts in the species composition of biomes. Alteration of several fundamental ecosystem processes, such as nutrient cycles, is to be expected. An increase in intensity and spread of wildfires due to higher midsummer dryness might occur. Moreover, plant diseases and insect outbreaks severe enough to kill

---

[6] It is worthwhile to note that most agriculturally productive areas of the world are classed as "moist forest" (although the forests are no longer there). In the cool temperature zone (Western Europe, northcentral and northeastern North America, Japan) about 40% of the "moist forest" area would be expected to shift to a different Holdridge life-zone. Most likely the shift would not be in the direction of cooler, but warmer and/or wetter or drier. In the warm temperate zone (mid-central and mid-eastern North America, central China, southern Brazil and Uruguay) about 60% of the area would be expected to shift to another zone: warmer and/or wetter/drier. In the subtropics, (southeastern U.S., southern China and Indo-China, most of central Africa, much of Brazil, parts of India, and most of Central America) around 45% of the area is likely to change to another zone, either wetter or drier.

dominant species over large areas may occur with greater frequency (Batie and Shugart, 1989).

Again, the importance of disaggregation seems clearly evident. Meaningful policy responses with regard to either abatement or adaptation depend on details that will be lost in aggregative studies. Moreover, the potential of spillover (and even more complex interactive) effects between receptor systems are likely to be significant and cannot be omitted.

The problem of global warming has typically been addressed from the perspective of incremental changes in national income, measured for the U.S. and other industrialized countries according to the traditional national accounts perspective, and then extrapolated to the rest of the world (where detailed statistical data are lacking). One of the most glaring faults in the System of National Accounts (SNA) is the failure to reflect natural resource stocks (and/or changes in stocks) in the accounts. Thus, the income generated by exploiting a natural resource (such as a mine, a forest or other environmental receptors) is not adjusted in the SNA to reflect declines in the asset values of such resources. It follows, of course, that aggregates such as GNP or GNI based on the SNA cannot be relied upon to give realistic measures of climatic impacts on natural systems. This may not introduce a large percentage error in the case of the U.S., Japan or Germany, but the error is much greater for a country that is more dependent on agriculture or forestry.[7] The problem is compounded by the use of market exchange rates (even if adjusted for so-called purchasing power parity) to compare GNP or GNI across countries. For purposes of international trade, such exchange rates are reasonable enough proxies for underlying values of traded commodities, but for purposes of computing damages to natural systems in countries deriving a large percentage of national income from such systems, they are inadequate.

For instance, Nordhaus (1989) estimated potential damage due to sea level rise (SLR) on the basis of using current land prices in Bangladesh converted to dollar equivalents based on the prevailing exchange rate. In effect, this method values loss of land in Bangladesh capable of producing three crops a year at around one sixth the value of an equal amount of land in the Netherlands producing only one crop per year. We argue that this sort of monetary aggregation is inappropriate and seriously misleading.

The pursuit of rigor is admirable, but it often leads to the inclusion of that which can be precisely quantified and the omission of that which cannot: This may lead policymakers to consider the latter category as having zero value. An alternative is to enumerate receptors to be damaged (or enhanced) by global warming, and provide a range of possible values or a qualitative assessment. A broader accounting framework (see, e.g., Repetto, 1987; Bartelmus *et al.* 1989) would help overcome the tendency to disregard environmental damages where they are especially difficult to estimate. It can readily be partitioned according to the degree of accuracy of the account categories, thus maintaining the "integrity" of the quantifiable balances. At the same time, it reveals that these balances are not really totals.

---

[7] See, e.g., Goldemberg *et al.* (1987), Chandler *et al.* (1990) and Mills *et al.* (1991).

An Expanded Research Agenda — Individual Receptors

1. Improve the basis of damage estimates by capturing more non-linearities.
2. Accelerate analyses of regional and seasonal variability in global warming.
3. Improve studies of natural adaptation to global warming.
4. Develop new methods of economic valuation of damages due to climate change, including losses due to problems in adaptation.
5. Include a broader range of receptors in damage assessments.
6. Expand national economic accounting to include natural resource stocks, including the environment.

REGIONAL ECONOMIC ANALYSIS

As noted in the previous section, differentials in the spatial distribution of impacts of the greenhouse effect may be very pronounced. The same is true of policies. A major example is the recent revision of the Clean Air Act and its implications for low-sulfur vs. high-sulfur coal producing regions, as well as regions that produce competing fuels. Policies to promote ethanol as a gasoline substitute would obviously be a boon to the Midwest and a blow to the Gulf States. Overall, even the so-called "costless" range of energy conservation, so often touted as a major tactic to combat global warming (see, e.g., Chandler *et al.*, 1990), could lead to significant unemployment in energy-producing regions. Moreover, such policies may have significant market and spillover effects on other regions (see, e.g., the study on regional shifts associated with changing energy prices by Miernyk *et al.* 1977).

The fact that some parts of a nation may gain as a result of an environmental policy while others stand to lose, or that losses may be spatially uneven within a single nation's borders, have important implications. This is not to suggest that pollution abatement policies should not be pursued, but it is a warning signal of the need to pay some attention to the full range of impacts. We do not suggest tolerating unfettered parochialism, but rather make a plea for avoiding policies based primarily on an averaging of differential impacts.

The major reason why regional impacts of global warming policy are important comes from the realities of the policy-making process in many nations. Regions that are negatively impacted can form very powerful coalitions that are able to stymie the forward progress of national policy. This has been evident in the air pollution control legislation in the U.S., either explicitly, as in delaying the 1990 Clean Air Act Amendment acid rain provisions, or implicitly, as in the 1977 Amendments, where the veil of the national interest was invoked to promote the regional self-interest of high-sulfur coal producers (see Ackerman and Hassler, 1981).

Mechanisms for overcoming regional obstacles to sound national policy need to be explored. An example is compensating regions negatively impacted. The funds could come from the taxes that many have suggested be applied to fossil fuels in order to make them reflect their full social costs. Compensation could take the form of unemployment insurance, retraining and relocation benefits for workers, and/or foregone royalty payments to property owners. Many compensatory policies can also be justified on the grounds of increased vertical equity (to be discussed below).

The above would at best be a short-run solution. For the longer term, an efficient and equitable solution would require additional features. These might begin with consistent long-run national energy and environmental policies to serve as a reference point for individuals to make investment, location, and career plan decisions. Second, individual and communities need to be made aware of personal responsibility for their actions. In the face of a severe limit on, say, coal burning in the year 2020, it would not be unreasonable, given the long lead time, to ask those affected to make the proper adjustment with minimal if any compensation. Third, there is a need for a regional planning apparatus to cope with these adjustments. This would benefit from more comprehensive analyses of non-renewable resources extraction in regional economies (Howe, 1987), setting aside a legacy for the future by investing proceeds of extraction in a sustainable manner (Mikesell, 1989), and diversifying the economy (Rose *et al.*, 1982), including opportunities for going into the renewable energy business.[8]

An Expanded Research Agenda — Regional Impacts

1. Improve the modeling capability to analyze regional and interregional impacts of global warming policy.
2. Study the role of regions in forming national energy and environmental policy, including the role of coalitions and the merits of interregional compensation.
3. Study means for minimizing the negative impacts of dislocations on regional economies.
4. Improve the planning apparatus for regions to promote sustainable economic development.

SOCIOECONOMIC DISAGGREGATION

The same criticism leveled at national level impact analyses—that they are too aggregative—can also be applied to many studies at the regional level. In democratic societies, the ultimate level at which welfare is measured is that of the individual. Matters of equity, or distributive justice, are best addressed at this level.[9] A disproportionate number of the economically disadvantaged are inhabitants of flood plains and low lying coastal areas, especially in lesser developed countries. The poor are also more vulnerable to damage to their own food production or to the rising prices that global warming induced drought would bring about. The residents of (energy) mining communities whose jobs and incomes are directly and

---

[8]There is a great deal of overlap between the issue of global warming and that of sustainable development, which refers to the continuous improvement in the standard of living without degrading the environment (see, e.g., Pearce, 1990). Sustainable development is thus the broader theme, and its usefulness for global warming policy should be explored, a case in point being lagging regions.

indirectly tied to the primary sector are likely to be hard hit by global warming policies. Given the significantly lower per capita incomes and higher poverty rates found in areas dependent on extractive industries (Elo and Beale, 1984), this will lead to even wider disparities of income in countries such as the U.S.

Similar disparities may relate to ethnic and racial minorities, in general, and because of their geographic concentration or overlap with low income groups. Recent studies have shown that these groups are especially sensitive to the direction of U.S. energy policy (see, e.g., Rose *et al.,* 1990).

Some of the leading policies to mitigate global warming may have undesirable distributional impacts. A great emphasis is placed on full-cost pricing of fossil fuels, but while this will promote an efficient allocation of resources, the market, or pricing in general, is blind to equity. For example, it is well documented that a gasoline tax in the U.S. would be regressive. It is often suggested that tax proceeds be channeled into the development of renewable resources. If equity is to be taken seriously, however, some of these proceeds might be used as compensation (e.g., energy tax credits for the disadvantaged).

Simply arriving at the conclusion that redistribution policies may be needed, however, is only a first step. There have been inherent difficulties in implementing such measures in the past, since the process usually involves coordination between two or more levels of government. That is, governmental specialization in most industrialized nations designates one agency as responsible for energy, another responsible for taxation, and another for welfare policies. One of the major themes of sustainable development is the emphasis on the interconnection of the economic, the social, and the environmental realms. But there is still a need for government to respond with appropriate institutions capable of properly implementing a comprehensive strategy.

Another use of distributional impact analysis is to provide the individual with better information on how he or she will be affected. Most impact studies or cost-benefit analyses provide rather aggregative results, typically on how the entire community, or even nation, will be affected. Expecting individuals to base their decision solely on this information imparts a degree of altruism upon the individual far beyond that which economists usually concede. Participatory democracy is only effective if the electorate is well informed.[10] Thus, distributional information is a must for the public choice process. Distributional impact models have recently been

---

[9]Pearce *et al.* (1990) emphasizes the concern for justice associated with the sustainable development theme in terms of the "socially disadvantaged" as well as between generations and with respect to nature. The Brundtland Report (1987) notes that "poverty is a major cause and effect of global environmental problems." Hence, it is desirable to analyze the distribution of policy impacts across income groups. Note, also that the version of equity discussed in this section differs from two others that have received attention: (1) intergenerational equity (see, e.g., Nordhaus, 1981; Crosson, 1989) and (2) international equity (see, e.g., d'Arge, 1989; Rose, 1990; Chapman and Drennen, 1990; Solomon and Ajuja, 1991; Burtaw and Toman, 1991).

[10]The enhancement of growth and equity through increased public participation in policy decisions is also a major theme of sustainable development.

devised and successfully applied to related resource/environmental issues (see, e.g., Rose *et al.*, 1988; 1990). Their applicability to global warming policy needs to be examined.

Again, we are not suggesting that individuals do or should make their decisions based solely on narrow personal gain or purely altruistic equity grounds. We are suggesting that there is a need for policymakers to inform them of the range of impacts. At the same time, policymakers will find it useful to gauge public attitudes, even in the context of efficiency-based incentive systems.

Distributional impacts of public policy are often neglected. Two of the main reasons include the difficulty in measuring them and their association with normative judgements. An acceleration of the development of operational distributional impact models and the positive economic implications of the disbursement of gains and losses should help overcome this neglect. In addition, the magnitude of the impacts of global warming will stimulate discussions about distributional justice.

Another concern is that attention to equity will lead to the diminution of economic efficiency, considered by most economists to be the highest priority objective. Fortunately, some policy instruments, such as tradeable carbon emission permits, allow for the separation of equity and efficiency considerations and thus avoid the equity-efficiency tradeoff. At the same time, decisions on the initial allocation of the permits makes it impossible to avoid value judgements relating to equity (see Barret *et al.*, 1992).

An Expanded Research Agenda — Socioeconomic Impacts

1. Improve modeling capabilities to analyze the impacts of global warming policy across income and racial/ethnic groups.
2. Analyze the impact of mitigation, damage and adaptation across socioeconomic groups.
3. Identify ways to minimize the regressivity of these impacts through redistribution and other means.
4. Explore how the results of distributional analysis can be used to inform citizens more effectively about the global warming issue.
5. Measure public attitudes in response to individual and aggregate impacts.

CONCLUSION

The several assessments of previous research on the economics of global warming and the establishment of research priorities to address it represent a good start, but practically ignore several key research areas associated with the issue. We have identified four areas that would provide a higher resolution (finer level of detail) necessary for policymaking. At the same time, it should be emphasized that we do not contend that they are the only four areas worth adding to the priority list.

Some common themes run across these technological, natural receptor, regional, and socioeconomic considerations. They pose some relatively difficult conceptual problems, are often hard to measure empirically, and are less amenable to the tools favored by most of the economics profession. Some or all of these considerations may help explain why they have been neglected thus far. Still, it is our hope that in setting future research priorities, both funding agencies and economists will realize the importance of requiring higher resolution in devising the best possible response to the global warming issue.

REFERENCES

Ackerman, B. and W. Hassler, 1981. *Clean Coal/Dirty Air.* New Haven: Yale University Press.

Ahmad, J., S. El Serafy and E. Lutz (eds.) 1990. *Environmental Accounting for Sustainable Development.* Washington, DC: World Bank.

Ayres, R.U. 1978. *Resources, Environment and Economics.* New York: Wiley.

Ayres, R.U. 1990. "Energy Conservation in the Industrial Sector," in J. Tester *et al.* (eds.), *Energy and the Environment in the 21st Century.* Cambridge, MA: MIT Press.

Ayres, R.U. 1991. "Eco-Restructuring: Managing the Transition to an Ecologically Sustainable Economy," paper presented at the American Association for the Advancement of Science Meetings, Washington, DC.

Barrett, S., M. Grubb, K. Roland, A. Rose, Richard Sandor and T. Tietenberg. 1992. *A Global System of Combating Global Warming for Tradeable Carbon Emission Entitlements.* Geneva: UNCTAD.

Bartlemus, P., Stahmer and J. Van Tongeren. 1989. "SNA Framework for Integrating Environmental and Economic Accounting," International Association for the Review of Income and Wealth 21st General Conference.

Batie, S. and H. Shugart, 1990. "The Biological Consequences of Climate Changes; An Ecological and Economic Assessment," in N. Rosenberg *et al.* (eds.) *Greenhouse Warming: Abatement and Adaptation.* Washington, DC: Resources for the Future.

(Brundtland Report). 1987. World Commission on Environment and Development. *Our Common Future.* New York: Oxford University Press.

Burtraw, D. and M. Toman, 1991. "Equity and International Agreements for CO_2 Containment." Discussion Paper ENR 91-07. Washington, DC: Resources for the Future.

Chandler, W. (ed.). 1990. *Carbon Emissions Control Strategies.* Washington, DC: World Wildlife Fund and The Conservation Foundation.

Chapman, D. and T. Drennan. 1989. "Economic Dimensions of CO_2 Treaty Proposals." *Cornell Agricultural Economics Staff Paper* No. 89-26.

Crosson, P. 1989. "Climate Change: Problems of Limits and Policy Responses." in N. Rosenberg *et al.* (eds.). *Greenhouse Warming: Abatement and Adaptation.* Washington, DC: Resources for the Future.

d'Arge, R. 1989. "Ethical and Economic Systems for Managing the Global Commons," in D. Botkin *et al.* (eds.), *Changing the World Environment.* New York: Academic Press.

Darmstadter, J. 1991. "The Economic Cost of CO_2 Mitigation: A Review of Estimates for Selected World Regions," RFF Discussion Paper, Washington, D.C.: Resources for the Future.

Duchin, F. 1990. "Evaluating Strategies for Environmentally Sound Economic Development: An Input-Output Approach," New York: Institute for Economic Analysis.

Duchin, F. 1991. "Industrial Input-Output Analysis: Implications for Industrial Ecology," Prepared for the U.S. National Academy of Sciences Colloquium on Industrial Ecology.

Duchin, F. 1991. "Prospects for Environmentally Sound Economic Development in the North, in the South, and in North-South Economic Relations: the Role for Action-Oriented Analysis," Prepared for the United Nations Conference on Environment and Development.

Economic Report of the President. 1990. Washington, DC: USGPO.

Elo, I.T. and C.L. Beale. 1984. "Natural Resources and Rural Poverty," Rural Development, Poverty and Natural Resources Workshop Paper Series. Washington, DC: Resources for the Future.

Emanuel, W.R., H.H. Shugard, and M.L. Stevenson. 1985: "Climate Change and the Broad-Scale Distribution of Terrestrial Ecosystem Complexes," *Climatic Change* 7:29-43.

Frederick, K. and P. Gleick, 1989. "Water Resources and Climate Change," in N. Rosenberg *et al.* (eds.), *Greenhouse Warming: Abatement and Adaptation.* Washington, DC: Resources for the Future.

Goldemberg, J. *et al.* 1987a. *Energy for Development.* Washington, DC: World Resources Institute.

Goldemberg, J. *et al.* 1987b. *Energy for a Sustainable World.* Washington, DC: World Resources Institute.

Grubb, M. 1989. *The Greenhouse Effect: Negotiating Targets.* London: The Royal Institute of International Affairs.

Hall, D. 1990. "Preliminary Estimates of Cumulative Private and External Costs of Energy," *Contemporary Policy Issues* 8: 283-307.

Heuting, R. 1980. *New Scarcity and Economics Growth: More Welfare through Less Production?* Amsterdam: North-Holland.

Hogan, W. and D. Jorgenson, 1991. "Productivity Trends and the Cost of Reducing CO_2 Emissions," *Energy Journal* 12: 67-85.

Howe, C.W. 1987. "On the Theory of Optimal Regional Development Based on an Exhaustible Resource," *Growth and Change* 18: 53-68.

Lashof, D. and D. Ahuja, 1990. "Relative Global Warming Potentials of Greenhouse Gas Emissions." *Nature.*

Leemans, R. and W. Cramer, 1991. "The IIASA Database for Mean Monthly Values of Temperature, Precipitation, and Cloudiness on a Global Terrestrial Grid," IIASA Research Report (forthcoming).

Lovins, A. *et al.* 1981. *Least-Cost Energy: Solving the CO_2 Problem*. Andover, MA: Brickhouse.

Manne, A. and R. Richels, 1990. "CO_2 Emission Limits: An Economic Cost Analysis for the USA," *Energy Journal* 11: 51-74.

Manne, A. and R. Richels. 1991. "Global CO_2 Emission Reductions - the Impacts of Rising Energy Costs," *Energy Journal* 12: 87-107.

Miernyk, W., F. Giarratani and K. Socher. 1977. *The Regional Impacts of Rising Energy Prices.* Cambridge, MA: Ballinger.

Mikesell, R. 1989. "Depletable Resources, Discounting and Intergenerational Equity," *Resources Policy* 15: 292-296.

Miller, J. 1989. "Chinese Bring Chill to Backers of Ozone Protocol," *New Scientist* 11: 28.

Mills, E., D. Wilson and B. Johansson. 1991. "Getting Started: No Regrets Strategies for Reducing Greenhouse Gas Emission," *Energy Policy* (forthcoming).

(NOAA) National Oceanographic and Atmospheric Administration. 1991. *Economics and Global Change.* Washington, DC.

Nordhaus, W. 1982. "How Fast Should We Graze the Global Commons?" *American Economic Review* 72: 242-246.

Nordhaus, W. 1989. "The Economics of the Greenhouse Effect," paper presented at the International Energy Workshop. Laxenburg, Austria: International Institute for Applied Systems Analysis.

Nordhaus, W. 1990. "Economic Policy in the Face of Global Warming," in J. Tester, D. Woods, and N. Ferrari (eds.), *Energy and the Environmental in the 21st Century.* Cambridge, MA: MIT Press.

Nordhaus, W. 1991. "The Cost of Slowing Climate Change: a Survey," *Energy Journal* 12: 37-65.

Nordhaus, W. and J. Tobin. 1972. *"Is Growth Obsolete?" Economic Growth.* New York: National Bureau of Economic Research, General Series 96.

Pearce, D., E. Barbier, and A. Markandya. 1990. *Sustainable Development.* Aldershot, UK: Edward Elgar.

Peskin, H. 1972. "National Accounting and the Environment," Artikler Fra Statistik Seulvalburos (50), Oslo, Norway.

Peskin, H. 1989. "A Proposed Environmental Accounts Framework," in Y. Ahwad, S. El Savaty and E. Lutz (eds.). *Environmental Accounting for Sustainable Development.* Washington, DC: World Bank.

Peskin, H. and E. Lutz. 1990. "A Survey of Resource and Environmental Accounting in Industrialized Countries," Environmental Working Paper 37. Washington, DC: World Bank.

Repetto, R. 1985. "Natural Resource Accounting in a Resource-Based Economy: An Indonesian Case Study," paper presented at the Third Environmental Accounting Workshop. Paris: UNEP and World Bank.

Rose, A. 1990. "Reducing Conflict in Global Warming Policy: Equity as a Unifying Principle," *Energy Policy* 18: 927-935.

Rose, A. and C.Y. Chen. 1991. "Sources of Change in Energy Use in the U.S. Economy: A Structural Decomposition Analysis," *Resources and Energy* 13:1-21.

Rose, A., D. Hurd and M. Sheehan. 1982. "Non-Renewable Resources and the Development of Arid Lands: A Planning Approach," in *Alternative Strategies for Desert Development and Management*. Vol. 4, UNITAR, Elmsford, NY: Pergamon Press.

Rose, A., B. Stevens, and G. Davis. 1988. *Natural Resource Policy and Income Distribution*. Baltimore: Johns Hopkins University Press.

Rose, A., S. Seninger, and G. Davis. 1990. "Minority Employment in the Energy Sectors of the U.S. Economy: Projections for the 1990s," in *Proceedings of the Socioeconomic Energy Research and Analysis Conference*. Washington, DC: U.S. Department of Energy.

Schneider, S. 1989. "The Changing Climate," *Scientific American* 216: 38-47.

Schneider, S. and N. Rosenberg. 1989. "The Greenhouse Effect: Its Causes, Possible Impacts and Associated Uncertainties," in N. Rosenberg et al. (eds.) *Greenhouse Warming: Abatement and Adaptation*. Washington, DC: Resources for the Future.

Sedjo, R. and M. Solomon. 1989. "Climate and Forests," in N. Rosenberg et al. (eds.), *Greenhouse Warming: Abatement and Adaptation*. Washington, DC.

Solomon, B. and D. Ahuja. 1991. "An Equitable Approach to International Reductions of Greenhouse Gas Emissions," Global Climate Change Program, US EPA.

Solomon, M. and D.C. West. 1985. "Potential Responses of Forests to Carbon Dioxide Induced Climate Change," in *Characterization of Information Requirements for Studies of Carbon Dioxide Effects*. U.S. Department of Energy, Washington, DC.

Tester, J., D. Wood and N. Ferrari (eds.), 1990. *Energy and the Environmental in the 21st Century*. Cambridge, MA: MIT Press.

Wood, D. and Y. Kaya (eds.) 1990. *Proceedings of the Workshop on Economic/Energy/Environmental Modeling for Climate Policy Analysis*. Washington, DC, October 22-23, 1990, Cambridge, MA: MIT Center for Energy Policy Research (and Tokyo: University of Tokyo Global Environmental Study Laboratory).

Global Climate Change: Implications, Challenges and Mitigation Measures. Edited by S.K. Majumdar, L.S. Kalkstein, B. Yarnal, E.W. Miller, and L.M. Rosenfeld. © 1992, The Pennsylvania Academy of Science.

Chapter Twenty-Four

PREDICTED EFFECTS OF CLIMATE CHANGE ON AGRICULTURE:
A Comparison of Temperate and Tropical Regions

CYNTHIA ROSENZWEIG[1] and DIANA LIVERMAN[2]

[1]GISS
2880 Broadway
New York, NY 10025
and
[2]The Pennsylvania State University
Department of Geography
University Park, PA 16802

INTRODUCTION

Agriculture is the basic activity by which humans live and survive on the earth. Assessing the impacts of climate change on agriculture is a vital task. In both developed and developing countries, the influence of climate on crops and livestock persists despite irrigation, improved plant and animal hybrids and the growing use of chemical fertilizers. The continued dependence of agricultural production on light, heat, water and other climatic factors, the dependence of much of the world's population on agricultural activities, and the significant magnitude and rapid rates of possible climate changes all combine to create the need for a comprehensive consideration of the potential impacts of climate on global agriculture.

This chapter compares the effects of climate change on agriculture in temperate and tropical regions. These regions differ significantly in their biophysical characteristics of climate and soil, and in the vulnerability of their agricultural systems and people to climate change. The focus is on the impacts of the climate changes projected to occur as a result of higher levels of carbon dioxide in the atmosphere and associated global warming. The projected climate changes for the temperate and tropical areas differ in that climate models project larger temperature increases in temperate regions than in tropical regions. The projections of changes in the hydrological cycle are more similar but rather uncertain, showing a mixed picture of regional precipitation increases and decreases in both areas.

COMPARISON OF TEMPERATE AND TROPICAL AGRICULTURE

The tropics are defined as the geographical area lying between 23.5°N and 23.5°S latitude, while the temperate regions are found above these parallels. Climatologically, the tropics are characterized by high year-round temperatures and weather is controlled by equatorial and tropical air masses. Tropical precipitation is primarily convective. In the more humid tropical regions, annual rainfall is often above 2000 mm and falls in almost all months of the year. In the drier tropics, rainfall can fall below 50 mm, and be very seasonal. The remainder of the region lies between these precipitation regimes, with distinct wet and dry seasons. Agriculture is frequently limited by the seasonality and magnitude of moisture availability.

In the mid-latitude temperate zone, weather is controlled by both tropical and polar air masses. Precipitation here occurs along fronts within cyclonic storms. The temperate region also has many different climate regions with warmer and cooler temperatures and seasonal rainfall. Temperate agriculture is often characterized as predominantly limited by seasonally cooler temperatures.

Reported experiments have shown that even though yield per day is often higher in the tropics, total crop growth season is shorter (Haws *et al.*, 1983). Leaf area expansion and phasic development are faster in the tropics because of higher temperatures during vegetative growth. Nevertheless, crop yields are consistently found to be higher in temperate regions than in the tropics (see Table 1) (FAO, 1990). Numerous factors contribute to this result. Soils in the humid tropics tend to be highly leached of nutrients and are therefore unproductive because of high temperatures, intense rainfall, and erosion. Soils in the drier tropics are often hampered by accumulations of salt and lack of water (Barrow 1987). Temperate soils are generally viewed as more favorable to agriculture than tropical soils because of higher nutrient levels. However, there are exceptions in both regions, with highly productive volcanic and fluvial soils found in the tropics, and poorly developed and infertile soils in temperate regions.

Agricultural production is also severely limited in many humid tropical regions by the wide range of weeds, pests, and diseases that flourish in consistently warm and moist climates. The growth of some crops and varieties, which require long

TABLE 1
Average Yields in Temperate and Tropical Areas

CROP	TEMPERATE (kg/ha)	TROPICAL (kg/ha)	%
Rice	4109	1958	48
Wheat	2984	1363	46
Maize	3993	1351	34
Sorghum	2270	1249	55
Soybean	1620	1038	64
Beans (dry)	1079	640	59
Groundnut	1667	1036	62
Potato	18056	8704	48
Sweet potato	13594	6881	51
Cassava	11844	9103	77
Sugarcane	61190	53328	87

Source: Haws *et al.*, 1983

hours of daylight to reach maturity, is also limited by the invariable day lengths of the tropics. Solar radiation, which is critical to plant growth, and whose intensity is controlled by the angle of the sun, daylength, and cloudiness, is lower in winter and higher in summer in temperate zones. In the tropics, solar radiation is often limited by cloudiness during the rainy seasons.

Agricultural crops and cropping systems have been developed for, and adapted to, these varied regimes of climate, soil, diseases and pests (Haws *et al.*, 1983). The main commercial agricultural crops and their adaptations include:

(a) Cassava and sugarcane, which only grow in tropical areas and have a crop duration of one year or longer. Cassava is drought resistant, but sugarcane requires irrigation in dry areas.
(b) Sorghum, groundnut, and sweet potato, which grow in both tropical and subtropical regions in relatively dry seasons.
(c) Rice, which is mainly grown in tropical and subtropical zones in the rainy season or with irrigation.
(d) Maize and field beans grow in both zones, preferably with seasons with enough rain.
(e) Wheat, soybean, and potato are crops of the subtropical and temperate zones and grow in the tropics at high (cooler) elevations.
(f) Sugarbeet is grown only in the temperate zone.

A number of "luxury" agricultural crops, especially fruit (bananas, pineapples), stimulants (coffee, tea) and spices grow only, or best, in the tropics. Tropical regions are also important in providing winter season produce for temperate zones.

In temperate agriculture, plant breeding and fertilizer use produced dramatic yield increases for many crops early in the twentieth century. Similar increases occurred more recently in tropical regions for crops such as wheat, maize and rice, which benefited from the technological package of improved seeds, fertilizer, mechanization and pesticides known as the Green Revolution.

In both temperate and tropical regions, irrigation has been developed in areas where dry seasons exist and adequate water can be reserved from other seasons or brought in from adjacent regions. Irrigation is an important buffer against climate variability and climate change. About 20% of the world's cropland is irrigated, mostly in Asia, producing about 40% of the annual crop production.

Differences in farming systems, technology and economics also contribute to the yield differences in temperate and tropical regions. Agriculture in temperate regions is characterized by high levels of inputs (quality seed stock, fertilizer, herbicides and pesticides), and a high degree of mechanization and capitalization. However, there are wide variations in the use of technology, European agriculture being particularly intensive. In tropical regions, many farmers cannot afford inputs, and governments cannot afford to subsidize them. In some parts of the tropics, traditional technologies, such as multiple cropping and terracing, act to buffer the system against climate variability, conserve soil fertility, and increase yields.

In some senses, the tropics are more dependent on agriculture, and therefore more vulnerable to climatic change, than the temperate regions. As much as 75% of the world's population live in the tropics, and two thirds of these people are reliant on agriculture for their livelihoods. With low levels of technology, land degradation, unequal land distribution, and rapid population growth, many tropical regions are near or exceeding their capacity to feed themselves (Table 2).

Indeed, some authors have argued that the unequal social structures and international position of many tropical countries also increase vulnerability to climate change. Unequal land tenure, high numbers of landless rural dwellers, low incomes and high national debts exacerbate the negative impacts of climate variability, as some people have no extra land, job, savings or government assistance to see them through droughts or other climatic extremes (Jodha 1989). When the economic system is oriented towards export rather than subsistence agriculture, climatic change (as well as low export prices) may threaten the whole national economy and food system. Those regions that cannot feed their populations depend on cereal imports from the major cereal exporters such as the USA, France, Canada,

TABLE 2

The Capacity of Lands in the Developing World to Support Their Populations At Different Levels of Technology

1975 Population Levels, Actual Land Area, 1/3 in Non-Food Crops

	LOW TECHNOLOGY	HIGH TECHNOLOGY
AFRICA	0.4	4.5
ASIA	0.7	3.1
SW ASIA	0.8	1.9
SOUTH AMERICA	0.6	5.8
CENTRAL AMERICA	0.5	3.7

Number is the number of times the region could support its 1975 population e.g. 2 = 2 times population.

Source: FAO, 1984.

Australia, Argentina, and Thailand (FAO 1990). All except Thailand would be defined as temperate agricultural producers.

THE INFLUENCE OF CLIMATE CHANGE ON CROP PRODUCTION

At the basis of any understanding of climate impacts on agriculture lies the biophysical sciences. The rates of most biophysical processes are highly dependent on climate variables such as radiation, temperature, and moisture, that vary regionally. For example, rates of plant photosynthesis depend on the amount of photosynthetically active radiation and levels of atmospheric carbon dioxide (CO_2). Temperature is an important determinant of the rate at which a plant progresses through various phenological stages towards maturity. The accumulation of biomass is constrained by the availability of moisture and nutrients to a growing plant.

Numerous studies have examined the impacts of past climatic variations on agriculture using case studies, statistical analyses and simulation models (e.g. Nix 1985; Parry 1978; Thompson 1975; World Meteorological Organization 1979). Such studies have clearly demonstrated the sensitivity of both temperate and tropical agricultural systems and nations to climatic variations and changes. In the temperate regions, the impacts of climate variability, particularly drought, on yields of grains in North America and the Soviet Union have been of particular concern because of their effects on world food security. In the tropics, drought impacts on agriculture and resulting food shortages have been widely studied, especially when associated with the failure of the monsoon in Asia or the rains in Sudano-Sahelian Africa. In the temperate regions, climatic variations are associated with economic disruptions; in the tropics, droughts bring famine and widespread social unrest (Pierce 1990).

THE BIOPHYSICAL IMPACT OF CLIMATIC CHANGES ASSOCIATED WITH GLOBAL WARMING

It is frequently assumed that global change will bring higher temperatures, altered precipitation, and higher levels of atmospheric CO_2 (IPCC 1990a). What might these changes mean for the biophysical response of agricultural crops?

Interactions with thermal regimes. Higher temperatures in general hasten plant maturity in annual species, thus shortening the growth stages during which pods, seeds, grains or bolls can absorb photosynthetic products. This is one reason yield are lower in the tropics. Because crop yield depends on both the rate of carbohydrate accumulation and the duration of the filling periods, the economic yields of both temperate and tropical crops grown in a warmer and CO_2-enriched environment may not rise substantially above present levels, despite increases in net photosynthesis (Rose 1989).

Because temperature and tropical regions differ in both current temperature and the temperature rise predicted for climate change, the relative magnitudes of combined CO_2 and temperature effects will likely be different in the different regions. In the mid-latitudes, higher temperatures may shift biological process rates toward optima, and beneficial effects are likely to ensue. Increase in temperature will also lengthen the frost-free season in temperate regions, allowing for longer duration crop varieties to be grown and offering the possibility of growing successive crops (moisture conditions permitting). In tropical locations where increased temperatures may move beyond optima, negative consequences may dominate.

Both the mean and extreme temperatures that crops experience during the growing season will change in both temperate and tropical areas. Extreme temperatures are important because many crops have critical thresholds both above and below which crops are damaged. Prolonged hot spells can be especially damaging (Mearns *et al.*, 1984). Critical stages for high temperature injury include seedling emergence in most crops, silking and tasseling in corn (Shaw, 1983), grain filling in wheat (Johnson and Kanemasu, 1983), and flowering in soybeans (Mederski, 1983). In general, higher temperatures should decrease cold damage and increase heat damage. Agro-climatic zones are expected to shift poleward as lengthening and warming growing seasons allow new or enhanced crop production (soil resources permitting) (Rosenzweig, 1985).

Changes in hydrological regimes. The hydrological regimes in which crops grow will surely change with global warming. While all GCMs predict increases in mean global precipitation (because a warmer atmosphere can hold more water vapor), decreases are forecast in some regions and increases are not uniformly distributed. The crop water regime may further be affected by changes in seasonal precipitation, within-season pattern of precipitation, and interannual variation of precipitation. Increased convective rainfall is predicted to occur, particularly in the tropics, caused by stronger convection cells and more moisture in the air.

Too much precipitation can cause disease infestation in crops, while too little can be detrimental to crop yields, especially if dry periods occur during critical development stages. For example, moisture stress during the flowering, pollination, and grain-filling stages is especially harmful to maize, soybean, wheat and sorghum (Decker *et al.*, 1986).

The amount and availability of water stored in the soil, a crucial input to crop growth, will be affected by changes in both the precipitation and seasonal and annual evapotranspiration regimes. Some GCMs predict mid-continental drying the Northern Hemisphere (Manabe and Wetherald, 1986; Kellogg and Zhao, 1988) and other GCM predictions have been interpreted to suggest that the rise in potential evapotranspiration will exceed that of rainfall resulting in drier regimes throughout the tropics and low to mid-latitudes (Rind *et al.*, 1990). Because the soil moisture processes are represented so crudely in the current GCMs, however, it is difficult to associate much certainty with these projections (IPCC, 1990a).

Global climate change is likely to exacerbate the demand for irrigation water (Adams *et al.*, 1990). Higher temperatures, increased evaporation, and yield decreases contribute to this projection. However, supply of needed irrigation water under climate change in uncertain. Where water supplies are diminishing, such as the Ogallala Aquifer in the United States, extra demand might require that some land be withdrawn from irrigation (Rosenzweig, 1990).

Physiological effects of CO_2. The study of agricultural impacts of trace gas induced climate change is complicated by the fact that increasing atmospheric CO_2 has other effects on crop plants besides its alteration of their climate regime. These are often called "fertilizing" effects, because of their perceived beneficial physiological nature. Specifically, most plants growing in enhanced CO_2 exhibit increased rates of net photosynthesis. The higher photosynthesis rates are then manifested in higher leaf area, dry matter production, and yield for many crops (Kimball, 1983; Acock and Allen, 1985; Cure, 1985). In several cases, high CO_2 has contributed to upward shifts in temperature optima for photosynthesis (Jurik *et al.*, 1984) and to enhanced growth with higher temperatures (Idso *et al.*, 1987); other studies, however, have not shown such benefits (Jones *et al.*, 1985; Baker *et al.*, 1989).

CO_2 enrichment also tends to close plant stomates, and by doing so, reduces transpiration per unit leaf area while still enhancing photosynthesis. The stomatal conductances of 18 agricultural species have been observed to decrease markedly (by 36%, on average) in an atmosphere enriched by doubled CO_2 (Morison and Gifford, 1984). However, crop transpiration per ground area may not be reduced commensurately, because decreases in individual leaf conductance tend to be offset by increases in crop leaf area (Allen *et al.*, 1985). In any case, higher CO_2 often improves water-use efficiency, defined as the ratio between crop biomass accumulation or yield and the amount of water used in evapotranspiration. Increases in photosynthesis and resistance with higher CO_2 have been shown to occur at less than optimal levels of other environmental variables, such as light, water, and some of the mineral nutrients (Acock and Allen, 1985).

Temperate crops may benefit more from increasing CO_2 than tropical crops. In crop species with the C3 pathway characteristic of non-tropical plants (e.g., wheat, soybean, cotton) CO_2 enrichment has been shown to decrease photorespiration, the rapid oxidation of recently formed sugars in the light, a process which lowers the efficiency of overall photosynthesis. C4 crops, which are particularly characteristic of tropical and warm arid regions (e.g., maize, sorghum, and millet), are more efficient photosynthetically under current CO_2 levels than C3 plants (because they fix CO_2 into malate in their mesophyll cells before delivering it to the RuBP enzyme in the bundle-sheath cells). Because of this CO_2-concentrating and photorespiration-avoiding mechanism, experimental data show that C4 plants are less responsive to CO_2 enrichment (Acock and Allen, 1985).

The physiological effects of high levels of atmospheric CO_2 described above have been observed under controlled experimental conditions. In the open field, however, their magnitude and significance are still largely untested, and their importance relative to the predicted large-scale climatic effects uncertain. Greenhouse and field-chamber environments tend to be much smaller, less variable, and more protected from wind than field conditions. Furthermore, physiological feedback mechanisms such as starch accumulation or lack of sink (that is, growing, storing, or metabolizing tissue) for the products of photosynthesis may limit the extent to which the "fertilizing" CO_2 effects may be realized. Finally, if trace gas emissions continue to grow unchecked, their climate warming effect is projected to continue even up to 2000 ppm (Manabe and Bryan, 1985), but the beneficial boost to photosynthesis appears to level off at about 400 ppm for C4 crops and about 800 ppm for C3 crops (Akita and Moss, 1973).

Soils. Climate change will also have an impact on the soil, a vital element in agricultural ecosystems. Higher air temperatures will cause higher soil temperatures, which should generally increase solution chemical reaction rates and diffusion-controlled reactions (Buol *et al.,* 1990). Solubilities of solid and gaseous components may either increase or decrease, but the consequences of these changes may take many years to become significant (Buol *et al.,* 1990). Furthermore, higher temperatures will accelerate the decay of soil organic matter, resulting in release of CO_2 to the atmosphere and decrease in carbon/nitrogen ratios, although these two effects should be offset somewhat by the greater root biomass and crop residues resulting from plant responses to higher CO_2.

In temperate countries where crops are already heavily fertilized, there will probably be no major changes in fertilization practices, but alterations in timing and method (e.g., careful adjustment of side-dress applications of nitrogen during vegetative crop growth) are expected with changes in temperature and precipitation regimes (Buol *et al.,* 1990). In tropical countries, where fertilization level is not always adequate, the need for fertilization will probably increase.

Sea level rise, another predicted effect of global warming, will caused increased flooding, salt-water intrusion, and rising water tables in agricultural soils located near coastlines. This is particularly crucial in tropical countries such as Bangladesh, with large agricultural regions and high rural population located near current sea level.

Pests. Pests are organisms that affect agricultural plants and animals in ways considered unfavorable. They include weeds, and certain insects, arthropods, nematodes, bacteria, fungi, and viruses. Because climate variables (especially temperature, wind and humidity) control the geographic distribution of pests, climate change is likely to alter their ranges. Insects may extend their ranges where warmer winter temperatures allow their overwintering survival and increase the possible number of generations per season (Stinner, *et al.,* 1989). Pests and diseases from low latitude regions, where they are much more prevalent (see Table 3) may be introduced at higher latitudes. As a consequence of pest increase, there may be a substantial rise

TABLE 3
Crop disease in temperate and tropical regions

| | Number of diseases reported | |
Crop	Temperate	Tropical
Rice	54	500-600
Maize	85	125
Citrus	50	248
Tomato	32	278
Beans	52	250-280

(Source: Swaminathan, 1986)

in the use of agricultural chemicals in both temperate and tropical regions to control them.

SPECIFIC STUDIES IN TEMPERATE AND TROPICAL REGIONS

Very few integrated regional studies have been completed on climate change impacts on agriculture. Parry *et al.* (1988a) report on integrated agricultural sector studies in high-latitude regions in Canada, Iceland, Finland, USSR, and Japan, concluding that warmer temperatures may aid crop production by lengthening the growing season, but that potential for higher evapotranspiration and drought conditions may be detrimental.

Adams *et al.* (1990) conducted an integrated study for the US linking models from atmospheric science, plant science, and agricultural economics. While the outcomes depend on the severity of climate change and the compensating effects of carbon dioxide on crop yields, the simulations suggest that irrigated acreage will expand and that regional patterns of US agriculture will shift with predicted global warming. With the more severe climate change scenario tested, the movement of US production into export markets is substantially reduced.

The Missouri, Iowa, Nebraska, and Kansas (MINK) study integrated both within the agricultural sectors and across other sectors (Rosenberg and Crosson, 1990). The study incorporated both the physiological effects of CO_2 and adaptation by farmers to the climatic conditions of the 1930s. Even with the relatively mild warming (1.1°C) of the 1930s and with farmer adaptation and CO_2 physiological effects taken into account, regional production declined by 3.3%. Given the IPCC (1990a) estimate of 2.5°C warming for doubled CO_2 conditions, these results imply agricultural losses of about 10% for equilibrium warming of doubling of carbon-dioxide-equivalent (Cline, 1991).

The second volume of Parry *et al.* (1988b) includes studies of the impact of climate changes on agriculture in Kenya, Brazil, Ecuador, India, and Australia. These case studies used the impacts of past climatic variations, rather than projections of future climate, to provide insights into the sensitivity of agriculture to climate change.

Liverman (1991) and Liverman and O'Brien (1991) have described how global warming may affect Mexican agriculture, using GCM output to project declines in moisture availability and maize yields at several sites in Mexico.

A number of conferences and publications have recently raised concern about the possible impacts of global warming in tropical regions, but they have not included explicit analyses of how global warming may affect tropical agriculture (Fundacion Universo Veintinuo 1990; HARC 1991; Suliman 1990; Universidad de Sao Paulo 1990).

GLOBAL ESTIMATES OF AGRICULTURAL IMPACTS

Global estimates of agricultural impacts have been fairly rough to date, because of lack of consistent methodology and uncertainty about the physiological effects of CO_2. General studies of how climate change might affect agriculture include those of the National Defense University (1983), Liverman (1986), and Warrick (1988). Kane *et al.* (1989) broadly predicted improvements in agricultural production at high latitudes and reductions in northern hemisphere mid-continental agricultural regions. The IPCC (1990b) concluded that while future food production should be maintained, negative impacts were likely in some regions, particularly where present-day vulnerability is high.

An international project of the US Environmental Protection Agency (EPA), "Implications of Climate Change for International Agriculture: Global Food Trade and Vulnerable Regions," has been established to estimate the potential effects of greenhouse gas-induced climate change on global food trade, focusing on the distribution and quantity of production of the major food crops for a consistent set of climate change scenarios and CO_2 physiological effects. Other goals of the project are to determine how currently vulnerable, food-deficit regions may be affected by global climate change; to identify the future locations of those regions and the magnitudes of their food-deficits; and to study the effectiveness of adaptive responses, including the use of genetic resources, to global climate change.

As part of the EPA project, crop specialists are estimating yield changes at over 100 sites in over 20 countries (see Figure 1), under common climate change scenarios using compatible crop growth models. The focus is on staple food crops: wheat, rice, maize, and soybeans. The crop models are those developed by the International Benchmark Sites Network for Agrotechnology Transfer (IBSNAT, 1990) — a global network of crop modelers funded by the U.S. Agency for

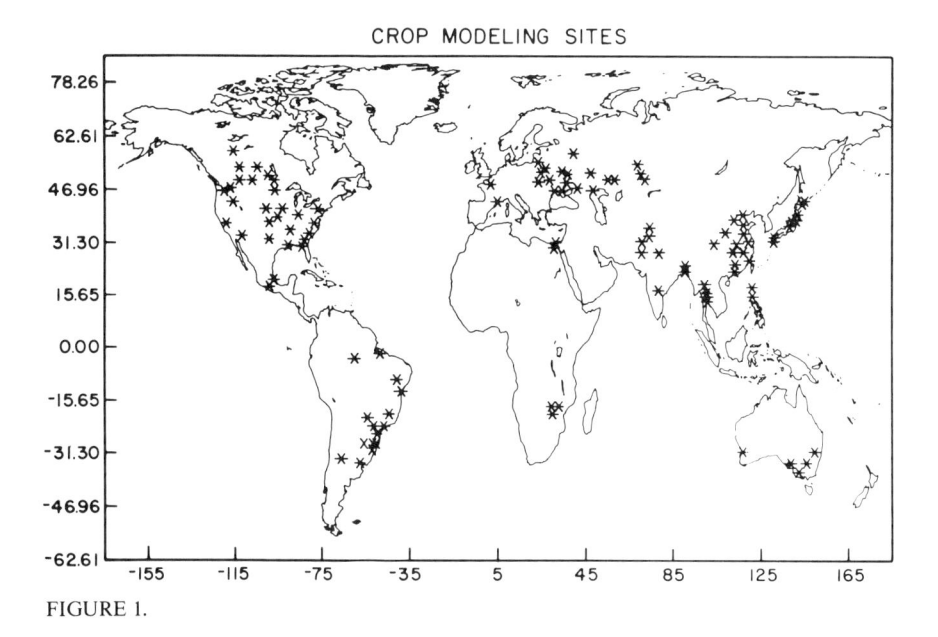

FIGURE 1.

International Development. The choice of the IBSNAT crop models was based on several criteria. First, the models simulate crop response to the major climate variables of temperature, precipitation, and solar radiation, and include the effects of soil characteristics on water availability for crop growth. Second, the models have been validated for a range of soil and climate conditions. Third, the models are developed with compatible data structures so that the same soil and climate data bases could be used with all crops.

Preliminary national production changes for wheat based on IBSNAT crop model results are shown in Table 4 (Rosenzweig and Iglesias, *et al.*, 1992). Results from individual sites have been aggregated according to rainfed and irrigated practice and contribution to regional and national production. The table shows national production changes for the three climate change scenarios with (555 ppm) and without (330 ppm) the physiological effects of CO_2 on crop growth.

In general, these results show that the climate change scenarios without the physiological effects of CO_2 cause decreases in estimated national production, while the physiological effects of CO_2 mitigate the negative effects. Production declines

TABLE 4
2xCO$_2$ GCM Scenario Production Changes

WHEAT SCENARIO

COUNTRY/	Yield T/Ha	Area Ha x 1000	Production T x 1000	GISS 330 T/Ha %	GISS 555 T/Ha %	GFDL 330 T/Ha %	GFDL 555 T/Ha %	UKMO 330 T/Ha %	UKMO 555 T/Ha %
AUSTRALIA	1.38	11,546	15,574	-18	8	-16	11	-14	9
BRAZIL	1.31	2,788	3,625	-51	-33	-38	-17	-53	-34
CANADA	1.88	11,365	21,412	-12	27	-10	27	-38	-7
CHINA	2.53	29,092	73,527	-5	16	-12	8	-17	0
EGYPT	3.79	572	2,166	-36	-31	-28	-26	-54	-51
FRANCE	5.93	4,636	27,485	-12	4	-28	-15	-23	-9
INDIA	1.74	22,876	39,703	-32	3	-38	-9	-56	-33
JAPAN	3.25	237	772	-18	-1	-21	-5	-40	-27
PAKISTAN	1.73	7,478	12,918	-57	-19	-29	31	-73	-55
URUGUAY	2.15	91	195	-41	-23	-48	-31	-50	-35
USSR ww	2.46	18,988	46,959	-3	29	-17	9	-22	0
USSR sw	1.14	36,647	41,959	-12	21	-25	3	-48	-25
USA	2.72	26,595	64,390	-21	-2	-23	-2	-33	-14
WORLD	2.09	231*	482*	-16	11	-22	4	-33	-13

C. Rosenzweig, *et al.*; World weighted by country production. *x1,000,000.

U.S. Environmental Protection Agency, Global Climate Change Division. *Climate Change and International Agriculture: Crop Modeling Study* (in preparation).

occur in many locations, however, even with the compensating CO_2 effects. Production changes tend to be less negative and even positive in some cases in countries in mid and high latitudes, while simulations in countries in the low latitudes indicate more detrimental effects of climate change on agricultural production. The UKMO climate change scenario (mean global warming of 5.2°C) generally causes the largest production declines, while the GFDL and GISS (4.0 and 4.2°C mean global warming, respectively) production changes are more moderate.

When embedded in a global agricultural food trade model, the Basic Linked System (Fischer *et al.,* 1988), the production change estimates based on IBSNAT crop model results will allow for projection of potential impacts on food prices, shifts in comparative advantage, and altered patterns of global trade flows for a suite of global climate change, population, growth, and policy scenarios.

ADAPTATION TO CLIMATE CHANGE

The importance of farm and state level adaptations to climate change and variability has been demonstrated in a number of studies (e.g. Rosenberg *et al* 1989; Waggoner 1983; White 1974) Adaptations to climate change exist at the various levels of agricultural organization. In temperate regions, farm-level adaptations include changes in planting and harvest dates, tillage and rotation practices, substitution of crop varieties or species more appropriate to the changing climate regime, increased fertilizer or pesticide applications, and improved irrigation and drainage systems. Governments can facilitate adaptation to climate change through water development projects, agricultural extension activities, incentives, subsidies, regulations, and provision of insurance.

Similar adaptations could occur in tropical regions as shown in Table 5. In Mexico, preliminary results from the EPA International Agriculture project suggest that severe declines in maize yields under global warming could be mitigated if climate change is accompanied by increases in irrigation, fertilizer use and the use of drought resistant varieties. The major problem in the tropics, compared to most temperate

TABLE 5
Examples of Adaptations to Possible Negative Impacts of Global Warming (Warmer, Drier Climate) in the Tropics

Expand irrigation systems
Switch to new cultivars (e.g. maize to more drought tolerant sorghum)
Diversify activities (new crops, livestock, alternative employment)
Provide public relief and improved climate information
Increase food imports
Increase use of fertilizer and pesticides
Return to traditional water and soil conserving technologies (e.g. raised fields, agroforestry)
Migration and urbanization

Source: based on Jodha (1989) and Liverman (1991)

most temperate regions, is the relative lack of resources, institutions, and infrastructure to promote such adaptations.

CONCLUSIONS

In general, the tropical regions appear to be more vulnerable to climate change than the temperate regions for several reasons. On the biophysical side, temperate C3 crops are likely to be more responsive to increasing levels of CO_2. Second, tropical crops are closer to their high temperature optima and experience high temperature stress, despite lower projected amounts of warming. Third, insects and diseases, already much more prevalent in warmer and more humid regions, may become even more widespread.

Tropical regions may also be more vulnerable to climate change because of economic and social constraints. Greater economic and individual dependence on agriculture, widespread poverty, inadequate technologies, and lack of political power are likely to exacerbate the impacts of climate change in tropical regions.

In the light of possible global warming, plant breeders should probably place even more emphasis on development of heat- and drought-resistance crops. Research is needed to define the current limits to these resistances and the feasibility of manipulation through modern genetic techniques. Both crop architecture and physiology may be genetically altered to adapt to warmer environmental conditions. In some regions it may be appropriate to take a second look at traditional technologies and crops as ways of coping with climate change.

At the regional level, those charged with planning for resource allocation, including land, water, and agriculture development should take climate change into account. In coastal areas, agricultural land may be flooded or salinized; in continental interiors and other locations, droughts may increase. These eventualities can be dealt with more easily if anticipated.

As climatic factors change, a host of consequences will ripple through the agricultural system, as human decisions involving farm management, grain storage facilities, transportation infrastructure, regional markets, and trade patterns respond. For example, field-level changes in thermal regimes, water conditions, pest infestations, and most importantly, quantity and quality of yields, may lead to changes in farm management decisions based on altered risk assessments. Consequences of these management decisions could result in local and regional alterations in farming systems, land use, and food availability. Ultimately, impacts of climate change on agriculture may reverberate throughout the international food economy and global society.

At the national and international levels, the needs of regions and people vulnerable to the effects of climate change on their food supply should be addressed. In many cases, reducing vulnerability to current climate variability should also serve to mitigate the impacts of global warming.

It is important to ask, "What will or should agriculture be like in the next century?" Even if the answer is unknown, the flexibility gained in attempting to imagine the agricultural future should be a useful tool for adaptation to climate change.

REFERENCES

Acock, B. and L.H. Allen, Jr. 1985. Crop responses to elevated carbon dioxide concentrations. In B.R. Strain and J.D. Cure (eds.), *Direct Effects of Increasing Carbon Dioxide on Vegetation,* DOE/ER-0238, U.S Dept. of Energy, Washington, D.C. pp. 53-97.

Adams, R.M., C. Rosenzweig, R.M. Peart, J.T. Ritchie, B.A. McCarl, J.D. Glyer, R.B. Curry, J.W. Jones, K.J. Boote, and L.H. Allen, Jr. 1990. Global climate change and US agriculture. *Nature* 345 (6272): 219-224.

Akita, S. and D.N. Moss. 1973. Photosynthetic responses to CO_2 and light by maize and wheat leaves adjusted for constant stomatal apertures. *Crop Sci.* 13:234-237.

Allen, L.H., Jr., P. Jones, and J.W. Jones. 1985. Rising atmospheric CO_2 an evapotranspiration. In *Advances in Evapotranspiration. Proceedings of the National Conference on Advances in Evapotranspiration.* December 16-17, 1985. American Society of Agricultural Engineers. St. Joseph, Michigan. pp. 13-27.

Baker, J.T., L.H. Allen, Jr., K.J. Boote, P. Jones, and J.W. Jones. 1989. Response of soybean to air temperature and carbon dioxide concentration. *Crop Sci.* 29:98-105.

Barrow, C. 1987. *Water Resources and Agricultural Development in the Tropics.* Longman, London UK. 356 pp.

Buol, S.W., P.A. Sanchez, S.B. Weed, and J.M. Kimble. 1990. Predicted impact of climatic warming on soil properties and use. In Kimball, B.A., N.J. Rosenberg, and L.H. Allen, Jr. (eds.) *Impact of Carbon Dioxide, Trace Gases, and Climate Change on Global Agriculture.* ASA Special Publication Number 53. pp. 71-82.

Cline, W.R. 1991. *Estimating the Benefits of Greenhouse Warming Abatement.* Institute for International Economics. Washington, DC. 92 pp. (in press).

Cure, J.D. 1985. Carbon dioxide doubling responses: A crop survey. In B.R. Strain and J.D. Cure (eds.), *Direct Effects of Increasing Carbon Dioxide on Vegetation,* DOE/ER-0238, US Dept. of Energy, Washington, D.C. pp. 99-116.

Decker, W.L., V.K. Jones, and R. Achutuni. 1985. The impact of CO_2-induced climate change on US agriculture. In M.R. White (ed.), *Characterization of Information Requirements for Studies of CO_2 Effects: Water Resources, Agriculture, Fisheries, Forests and Human Health.* US Dept. of Energy, DOE/ER-0236, Washington, D.C.

Fischer, G., K. Frohberg, M.A. Keyzer, and K.S. Parikh. 1988. *Linked National Models: A Tool for International Food Policy Analysis.* Kluwer. Dordrecht.

Food and Agriculture Organization of the United Nations. 1984. *Land, Food and People.* FAO. Rome.

Food and Agriculture Organization. 1990. FAO Yearbook. Production. Vol. 44. Food and Agriculture Organization of the United Nations. Rome, 1991.

Fundacion Universo Veintiuno. 1990. *Memoria del Seminario Internacional sobre Calentamiento Global: Una Vision Latinoamericana.* Fundacion Universo Veintiuno. Mexico D.F.

HARC (Houston Area Research Center). 1991. *The Regions and Global Warming.* Draft Woodlands Conference papers. HARC. Houston TX.

Haws, L.D., H. Inoue, A. Tanaka, and S. Yoshida. 1983. Comparison of crop productivity in the tropics and temperate zone. In *Potential Productivity of Field Crops Under Different Environments.* International Rice Research Institute. Los Banos, Philippines. pp. 403-413.

Idso, S.B., B.A. Kimball, M.G. Anderson, and J.R. Mauney. 1987. Effects of atmospheric CO_2 enrichment on plant growth: the interactive role of air temperature. *Agriculture, Ecosystems and Environment* 20:1-10.

Intergovernmental Panel on Climate Change. 1990a. *Climate Change: The IPCC Scientific Assessment.* J.T. Houghton, G.J. Jenkins, and J.J. Ephraums (eds.). World Meteorological Organization and United Nations Environmental Program. Cambridge University Press. Cambridge. 365 pp.

Intergovernmental Panel on Climate Change. 1990b. Climate Change: The IPCC Impacts Assessment. Report prepared for IPCC by Working Group II. Eds. W.J. McG. Tegart, G.W. Sheldon, and D.C. Griffiths. Australian Government Publishing Service. Canberra.

International Benchmark Sites Network for Agrotechnology Transfer. 1990. *Proceedings of IBSNAT Symposium: Decision Support Sustem for Agrotechnology Transfer.* Held at the 81st Annual Meeting of the American Society of Agronomy, Las Vegas, Nevada. October 17, 1989. University of Hawaii, Honolulu.

Jodha, N.S. 1989. Potential Strategies for Adapting to Greenhouse Warming: Perspectives from the Developing World. In N.J. Rosenberg, W.E. Easterling III, P.R. Crosson, and J. Darmstadter (eds.) *Greenhouse Warming: Abatement and Adaptation.* Resources for the Future. Washington. pp. 147-158.

Johnson, R.C. and E.T. Kanemasu. 1983. Yield and development of winter wheat at elevated temperatures. *Agron. J.* 75:561-565.

Jones, P., L.H. Allen, Jr., and J.W. Jones. 1985. Responses of soybean canopy photosynthesis and transpiration to whole-day temperature changes in different CO_2 environments. *Agron. J.* 77:242-249.

Jurik, T.W., J.A. Weber, and D.M. Gates. 1984. Short-term effects of CO_2 on gas exchange of leaves of bigtooth aspen *(Populus grandidentata)* in the field. *Plant Physiol.* 75:1022-1026.

Kane, S., J. Reilly, and R. Bucklin. 1989. Implications of the greenhouse effect for world agricultural commodity markets. US Department of Agriculture. Washington, DC.

Kellogg, W.W. and Z.-C. Zhao. 1988. Sensitivity of soil moisture to doubling of carbon dioxide in climate model experiments, I, North America. *J. Clim.* 1:348-366.

Kimball, B.A. 1983. Carbon dioxide and agricultural yield: an assemblage and analysis of 430 prior observations. *Agron. J.* 75:779-788.

Liverman, D.M. 1986. The Sensitivity of Global Food Systems to Climatic Change. *Journal of Climatology.* 6:355-373.

Liverman, D.M. 1991. The potential impacts of global warming in Mexico: some preliminary results. In J. Reilly ed. *The Impacts of Climate Change on Agriculture and Forests.* Westview. Boulder. (In press.)

Liverman, D.M. and K. O'Brien. 1991. The impacts of global warming in Mexico. *Global Environmental Management.* Forthcoming December 1991.

Manabe, S. and K. Bryan, Jr. 1985. CO$_2$-Induced change in a coupled ocean-atmosphere model and its paleoclimatic implications. *J. of Geophysical Research* 90 (C6):11, 689-11,707.

Manabe, S. and R.T. Wetherald. 1986. Reduction in summer soil wetness induced by an increase in atmospheric carbon dioxide. *Science* 232:626-628.

Mearns, L.O., R.W. Katz, and S.H. Schneider. 1984. Extreme high temperature events: Changes in their probability with changes in mean temperature. *J. of Climate and Applied Meteorology* 23:1601-1613.

Mederski, H.J. 1983. Effects of water and temperature stress on soybean plant growth and yield in human temperature climates. In C.D. Raper and P.J. Kramer (eds.). *Crop Reactions to Water and Temperature Stresses in Humid, Temperate Climates*. Westview Press. Boulder. pp. 35-48.

Morison, J.I.L. and R.M. Gifford. 1984. Plant growth and water use with limited water supply in high CO$_2$ concentrations. I. Leaf area, water use and transpiration. *Austral. J. Plant Physiol.* 11:361-374.

National Defense University. 1980. *Crop Yields and Climate Change to the Year 2000. Vol. 1.* J. McNair, Washington DC.

Nix, H.A. 1985. Agriculture. In R.W. Kates, J.H. Ausubel, and M. Berberian eds. *Climate Impact Assessment*. SCOPE 27. Wiley. New York.

Parry M.L. 1978. *Climatic Change, Agriculture and Settlements.* Dawson. Folkestone UK.

Parry, M.L., T.R. Carter, and N.T. Konijn (eds.). 1988a. *The Impact of Climatic Variations on Agriculture. Volume I: Assessments in Cool Temperature and Cold Regions.* Kluwer Academic Publishers. Dordrecht. 876 pp.

Parry, M.L., T.R. Carter, and N.T. Konijn (eds.). 1988b. *The Impact of Climatic Variations on Agriculture. Volume I: Assessments in Semi-Arid Regions.* Kluwer Academic Publishers. Dordrecht. 764 pp.

Pierce, J.T. 1990. *The Food Resource.* Longman Publishing. New York. 334 pp.

Rind, D., R. Goldberg, J. Hansen, C. Rosenzweig, and R. Ruedy. 1990. Potential evapotranspiration and the likelihood of future drought. *J. Geophys. Res.* 95(D7):9983-10, 004.

Rose, E. 1989. Direct (physiological) effects of increasing CO$_2$ on crop plants and their interactions with indirect (climatic) effects. In J.B. Smith and D. Tirpak (eds.) *The Potential Effects of Global Climate Change on the United States. Report to Congress.* Appendix C-2. EPA-230-05-89-053. U.S. Environmental Protection Agency. Washington, D.C. pp. 7-1 - 7-37.

Rosenberg, N.J., W.E. Easterling III, P.R. Crosson, and J. Darmstadter (eds.) 1989. *Greenhouse Warming: Abatement and Adaptation.* Resources for the Future. Washington.

Rosenberg, N.J. and P.R. Crosson. 1990. *Processes for Identifying Regional Influences of the Responses to Increasing Atmospheric CO$_2$ and Climate Change: the MINK Project, An Overview.* Resources for the Future. Washington, DC.

Rosenzweig, C. 1985. Potential CO$_2$-induced climate effects on North American wheat-producing regions. *Climatic Change* 4:239-254.

Rosenzweig, 1990. Crop response to climate change in the southern Great Plains: A simulation study. The Professional Geographer 42(1):20-37.

Rosenzweig, C., A. Iglesias, B. Baer, W. Baethgen, M. Brklacich, T.Y. Chou, B. Curry, R. Delecolle, H.M. Eid, C.R. Escano, J. Jones, Z. Karim, L. Koval, D. Liverman, G. Menzhulin, W.S. Meyer, P. Muchena, A. Qureshi, G. Rao, J.T. Ritchie, O. Sala, H. Seino, S. J.F. de Siqueria, M.L.C. Tongyai, and J. Zhiqing. 1991. *Climate Change and International Agriculture: Crop Modeling Study.* U.S. Environmental Protection Agency. Washington, DC. (in preparation).

Shaw, R.H. 1983. Estimates of yield reductions in corn caused by water and temperature stress. In C.D. Raper and P.J. Kramer (eds.). *Crop Reactions to Water and Temperature Stresses in Humid, Temperate Climates.* Westview Press. Boulder. pp. 49-66.

Stinner, B.R., R.A.J. Taylor, R.B. Hammond, F.F. Purrington, D.A. McCartney. 1989. Potential effects of climate change on plant-pest interactions. In Smith, J.B. and D.A. Tirpak (eds.). *The Potential Effects of Global Climate Change on the United States.* Appendix C-2. US Environmental Protection Agency. Washington, DC. pp. 8-1 to 8-35.

Suliman, M. ed. 1990. *Greenhouse Effect and its Impact on Africa.* Institute for African Alternatives. London. 90pp.

Swaminathan, M.S. 1986. Building National and Global Security Systems. In M.S. Swaminathan and S.K. Sinha (eds.). *Global Aspects of Food Production.* Tycooly Press. pp. 417-449.

Thompson, L.M., 1975. Weather variability, climate change and food production. *Science* 188:534-541.

Universidad de Sao Paulo and Woods Hole Research Center. 1990. *Regional Conference on Global Warming and Sustainable Development.* Sao Paulo, Brasil June 18-20 1990. Conference Statement.

Waggoner, P.E. 1983. Agriculture and a climate changed by more carbon dioxide. In *Changing Climate.* National Academy of Sciences Press. Washington, D.C. pp. 383-418.

Warrick, R.A. 1988. Carbon Dioxide, Climatic Change and Agriculture. *The Geographical Journal.* 154(2):221-233.

White, G.F., (ed.) 1974. *Natural Hazards: Local, National, Global.* Oxford University Press, New York.

World Meteorological Organization. 1979. *Proceedings of the World Climate Conference.* WMO. Geneva.

Global Climate Change: Implications, Challenges and Mitigation Measures. Edited by S.K. Majumdar, L.S. Kalkstein, B. Yarnal, E.W. Miller, and L.M. Rosenfeld. © 1992, The Pennsylvania Academy of Science.

Chapter Twenty-Five

IMPACTS OF GLOBAL CLIMATE CHANGE ON HUMAN HEALTH: Spread of Infectious Disease

ROBERT E. SHOPE
Yale Arbovirus Research Unit
Box 3333
New Haven, CT 06510

INTRODUCTION

It has long been known that infectious diseases are indigenous to one or another part of the globe and can be spread. The historic spread of human infectious diseases includes syphilis to Europe in 1494, possibly from the Americas and resulting in rapid venereal spread through Europe; smallpox from Europe in 1518 to the New World where it virtually annihilated the Aztec empire; cholera from Asia to Europe, Africa, and sometimes the Americas in seven pandemics since 1800;[1] and most recently AIDS, probably from Africa to N. America and then throughout the world in the last two decades.

As far as we know, global climate change was not responsible for spread of these diseases in the past. More likely they spread as a result of human social and behavioral change. Nevertheless, spread of a very special set of diseases is apt to occur with climate change. These are diseases transmitted by arthropods—mosquitoes, sand flies, midges, and ticks—and other diseases transmitted from animals to people, which we call zoonoses. Climate change may also affect diseases spread by snails or by water, such as schistosomiasis and cholera, because changes in rainfall will have an impact on flow of rivers and levels of lakes; melting of polar ice may raise the sea level and inundate coastal and delta regions.

The infections that will spread with climate change have some commonalities.[2] They are focal, and their distribution is limited by the ecology of their reservoir, be it arthropod, snail, or water. They usually have a two- or three-host life cycle, meaning that in addition to infecting people, they infect a vector and frequently also a wild vertebrate animal host. Either the vector or the host, or both, are the reservoir. The range of the reservoir is delineated by temperature and sometimes water. In order to survive global climate change (and some of these infectious agents will not survive) the agents will need to have reservoirs that will survive; they will probably survive by moving in a polar direction, north in the Northern Hemisphere, in order to find a temperature range that is ecologically permissible. If the agent and reservoir are successful in the newly warmer climate, the agent can be expected to multiply more rapidly, and if the reservoir is an arthropod or snail, it too will develop more rapidly (it may also have a shorter life). It will be obvious to the reader, that the reservoir must survive the change, the agent must be able to move if the reservoir is translocated, and the reservoir must be able to adapt to conditions in the new ecologic zone.

Among the diseases that have been predicted to be more severe or move into more populated areas of North America are the mosquito-borne viral diseases dengue, St. Louis encephalitis, and yellow fever; the sand fly-borne protozoal disease leishmaniasis; the water-associated disease cholera; and the bat-borne vampire bat rabies.[2,3] This chapter will illustrate how one of these, dengue, may spread, and it will suggest how some other diseases of Africa and Asia may extend their geographic range and cause more serious human illness than currently encountered.

GLOBAL CLIMATE CHANGE AND THE SPREAD OF DENGUE

Dengue is a human disease caused by any one of four serotypes of dengue virus. Dengue fever is a common affliction throughout the tropics, characterized by fever, rash, muscle and joint pains, and severe prostration especially in adults, but usually the disease is followed by complete recovery. Dengue hemorrhagic fever is a severe form found usually in children and most commonly in Asia. It is complicated by hemorrhage, shock and sometimes death. Dengue hemorrhagic fever cases in Thailand in 1990 numbered more than 300,000.[4] Dengue hemorrhagic fever usually occurs in children having a second infection with a serotype different from the first infecting virus. Small amounts of residual antibody after the first infection, enhance the infectivity of the second dengue virus and hence, for reasons that are not completely understood, enhance the severity of the second infection. Dengue virus is in the family Flaviviridae which in addition to dengue is comprised of yellow fever, Japanese encephalitis and viruses causing other serious arthropod-borne diseases of people and domestic animals.

Dengue viruses are transmitted by mosquitoes. The virus replicates in the mosquito, a process that takes one to two weeks from the time a mosquito imbibes infected human blood until the virus reaches the salivary gland of the mosquito and replicates to titer high enough to be transmitted. This incubation period in the mosquito is shorter if the ambient temperature is warmer. The most common vector is *Ae. aegypti*, a mosquito that is domesticated and breeds in and around human dwellings; in flower pots, water storage jars, cisterns, metal cans, discarded tires, and any other fresh water containers that people leave standing. The mosquito does not survive freezing weather and, thus, is primarily a tropical vector. Historically, *Ae. aegypti* was transported long distances in drinking water aboard sailing ships. One of the classic descriptions of a dengue epidemic is that of Benjamin Rush in Philadelphia in 1780. The mosquito and dengue virus were imported to Philadelphia during the summer, but the mosquito did not live over the winter and the disease disappeared. Sporadic cases of dengue continue to occur in the northern U.S. in travellers from the tropics, but the virus is not now transmitted outside of the tropics and subtropics because of the lack of the vector.

What is it about dengue that makes it so likely to spread with global climate change? The dengue viruses are distributed wherever *Ae. aegypti* mosquitoes occur. The mosquito lives in the tropics wherever people live. The mosquito thrives in areas of high rainfall, but it also paradoxically thrives in desert areas because people often provide water containers for breeding. *Ae. aegypti* does not stand freezing weather, thus, it is limited to tropical and subtropical regions. Two aspects of global warming are particularly worrisome. Firstly, the warming is expected to be greater in temperate zones than in the tropics, and secondly, the warming is expected to be more marked at night than during the day. Both conditions favor the spread of the mosquito into the temperate areas. The warming at night is especially favorable to *Ae. aegypti* because it is the extreme temperature, i.e. freezing which occurs at night, that is most deleterious to the mosquito.

Another vector of dengue virus, *Ae. albopictus*, is widely distributed in Asia. This species has evolved in its more northerly distribution in Asia in a manner that permits it to withstand freezing. The northern form of the mosquito is triggered by shortened periods of sunlight to enter diapause, a physiological state of the egg that makes the egg resist cold temperature and delays hatching until the spring. Thus, the mosquito is able to survive freezing. *Ae. albopictus* was introduced during the 1980's into the Americas and now represents a potential risk for transmission of dengue in the United States. The mosquito that established itself in the United States is the diapausing form, and it appears to have adapted well to the more northern states. Now in both Asia and North America, there is the potential with global climate change for the vectors of dengue to move further north, perhaps much further north than today.

The risk of explosive epidemics is enhanced because of two other properties of this vector-virus relationship. Firstly, within limits dengue viruses multiply more rapidly in mosquitoes at high temperatures than at low ones. Secondly, the mosquito also develops more rapidly at high temperatures than at low ones. This combination is conducive to a very short incubation period in the mosquito and rapid mosquito population increase. A short incubation period in the mosquito along with rapid population increase in turn can lead to more rapid, and sometimes explosive transmission in the human population. This prediction, however, should be accompanied by a caution. Warmer temperatures also lead to a shorter life span of the mosquito, and shorter life means less time to transmit the virus to another person. Some scientists hold that the short life span will favor less transmission even though the amplification of dengue virus will proceed faster at high temperatures. It remains to be seen which parameter will prevail.

A further concern is the rapid increase of the human population and the continued growth of cities in the twenty-first century. As the earth warms, and if conditions in the vast sparsely inhabited land masses in Canada, Alaska, and Siberia become more suitable for human dwelling, it is possible that new cities will spring up. The *Aedes* mosquitoes and dengue could follow if conditions are favorable.

GLOBAL CLIMATE CHANGE AND SCHISTOSOMIASIS

Schistosomiasis is a parasitic disease caused by trematodes of three major species, *Schistosoma mansoni, S. haematobium,* and *S. japonicum.* The disease afflicts about 600,000,000 people in 79 countries of Africa, South America, and eastern Asia.[5] When first infected the patient may be acutely ill with skin rash or fever, abdominal pain, and malaise. Later the chronic phase manifests itself as urinary tract, liver, lung, or intestinal disease and varies according to the species of schistosome. Patients are chronically ill and after several years, the body's reaction to schistosomiasis may lead to death, usually by urinary tract obstruction, carcinoma of the bladder, portal or pulmonary hypertension, or some complication of these.

The life cycle of the parasite is complex involving snails, water, and human beings. The cycle is susceptible to environmental change, especially in water-associated stages. The cercariae (larvae) emerge from infected snails into water. These cercariae penetrate the intact human skin, and migrate through the tissues to find target sites in the human host. There they mature into adult worms, mate, and the female deposits eggs in the venous systems of the bladder or liver. The eggs migrate to the ureters or intestines, are excreted in urine or feces, and hatch in the water as miracidia that swim to find and enter a snail host. In the snail, the parasite undergoes two generations of sporocysts and emerges as the cercariae.

The snail hosts of schistosomes differ for each of the major species of parasite. The host of *S. hematobium* is the genus *Bulinus*; that of *S. mansoni* is *Biomphalaria*, and that of *S. japonicum* is *Oncomelania*. The ecology of each genus of the parasite and snail differs, but some generalities hold. A major determinant of schistosome distribution is the distribution of the snail host. Snail populations are dependent on temperature, water, and water currents. The ecological conditions needed by snails for survival have been carefully studied. Snails of the species transmitting *S. mansoni* in Brazil had an optimum temperature of 25° to 28°C and lived at 7°C for several days; they died after 2 hours at 42°C. The shedding of cercariae stopped below 13°C and above 41°C. Most of these snails died when removed from water, although a few survived if humidity was high enough.[6] Snails of the species transmitting *S. japonicum* in contrast, survived dessication for several weeks, and died at temperatures above 30°C and below 0 to −5°C.[7] The optimum temperature of snail reproduction was 26°C. These studies established that ambient temperature is an important limiting factor of the survival of snails and of the shedding of cercariae.

Aquatic birds have been implicated in the distribution of *Biomphalaria* and *Bulinus* snails to new areas. This accidental airborne transport is very effective in seeding new sites.[8] In addition, these snails are hermaphroditic in nature and can self-fertilize and increase in numbers rapidly, once transported. Thus, the major question associated with global climate change is not how the snail will be transported and established, but rather, whether the temperature, water, and other conditions to support the snail and the schistosome parasite are adequate.

If the temperature in the tropics of Africa and South America rises sufficiently, it is likely that some of the present foci of schistosomiasis will be too hot to support the parasite. Areas of Africa on the east and west coasts already have high temperatures, and it has been suggested that this is the reason why *Biomphalaria* has not colonized these zones.[9] On the other hand, other areas in Europe, Asia, and the Americas, now too cold to support the host snails, can be expected in the future to be favorable ecologically for schistosomiasis.

RIFT VALLEY FEVER AND GLOBAL CLIMATE CHANGE

Rift Valley fever is a disease with a track record of appearing in 1977 in epidemic form in Egypt where no trace of prior infection could be found, and again in 1987 in Mauritania of causing an epidemic where it had previously caused only inapparent infections. These two epidemics accompanied ecological change following the completion of dams; such ecological change may differ from that of global climate change, nevertheless I believe this dramatic disease is a prime candidate to be affected by global climate change.

The 1987 epidemic in southern Mauritania confirmed that Rift Valley fever had potential to cause serious human disease.[10-12] About 43% of cases presented with fever without life-threatening disease, but an additional 48% had jaundice and of these, 18% presented with hemorrhage. About half of these latter patients died. Another 5% had encephalitis and about 2% had retinitis that led to blindness. The disease also affects sheep, cattle, and other domestic animals causing abortion and fatal hepatitis. Mortality of nearly 100% is sometimes observed in newborn sheep.

Rift Valley fever is caused by a virus in the family Bunyaviridae, transmitted by mosquitoes and also through inhalation of aerosol from the infected blood and afterbirths of sick animals. The disease derives its name from the historic epizootics that occurred periodically after 1910 in sheep and cattle in Africa's Rift Valley that extends from South Africa through Tanzania, Kenya, Uganda, and Sudan. In East and South Africa, the disease appears in epizootics in the grasslands. These outbreaks follow heavy rains and the consequent emergence of large numbers of *Aedes* and *Culex* mosquitoes. Until 1977 it had never been reported north of the Sahara.

In 1977, there was an explosive outbreak in the Nile delta of Egypt. An estimated

200,000 human cases were observed with at least 598 deaths.[13] Studies of human sera collected prior to 1977 showed that this was a new disease to Egypt. Since the infection is ecologically restricted, its appearance in Egypt is believed to be a consequence of ecological change, perhaps linked to the building of the Aswan High Dam and inundation of 800,000 hectares for agricultural development between 1970 and 1977. There is no doubt that the virus was transported, probably from the Sudan where an epizootic was recorded in 1976. It could have been transported as a viremic animal or human, or an infected insect blown northward by the wind.[14]

Curiously, Rift Valley fever failed to establish itself in Egypt. The epidemic and epizootic swept through Egypt during 1977 and 1978, with sporadic infections detected until 1980, then there was no more evidence of the virus. The reservoir of the Rift Valley fever virus in sub-Saharan Africa has been a mystery. Recently, scientists believe they have discovered the secret through studies in Kenya. After heavy rains, epizootics start, involving sheep and cattle pastured in the grasslands. The rains flood large depressions, called "dambos" and from these depressions hatch massive number of *Aedes* mosquitoes. Larvae and pupae of *Aedes lineatopennis* collected in flooded dambos were raised to adults; both male and female adults were found infected with Rift Valley fever virus.[15] The most likely explanation for this finding is that the virus is in the mosquito egg and is passaged transovarially to the emerging mosquito. The mosquito itself appears to be the reservoir and, since rain is needed to hatch the eggs and heavy rain only occurs every few years, this phenomenon may explain the long periods between epizootics. It also offers an explanation why Rift Valley fever was not established permanently in Egypt. Possibly an appropriate mosquito capable of transovarial transmission is missing.

The 1987 epidemic in the Senegal River Valley of Mauritania illustrates a second principle of ecological change. Whereas in Egypt, the virus was transported to virgin territory, in Mauritania, Rift Valley fever virus was already there, although no human cases had been recognized in the sparse settlements. The Diama Dam was completed in 1986 near the mouth of the Senegal River. Ecological impact studies were commissioned both by the U.S. Government and by the French Government. Scientists at the Institut Pasteur warned "The existence of an important focus of Rift Valley fever virus in the south of Mauritania, and in proximity to the Senegal River and in particular to the Diama and Manantali Dams, constitutes a potentially important risk of amplification of the virus in relation to the migration of domestic animal herds and human populations which pass through this focus."[16] Shortly after the warning, the epidemic began near Rosso, a community that was a focus of herdsmen and their animals, some of them migrating into the area to seek the agricultural benefits of the Diama Dam. Over a thousand Rift Valley fever cases occurred in the town of Rosso alone with an estimated 47 deaths.[17]

Why is Rift Valley fever such a threat? Firstly, the virus is capable of infecting and being transmitted by a wide variety of mosquitoes and can maintain itself in non-immune populations of sheep and cattle, infecting people at the same time.[13] Secondly, it has demonstrated the ability to be transported. Thirdly, it can apparently adapt easily to ecological change. The two major epidemics in 1977 and 1987 were associated with localized ecological change. With global climate change we may well be embarking not on localized, but rather generalized change. For a disease that seems to have capitalized on two isolated opportunities, can one imagine the possible consequences when offered almost unlimited opportunities for spread to areas of ecological change?

DISCUSSION AND CONCLUSIONS

Dengue, schistosomiasis, and Rift Valley fever are only three examples of major human diseases that can be expected to be influenced by global climate change. There are experimental vaccines for dengue and Rift Valley fever, and drugs for treatment of schistosomiasis. We can combat all three diseases with environmental sanitation and health education. In spite of these measures, we have not been successful in controlling them and we can expect local and world changes in temperature and rainfall to make their control more difficult.

Fortunately, the changes will happen gradually and if we act now, we have time to learn more about the epidemiology and ecology of the vector-borne and zoonotic diseases. We also have time to devise better control and prevention strategies. These studies will require interdisciplinary research. The trend today in graduate education and in university and government research is to specialization, and in infectious diseases the trend is to specialization at the molecular level. This trend is laudable to a point; many of our solutions will require understanding at the molecular level. However, this particular problem will also require training in more general and interdisciplinary fields including field ecology, general medicine, epidemiology, forestry and botany, entomology, climatology, and zoology to name a few.

We should aim to devise better direct intervention measures for these diseases. We also need more information about transport of agents, modes of transmission, their reservoirs, and the effect of temperature, rainfall, and other climate-related parameters on the vectors, vertebrate hosts, and the agents of disease themselves. Studies of ecology at the periphery of the ranges of the agents and their reservoirs would be especially valuable. Such information could be used to predict more accurately which of the diseases to target as threats, and which will be less likely to spread and/or become more severe.

REFERENCES

1. McNeill, W.H. 1977. *Plagues and Peoples.* Anchor Books, Doubleday, New York, NY.
2. Shope, R.E. 1990. Infectious diseases and atmospheric change, pp. 47-54. In: J.C. White (Ed.) *Global Atmospheric Change and Public Health.* Elsevier, New York, NY.
3. Shope, R.E. 1991. Global climate change and infectious diseases. *Environmental Health Perspectives* 96:171-174.

4. Chunsuttiwat, S. 1990. Epidemiology and control of dengue hemorrhagic fever in Thailand. *Southeast Asian J. Trop. Med. and Public Health* 21:684-685.
5. Iarotski, L.S. and A. Davis. 1981. The schistosomiasis problem in the world: Results of a WHO questionnaire survey. *Bull. WHO* 59:114-127.
6. Barbosa, F.S. and L. Olivier. 1958. Studies on the snail vectors of bilharziasis mansoni in Northeastern Brazil. *Bull. WHO* 18:895-908.
7. Iijima, T. and S. Sugiura. 1962. Studies on the temperature as a limiting factor for the survival of *Oncomelania nosophora*, the vector snail of *Schistosoma japonicum* in Japan. *Jap. J. Med. Sci. Biol.* 15:221-228.
8. Burch, J.B. 1975. Freshwater molluscs, pp. 311-321. In: N.F. Stanley and M.P. Alpers (Ed.) *Man-Made Lakes and Human Health*. Academic Press, New York, NY.
9. Sturrock, R.F. 1965. The development of irrigation and its influence on the transmission of Bilharziasis in Tanganyika. *Bull. WHO* 32:225-236.
10. Jouan, A., B. Philippe, O. Riou, I. Coulibaly, B. Leguenno, J. Meegan, M. Mondo and J.P. Digoutte. 1989. Les formes cliniques benignes de la fievre de la Vallee du Rift pendant l'epidemie du Mauritanie. *Bull. Soc. Pathol. Exot. Filiales.* 82:620-627.
11. Philippe, B., A. Jouan, O. Riou, I. Coulibaly, B. Leguenno, J. Meegan, M. Mondo and J.P. Digoutte. 1989. Les formes hemorrhagiques de la fievre de la Vallee du Rift en Mauritanie. *Bull. Soc. Pathol. Exot. Filiales.* 82:611-619.
12. Riou, O., B. Philippe, A. Jouan, I. Coulibaly, M. Mondo and J.P. Digoutte. 1989. Les formes neurologiques et neurosensorielles de la fievre de la Vallee du Rift en Mauritanie. *Bull. Soc. Pathol. Exot. Filiales.* 82:605-610.
13. Meegan, J.M. and R.E. Shope. 1981. Emerging concepts on Rift Valley fever, pp. 267-282. In: M. Pollard (Ed.) *Perspectives in Virology XI*. Alan R. Liss, Inc., New York, NY.
14. Sellers, R.F. 1981. Rift Valley fever, Egypt 1977: Disease spread by windborne insect vectors? *Veterinary Record* 110:73-77.
15. Linthicum, K.J., F.G. Davies, A. Kairo and C.L. Bailey. 1985. Rift Valley fever virus (family Bunyaviridae, genus *Phlebovirus*). Isolations from Diptera collected during an inter-epizootic period in Kenya. *J. Hyg. Camb.* 95:197-209.
16. Digoutte, J.P. and C.J. Peters. 1989. General aspects of the 1987 Rift Valley fever epidemic in Mauritania. *Res. Virol.* 140:27-30.
17. Jouan, A.F. Adam, O. Riou, B. Philippe, N.O. Merzoug, T. Ksiazek, B. Leguenno and J.P. Digoutte. 1990. Evaluation des indicateurs de sante dans la region du Trarza lors de l'epidemie de fievre de la Vallee du Rift en 1987. *Bull. Soc. Pathol. Exot. Filiales.* 83:621-627.

Global Climate Change: Implications, Challenges and Mitigation Measures. Edited by S.K. Majumdar, L.S. Kalkstein, B. Yarnal, E.W. Miller, and L.M. Rosenfeld. © 1992, The Pennsylvania Academy of Science.

Chapter Twenty-Six

IMPACTS OF GLOBAL WARMING ON HUMAN HEALTH:
Heat Stress-Related Mortality

LAURENCE S. KALKSTEIN
Center for Climatic Research
Department of Geography
University of Delaware
Newark, Delaware 19716

INTRODUCTION

With the threat of a human-induced global warming, interest in climate/human health studies has dramatically increased, and at least three comprehensive reports summarizing most of this research have appeared in recent years![1,2,3] The impact of weather on human well-being goes beyond mortality; even birth rates and sperm counts appear to be affected by climatological phenomena.[4,1] However, the majority of the climate/human health evaluations have concentrated on mortality, and most all of the studies correlate a number of climate variables with daily or weekly mortality statistics. These data are not difficult to obtain for the United States or Canada, and they are much easier to interpret than most sources of general morbidity data, such as hospital admissions tallies or visits to the doctor.

There is disagreement among researchers concerning the impact of weather on human mortality. For example, medical researchers have noted that mortality attributed to weather seems to vary considerably with age, sex, and race, but there is disagreement in defining the most susceptible population group. Other issues of contention exist, including the role of air conditioning in mitigating heat-related mortality and whether small, irregular aberrations in weather or very large fluctuations have a greater impact on mortality.

This essay will evaluate these questions and others. First, historical weather/mortality relationships will be discussed, as it is impossible to determine the impact of global warming on human health without understanding present-day and past relationships. Most of this work concentrates on the United States, but results from a pilot study in China will be briefly presented. Second, and most important, the potential impact of a large-scale global warming on human mortality will be discussed, with allowances for possible human acclimatization to the expected increasing warmth.

HISTORICAL RELATIONSHIPS

Although a few studies contend that mostly long term (i.e., monthly and annual) fluctuations in temperature affect mortality,[5] and only small aberrations can be explained by daily temperature variability,[6] most researchers insist that hot weather extremes have a more substantial impact on human mortality. Thus many "heat stress" indices have been developed to help assess the degree of impact.[7,8,2] Most research on heat/mortality relationships indicate that high temperature alone has a very dramatic effect on mortality, and other factors such as humidity and wind are essentially unimportant.[9,10] In fact, the notion of a "temperature threshold" is quite common in certain studies, and can be demonstrated when daily mortality in New York City and Shanghai, China are compared to maximum temperature (Figure 1). Particularly interesting is the *lack* of a weather/mortality relationship at temperatures below the threshold, indicating that only the warmest 10-15 percent of all days in summer have an impact on human mortality.[11] For example, the five cities which demonstrate the greatest rise in mortality during very hot conditions are New York, Chicago, Philadelphia, Detroit, and St. Louis, respectively.[12]

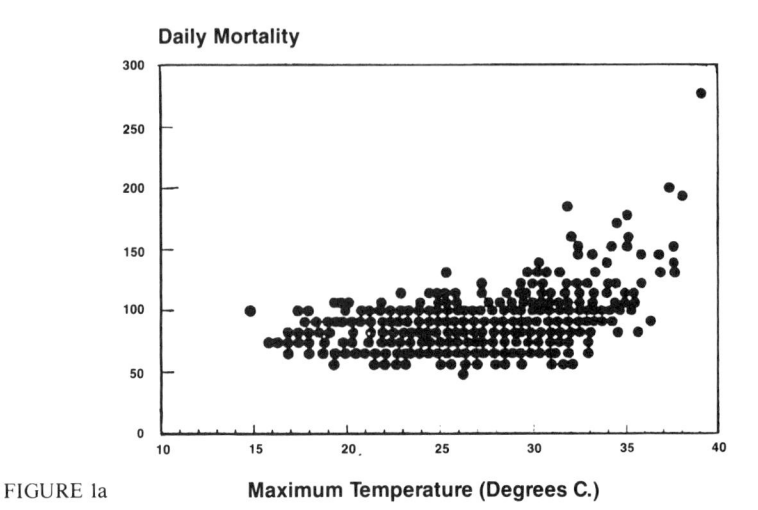

FIGURE 1a   Maximum Temperature (Degrees C.)

FIGURE 1 *continued* . . .

Daily Mortality

FIGURE 1b   Maximum Temperature (Degrees C.)

Daily Mortality

FIGURE 1c   Maximum Temperature (Degrees C.)

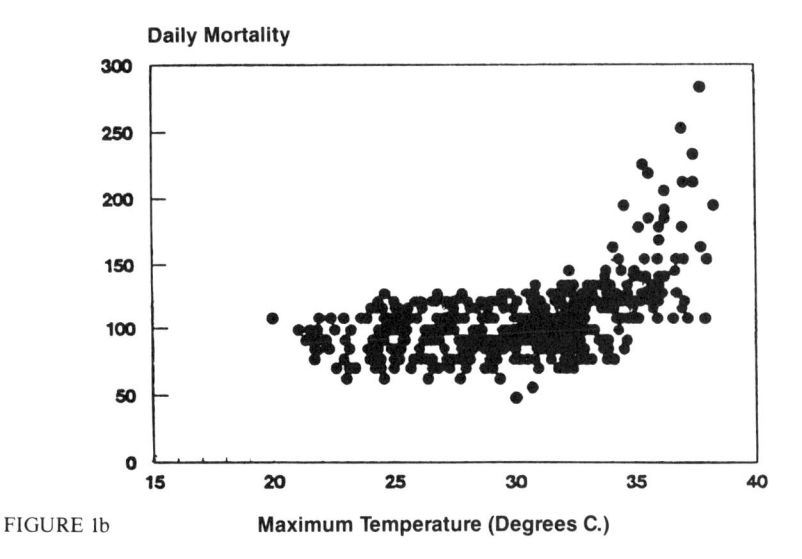

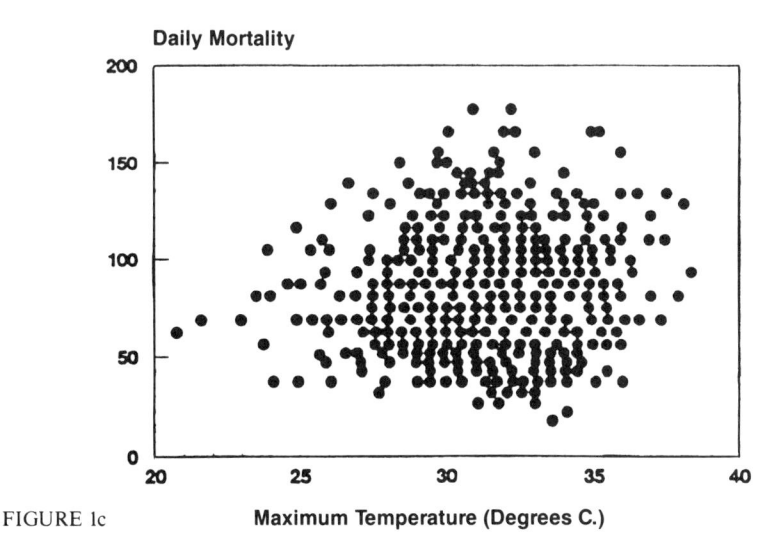

FIGURE 1. Relationship between summer maximum temperature and daily mortality at New York (Figure 1a), Shanghai (Figure 1b), and Jacksonville (Figure 1c). Sources: Kalkstein and Davis, 1989; Tan, in press.

The threshold temperatures for these cities are: New York: 33°C; Chicago: 32°C; Philadelphia: 33°C; Detroit: 32°C; and St. Louis: 36°C. The magnitude of the threshold temperature appears to be related to its frequency across these cities;

for example, a maximum temperature of 36°C in St. Louis occurs with approximately the same frequency as a maximum temperature of 32°C in Detroit. This strongly suggests that the notion of a "heat wave" is relative on an inter-regional scale, and is dependent upon the frequency of a given maximum temperature.

However, there is also evidence to refute this notion. Mortality rates in the southern and southwestern United States do not seem to be affected by weather in the summer, no matter how high the temperature. For example, the cities of Dallas, Atlanta, New Orleans, Oklahoma City, and even Phoenix show little change in mortality even during the hottest weather. A similar phenomenon is noted in China, where Guangzhou, with a summer climate similar to New Orleans, demonstrates much less day-to-day variation in mortality than Shanghai, which is located at a higher latitude.[13] In fact, threshold temperatures for these hotter cities are virtually impossible to define. One posible explanation may involve the variance in summer temperatures between the two regions. In the northern and midwestern cities, the very hot days or episodes are imbedded within periods of cooler weather. Thus, the physiological and behavioral "shock value" of a very high temperature episode is quite high. This point is further substantiated by the fact that most of the high mortality days occur during hot weather *early* in the summer season. Thus the first or second heat episodes of June or early July are much more critical than a comparative episode in August. In the southern cities, the hottest periods are less unique, as they do not vary as much from the mean. This seems to play a role in diminishing the impact of a very hot episode on human mortality.

Thus, a most interesting finding relating to heat-related mortality which potentially has a profound influence on such deaths in a warmer world is the large degree of inter-regional response. A recent EPA-sponsored study demonstrated that many cities in the northeastern and midwestern United States show a sharp rise in total mortality during unusually hot weather conditions, and in some cases, daily mortality can be more than double baseline levels when the weather is oppressive.[12]

The use of a new automated air mass-based synoptic procedure to evaluate weather/mortality relationships has supported and expanded the findings from the threshold temperature research (refer to Kalkstein *et al.*, 1990[14] for a discussion of synoptic index development). The synoptic procedure is designed to classify days which are considered to be meteorologically homogeneous into air mass categories. Thus, the synergistic relationships that exist between numerous weather elements which comprise an air mass can be evaluated simultaneously, representing a significant improvement over an individual weather element approach.

The synoptic procedure was applied to ten U.S. cities in different climates to determine inter-regional mortality/weather sensitivities: Atlanta, Boston, Chicago, Dallas, Memphis, New York, Philadelphia, St. Louis, San Francisco, and Seattle. For many of these cities, a single offensive summer synoptic category was noted, which possessed a much higher mean mortality than the other categories (refer to Table 1; St. Louis example). Although category 6 in St. Louis was only slightly cooler than category 9, it appears that category 9 alone exceeded an oppressive threshold associated with very high mortality.

A comparison of results from the 10 cities is very instructive (Table 2). Most of the seven cities with a moderate or strong weather/mortality signal (as determined by the existence of an oppressive synoptic category associated with unusually high mortality) were located in the Northeast or Midwest (Memphis is a notable exception). The San Francisco area, which experiences infrequent but sometimes

TABLE 1
Mean daily mortality for each summer synoptic category for St. Louis[a]

Category number	Whites	Non-Whites	Elderly	Total[b]
1 (cool, continental)	69	25	60	94
2 (transitional)	71	29	63	100
3 (maritime)	71	27	63	98
4 (overrunning)	68	25	59	93
5 (maritime tropical, cloudy)	72	28	64	100
6 (maritime tropical, sunny)	73	30	64	103
7 (transitional to maritime)	70	27	60	97
8 (frontal wave)	70	25	60	95
9 (oppressive, tropical)	88	41	84	129
10 (cold front passage)	68	25	59	93
Overall mean excluding category 9	70	27	62	97
Percent that category 9 exceeds overall mean	26	52	36	33

[a] Values represent a one-day lag between synoptic category occurrence and mortality response.
[b] Total equals the sum of the white and non-white values.

TABLE 2
A comparison of the relative impacts of weather on mortality for ten selected cities.

STRONG WEATHER/MORTALITY SIGNAL[a]
Boston
Memphis
New York City
Philadelphia
St. Louis
MODERATE WEATHER/MORTALITY SIGNAL
Chicago
San Francisco
NO WEATHER/MORTALITY SIGNAL
Atlanta
Dallas
Seattle

[a] To have a moderate or high weather/mortality signal, a city must possess a synoptic category with unusually high mean daily mortality. In this analysis, this was determined subjectively.

persistent hot weather, is also associated with this group. Of the three cities with no weather/mortality signal, two are located in the South, and Seattle possessed virtually no air mass types associated with very hot weather. Thus, this inter-regional disparity in response is supported within the synoptic evaluation.

Interestingly, a majority of studies have found that most of the excess deaths that occurred during periods of intense heat were not attributed to causes traditionally considered to be weather-related (such as heat stroke), but were attributed to a broad range of illnesses, accidents, and even adverse effects of medicinal agents.[15,16] In fact, heat stroke and heat exhaustion comprise a very small proportion of the increase in deaths which traditionally occurs during very hot episodes. An evaluation of individual causes of deaths which seem to rise during hot weather produces some surprises, with complications of pregnancy, ischemic heart disease, and various injuries ranking high on the list.[16]

Most research indicates that mortality rates during extreme heat vary with age, sex, and race. Oechsli and Buechley (1970)[17] found that mortality rates during heat waves increase with age; this is supported by more recent work.[16] The elderly seem to suffer from impaired physiological responses and often are unable to increase their cardiac output sufficiently during extremely hot weather. In addition, sweating efficiency decreases with advancing age, and many of the medications commonly taken by the elderly have been reported to increase sensitivity to the heat.[15] Certain researchers have determined slight rises as well in the mortality of infants during heat waves.[18,19]

Although there is general agreement concerning the impact of heat on the elderly, there are conflicting results involving mortality/gender and mortality/race relationships. Several studies have noted increased mortality rates among females during hot weather;[20] this may be attributed to differences in dress among the sexes.[21] However, at least two studies have found higher heat-related mortality rates among men.[18,22] A study on race/heat-related mortality have found that blacks are more susceptible in St. Louis and whites are more susceptible in New York.[23] However, two other studies have discovered that white mortality rates are higher than black's under almost all examined conditions.[18,19] Rather than race, socioeconomic status may have an influence on weather/mortality relationships, and large numbers of deaths during heat waves are found among poor inner-city residents who have little access to cooler environments.

IMPLICATIONS FOR A LARGE-SCALE GLOBAL WARMING

A proper analysis which considers the impact of global warming on heat-related mortality should address all of the issues discussed above, including:

1. an understanding of thresholds or tolerance limits of humans;
2. the impact of possible acclimatization to warmer conditions;
3. variations in response among the sexes and races;
4. demographic changes expected to occur within the next century.

At present, we have some knowledge about the first of these issues and a lesser amount of information about the third. There is very little known about the second and fourth issues. Thus, any estimates of the impacts of global warming on human mortality must be treated with skepticism. However, there have been estimates offered for 15 U.S. cities, and they vary consideraly based on scenarios using different assumptions about the issues listed above.

Assuming that people react to weather in a warmer world much as they do today (implying little or no acclimatization), it is quite likely that deaths from heat stress-related causes will increase dramatically (refer to Table 3 for the full range of estimates). Thus, EPA has reported to Congress that we might expect a sevenfold increase in heat-related deaths by the middle of the 21st century if acclimatization does not occur.[2,24] This would render heat-related deaths as a major killer, rivaling the present number of deaths from leukemia. The greatest brunt of this increase would be borne in the northeastern and midwestern U.S., and it is estimated that the number of heat-related deaths in New York City would rise from the present 320 during a typical summer today to 1,743 during a typical summer in the mid 21st century.[24] Under such circumstances, the number of days exceeding the threshold temperature would show at least a two-fold increase, contributing to this dramatic rise.

Of course, a more likely scenario is that some degree of acclimatization will occur, as the gradual increase in warmth over the next 60 years would permit time for certain social and physiological adjustments. A satisfactory means to account for this acclimatization has alluded the scientific community, although an attempt was made in the EPA *Report to Congress* on the potential effects of global climate change.[24] *Analog cities,* which possess present day weather most duplicative of predicted mid-21st century weather for each evaluated city, were established to account for full acclimatization. For example, the use of a climate change scenario developed from a global circulation model to estimate weather in Atlanta in the mid-21st century will produce a regime which approximates the present weather in New Orleans.[25] Since New Orleans residents are fully acclimatized to this regime, the weather/mortality relationships that exist today for New Orleans can be utilized for the mid-21st century estimated climate in Atlanta to account for full acclimatization.

Employing this acclimatization procedure, new estimates were developed for the number of heat stress-related deaths in the mid-21st century, yielding very different results from the unacclimatized estimates. Rather than the sevenfold increase developed for the unacclimatized model, less than a two-fold increase results when accounting for acclimatization. In fact, many cities actually show a *decrease* in heat-related mortality if the population fully acclimatizes to a climage change. For example, the number of heat-related deaths which occur in an average summer today in Los Angeles is 84. If residents in Los Angeles fully acclimatize to the increased warmth expected in the mid-21st century, the number of heat-related deaths is estimated to drop to virtually zero.[24] Astoundingly, if residents of Los Angeles don't acclimatize at all, the number of heat-related deaths estimated for an average summer in the mid-21st century rises to 1570! Similar disparities between acclimatized and unacclimatized estimates are noted for many northern and midwestern cities such as Detroit, New York, and St. Louis.

TABLE 3
Estimates of Future Mortality in Summer Attributed to Weather[a]

City	Present Total[c]	GISS No Acclim	TRANS Partial Acclim	A[1b] Full Acclim[d]	GISS No Acclim	TRANS Partial Acclim	A[2] Full Acclim[d]	GISS No Acclim	2x Partial Acclim	CO₂ Full Acclim[d]	Change: Present to 2 × CO₂[e] No Acclim	Partial Acclim	Full Acclim
Atlanta													
Elderly	9	25	13	0	63	32	0	85	43	0	76	34	-9
Total	18	45	23	0	118	59	0	159	79	0	141	61	-18
Chicago													
Elderly	104	173	86	0	300	408	520	240	347	458	136	243	354
Total	173	295	145	0	511	725	940	412	622	835	239	449	662
Cincinnati													
Elderly	28	62	52	45	27	63	0	150	123	69	122	95	41
Total	42	93	83	72	195	97	0	226	195	116	184	153	74
Dallas													
Elderly	9	33	33	33	109	68	28	172	153	74	163	144	65
Total	19	61	61	61	197	213	67	309	244	179	290	225	160
Detroit													
Elderly	70	120	91	0	301	151	0	349	174	0	279	104	-70
Total	118	201	152	104	512	254	0	592	295	0	474	177	-118
Kansas City													
Elderly	20	20	24	27	41	16	49	38	66	18	93	46	-73
Total	31	33	40	48	66	68	71	60	100	138	29	69	107
Los Angeles													
Elderly	53	94	50	7	194	132	71	1026	513	0	973	460	-53
Total	84	153	81	11	313	205	116	1654	824	0	1570	740	-84
Memphis													
Elderly	13	18	9	0	47	24	0	106	52	0	93	39	-13
Total	20	28	14	0	78	39	0	177	88	0	157	68	-20
Minneapolis													
Elderly	30	65	32	0	118	59	0	95	110	126	65	80	96
Total	46	96	47	0	175	87	0	142	186	235	96	140	189

TABLE 3 continues on following page...

TABLE 3 CONT.

City	Present Total[c]	GISS No Acclim	TRANS Partial Acclim	A[1,b] Full Acclim[d]	GISS No Acclim	TRANS Partial Acclim	A[2] Full Acclim[d]	GISS No Acclim	2x Partial Acclim	CO[2] Full Acclim[d]	Change: No Acclim	Present to 2 x CO[2][e] Partial Acclim	Full Acclim
New Orleans													
Elderly	0	0	0	0	0	0	0	0	0	0	0	0	0
Total	0	0	0	0	0	0	0	0	0	0	0	0	0
New York													
Elderly	212	514	257	0	897	448	6	1139	580	17	927	368	-195
Total	320	777	386	0	1375	689	8	1743	880	23	1423	560	-297
Oklahoma City													
Elderly	0	0	4	8	0	9	19	0	16	31	0	16	31
Total	0	0	6	12	0	19	29	0	23	47	0	23	47
Philadelphia													
Elderly	91	182	89	0	489	244	0	590	590	285	499	499	194
Total	145	288	142	0	778	388	0	938	700	466	793	555	321
St. Louis													
Elderly	71	203	102	0	471	373	278	459	226	0	388	155	-71
Total	113	325	162	0	754	564	375	744	372	0	631	259	-113
San Francisco													
Elderly	17	28	14	2	45	23	4	156	129	103	139	112	86
Total	27	44	23	3	71	38	7	246	202	159	219	175	132
Sum[f]													
Elderly	727	1537	856	122	3202	2050	975	4605	3122	1256	3878	2395	529
Total	1156	2439	1365	311	5143	3445	1613	7402	4810	2198	6246	3654	1042

[a] Mortality estimates are offered for elderly (age greater than 65 years old) and total populations for each city.
[b] Three general circulation models are used to estimate mortality under potential future conditions: The Goddard Institute for Space Studies (GISS) transient model A[1], transient model A[2], and doubled CO[2] (2xCO[2]) models. The transient models assume gradual changes of atmospheric CO[2] conditions. Transient A[1] was used as the scenario to estimate mortality for the period 1994-2010. Transient A[2] estimates mortality for the period 2024-2040.
[c] Values represent estimates of present day heat-related mortality during a typical summer at each city.
[d] Values represent estimates of future mortality for each scenario considering no acclimatization, partial acclimatization, and full acclimatization.
[e] Values represent the difference between present day and 2xCO[2] mortality.
[f] Values represent summations for each column.

This wide variation has much to do with the technique used in this evaluation to account for acclimatization. Since the change analogs for the northern cities are other cities located further south, and considering the southern populations respond much less dramatically to heat, it is not surprising that the fully acclimatized mortality estimates are so low. This points to some major shortcomings within our acclimatization procedure. First, the only similarities drawn between the target and analog cities relate to climate alone, with no consideration of possible urban structural or architectural disparities. For example, much of the low income urban population in the South, who are especially vulnerable to heat stress due to a lack of air conditioning or other amentities, reside in small frame houses often referred to as "shotgun shacks".[26] Although far from luxurious, these residences are well-adapted to the rigors of the southern summer, as they are often light colored with metal (reflective) roofs and well-ventilated with windows or doors on four sides. In contrast, most northern urban poor live in tenement dwellings constructed of dark-colored brick, black tar roofs, and windows often on two sides associated with a row house motif. During very hot conditions, these northern dwellings suffer from poor ventilation and absorb considerably more radiation than their southern counterparts. These types of factors might partially explain the differential mortality response between northern and southern populations.

Assuming that people can physiologically adapt to the predicted increasing warmth, we can expect only a partial acclimatization to occur as it is highly unlikely that the architectural makeup of the urban area will change significantly over the next 50 to 75 years to account for warmer conditions. This is particularly true for dwellings of the poor, and it is unlikely that the tenements of northern cities will be leveled and replaced by other forms of housing, such as shotgun shacks, which are more amenable to hotter conditions.

An added shortcoming of the acclimatization procedure is underscored by the fact that the analog and target cities are also very different in their demographic composition, as the differential proportion of elderly (who are particularly vulnerable to heat-related mortality) and minorities has not been accounted for. Furthermore, no attempt has been made to determine what these proportions might be in the target city in the mid-21st century. The expected increase in the percentage of elderly people (greater than 65 years old) over the next 75 years might partially counter the dampening in heat-related mortality attributed to physiological acclimatization.

Thus, the *Report to Congress* also includes an estimate of mid-21st century mortality assuming partial acclimatization.[24] The assumption here is that the apparent full acclimatization which presently exists in the South will not occur in a warmer North for two reasons: (1) any architectural changes in dwellings will lag considerably behind the warming itself, rendering many residences unsuitable for the increasing warmth, and (2) the demographic composition of these cities will be comprised of a larger proportion of people more vulnerable to heat-related stresses.

Finally, it is possible that the expected 1.5°C to 4.0°C warming by the mid-21 century might actually increase temperatures in many southern and southwestern cities to levels which might be beyond the tolerances of our society. For example, the average annual number of days in Atlanta which exceed 32°C today is 17. This number is expected to reach 53 by the mid-21st century using the Goddard Institute for Space Studies (GISS) doubled CO_2 scenario.[25] Even more troubling is the number of days with temperatures exceeding 38°C. Presently, cities such as Memphis, Jackson, and Birmingham experience three such days during an average summer. This will increase to about 20 days with the expected temperature increase.[27] Considering similar increases in the number of very hot days in Phoenix and other southwestern cities, it is possible that certain areas might simply become uninhabitable in the warmer world of the mid-21st century. Of course, this is highly speculative, as no examples presently exist to support this notion.

To improve upon present attempts to account for acclimatization, the Climate Change Division of EPA has developed a global warming/human health program which places some emphasis on an improved understanding of human acclimatization in a warmer world. The program will concentrate on the social and cultural adjustments expected to occur, including possible demographic alterations that could take place because of migration and other factors attributed to global warming.[28] Contacts have been made with the Institute of Medicine at the National Academy of Sciences to help in the formulation and coordination of such a study.

Of course, skeptics contend that global warming will probably not occur, that the earth and atmospheric system have feedback mechanisms to prevent a human-induced climatic change from occurring, and that time and money may be wasted in developing mitigating policies for a possibility that may never happen. There is no claim in this essay that global warming will definitely occur, but the risks are high enough to support a program that attempts to estimate impacts and tries to develop mitigating action. Thus, improved evaluations of the possible impacts of climate change on human mortality are needed to guide policy decisions and to foster the international cooperation that is vital to deal with the possible threats effectively.

REFERENCES

1. White, M.R. and I. Hertz-Picciotto. 1985. Human Health: Analysis of climate related to health. *Characterization of Information Requirements for Studies of CO_2 effects: Water Resources, Agriculture, Fisheries, Forests and Human Health.* Washington: Department of Energy.
2. Kalkstein, L.S. and K.M. Valimont. 1987. Climate Effects on Human Health, pp. 122-152. In: *Potential Effects of Future Climate Changes on Forests and Vegetation, Agriculture, Water Resources, and Human Health,* EPA Science and Advisory Committee Monograph #25389.
3. World Health Organization. 1990. *Potential health effects of climatic change.* Geneva.
4. Calot, G. and C. Blayo. 1982. Recent course of fertility in Western Europe. *Population Studies* 36:345-372.
5. Sakamoto, M.M. and K. Katayama. 1971. Statistical analysis of seasonal variation in mortality. *Journal of the Meteorological Society of Japan* 49:494-509.

6. Persinger, M.A. 1980. *The Weather Matrix and Human Behavior.* New York: Praeger.
7. Quayle, R., and F. Doehring. 1981. Heat stress: A comparison of indices. *Weatherwise* 34:120-124.
8. Steadman, R.G. 1984. A universal scale of apparent temperature. *Journal of Climate and Applied Meteorology* 23:1674-1687.
9. Driscoll, D.M. 1971. The relationship between weather and mortality in ten major metropolitan areas in the United States, 1962-1965. *Journal of the International Society of Biometeorology* 15:23-40.
10. Ellis, F.P., F. Nelson, and L. Pincus. 1974. Mortality during heat wave in New York City, July 1972 and August and September 1973. *Environmental Research* 10:1-13.
11. Kalkstein, L.S. 1991. Bioclimatological Research and the Issue of Climatic Sensitivity. *Physical Geography.* 3:274-286.
12. Kalkstein, L.S. 1989. The impact of CO_2 and trace gas-induced climate changes upon human mortality, pp. 1-12-1-35. In: J.B. Smith and D.A. Tirpak (Ed.) *The potential effects of global climate change on the United States: Appendix G-Health.* Washington, DC: US Environmental Protection Agency.
13. Tan, G. 1991. Weather-related human mortality in Shanghai and Guangzhou, China. *Proceedings on the Tenth Conference on Biometeorology and Aerobiology.* American Meteorological Society, In Press.
14. Kalkstein, L.S., P.C. Dunne and R.S. Vose. 1990. Detection of climatic change in the western North American arctic using a synoptic climatological approach. *Journal of Climate* 3:1153-1167.
15. Jones, T.S., A.P. Liang, E.M. Kilbourne, M.R. Griffin, P.S. Patriarca, G.G. Wassilok, R.J. Mullan, R.F. Herricek, H.D. Donnel, Jr., K. Choi and S.B. Thacker. 1982. Morbidity and mortality associated with the July, 1980 heat wave in St. Louis and Kansas City. *Journal of the American Medical Association* 247:3327-3330.
16. Kalkstein, L.S. and R.E. Davis. 1989. Weather and human mortality: An evaluation of demographic and inter-regional responses in the United States. *Annals of the Association of American Geographers* 79:44-64.
17. Oechsli, F.W. and R.W. Buechley. 1970. Excess mortality associated with three Los Angeles September hot spells. *Environmental Research* 3:277-284.
18. Bridger, C.A., F.P. Ellis and H.L. Taylor. 1976. Mortality in St. Louis, Missouri, during heat waves in 1936, 1953, 1954, 1955 and 1966. *Environmental Research* 12:38-48.
19. Kalkstein, L.S. 1988. The Impacts of Predicted Climate Change on Human Mortality. In: J.R. Mather (Ed.) *Publications in Climatology*, pp. 110.
20. Applegate, W.B., J.W. Runyan, Jr., L. Brasfield, M.L. Williams, C. Konigsverg and C. Fouche. 1981. Analysis of the 1980 heat wave in Memphis. *Journal of the American Geriatrics Society* 19:337-342.
21. Rotton, J. 1983. Angry, sad, happy? Blame the weather. *U.S. News and World Report* 95:52-53.
22. Ellis, F.P. 1972. Mortality from heat illness and heat-aggravated illness in the

United States. *Environmental Research* 15:504-512.
23. Schuman, S.H. 1972. Patterns of urban heat wave deaths and implications for prevention: Data from New York and St. Louis during July, 1966. *Environmental Research* 5:58-75.
24. Smith, J.B. and D.A. Tirpak, eds. 1989. *The potential effects of global climate change on the United States.* Washington, D.C.: U.S. EPA Office of Policy, Planning and Evaluation.
25. Hansen, J.,I. Fung, A. Lacia, S. Lebedeff, D. Rind, R. Ruedy and G. Russell. 1988. Global climate changes as forecast by the GISS 3-D model. *Journal of Geophysical Research* 93:9341-9364.
26. Kniffen, F.B. 1965. Folk housing, key to diffusion. *Annals of the Association of American Geographers* 55:649-577.
27. Titus, J.G. 1989. Regional studies: southeast, pp. 3-23-3-58. In: J.B. Smith and D.A. Tirpak (Ed.) *The potential effects of global climate change on the United States.* Washington, D.C.: U.S. Environmental Protection Agency.
28. Kalkstein, L.S. 1990. A proposed global warming/health initiative. *Environmental Impact Assessment Review.* 10:383-392.

Global Climate Change: Implications, Challenges and Mitigation Measures. Edited by S.K. Majumdar, L.S. Kalkstein, B. Yarnal, E.W. Miller, and L.M. Rosenfeld. © 1992, The Pennsylvania Academy of Science.

Chapter Twenty-Seven

THE COSTS OF CLIMATE CHANGE TO THE UNITED STATES

JAMES G. TITUS
Office of Policy Analysis
U.S. Environmental Protection Agency
Washington, DC 20460

INTRODUCTION

Given the consequences that the other chapters of this book expect to result from global warming, it is hard to imagine that we would deliberately alter our planet in such a fashion. Yet, preliminary analyses on the subject generally conclude that the value to society of avoiding these consequences is not as great as the cost of decreasing emissions, especially when one "discounts" future benefits to present values (e.g. Nordhaus 1990). This paper, however, makes two departures from previous studies, and thereby reaches a much higher estimate of the cost of heating our planet.

First, we focus on the range of uncertainty, rather than merely the central estimate, of a given impact. Focusing only on central estimates understates the value of an environmental risk because (a) our uncertainty tends to be skewed; (b) the damage function is often nonlinear; and (c) people are risk-averse. Suppose, for example, that our uncertainty regarding the impact of climate change on a farm is lognormal with the most likely impact a loss of 9 acre feet with nine-fold uncertainty; that the cost to the farmer is equal to the number of acre feet lost, raised to the 1.5 power; and that people will pay 25 cents to avoid a risk that has a standard deviation of $1. Elementary probability theory shows that the expected decline in water would be 16.5 acre-feet; the expected cost would be $105, with a standard deviation of $395; hence, people would be willing to pay about $200 to avoid such a risk. By contrast, using only the best estimates and ignoring risk aversion would lead one to expect a loss of 9 acre-feet and a cost of only $27.

Second, we include environmental and other nonmarket impacts. Because estimates of these assets range from poor to nonexistent, *we assume that society will undertake the necessary costs to offset environmental effects of global warming.*

The following sections (1) develop equations that summarize the nationwide impact of effects that have already been quantified; (2) develop state-by-state equations for four areas that have not been previously estimated in detail; (3) project the costs of climate change for two alternate emissions scenarios; and (4) calculate the benefits of reducing emissions. We assume that the U.S. economy will grow 1.2 to 2.1 percent per year. For the most part, our estimates are based on the GISS (Goddard Institute for Space Studies) and GFDL (Princeton Geophysical Fluid Dynamics Laboratory) models whose implications are discussed in detail by Smith and Tirpak (1989). Our calculations suggest that a CO_2 doubling would cost the United States $37-351 billion per year, even up to $92-130 billion. At a 3% discount rate, the CO_2 from burning one gallon of gasoline will cause damages of 16-36 cents.

NATIONWIDE CALCULATIONS BASED ON SMITH AND TIRPAK (1989)

A previous EPA Report to Congress (Smith and Tirpak 1989) quantified costs for agriculture, energy consumption, and sea level rise. For those studies, our task was to interpret the results in a common framework that enables us to estimate the costs for different years and amounts of climate change. The EPA study also provided estimates of increased mortality, which can be readily monetized using existing estimates of the regulatory cost of saving lives.

Agriculture

Crop modelers have already examined many of the ways by which global warming could affect agriculture, including longer growing seasons in colder areas, heat stress in the south, increased evaporation, changes in precipitation, the CO_2 fertilization effect, and changes in pests. Based on available studies, Adams *et al.* (1989) estimated the impact for the year 2060, assuming that CO_2 doubles and that the climate reaches the equilibria implied by the general circulation models. They reported that the GISS and GFDL models imply annual losses of $7.45 and $42.4 billion ($1984), ignoring CO_2 fertilization; and −$13.4 (net benefit) and $12.25 billion including CO_2 fertilization.[2]

Given these estimates, we need equations that can generalize the results for different climate scenarios and different years. (We use the term "generalize" to refer to both interpolation of smaller changes and extrapolation to larger changes). Our calculations employ separate equations for the GISS and GFDL models; we will discuss only the former. Our low and high equations are as follows:

$Cost_{median} = 5.853\ T/T_{eq2} - 16.4\ CO_2{}^*$
$Cost_{low} = Cost_{median} - 3.7\ T/T_{eq2} - 6.129\ T/T_{eq2} - 4.1\ CO_2{}^*$
$Cost_{high} = Cost_{median} + 1\ T/T_{eq2} + 6.3\ T/T_{eq2} + 5.0\ CO_2{}^*$
where $CO_2{}^* = (CO_2 - 330)/330$ for $CO_2 < 660$ ppm
$= \ln(CO_2/330)/\ln(2)$ for $CO_2 > 660$ ppm
T = transient global temperature increase;
T_{eq2} = The model's estimated equilibrium warming for $2 \times CO_2$;
and: We multiply these results by 1.005 (year-1985) to account for population growth and by 1.263 to convert to $1990.

The median equation assumes that the climate impact is directly proportional to the change in global temperatures. Up to a doubling, we assume that the beneficial impacts of CO_2 are also linear. However, because crop modelers have noted that the beneficial impact of CO_2 eventually reaches a point of diminishing returns, we assume that the impact is only logarithmic past the doubling point (i.e. a CO_2 quadrupling has twice the beneficial impact of a CO_2 doubling). The low and high equations include Adams' yield uncertainties for climate and CO_2 fertilization.

Energy Consumption

Global warming would decrease energy consumption for space heating while increasing requirements for cooling. However, the cost of cooling a house one degree is greater than the cost of heating it one degree because (1) air conditioners require a more expensive form of energy (electricity) than home heating (mostly oil and natural gas); and (2) air conditioning takes place during peak hours while heating is mostly required at night.

Linder and Inglis (1989) developed a national model of how electricity consumption responds to temperature changes based on daily demand and weather data for five utility regions. They estimated that the increased consumption of electricity in the year 2060 implied by the GISS transient model would increase costs $37 billion with low economic growth and 53-81 billion ($1986) with high economic growth. Our generalizing equations are as follows:

Baseline = P growth 79.6
$electric_{median} = (1.016^T - 1)$ Baseline
$electric_{high} = electric_{median}\ \sigma$
$electric_{low} = electric_{median}\ /\ \sigma$
where

T is transient global temperature increase;
P accounts for increasing price, assuming that (real) electricity prices grow 1.37% through 2060 and are stable thereafter;
   growth accounts for growth in electricity consumption, assuming that in the absence of climate change, consumption grows at 90 percent of the rate of general economic growth; and
σ represents a four-fold and uncertainty for a given rate of growth.

These estimates fail to account for impacts of changing precipitation (e.g. pumping for irrigation) and adaptation (e.g. more insulation). They ignore the impact of global warming on the cost of meeting the increased demand, by assuming that it could be met by coal-fired utilities. If non-greenhouse gas sources are required, the unit generating costs may be greater; reduced availability for cooling water may also increase the costs of new capacity. We do not adjust these calculations downward to account for reductions in space heating.[3]

Sea Level Rise

Global warming could raise sea level by (1) expanding seawater, (2) melting mountain glaciers, causing the ice sheets in (3) Greenland and (4) Antarctica to melt or discharge ice into the oceans, and by (5) depleting groundwater tables. Our calculations consider only the first three categories.

We used the same equations as IPCC for mountain glaciers and Greenland.[4] For thermal expansion, we used the model developed by Hoffert *et al.* (1980) to specify these equations, with $\pi = 0$ for the low scenario and $\pi = 1$ for the high estimate (π is the ratio of polar bottom water warming to temperature increase at the surface of the ocean). In both cases, we ran the model for 500 years for an assumed warming of 2°C. In the low case, the rise was 18, 24, and 30 cm after 100, 200 and 500 years, respectively. For the high case, the rise was 21, 33, and 59 cm.

We then estimated the following regression equations:

$\text{Expand}_{low}(t)$ = 1.26T(t-1) - 1.24T(t-2) + 0.104T(t-3) + 0.0141T(t-4) + 0.003393T(t-5) + 1.712 Expand_{low}(t-1) - 0.721 Expand_{low}(t-2)

$\text{Expand}_{high}(t)$ = 1.24T(t-1) - 1.3383T(t-2) + 0.374T(t-3) + 0.023386T(t-4) + 0.009048T(t-5) + 0.003804T(t-6) + 1.796957 Expand_{low}(t-1) - 0.79832 Expand_{low}(t-2)

where T represents transient global temperature and both equations use time steps of 5 years. In each case the R-squared was greater than 0.9999.

Note that these equations use first and second order lagged dependant variables. The effect of doing so is to impose the assumption that after the first 5 or 6 time periods, the effect of a temperature increase diminishes as the sum of two declining exponentials. The adjustment times for the low and high equations are 100 and 600 years. Figure 1 shows the resulting sea level rise projections.

Using these projections, we estimate the costs of sea level rise for wetland loss, dike construction, land for dikes, loss of dry land, beach nourishment, and the cost of elevating infrastructure in low areas, based on Titus *et al.* (1991). For each of the cost categories, the study had reported low and high estimates for the 12 (baseline), 50, 100, and 200 cm scenarios. For sea level rise less than 200 cm, this analysis interpolates those estimates to calculate the total cost of sea level rise by a given year. For dry-land loss and dike costs, we assume that development in coastal areas will grow at the same rate as the general economy; but we assume that does not increase wetland loss or the cost of protecting barrier islands.

The slope of coastal land is generally much steeper above the one-meter contour than below it. Thus, for sea level rise greater than 200 cm, our calculations for elevating structures and loss of dry land assume that costs rise at the same rate as with a rise of 100 to 200 cm. We assume that a 3.3 m rise would inundate the remaining coastal wetlands, and interpolate linearly. For sand, we assume that the incremental costs stay the same. For dikes, we assume that costs rise with the 1.5 power of sea level rise.

Health

Heat- and cold-related mortality are the only health impacts of global warming that have been quantified. Kalkstein (1989) estimates that given today's population in the 15 largest cities, a CO_2 doubling would increase annual heat-related deaths by 529-3878 among the elderly and 513-2368 among other age groups, while reducing cold-related deaths by 59-123 and 25-68 among the two groups. Violette and Chestnut (1989) estimate that the value of reducing the risk of a statistical death is between $1.6 and 8.5 million; this estimate does not say how much a human life is worth, but rather the extent to which society currently invests resources to avoid deaths from pollution and accidents.

Increase in Temperature and Sea Level Over Time

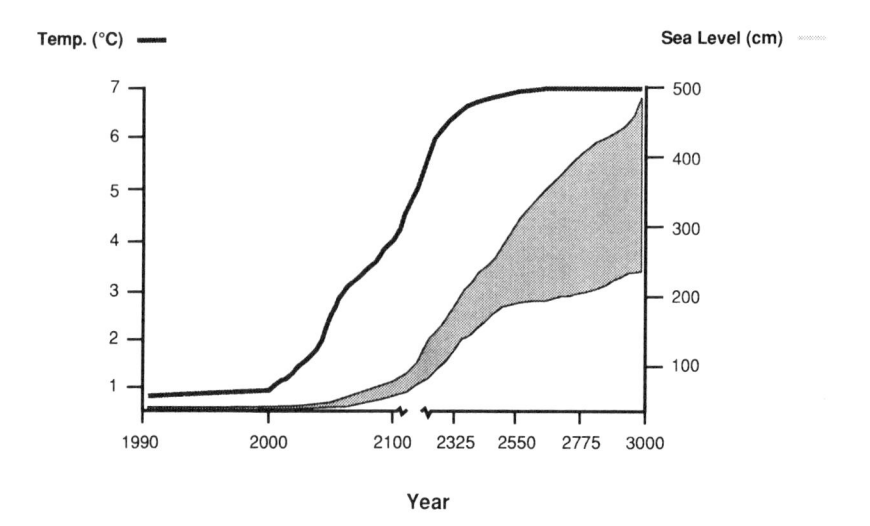

FIGURE 1. Sea level curves show the range of uncertainty given the trend for temperatures. The temperature curve starts at 0.5°C to account for past warming due to greenhouse gases.

We generalize the relationships as follows:

Deaths$_{high}$ = 6110 T
Deaths$_{low}$ = 1646 T
Cost$_{high}$ = 8,500,000 Deaths growth
Cost$_{low}$ = 8,500,000 Deaths growth

where growth represents economic growth as described above.

STATE BY STATE CALCULATIONS OF EFFECTS NOT PREVIOUSLY QUANTIFIED

For the impacts not previously quantified, calculating impacts on a state-by-state basis seemed to be an appropriate level of aggregation. The national level is too broad because it cannot capture all the ambiguities, such as some areas becoming wetter while others become drier; and a warmer or drier climate helping a cold area suffering from too much rainfall, while harming a warm area where water supplies are only barely adequate. In contrast, state-level analyses can capture these regional differences.

We generate our scenarios for annual and summer changes in climate by assuming that (1) the climate of a state is characterized by the climate of its capital; (2) the change in temperature for a CO_2 doubling is characterized by the difference between temperatures projected by the doubled CO_2 and control runs of the GISS and GFDL models; (3) the change in rainfall for a CO_2 doubling is characterized by the ratio of changes implied by the doubled CO_2 and control runs; and (4) the change in rainfall or temperature by a particular year is proportional to the change in global temperatures.

Automobile Air Conditioners

Warmer temperatures will lead motorists to run their air conditioners for a greater percentage of the time. General Motors estimates that the average automobile in the United States uses 20 gallons of gasoline to run its air conditioner for every 10,000 miles driven. It also estimates that on days with a daily average temperature of 50, 60, and 90°F, 0, 30, and 90 percent of all miles driven by automobiles with air conditioners are driven with the air conditioners on.[5] These assumptions imply that the U.S. currently consumes 3.84 billion gallons per year on automobile air conditioning.

Our calculations use those results along with the assumption that fuel consumption is proportional to the difference between the daily average temperature and 50°F.

Air Pollution: Ozone

Climate plays an important role in determining whether air pollutants are transported out of harm's way or linger enough to threaten people's health. Of the many ways by which global warming could change air quality, the only quantified so far is the impact of warmer temperatures on the formation of tropospheric ozone.

Gerry (1987) estimated the possible impacts of 2 to 5°C warmings on ozone concentrations in Los Angeles, New York, Philadelphia, and Washington,[6] with ozone concentrations increasing 1.6875 ± .3684% per degree warming. Morris *et al.* (1988) applied a regional transport model to central California and the part of the United States (approximately) between 78 and 98°N longitude (that is, west of central Maryland and east of San Antonio, Texas.) They found that maximum ozone concentrations would increase 1.4525 ± .497 percent per degree warming.[7] Because Gerry and Morris *et al.* reach similar results, we assume that ozone concentrations increase 1.5 ± 0.5% per degree warming.

Estimating the cost of preventing such an increase requires us to consider (a) the percentage reduction in emissions, (b) the absolute level of emissions that would otherwise occur, and (c) the cost of reducing emissions by one ton. Smith and Tirpak (1988) estimate that the 6% increase in ozone concentrations implied by Morris *et al.* would require emissions of volatile organic compounds (VOC) to be reduced 11.7%; hence we assume that the reduction in VOC would be 1 to 2% for every 1% increase in ozone concentrations.

Pechan Associates (1990) estimates that the baseline emissions for 2005 would be approximately 14 million tons, and that emissions are increasing by about 3% per year. They also estimate that the incremental cost of reducing emissions ranges from $1700 to $5000. Smith and Tirpak state that the cost is $5000. EPA's Office of Air Quality Planning and Standards told us that even at the $5000/ton cost, the available emission reductions are not unlimited. Nevertheless, we assume that emissions can be controlled for $1700-5000 per ton.

WATER RESOURCES

Rising temperatures and reduced rainfall could increase the demand for irrigation water and diminish the supply of water for all uses. The resulting reductions in river flows would reduce hydropower production and worsen water quality (unless the discharge of pollutants was also reduced).[8]

Baseline Conditions

Our starting points are USGS data on withdrawals of ground and surface water by the agriculture, residential, and industrial sectors of each state. Based on Gibbons (1988), we assume that the price of surface water is currently $35-85 per acre foot, and that the elasticity of demand for water is between 0.5 and 1.0.[9] We assume the same range for elasticity of supply, which is probably optimistic because in the west current supplies are already oversubscribed, and in the east, water experts generally doubt that any more major dams will be built due to the adverse environmental effects. We assume that residential and industrial demand will increase by the same rate as economic growth, but that irrigation demand does not increase. As Table 2 shows, our baseline water prices range from $60 to $250 per acre foot by 2060.

Our approach is similar for groundwater. However, in those regions where the ratio of groundwater overdraft to recharge is greater than 0.2, we assume that the current price of water is the same as the surface price, and that there is a near-fixed supply elasticity of 0.1. Where there is no overdraft, we assume that unlimited water can be pumped for $10 per acre-foot. For intermediate cases, we interpolate between these two assumptions.

Changes in Supply and Demand

Previous studies have estimated changes in water requirements (per acre) for a dozen or so sites that are already irrigated in the Southeast (Peart *et al.* 1989), Great Plains (Rosensweig), and the Great Lakes Region (Ritchie).[10] We estimated summary regression equations, shown in Table 1.[11] We use the state-specific results of Paterson and Keller (1990) low and high estimates of the elasticity of irrigated acreage with respect to temperature and precipitation.

Finally, residential lawn watering and other outdoor uses would also change. Because only 25 percent of residential use is consumptive, we assume that it would increase by 25 percent of the fraction by which irrigation per acre increases in a given state. We include this impact within irrigation demand. We calculate the increased cost of delivering more water to residential users, based on the current difference between residential and market prices.

The supply of water could change for two reasons: (1) more (or less) rainfall would increase (decrease) the amount of water flowing in rivers and recharging aquifers; and (2) higher temperatures increase evaporation and hence reduce the availability of water if nothing else changes. We use the simple model from Waggoner and Revelle (1990) which assumes that runoff decreases 2-4 percent if either precipitation decreases 1 percent or temperature increases 0.4°. We assume that in the west, river flow, groundwater, and surface supplies are available for withdrawal decline by the same fraction as runoff. For the east, our assumptions are the same, except that because flows are very large relative to withdrawals, we assume that runoff changes have no impact on surface supplies.[12]

If the change in climate is small, the cost to society can be calculated simply as the price of water times the sum of the increase in demand plus the decrease in supply. But for large changes, this assumption understates the impact because as prices rise, each additional shift in supply or demand costs more than the previous shift. Figure 2 illustrates the general case. As demand increases, more water must be delivered; prices will rise, which will lead some current users to conserve water and spend more on labor, fertilizer, or some other factor of production to achieve a harvest of a given size.

The cost of a reduction in supply can also be divided into increased extraction cost and the cost of conserving water. In Figure 2c, the left triangle shows the

TABLE 1
Regression Equations Predicting Irrigation Per Acre As A Function of Climate Change

SOUTHEAST (Results Assume 600 ppm CO_2 fertilization)

Δln(Soy Irrigation) = 0.05ΔT_jja - 0.619Δln(Pjja)
        (0.00365) (0.0891)

       R^2 = 0.745 S.E. = 0.087

Δln(Corn Irrigation) = 0.0200ΔTjja - 0.592Δln(Pjja)
         (0.00372) (0.0808)

       R^2 = 0.9400 S.E. = 0.0372

GREAT PLAINS (Results Ignore CO_2 fertilization)

Δln(Corn Irrigation) = 0.0172ΔTjja - 0.45567Δln(Pjja)
         (0.0921) (0.00252)

       R^2 = 0.833 S.E. = .0387

Δln(Wheat Irrigation) = 0.1763ΔT -1.25Δln(P) + 0.00918Δln(Yield)
         (0.116) (0.250)   (0.001911)

       R^2 = 0.551 S.E. = 0.0348

GREAT LAKES

With 600 ppm CO_2 fertilization

Δ%Soy Irrigation = -0.684 + 0.193ΔTjja -0.687Δ%Pjja +0.165Δ%Yield
         (0.0387) (0.467)   (0.0432)

       R^2 = 0.849 S.E. = 0.226

Δ%Corn Irrigation = -1.05 + 0.0712ΔTjja + 0.408Δ%Yield
         (1.23)   (0.127)

       R^2 = 0.511 S.E. = 0.125

Without CO_2 fertilization

Δ%Soy Irrigation = 0.0855ΔTjja -1.5523Δ%Pjja +0.360Δ%Yield
        (0.0137)   (0.4113) (0.1253)

       R^2 = 0.659 S.E.= 0.3186

Δ%Corn Irrigation = 0.0697ΔTjja - 1.539 Δ%Pjja + 1.0012Δ%Yield
         (0.017)   (0.127)   (0.2445)
       R^2 = 0.5353 S.E.= 0.331

Combined

Δ%Soy Ir = 0.0900ΔTjja -1.5817Δ%Pjja +0.202Δ%Yield - 0.000763ΔCO2
     (0.0111)   (0.2647)   (0.0484)   (0.00022583)

       R^2 = 0.736 S.E.= 0.2841

Δ%Corn Ir = 0.0641ΔTjja - 0.729 Δ%Pjja + 0.562Δ%Yield -0.00322ΔCO2
      (0.0126)   (0.265)   (0.143)   (0.000201)

       R^2 = 0.8004 S.E.= 0.2848

TABLE 2
Change in Surface Water Supply and Demand for GISS CO_2 Doubling Low Scenario

	Base	Consumption Climate Change	P0	Price P1	% Change Irrigation	% Change Demand	% Change Supply	Cost Demand	Cost Supply
AL	17.52	15.63	85	96	215.33	.24	-20.64	3.59	.00
AZ	3.60	3.34	58	63	1.22	.97	-14.70	2.05	48.15
AR	4.76	4.46	80	94	110.41	10.64	-20.64	41.67	136.01
CA	29.97	27.85	60	76	24.91	17.21	-26.35	327.07	891.27
CO	16.11	16.63	60	72	34.11	23.65	-13.82	243.12	256.04
CT	2.09	1.82	85	98	89.54	.56	-24.70	1.00	.00
DE	.10	.10	84	81	46.21	-.47	6.43	-.04	.00
FL	5.54	4.86	75	92	37.86	7.91	-28.57	33.55	.00
GA	9.54	9.88	84	82	63.09	.83	6.43	6.67	.00
ID	24.57	26.40	66	59	-9.84	-4.43	20.84	-71.67	.00
IL	28.05	26.46	85	90	78.59	-.06	-11.00	-1.38	.00
IN	22.78	21.50	85	90	78.93	.03	-11.00	.59	.00
IA	4.02	3.80	85	90	130.56	.37	-11.00	1.28	58.05
KS	1.47	1.34	73	106	124.31	31.69	-37.18	36.68	85.92
KY	7.71	7.95	85	82	53.17	-.25	6.43	-1.64	.00
LA	18.60	17.10	82	97	159.06	8.29	-21.92	129.80	.00
ME	1.35	1.18	85	98	378.67	1.20	-24.70	1.39	.00
MD	1.70	1.75	85	82	82.31	-.26	6.43	-.37	.00
MA	3.67	3.19	85	98	91.92	.22	-24.70	.70	.00
MI	24.49	20.29	85	103	168.38	.65	-31.81	13.64	.00
MN	4.20	3.97	85	90	175.69	.48	-11.00	1.72	.00
MS	2.39	2.20	83	96	182.69	7.36	-20.64	14.86	.00
MO	11.21	10.00	85	96	138.39	.27	-20.64	2.55	309.08
MT	13.12	14.76	58	55	8.86	7.08	18.27	54.94	-217.84
NE	7.65	7.69	71	83	59.01	17.13	-13.82	97.95	136.76
NV	3.49	3.09	58	70	8.67	6.98	-26.35	14.52	91.06
NH	.56	.49	85	98	378.67	.96	-24.70	.46	.00
NJ	3.67	3.21	85	99	109.41	1.28	-24.70	4.03	.00
NM	2.43	2.43	56	66	18.88	16.83	-14.70	24.08	36.42
NY	12.61	11.02	85	99	431.80	1.47	-24.70	15.90	.00
NC	12.75	13.17	85	82	68.00	.27	6.43	2.88	.00
ND	15.91	14.86	85	92	311.99	1.19	-13.82	16.14	291.91
OH	22.79	18.83	85	103	167.98	.13	-31.81	2.57	.00
OK	1.26	1.04	80	106	102.45	9.64	-37.18	10.02	67.30
OR	6.93	7.64	59	54	.80	.44	20.84	1.80	-125.31
PA	26.23	21.86	85	104	422.65	1.83	-31.81	41.12	.00
RI	.24	.21	84	98	90.15	1.27	-24.70	.26	.00
SC	9.80	10.12	85	82	77.58	.16	6.43	1.35	.00
SD	.44	.65	59	102	214.37	153.78	-13.82	51.71	14.30
TN	16.83	15.00	85	96	230.71	.08	-20.64	1.13	.00
TX	9.18	8.31	71	96	70.14	22.61	-33.20	155.25	426.99
UT	4.34	4.62	62	56	-6.58	-4.21	18.27	-11.28	-69.09
VT	.53	.46	85	98	321.27	.43	-24.70	.19	.00
VA	9.11	9.39	85	82	31.32	-.08	6.43	-.61	.00
WA	9.47	10.68	61	54	-1.32	-1.13	28.45	-6.56	.00
WV	9.47	9.76	85	82	-2.57	-.09	6.43	-.73	.00
WI	9.12	8.60	85	90	284.09	.03	-11.00	.23	.00
WY	5.62	5.80	57	68	27.71	23.84	-13.82	80.74	84.45
US		(scaled and including price uncertainty)						36.5	551.5

Note: P0 and P1 = baseline and greenhouse-induced price of water ($/ac-ft)
Costs are in millions of dollars per year; quantities are in millions of acre-feet per year.

increased cost of supplying water from the preexisting pumps that continue to operate. The right triangle illustrates the difference between the value of using water for users who choose to consume at a higher price and the previous cost of pumping from wells that are closed due to the increased cost.

We do not allow for water being transported between states; nor do we allow for consumers to switch between ground and surface water. To prevent anomalies, in which dramatic supply and demand crunches cause implausible price rises, we assume that a state can always build a project that provides all of the increased water requirements for $300 to 700 per acre foot.

Water Pollution and Hydropower

We assume that the discharge of pollutants would change by the same proportion as does river flow, to maintain current water quality. EPA's 1990 Report to Congress on the Cost of a Clean Environment reports that state and local water pollution costs for point sources (mostly sewage treatment plants) which are growing

Cost of Increased Demand or Decreased Supply of Water

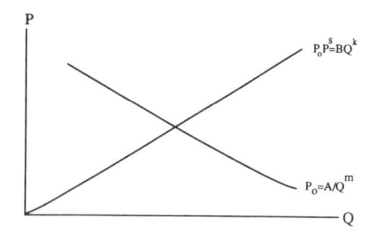

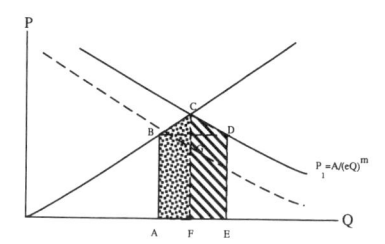

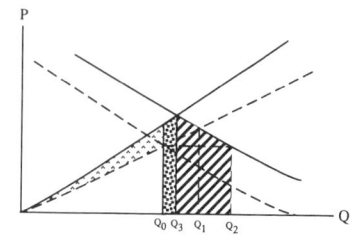

FIGURE 2.

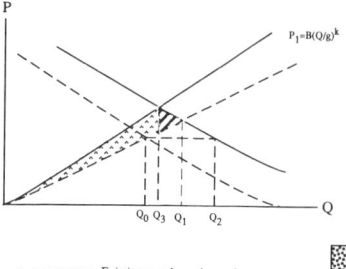

at 3.6 percent per year and will be $18.8 billion by the year 2000. Other costs (mostly industrial) are growing at 4.5 percent per year and will reach $45.3 billion by the year 2000. We assume that in the baseline, these costs would only grow in step with the general economy after 2000.

According to EPA Region III, assuming that one has already achieved secondary (85%) control, the cost elasticity for further control is about 1.0. For the last several years, each industry-specific water-quality regulation issued by EPA has been accompanied by a Regulatory Impact Assessment, which must estimate the marginal cost of controlling pollutants under the regulation, as well as an estimate of the marginal cost of the next more stringent technology. Based on these analyses, we assume elasticities of 1.0 to 1.73.

We assume that hydropower production declines in proportion to the change in runoff, using data on current hydropower from Edison Electric Institute (1985). We assume that there will be no increase in generating capacity.

Results

Tables 2 and 3 illustrate our intermediate calculations of surface water supply and demand, based on the GISS general circulation model. We display our results at the state level *only* to permit the reader to better understand the limitations of our calculations. Our nationwide results are presented both in $1990 and $1990 scaled downward to account for economic growth. Note that for surface water, the uncertainty surrounding our baseline assumptions is greater than the impact of climate change. Nevertheless, our equations imply that increased demand for irrigation water could raise the total demand for water by 10 percent for about 10 states. About three quarters of the states would experience declines in water supplies while the rest would have increased availability. The scaled cost of these changes would be $1-3 billion per year for the GISS scenario and $3-7 billion for GFDL.

Table 4 illustates our groundwater calculations for both the low and high scenarios. Although the low-growth scenario implies that none of the states would be paying over $200/acre foot, the high-growth scenario has 24 states doing so, with 13 states paying over $500/acre foot (at the margin). The scaled cost would be $1.1-4.6 billion per year for the GISS scenario and $1.5-6 billion for GFDL.

Finally, Table 5 illustrates our estimates for the other water resources and the total cost. The water *quality* problems of global warming would be more expensive than the water *quantity* problems. Under GISS, (scaled) pollution control costs would increase $15-52 billion (compared with a base of $64 billion), with the total water resource cost $21-60 billion. With GFDL, the cost is $31-87 billion, of which $21-67 billion is for pollution control. Nevertheless, some areas might have more favorable conditions. Under the GISS scenario, wetter conditions in the Pacific Northwest would increase hydropower production by 2-5 billion (unscaled) dollars, more than offsetting declines in California and Washington. However, the dry conditions of the GFDL scenario would lead hydropower production to decline in every state for a total loss of 4-13 billion (unscaled) dollars.

TABLE 3
Change in Surface Water Supply and Demand for GISS CO_2 High Scenario

	Consumption Base	Climate Change	Price PO	Price P1	% Change Irrigation	% Change Demand	% Change Supply	Cost Demand	Cost Supply
AL	24.42	19.95	259	389	232.94	.15	-33.35	9.82	.00
AZ	3.94	3.44	75	100	1.27	.85	-24.43	2.52	60.24
AR	6.48	5.49	217	351	148.08	7.71	-33.35	110.68	472.60
CA	34.30	28.12	88	174	27.00	14.71	-41.41	464.25	1530.54
CO	18.30	17.55	86	133	36.20	19.49	-23.06	321.94	385.16
CT	2.91	2.30	256	432	156.21	2.52	-39.13	18.96	.00
DE	.14	.14	252	223	50.53	-1.14	11.77	-.39	.00
FL	5.54	4.86	75	92	37.86	7.91	-28.57	33.55	.00
GA	9.54	9.88	84	82	63.09	.83	6.43	6.67	.00
ID	24.57	26.40	66	59	-9.84	-4.43	20.84	-71.67	.00
IL	28.05	26.46	85	90	78.59	-.06	-11.00	-1.38	.00
IN	22.78	21.50	85	90	78.93	.03	-11.00	.59	.00
IA	4.02	3.80	85	90	130.56	.37	-11.00	1.28	58.05
KS	1.47	1.34	73	106	124.31	31.69	-37.18	36.68	85.92
KY	7.71	7.95	85	82	53.17	-.25	6.43	-1.64	.00
LA	18.60	17.10	82	97	159.06	8.29	-21.92	129.80	.00
ME	1.35	1.18	85	98	378.67	1.20	-24.70	1.39	.00
MD	1.70	1.75	85	82	82.31	-.26	6.43	-.37	.00
MA	3.67	3.19	85	98	91.92	.22	-24.70	.70	.00
MI	24.49	20.29	85	103	168.38	.65	-31.81	13.64	.00
MN	4.20	3.97	85	90	175.69	.48	-11.00	1.72	.00
MS	2.39	2.20	83	96	182.69	7.36	-20.64	14.86	.00
MO	11.21	10.00	85	96	138.39	.27	-20.64	2.55	309.08
MT	13.12	14.76	58	55	8.86	7.08	18.27	54.94	-217.84
NE	7.65	7.69	71	83	59.01	17.13	-13.82	97.95	136.76
NV	3.49	3.09	58	70	8.67	6.98	-26.35	14.52	91.06
NH	.56	.49	85	98	378.67	.96	-24.70	.46	.00
NJ	3.67	3.21	85	99	109.41	1.28	-24.70	4.03	.00
NM	2.43	2.43	56	66	18.88	16.83	-14.70	24.08	36.42
NY	12.61	11.02	85	99	431.80	1.47	-24.70	15.90	.00
NC	12.75	13.17	85	82	68.00	.27	6.43	2.88	.00
ND	15.91	14.86	85	92	311.99	1.19	-13.82	16.14	291.91
OH	22.79	18.83	85	103	167.98	.13	-31.81	2.57	.00
OK	1.26	1.04	80	106	102.45	9.64	-37.18	10.02	67.30
OR	6.93	7.64	59	54	.80	.44	20.84	1.80	-125.31
PA	26.23	21.86	85	104	422.65	1.83	-31.81	41.12	.00
RI	.24	.21	84	98	90.15	1.27	-24.70	.26	.00
SC	9.80	10.12	85	82	77.58	.16	6.43	1.35	.00
SD	.44	.65	59	102	214.37	153.78	-13.82	51.71	14.30
TN	16.83	15.00	85	96	230.71	.08	-20.64	1.13	.00
TX	9.18	8.31	71	96	70.14	22.61	-33.20	155.25	426.99
UT	4.34	4.62	62	56	-6.58	-4.21	18.27	-11.28	-69.09
VT	.53	.46	85	98	321.27	.43	-24.70	.19	.00
VA	9.11	9.39	85	82	31.32	-.08	6.43	-.61	.00
WA	9.47	10.68	61	54	-1.32	-1.13	28.45	-6.56	.00
WV	9.47	9.76	85	82	-2.57	-.09	6.43	-.73	.00
WI	9.12	8.60	85	90	284.09	.03	-11.00	.23	.00
WY	5.62	5.80	57	68	27.71	23.84	-13.82	80.74	84.45
US	Adding in the price uncertainty:							1763.7	2715.6

Note: PO and P1 = baseline and greenhouse-induced price of water ($/ac-ft)
Costs are in millions of dollars per year; quantities are in millions of acre-feet per year.

TABLE 4
Change in Groundwater Supply and Demand for GISS CO_2 Doubling

State	Overdraft L	Overdraft H	Elasticity L	Elasticity H	P0 L	P0 H	P1 L	P1 H	Demand Cost Low	Demand Cost High	Supply Cost Low	Supply Cost High
AL	.18	.42	∞	.1	50	50	500	1007	.8	3.2	.0	100.9
AZ	.65	.88	.1	.1	61	72	93	150	1.6	3.7	84.9	132.6
AR	1.68	2.60	.1	.1	67	135	131	598	165.4	472.6	205.6	741.7
CA	.25	.43	.1	.1	71	106	156	450	117.4	333.7	677.2	2291.8
CO	.00	.00	∞	∞	10	10	10	10	16.5	10.3	.0	.0
CT	.00	.00	∞	∞	10	10	10	10	.0	.1	.0	.0
DE	.02	.04	∞	∞	12	12	16	16	.0	.0	.0	.0
FL	.16	.36	∞	.1	45	45	374	1097	27.2	115.2	.0	1460.4
GA	.12	.20	∞	.1	35	35	440	394	7.2	34.1	.0	-77.5
ID	.11	.17	∞	.3	34	34	187	121	-14.0	-6.7	.0	-657.8
IL	.00	.00	∞	∞	10	10	10	10	-.1	.2	.0	.0
IN	.00	.00	∞	∞	10	10	10	10	1.3	.9	.0	.0
IA	.68	1.40	.1	.1	117	136	500	741	3.5	20.2	30.7	138.8
KS	.90	1.47	.1	.1	61	174	88	1045	299.3	618.7	461.9	2084.1
KY	.00	.00	∞	∞	10	9	10	10	.0	.0	.0	.0
LA	.29	.51	.1	.1	89	156	315	969	58.0	211.8	106.8	556.5
ME	.00	.00	∞	∞	10	10	10	10	.0	.0	.0	.0
MD	.02	.04	∞	∞	12	12	17	17	.1	.0	.0	.0
MA	.00	.00	∞	∞	10	10	10	10	.0	.2	.0	.0
MI	.07	.17	∞	.3	23	23	304	803	3.3	26.4	.0	243.2
MN	.00	.00	∞	∞	10	10	10	10	2.0	1.9	.0	.0
MS	.30	.50	.1	.1	84	162	268	964	67.5	245.5	912.9	456.0
MO	.76	1.69	.1	.1	109	150	500	#	4.9	47.5	26.7	158.6
MT	.38	.58	.1	.1	93	80	353	220	.2	1.9	-23.6	-26.8
NE	.64	1.27	.1	.1	105	128	499	824	38.9	200.4	268.7	1434.2
NV	.83	1.51	.1	.1	73	101	176	460	1.7	6.2	27.5	99.8
NH	.00	.00	∞	∞	10	10	10	10	.0	.0	.0	.0
NJ	.03	.08	∞	.8	15	15	79	116	.9	3.0	.0	56.9
NM	.45	.68	.1	.1	65	84	117	238	11.1	41.1	45.7	98.8
NY	.03	.08	∞	.8	15	15	82	121	1.3	3.8	.0	59.1
NC	.13	.25	∞	.1	39	39	500	419	.7	3.3	.0	-57.2
ND	.73	1.53	.1	.1	119	145	500	818	5.3	22.1	38.5	183.8
OH	.00	.00	∞	∞	10	10	10	9	.1	.3	.0	.0
OK	2.12	4.35	.1	.1	78	167	215	#	26.8	105.6	76.9	481.7
OR	.10	.14	∞	.5	31	31	97	69	.0	1.0	.0	-38.8
PA	.04	.09	∞	.7	16	16	100	173	1.0	2.9	.0	121.1
RI	.00	.00	∞	∞	10	9	10	9	.0	.0	.0	.0
SC	.13	.24	∞	.1	39	39	500	417	.2	.6	.0	-16.3
SD	.79	1.31	.1	.1	92	155	346	867	13.4	54.1	15.3	70.2
TN	.00	.00	∞	∞	10	10	10	10	1.1	1.0	.0	.0
TX	2.14	3.71	.1	.1	70	138	148	751	222.6	588.0	611.8	2584.1
UT	.44	.53	.1	.1	69	56	142	80	-2.0	-5.7	-56.7	-31.3
VT	.00	.00	∞	∞	10	10	10	10	.0	.0	.0	.0
VA	.02	.04	∞	∞	12	12	17	17	.0	.0	.0	.0
WA	.12	.20	∞	.1	36	36	434	260	-.6	71.1	.0	-198.5
WV	.00	.00	∞	∞	10	10	10	9	.0	.0	.0	.0
WI	.00	.00	∞	∞	10	10	10	10	1.5	.9	.0	.0
WY	.43	.64	.1	.1	67	89	127	260	4.1	13.9	11.9	26.7
US	(scaled, including price uncertainty)								222.5	1263.4	638.6	4478.5

NOTE # Supply curve shift is unrealistically costly; hence we assume that project is built that delivers water for $800/acre-foot.
See Table 2 for explanation of other units.

TABLE 5
Other Water Related Costs for GISS CO_2 Doubling
($millions, scaled for economic growth)

State	Residential Low	Residential High	Pollution Low	Pollution High	Hydropower Low	Hydropower High	Total Water costs ($billions) Low	median	high	mean
AL	13.5	33.8	277.2	858.9	39.2	114.1	.377	.615	1.004	.693
AZ	.1	.4	152.1	478.1	40.9	120.6	.245	.405	.669	.459
AR	3.2	16.0	162.3	502.8	9.9	28.8	.333	.635	1.211	.783
CA	15.6	77.5	2708.7	8275.9	198.7	571.8	3.764	6.281	10.482	7.161
CO	4.1	20.3	133.2	419.8	61.9	183.0	.390	.583	.873	.633
CT	-9.7	28.0	281.0	862.1	1.6	4.6	.305	.516	.874	.593
DE	.4	1.1	-30.6	-23.2	.0	.0	-.009	-.017	-.029	-.020
FL	1.5	3.8	1335.5	4058.5	1.1	3.0	1.606	2.642	4.345	2.990
GA	6.5	16.3	-294.3	-222.6	-4.8	-14.8	-.085	-.159	-.297	-.193
ID	.0	-.1	-130.4	-98.0	-50.7	-162.3	-.172	-.276	-.444	-.309
IL	-39.2	111.5	357.1	1132.8	.2	.7	.388	.679	1.191	.796
IN	-10.5	31.7	171.0	542.5	.9	2.5	.187	.323	.560	.376
IA	1.5	7.5	87.0	276.1	1.8	5.4	.138	.238	.410	.276
KS	3.3	16.7	418.4	1245.4	.0	.1	.556	1.176	2.489	1.558
KY	-6.9	19.8	-173.1	-130.9	-4.1	-12.7	-.052	-.097	-.182	-.118
LA	8.3	20.9	324.0	1000.7	.0	.0	.453	.794	1.393	.930
ME	-4.8	16.7	105.3	322.9	8.7	25.2	.125	.211	.354	.241
MD	7.4	20.3	-241.3	-182.6	-2.4	-7.3	-.068	-.125	-.231	-.151
MA	-20.0	57.8	512.5	1572.0	1.0	2.9	.554	.939	1.593	1.080
MI	-40.0	128.7	1186.8	3578.1	5.3	15.0	1.313	2.201	3.689	2.515
MN	-13.9	36.6	132.6	420.5	1.6	4.8	.145	.255	.445	.298
MS	1.2	2.9	177.1	548.8	.0	.0	.247	.448	.810	.534
MO	16.2	88.7	347.5	1076.7	5.8	16.8	.646	1.004	1.560	1.106
MT	.2	1.6	-93.6	-70.5	-37.3	-118.7	-.092	-.161	-.283	-.189
NE	.8	4.2	64.6	203.5	3.3	9.8	.166	.417	1.045	.636
NV	.3	1.6	100.5	306.9	25.8	74.4	.183	.293	.468	.327
NH	-2.6	9.0	94.0	288.3	4.8	13.7	.108	.181	.303	.206
NJ	11.9	30.3	671.7	2060.5	.0	.0	.744	1.242	2.072	1.416
NM	.1	.6	65.4	205.5	.3	1.0	.101	.172	.295	.199
NY	88.5	225.2	1557.4	4777.5	115.0	332.1	1.944	3.192	5.243	3.611
NC	6.7	16.9	-300.8	-227.5	-7.4	-23.0	-.099	-.176	-.314	-.208
ND	1.3	6.5	26.6	84.0	5.8	17.1	.255	.352	.486	.371
OH	-48.5	142.7	1393.5	4201.6	.9	2.6	1.501	2.520	4.231	2.882
OK	4.5	22.9	544.5	1620.5	14.9	42.1	.651	1.140	1.995	1.333
OR	.0	.3	-359.8	-270.5	-179.0	-573.1	-.354	-.584	-.965	-.662
PA	56.9	142.6	1541.2	4646.9	8.0	22.7	1.761	2.902	4.784	3.288
RI	-3.6	10.3	86.1	264.2	.0	.0	.093	.158	.268	.182
SC	4.2	10.5	-161.0	-121.8	-3.7	-11.3	-.051	-.091	-.164	-.108
SD	.8	3.9	28.7	90.3	14.0	41.3	.072	.127	.223	.149
TN	10.1	25.8	330.6	1024.4	37.0	107.8	.416	.686	1.131	.777
TX	18.4	91.8	2323.3	6981.3	5.9	16.8	2.894	5.032	8.750	5.864
UT	-.2	-.9	-197.8	-148.8	-4.7	-14.9	-.101	-.167	-.274	-.189
VT	-1.7	5.8	47.9	146.8	3.8	11.0	.057	.095	.159	.108
VA	3.8	10.8	-278.9	-211.0	-1.3	-4.2	-.085	-.152	-.272	-.180
WA	-33.8	56.0	-771.9	-578.7	-439.5	-1432.9	-.448	-.960	-2.056	-1.283
WV	-2.0	3.1	-86.8	-65.6	-.5	-1.6	-.028	-.050	-.089	-.059
WI	-14.4	48.1	149.2	473.2	4.1	12.1	.169	.293	.510	.342
WY	.4	2.0	19.0	59.8	3.2	9.3	.078	.119	.180	.129
US	40.	1629.	14741.	52244.	-112.	-572.	21.31	35.83	60.46	41.15

NOTE: These results include a downward scaling by a factor of 3.4, so that projected economic growth does not make them look too high in relation to today's economy. Tables 2, 3, and 4 omit this scaling so that the reader can better understand the intermediate calculations that they represent.

FORESTS

Although no one has estimated the nationwide loss of forests resulting from a change in climate, several researchers have estimated the decline in forest biomass. Solomon (1986) applied a forest stand simulation model to 24 sites in eastern Canada and the United States, using climate projections from Mitchell (1983). In an EPA Report to Congress, Urban and Shuggart (1989) and Botkin *et al.* (1989) examined sites in the southeastern United States and the Great Lakes region, using the GISS, GFDL, and Oregon State University Models. Figure 3 illustrates the results of the three studies.[13]

Our approach was to (1) develop generalizing regression equation that express the change in biomass as a function of temperature and precipitation; and (2) use rough estimates of the value of forests to estimate the value of the forest changes. We limited our efforts to the 30 states east of, or bordering, the Mississippi River, excluding Florida.

Projected Declines in Forests Resulting from CO_2 Doubling

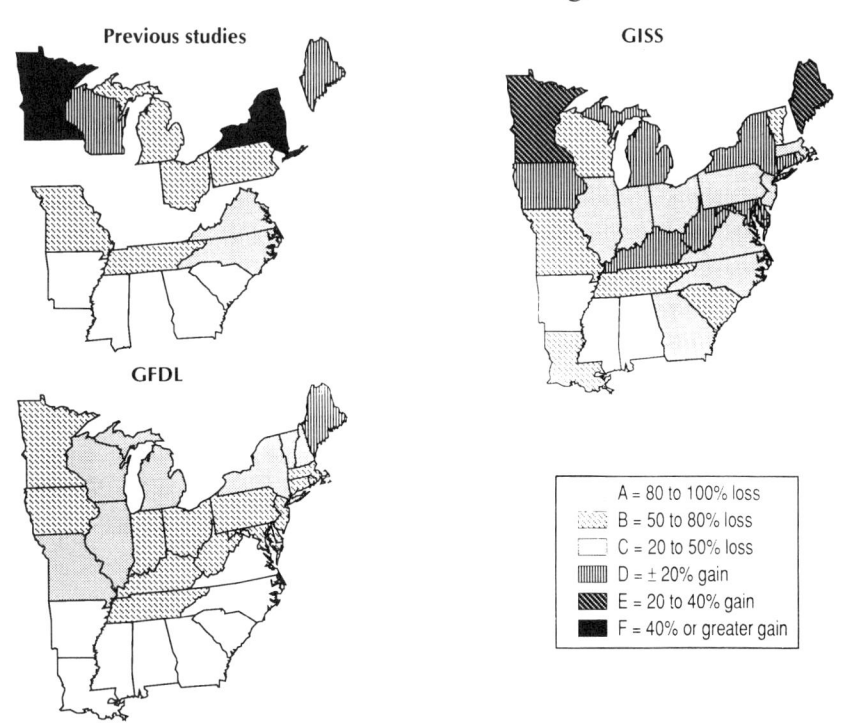

FIGURE 3. Estimates from previous studies reported by Soloman (1986), Botkin *et al.* (1989), and Urban and Shuggart (1989); states not shown were not examined in those studies. The projections for GISS and GFDL were based on the equations shown in Table 6.

Change in Biomass

We characterize the Solomon and EPA [studies] separately, given the differences in approach, climate assumptions, and quality of reported results.[14] In specifying our regression equations, we focused primarily on two alternate formulations: (1) modeling biomass as a function of climate and (2) modeling the change in biomass as a function of both climate and the change in climate.

Table 6 shows our equations. The logarithmic equations treat biomass as extremely sensitive to changes in precipitation (elasticities of 12 and 14), because this functional form places excess emphasis on accuracy in cases where the loss is near 100%.[15] Although an elasticity of 4-12 is too sensitive for small changes in precipitation, it understates the sensitivity for large declines in precipitation by a similar percentage.

Temperature was the primary driving factor for all of the modeling studies we used. Our regression equations mostly suggest that the optimal biomass occurs with an annual average temperature of around 8 to 10 degress, depending on rainfall.

We see no compelling reason to favor one equation over the other; so we treat them as equally valid generalizations of the modeling studies. Because biomass can not be negative, we characterize our uncertainty with a log normal distribution. (We remind the reader that a best estimate in which biomass will decline can still imply that the mean estimate is for an increase in biomass.)[16] Table 7 illustrates summary statistics of the simulations[17]; Figure 3 compares our best estimates with the estimates of previous studies. We assumed that our simulations had two types of sytematic error: (1) for a given state the four projections have a correlation of 0.5 with one another;[18] and (2) for a given equation, the state-specific projections have a correlation of 0.5.[19]

Value of Forest Changes

Most economic information on the value of forests concerns the value of timber. Very little has been done to estimate the values of habitat, recreation, natural recharge of water supplies, reduction in air pollution, scenic vistas, screening noise and unsightly infrastructure, and providing shade for pedestrians, parked vehicles, and buildings. Moreover, even if we had such studies, they probably would focus on the value at the margin, not the total value. The proper question for someone in Mississippi is not so much "How much would you pay to keep this forest alive" as it is "How much is it worth for the whole region to not look like West Texas?" In the past, these decisions have not confronted us — hence existing analyses would probably understate the value of forests.

Our baseline price assumption considers the observation that forests generally sell for $300 to $1000 per acre less after being logged. Assuming a 33% tax rate and a 10% rate of return implies that the forests are worth $45 to $150 per acre per year. We assume that the elasticity of demand for forest services is unity.

Our cost estimates ignore "substitution opportunities" across states. To the extent that timber is the resource being valued, this assumption clearly tends to overstate the cost. However, most of the other forest values are very site specific; so substitution is not really an option in the short run. To constrain the calculations from implying absurdly high costs for a given state, we assume that the total cost per acre can not exceed five times the initial value, that is $225 to 750/acre. On engineering grounds, this assumption seems reasonable since it would probably be possible to irrigate forests at that price.[20]

As Table 7 shows, the projected losses in the southeast area are more than enough to offset any possible gains elsewhere. Nevertheless, the reader might logically ask: How could we possibly lose $24 to 60 billion in forests? As a rough check, we note that even the partial studies of forest values are in the same league. For example, annual consumer surplus associated with nonwater recreation in U.S. land is $81 billion (Bergstrom and Cordell, 1991). Although not all of it is associated with forests (e.g. historic sites), surely a large fraction is.

Kielbaso (1989) estimates that the value of trees along streets is $25 billion. The Council of Tree and Landscape Appraisers estimates that street trees represent one-tenth the value of all urban trees (e.g. backyards and parks), which implies that the total value of urban trees is $250 billion. Assuming a real estate cost of capital of

TABLE 6

Regression Equations Summarizing Forest Modeling Results

Using the results of Botkin *et al.* and Urban & Shuggart

1. $\ln(B+1) = -9.37 + 0.659T - 0.0396T^2 + 9.121 \ln(P)$
    $(0.223)\quad(0.00735)\quad(2.16)$

    $+ 3.58 \ln(Pjja)$ S.E. = 1.29 R^2 = 0.557
    (1.07) T_{max} = 8.2 D_{15} = 4 to 6

2. $\triangle\%B = (0.238 + 0.0686P - 0.0313T) \wedge T + 3.2449 \wedge \%P$
    $(0.108)\quad(0.0294)\quad(0.00317)\quad\quad(0.956)$

    S.E. = .350 R^2 = 0.756 D_{15} = 0.9 T_{max} = 7.6 + 2.2P

Using Solomon's result

3. $\ln(B+1) = -4.08 + 0.547T - 0.0262T^2 + 4.39 \ln(Pjja)$
    $(0.167)\quad(0.00617)\quad(1.79)\quad(1.07)$

    S.E. = 0.819 R^2 = 0.634 D_{15} = 5.3 T_{max} = 16.2

4. $\triangle\%B = (-0.244 + 0.117P - 0.0146T) \wedge T + 0.373 \wedge \%P$
    $(0.111)\quad(0.0371)\quad(0.00259)\quad\quad(0.239)$

    S.E. = .166 R^2 = 0.49 D_{15} = 29 T_{max} = -16.7 + 8P

Note. B = biomass, T = temperature (°C) P = rain (mm/day)
Annual values unless subscripted JJA. T_{max} is the temperature at which the model predicts maximum biomass. D_{15} is the ratio of the sensitivity to a one degree rise in temperature to a 1% increase in rainfall.

10% implies that the annual services of urban trees is $25 billion. Given that these studies examine all but two of the many nontimber uses of forests, it seems reasonable that a large scale loss of forest might be valued in the tens of billions of dollars.

Summary of Results for Doubled CO_2 by 2060

Table 8 summarizes our calculations for a CO_2 doubling by 2060. The GISS scenario implies a (scaled) cost of $37-229 billion; the GFDL scenario, $48-351 billion.

Our estimates for electricity and agriculture are lower than the Smith and Tirpak estimates primarily because they reported actual costs estimated for 2060, while we have scaled these estimates downward to account for economic growth; in addition, we calculate the benefits of CO_2 fertilization assuming a concentration

TABLE 7
Costs of Forest Decline from a GISS Doubled CO_2

State	Current % Forest Acres million	Change in Biomass Low	Medium	High	Cost Divided By Current Value Low	Medium	High	Cost ($ billions) Low	High
AL	21.73	-80.9	-93.6	-97.9	1.7	2.8	3.9	2.3	8.8
AR	16.99	-68.9	-89.9	-96.7	1.2	2.3	3.4	1.3	5.9
CT	1.82	31.0	-26.0	-58.2	-.3	.3	.9	.0	.1
DE	.40	66.4	-6.3	-47.2	-.5	.1	.6	.0	.0
GA	23.91	-1.2	-50.4	-75.1	.0	.7	1.4	.0	3.1
IL	4.26	5.5	-52.5	-78.6	-.1	.7	1.5	.0	.6
IN	4.44	21.3	-32.7	-62.7	-.2	.4	1.0	-.1	.4
IA	1.56	65.7	-36.9	-76.0	-.5	.5	1.4	-.1	.2
KY	12.26	53.0	-14.6	-52.3	-.4	-2	.7	-.4	.8
LA	13.88	-41.4	-76.8	-90.8	.5	1.5	2.4	.5	3.3
ME	17.71	85.9	9.6	-35.4	-.6	-.1	.4	-.9	.6
MD	2.63	49.6	-17.2	-54.1	-.4	.2	.8	-.1	.2
MA	3.10	23.7	-30.5	-60.9	-.2	.4	.9	-.1	.3
MI	18.22	65.9	-26.8	-67.7	-.5	.3	1.1	-.7	1.8
MN	16.58	224.5	5.3	-65.8	-1.2	-.1	1.1	-1.6	1.5
MS	16.69	-79.6	-93.3	-97.8	1.6	2.7	3.8	1.7	6.7
MO	12.52	-11.1	-74.8	-92.9	.1	1.4	2.6	.1	.3.1
NH	5.02	72.0	.0	-41.8	-.5	.0	.5	-.2	.2
NJ	1.99	-2.1	-48.4	-72.8	.0	.7	1.3	.0	.2
NY	18.77	40.5	-19.6	-54.0	-.3	.2	.8	-.5	1.3
NC	18.89	3.7	-46.0	-71.9	.0	.6	1.3	-.1	2.2
OH	7.31	-12.3	-57.2	-79.2	.1	.8	1.6	.1	1.1
PA	17.00	-14.1	-57.4	-78.9	.2	.9	1.6	.2	2.5
RI	.40	22.1	-31.4	-61.5	-.2	.4	1.0	.0	.0
SC	12.26	-27.1	-64.8	-83.0	.3	1.0	1.8	.3	2.1
TN	13.26	-48.7	-75.7	-88.5	.7	1.4	2.2	.6	2.9
VT	4.48	87.4	5.7	-40.4	-.6	-.1	.5	-.2	.2
VA	15.97	22.0	-34.2	-64.4	-.2	.4	1.0	-.3	1.5
WV	11.94	37.6	-24.2	-58.3	-.3	.3	.9	-.3	.9
WI	15.32	171.0	17.9	-48.7	-1.0	-.2	.8	-1.3	.8
US	(including price uncertainty)							-0.3	57.4

of 600 ppm, whereas Smith and Tirpak used 330 and 660 ppm.[21]

Our calculations suggest that the direct economic effects considered by Smith and Tirpak would be overshadowed by the environmental and "quality of life" factors. The costs associated with air pollution, water pollution, lost forests, and health account for 80 percent of the total.

THE BENEFITS OF SLOWING THE CHANGE IN CLIMATE

The preceding estimates illustrate our assumed sensitivity of various sectors to a change in climate. But policymakers need to know the benefits of particular policies, which requires examining how the costs will rise through time with no policy and the benefit (reduction in costs) of implementing a particular policy.

Costs of Climate Change Through Time

Like the IPCC, we rely on EPA's Atmospheric Stabilization Framework for estimates of concentrations and temperatures through 2100,[22] which assumes a three-degree temperature sensitivity, and hence, less warming than the Smith and Tirpak study. The model projects a global warming of 4.6 degrees by 2100, with a change in concentrations sufficient to eventually cause a 6.8 degree warming. We assume that concentrations remain constant after 2100, and that global temperatures approach the equilibrium with an e-folding time of 40 years. Figure 1 illustrates

TABLE 8
Scaled Cost in the Year 2060

	GISS				GFDL			
	Low	Median	High	Mean	Low	Median	High	Mean
Agriculture	-7.6	-2.8	1.4	-3.1	-3.9	5.1	14.0	5.1
Electricity	2.4	5.6	13.3	8.1	2.4	5.6	13.3	8.1
Sea Level	1.7	5.7	19.1	11.8	1.7	5.7	19.1	11.8
Ozone	9.5	21.8	50.3	30.9	14.1	32.6	75.4	46.3
Mobile A/C	1.1	1.8	3.1	2.1	2.3	3.1	4.3	3.3
Health	2.5	9.9	31.8	20.0	2.5	8.9	31.8	20.0
Forests	-0.3	28.5	57.4	30.4	4.2	58.7	113.2	58.7
Water Resources	21.3	35.9	60.5	41.1	31.1	52.0	87.2	59.5
Surface Water								
Demand	0.04	0.25	1.8	1.67	1.06	2.33	5.17	2.33
Suply	0.64	1.68	4.5	1.68	0.89	2.28	5.81	3.53
Ground Water								
Demand	0.23	0.53	1.3	0.77	0.27	0.73	1.97	1.19
Supply	0.63	1.68	4.5	2.71	0.89	2.28	5.81	3.53
Water Pollution								
Public	2.19	5.81	15.4	9.3	3.71	8.42	19.1	11.8
Industrial	12.5	21.5	36.7	24.8	18.06	29.4	47.9	33.1
Hydropower	-0.11	-.35	-0.57	-.35	1.3	2.21	3.77	2.55
Residential	0.04	0.25	1.6	1.38	0.1	0.5	2.5	2.21
Total	37	92	229	139	48	130	351	212

the resulting estimates of temperature and sea level; note that substantial warming has already occurred.

We estimate transient regional scenarios of temperature and precipitation by interpolating and extrapolating the GISS and GFDL estimates according to the ratio of the transient temperature to the equilibrium temperature *of the model*. Thus, we remove any (perceived) upward bias associated with these two models; we only use GISS and GFDL to allocate climate change across regions, not to estimate global warming.

Figure 4 shows GISS transient estimates of the (scaled) costs of climate change. As expected, sea level costs peak after the year 2100 because by this time most wetlands have been inundated; a second peak occurs a century later when the substantial concentration of development between 1 and 2 meters is inundated.[23] Under the GFDL scenario, agricultural costs diminish at first, as agriculture becomes a smaller portion of the total economy; they later increase as the adverse impacts begin to grow more rapidly than the general economy. By contrast, the GISS scenario shows substantial near-term benefits from CO_2 fertilization. The other impacts rise with the change in climate.

Impacts of an Example Policy

For illustrative purposes, the easiest policy to consider is a small temporary reduction in emissions. We assume that during the decade 1995-2005, emissions of CO_2

Annual Cost of Climate Change Over Time

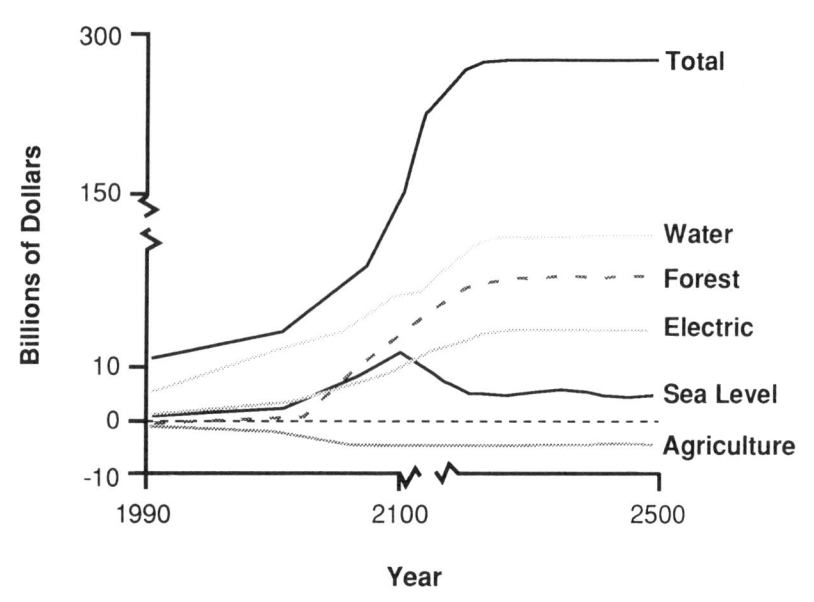

FIGURE 4.

are reduced by 10 billion metric tons of carbon, but that emissions are the same after 2005 as in the baseline.[24] As Figure 5 shows, the impact of such a reduction reaches a peak of 0.044°C a few decades later. However, because CO_2 remains in the atmosphere for centuries, the impact declines very slowly. Therefore, any reduction in emissions today would yield benefits well into the next millennium. The annual benefits would be about $1-2 billion during much of the next century.

Discounting

How much should we be willing to pay today to save a billion or so dollars per year for the next several centuries? Ultimately, the answer depends on whether or not we care about future generations. Economists generally assume that public policies should make no distinction between generations; i.e. that we care as much about a future generation as about our own. However, a dollar invested today will yield a return of several percent per year. If an emissions policy yields a lower return, the reasoning goes, future generations will be better off if this policy is not implemented.

The problems with this approach are that we do not know the extent to which the funds will come from investment or current consumption and to the extent that they come out of investment, we do not know what the foregone investment would have otherwise yielded. The theory of portfolio management offers the most

Long Term Impact If Carbon Emissions Are Reduced by 10 Billion Tons in the Next Decade

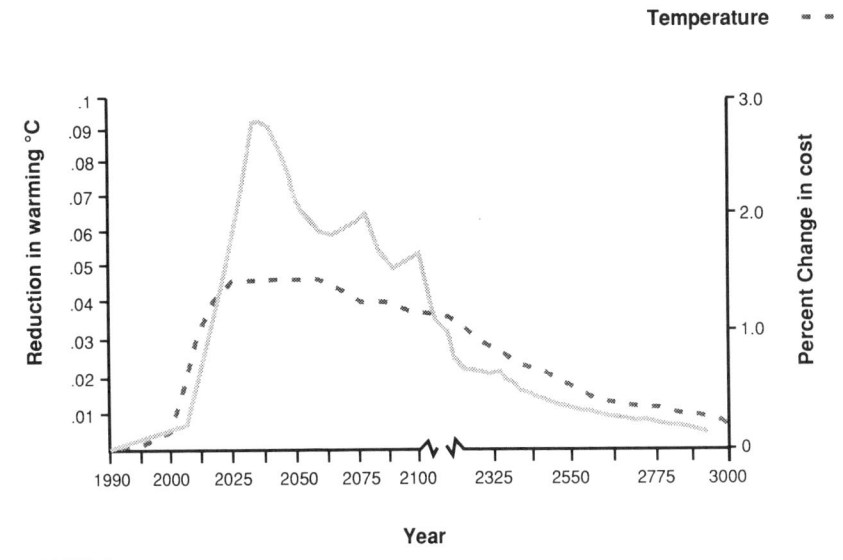

FIGURE 5.

thorough and tested approach for deciding on the required return of an investment. The capital asset pricing model shows that the required return is a linear function of the extent to which the investment increases the overall riskiness of the investor's portfolio. If there is no risk (e.g., U.S. Treasury Bill), we assume that the investment has a pretax inflation adjusted return of 4 percent. Financial analysts generally agree that a fully diversified portfolio (annual risk of about 15 percent) requires an extra 4-1/2 percent return to account for the risk,[25] i.e. 8-1/2 percent. A stock that rises or falls with, but twice as much as, the market would require twice the risk premium (9 percent) for a total return of 13 percent. Selling short on the same investment implies that when the market falls 1 percent, the investment rises 2 percent; thus, there is a risk premium of −9 percent, implying a required return of −5 percent per year.

The returns of environmental investments are probably not correlated with the stock market; the scientific uncertainties have nothing to do with Wall Street. But extending the analysis to consider other principal components of societal wealth suggests that the return from efforts to stop global warming are negatively correlated with the state of the environment. Thus, portfolio theory leads us to either (1) use a discount rate less than the risk-free rate; or (2) estimate the value of reducing the uncertainty, and discount the result using the risk-free rate. We opted for the latter approach. Our calculations already estimate the standard deviation of our uncertainty for a given model. We assume that model error and our ignorance about the future triples that uncertainty, and we apply the risk/return tradeoff implied by the capital asset pricing model.

Over the last century, the average pretax rate of return on Treasury Bills has been about 2 percent, when adjusted for inflation. The last few decades, however, has seen a rate closer to 5 percent. Moreover, taxes distort the picture: the after-tax rate may reflect how people weigh the present against the future. Nevertheless, because the total cost to society of forgoing an investment must include the taxes that are lost as well, we use 2, 3, and 4 percent.

Table 9 illustrates our calculations, with the results expressed in terms of cents

TABLE 9
Marginal Costs of Climate Change from Burning One Gallon of Gasoline
(present value in cents per gallon)

	Discount Rate		
	2%	3%	4%
Model			
GISS			
Before 2100	28.7	16.5	10.3
Long Run	97.0	24.7	12.0
GFDL			
Before 2100	42.6	25.2	14.0
Long Run	135.6	36.5	18.6

NOTE: These calculations assume that the ratio of worldwide to U.S. damages will be the same as the ratio of emissions (i.e. five). They also assume that the baseline concentration of greenhouse gases does not increase after the year 2100, implying an equilibrium warming of 6.8 degrees.

of damages per gallon of gasoline. Our results are consistent with the hypothesis of Cline (1991) that the impacts of global warming are understated unless one looks at it in the long run. With a 4 percent discount rate, most of the costs occur before the year 2100. By contrast, with a 2 percent rate, the costs are 9-43 cents per gallon through 2100, but the total cost is 97-135 cents per gallon. While we assume that concentrations are stable after 2100, Cline (1991) assumed that they will continue to increase for another century. Had we adopted his assumption, the additional costs of burning a gallon of gasoline would have appeared to be even greater.

REFERENCES

Adams, R.M., J.D. Glyer and B.A. McCarl. 1989. "The Economic Effects of Climate Change on U.S. Agriculture." In Smith and Tirpak, op cit.

Bergstrom, C. and H.K. Cordell. 1991. "An Analysis of the Demand for and Value of Outdoor Recreation in the United States." *Journal of Leisure Research.* 23:1:79.

Botkin, D.B., R.A. Nisbet and T.E. Reynales. 1989. "Effects of Climate Change on Forests of the Great Lake States." In Smith and Tirpak, op. cit.

Cline, W.R. "Estimating the Benefits of Greenhouse Warming Abatement" (Draft). Institute for International Economics: Washington, D.C.

Edison Electric Institute. 1985. *Statistical yearbook of the Electric Utility Industry.*

Environmental Protection Agency. 1990. *Environmental Investments: The Cost of a Clean Environment.* Report of the Administrator of the Environmental Protection Agency to the Congress of the United States. Washington, D.C.: Environmental Protection Agency.

Gery, M.W., R.D. Edmond and G.Z. Whitten. 1987. "Tropospheric Ultraviolet Radiation: Assessment of Existing Data and Effect on Ozone Formation." Research Triangle Park, N.C.: Environnmental Protection Agency.

Gibbons, D.C. 1986. *The Economic Value of Water.* Washington, D.C.: Resources for the Future.

Hoffert, M.I., A.J. Callegari and C-T Hsieh. 1980. "The Role of Deep Sea Heat Storage in the Secular Response to Climatic Forcing" *Journal of Geophysical Research* 85:C11:6667-6679.

Linder, K.P. and M.R. Inglis. 1989. "The Potential Effects of Climate Change on Regional and National Demands for Electricity." In Smith and Tirpak, op. cit.

Kalkstein, L. 1989. "The Impact of CO_2 and Trace Gas-Induced Climate Change upon Human Mortality." In Smith and Tirpak, op. cit.

Kielbaso, J.J. 1989. "City Tree Care Programs" In G. Moll and S. Ebenreck (eds). *Shading Our Cities* Washington, D.C.: Island Press.

Nordhaus, W.D. 1900. "To Slow or Not the Slow: The Economics of the Greenhouse Effect (February 5 draft). New Haven: Yale University.

Mitchell, JFB and G. Lupton. "The Seasonal Response of a General Circulation Model to Changes in CO_2 and Sea Surface Temperatures." *QJR Meteorol. Soc.* 109:113-152.

Morris, R.E., M.W. Gery, M-K Liu, G.E. Moore, C. Daly and S.M. Greenfield. 1989. "Sensitivity of a Regional Oxidant Model to Variations in Climate Parameters." In Smith and Tirpak, op. cit.

Pechan, E.H. and Associates. *Ozone Nonattainment Analysis: A Comparison of Bills.* U.S. Environmental Protection Agency, Office of Air and Radiation. 1990.

Peart, R.M., J.W. Jones, R.B. Curry, K. Boote and L.H. Allen. 1989. "Impact of Climate Change on Crop Yield in the Southeastern USA. In Smith and Tirpa, op. cit.

Peterson, D.F. and A.A. Keller. 1990. "Irrigation." In Waggoner (ed) op. cit.

Ritchie, J. B.D. Baer and T.Y. Chou. 1989. "Effect of Global Climate Change on Agriculture: Great Lakes Region." In Smith and Tirpak, op. cit.

Rosenzwieg, C. 1989. "Potential Effects of Climate Change on Agricultural Production in the Great Plains." In Smith and Tirpak, op. cit.

Solomon, A.M. 1986. "Transient response of forests to CO_2 induced climate change: simulation modeling experiments in eastern North America." *Oecologia* 68:567-579.

Smith, J. and D. Tirpak (eds). 1989. *Potential Impacts of Global Climate Change on the United States.* Washington, D.C.: Environmental Protection Agency.

Titus, J.G., R. Park, S. Leatherman, R. Weggel, M. Greene, P. Mausel, M. Treehan, S. Brown, C. Gaunt and G. Yohe. 1991. "Greenhouse Effect and Sea Level Rise: The Cost of Holding Back the Sea". *Coastal Management.* 19:3.

Urban, D.L. and H.H. Shygart. 1989. "Forest Response to Climate Change: A Simulations Study for Southeastern Forests." In Smith and Tirpak, op. cit.

Viole He, D.M. and L.G. Chestnut. 1989. *Valuing Risks* Washington, D.C.: Environmental Protection Agency.

Waggoner, P.E. and R.R. Revelle. 1990. "Summary". In Waggoner, P.E. (ed). *Climate Change and U.S. Water Resources.* New York: John Wiley and Sons.

1. See parallel effort by Cline (1991).
2. We have adjusted their figures to 1990 dollars.
3. Nordhaus (1990) estimates that the savings would be about $1.6 billion per year.
4. The IPCC model probably overstates the short-term contribution of mountain glaciers, but is otherwise superior to other efforts because the rate of melting depends upon the amount of snowcover remaining.
5. Even on a 90 degree day, some people will drive at night when it is cooler and hence not need their air conditioner; by contrast, even on a 60 degree day, a large portion of the driving will occur during the day.
6. We simply calculate the mean and standard deviation of the samples and divide the latter by the square root of the sample size.
7. Note, that Smith and Tirpak's summary of these results indicates that ozone concentrations will rise about 10 percent.
8. For more details on our water resource and forest calculations, detailed appendices are available from the author.
9. Elasticity refers to the percent change in quantify supplied or demanded for a 1 percent change in price.
10. The referenced publications describe the analysis undertaken and some of the general results. However, only Rosensweig published the site-specific results.
11. We matched the Great Lakes equations with USGS water regions 1, 4, 5, 7, and 9, as well as the state of Washington; the southeast with regions 2, 3, 6, and 8; and the Great Plains results with the rest of the nation.

12. This distinction was necessary because the relatively fixed supply of dams in the midst of plentiful surface water in the east implies that supply is inelastic when demand shifts but effectively elastic when runoff changes.
13. Note that the figures for Solomon are estimates based on reading the graphs published in the article. Dr. Solomon has changed jobs and thrown away the raw data on which the figures were based. We also had to estimate the climate change that Dr. Solomon had used in the model runs; the results were based on a 1983 run of the United Kingdom Meteorological Office's model. Dr. Mitchell of the UKMO told us that they had thrown away their raw data as well. Finally, the reader may note that Botkin failed to report a number of his results. We attempted to secure from Botkin his unreported results, but he declined to cooperate out of a conviction that other forest modelers—rather than interlopers such as this author—should be doing this type of work.
14. We relied on historic climate data provided by Roy Jenne and Dennis Joseph of the National Center for Atmospheric Research. We had to omit a few of the Solomon sites from our analysis because NCAR had no available weather data at those sites.
15. The act of minimizing a sum of squares treats, the difference between a 10 and 50 percent loss in biomass as less than the difference between a 98 and a 99 percent loss, even though for most purposes we would be more interested in the former distinction. On the other hand, the arithmetic form would treat the first 50% loss as the same as going from 50% to zero. When considering the value of losing a forest, the last increment is particularly important. The problem is that the model accuracy is probably closer to arithmetic; hence focussing on the loss of that final 5% is pointless since the models we are trying to mimic are not that accurate anyway. One of our reasons for using arithmetic percentage decline and logarithms is that neither form is ideal, but their weaknesses are complementary.
16. Statistical theory shows that the variance of a mean which declines with the sample size if the observations are independent. We assumed that the projections for a given state had 0.5 correlations with one another.
17. These statistics are for exposition and model evaluation only. As mentioned in the previous paragraph, our cost estimates were based solely on the mean and variances of the logarithm of the change in biomass for each state.
18. On statistical grounds, the R-square values approaching 0.75 make a good case for a correlation of 0.5. According to regression theory, oversimplifications and omitted explanatory variables only cause systematic error if those variables are correlated with the variables included in the model. It seems likely that some of those variables are fairly random, while others are systematic.
19. Given that each equation accounts for 1/4 of the projection, it is straightforward to show that this assumption implies that a correlation of 0.125 between the projections for different states. The implied variance is thus:
variance = $0.125*(\Sigma\sigma)^2 + .875*\Sigma\sigma^2$
20. We may be assuming an artificially low price of water that climate change will invalidate. Future studies should probably calculate the amount of water necessary to offset the impact of climate change and use the water price assumed to result from climate change.
21. IPCC projects the CO_2 equilibrium climate to occur somewhat after 2060; by that time, it projects a CO_2 concentration of 600 ppm.
22. We are grateful to Bill Pepper of ICF Incorporated, the model's principal investigator.
23. An additional 60 cm of baseline sea level rise also occurs by this time.
24. Such a reduction is equivalent to 12 percent of projected emissions for that period, or about 80 billion barrels of oil.
25. E.G. Sharp, W. 1990. *Investments*, Chapter 23. New York: Prentice-Hall.

Global Climate Change: Implications, Challenges and Mitigation Measures. Edited by S.K. Majumdar, L.S. Kalkstein, B. Yarnal, E.W. Miller, and L.M. Rosenfeld. © 1992, The Pennsylvania Academy of Science.

Chapter Twenty-Eight

ENERGY SUPPLY TECHNOLOGIES FOR REDUCING GREENHOUSE GAS EMISSIONS

BARRY D. SOLOMON*
U.S. Environmental Protection Agency
6204-J, 401 M Street, S.W.
Washington, D.C. 20460

INTRODUCTION

This chapter reviews energy supply technologies for reducing CO_2 and CH_4 emissions. The technologies discussed are ones that are either economic today in many applications or are likely to become so within 10 years. These options include natural gas/methane, "clean" coal, nuclear power, and renewable energy sources. Later in the chapter, two EPA-sponsored case studies that considered these options (among others) for the United States are summarized. While increased use of energy-efficient technologies is the principal, cost-effective option for limiting energy-related greenhouse gas emissions, most observers agree that a long-term strategy must ultimately turn to renewable energy sources or some form of nuclear power.

After experiencing difficult times in the 1980s, renewable sources of energy, natural gas and nuclear power are experiencing renewed policy attention in the 1990s.

---

*Dr. Solomon is a senior energy policy analyst with the Office of Atmospheric and Indoor Air Programs at the U.S. Environmental Protection Agency, and was previously affiliated with the Climate Change Division. The views and opinions expressed herein are entirely his own and do not necessarily reflect the views or policies of the U.S. Environmental Protection Agency or any other agency of the U.S. Government.

Renewable sources of electricity, e.g., have received a boost from the Solar, Wind, Waste, and Geothermal Power Production Incentives Act and the Acid Rain Title of the Clean Air Act Amendments of 1990, while natural gas and renewable sources of motor fuels have been spurred by the Alternative Motor Fuels Act of 1988 as well as the Clean Air Act Amendments. These options, especially nuclear power, are being touted as part of President George Bush's National Energy Strategy of 1991! Fortuitously, many of the energy technologies that reduce greenhouse gas emissions also reduce emissions of acid rain precursors (though not always vice versa).

The total greenhouse gas emissions of a technology depend on the indirect and induced emissions from the fuel cycle and related industries, as well as the direct emissions. In the discussion that follows, I ignore the indirect and induced emissions out of convenience, but recognize that they are often significant. Thus, I assume that nuclear power and renewables have zero greenhouse gas emissions, even though nuclear fuel cycle emissions can in some cases be very significant.[2]

LOW AND NO-CARBON ENERGY SOURCES

"Clean" Coal, Natural Gas and Methane Recovery

Many of the world's largest nations such as the U.S., China, India, the CIS and Poland are heavily dependent upon coal consumption for electric power production. Refurbishment and life extension of existing coal-fired powerplants can increase their operating efficiency by 2-4% in North America and Europe, and by 10% in developing countries where powerplant heat rates are below those in the West.[3] The U.S. Government and electric utility industry have been sponsoring extensive research, development and demonstration of the so-called "clean" coal technologies, which are generally designed to burn coal with improved efficiency and much reduced environmental impact.[4] Increased use of such technologies was first envisaged in the early 1980s as a response to the acid rain problem. The Kalina Cycle, e.g., could be deployed with a 38-40% efficiency as compared to 35% in conventional plants. The Intergrated Gasification/Combined Cycle technology, a big success story at California's Cool Water facility, can be combined with pressurized fluidized bed coal combustion (PFBC) using a variety of technologies to increase energy conversion efficiencies and reduce CO_2 emissions by 10-20%. While there are several other more efficient coal combustion technologies that may become economical before the end of this decade, widespread use of clean coal could lead to rapid growth in overall coal consumption, potentially offsetting any gain in electric power production efficiency and reduction in CO_2 emissions.

Natural gas use is more promising than coal combustion for reducing emissions of CO_2 and many other pollutants. It is also increasingly used in motor vehicles in compressed form. While use of this premium fuel leads to just 55-60% CO_2 emission rate as compared to the same amount of coal consumption, its supply is scarcer, geographically concentrated, and more difficult to store. In Nigeria, the gas flaring and venting rate is still 2/3 of what is produced in the country.[5]

More than 2/3 of the world's proved gas resources are in the former USSR and the Middle East, with most of the gas being in just two countries (Russia and Iran).[6] While free trade between the U.S. and Canada should spur gas markets and pipeline construction somewhat, it is difficult to forsee a substantial increase in gas use in North America.[7]

The most promising electric power technology for efficient gas use in the aero-derivative combustion turbine. Because of advances in jet engine technology in the 1980s, including new materials and designs that enable combustion to occur at higher temperature, electricity can now be produced at up to 47% efficiency. The steam-injected gas turbine, or STIG (which has already been commercialized in California), for instance, takes steam not needed for process heat requirements and injects it back into the combuster for added power and efficiency. An intercooled STIG (or ISTIG) cools the compressor bleed air used for turbine blade cooling, thereby permitting a much higher inlet air temperature. The optimal scale of an ISTIG is roughly 110 MWe, and is available at a competitive capital cost of about $410/kw.[8] STIGs and ISTIGs, as is the case with many of the clean coal technologies, can also be developed with a variety of fuels as the feedstock, such as biomass sources.

A possible shortcoming of natural gas use from a greenhouse perspective is that it is mostly composed of CH_4, itself a very strong greenhouse gas if released to the atmosphere.[9] Thus it is crucial that methane leaks or losses from oil and gas production, pipeline transmission and distribution be recovered and used as energy. While natural gas leaks tend to be 1% or less of total gas production in the U.S. and Western Europe, they may be 3-5% of gas production in Russia where a large potential exists for cost-effectively reducing pipeline leaks. This can be done by replacing old cast iron pipes, sealing leaky pipe joints, and improving gas compressors in pipelines.[10]

Another attractive option for gas use is coalbed methane recovery, especially from deep underground coal seems that are mined with the more efficient longwall technology. Historically considered a safety problem, trapped coalbed methane is being increasingly recovered in several mines in OECD countries, such as the U.S., U.K., Germany and Australia. Preliminary estimates indicate that the largest potential resource base for coalbed methane recovery exists in the U.S., China, Russia, and Poland.[11] Coalbed methane recovery projects with in-mine boreholes, gob or vertical wells have also begun in these countries. Peak gas production would typically occur 2-3 years after the wells are in place.

Nuclear Fission Power

While at first glance nuclear power appears to offer a great potential for clean energy production without greenhouse gas emissions, the reality is something quite different. For example, as already alluded to, the nuclear fuel cycle may have large indirect CO_2 emissions, primarily from the electricity-intensive fuel enrichment stage, which is often based on coal. Secondly, nuclear plant construction activity has grinded to a halt in most of the world, with a de-facto moratorium existing in the U.S. for well over a decade. If the nuclear industry is to be revived in the 21st century, it will be based on new technologies. Leading candidates include Westinghouse's advanced "passive" pressurized water reactor (AP-600), General Electric's inherently safe, modular liquid metal cooled reactor (PRISM), and General Atomic's modular high temperature, gas-cooled reactor (MHTGR). Many obstacles will have to be surmounted, however, for nuclear power to make a comeback and to contribute toward a solution to the climate change problem. The key ones include economics, reactor licensing and construction, power plant safety, radioactive waste disposal, reactor decommissioning, and nuclear weapons proliferation. Each of these issues will be addressed in turn.

During the first two decades of the industry, nuclear power developed into a low-cost source of electricity. In the past 15 years, especially since the 1979 accident at Three Mile Island, however, the economics have been reversed. The cost to build and operate a nuclear power plant has escalated to the point that it would be more expensive than coal-fired power in most parts of the U.S. today.[12] Consequently, unless this situation is reversed again, capital invested in nuclear power would be much less cost-effective and slower than the equivalent amount of capital invested in energy efficiency in providing significant reductions or offsets in CO_2 emissions.[13] Nonetheless, optimistic claims have been made about cost reductions that can be achieved with advanced reactors that would make them much more competitive. For example, capacity factors of 80-90% or higher are often assumed as compared with about 65% in conventional U.S. nuclear power plants in recent years.[14] Similarly, nuclear reactor economics will be influenced by developments in power plant licensing and construction practices.

Concern has been expressed about the lengthy nuclear reactor licensing process in the U.S., which has been one of several factors that has contributed to cost escalation. The Nuclear Regulatory Commission's (NRC) new regulation of April 1989 (10 CFR Part 52) completely revises the licensing process, expediting the licensing of future reactors.[15] Three major revisions to current practice are affected by this rule. First of all, the rule allows a utility to get pre-approval for a site before a specific reactor design is chosen. When permit renewal procedures are considered, this site banking provision could result in the operation of a nuclear reactor at a site 80 years or more after the original site approval decision. The second revision allows the NRC to certify a nuclear reactor design by rule, which would generally prohibit them from imposing new requirements on the design. Finally, the new rule provides for the issuance of combined licenses for plant construction and operation based on final reactor design. While these reforms serve to expedite the nuclear power plant licensing process, in the process they severely limit public participation in the proceedings. Once a license is received, a utility could use modular construction techniques to lower the time required to build a nuclear plant perhaps by half or more, which currently is close to 10 years. However, in addition to regulatory hurdles, an electric utility will still have to contend with financing, labor, weather, and plant and materials related delays, which may be less controllable.

The new generation of nuclear power plants has been designed to achieve an extremely high degree of safety, and to minimize the safety-related role of operators. While the best test of plant safety will be commercial performance, it appears that new reactors may well be an improvement over existing nuclear reactors, and have been designed with greater attention to the most severe accidents. Plant safety is partly a perceptual issue, however, and the advanced reactor designs require detailed scrutiny, as has been provided in a study commissioned by the Union of Concerned Scientists.[16] The key weakness of these plants may involve reactivity excursion accidents, from control rod ejection or steam generator tube ruptures (MHTGR), or sodium boiling under unscrammed conditions (PRISM). Other problems are that some of the advanced reactors are being proposed without containment domes and offsite emergency planning, because of asserted enhanced accident prevention capabilities.

The disposal of radioactive wastes and spent fuel from nuclear power plants remains an unresolved problem. While low-level waste sites are operational and more are being (slowly) developed, deep underground repositories for high-level wastes remains elusive. The first commercial sites were to be built at Yucca Mountain, Nevada, and Gorleben, Germany, yet both sites have had major setbacks pushing potential commissioning dates back to about 2010.[17] Advanced nuclear reactors may rely on reprocessing, which raises special waste and nuclear weapons proliferation concerns in the U.S. If public trust in the radioactive waste solution is not achieved, future use of nuclear power may eventually halt.[18] A related problem is the decommissioning of old or retired reactors.[19] Whereas the first generation of smaller nuclear reactors are only beginning to be decommissioned, the cost, timing and waste issues pose significant uncertainties. Resolution of these problems will require consideration of their difficult social implications.[20]

The final hurdle faced by nuclear power if it is to make a comeback to help ameliorate the global climate problem is the potential for proliferation of nuclear weapons. This problem was underscored by the concern during the 1991 Persian Gulf War that both Iraq and Israel may have nuclear weapons capabilities. Williams and Feiveson have noted that a 3000 GW nuclear system based on plutonium-breeder reactors (which they assumed due to eventual shortage in the uranium supply) would place into global nuclear commerce about 5 million kilograms of separated plutonium a year.[21] The International Atomic Energy Agency's system of inspections and safeguards would be inadequate to the task. Internationalization of sensitive nuclear facilities and materials is called for, along with incentives for the industry to shift to technologies that are far less attractive sources of nuclear weapons-usable materials (i.e., without reprocessing and plutonium recycling). Such reforms would have to be incorporated into revisions to the Nuclear Non-Proliferation Treaty.

Renewable Energy Options

The most promising near-term renewable energy options for reducing greenhouse gas emissions include solar energy technologies, wind electric systems, and various forms of biomass energy. Hydroelectric power (along with biomass the most common form of renewable energy in use today) has only limited growth potential because of environmental and land use constraints, and often a shortage of available good sites to tap water power, especially in the U.S. Geothermal energy, another promising option, which is not a renewable resource because of the slow rate of heat transfer from the Earth's interior to the surface, but will be discussed later.

While solar and wind energy systems are used much less in the U.S. today than is hydroelectricity, most studies project these options as growing well beyond the potential of hydro power.[22] This is because of the vast solar resource base, which may be a couple hundred times our energy use rate, depending on the assumptions. The near-term direct solar technologies can be divided into three categories: solar buildings, solar-thermal collectors, and photovoltaic cells.[23] The use of solar energy in buildings encompasses active and passive space heating and cooling, and water heating. Passive solar energy systems are most cost-effective when incorporated into initial building design, adding about 10-12% to construction costs while lowering energy bills by 50-60%.[24] These systems range from proper site orientation, window glazing, sky lights, and trombe walls to underground housing. Despite the large economic potential, only a quarter million passive solar homes have been built in the U.S., and only a few non-residential solar buildings.[25] Active-solar systems require moving parts to collect, distribute, and store solar energy for water heating, space heating or space cooling. Typically, a black metal flat plate on a roof collects solar heat and transfers it through water or air into an insulated storage medium (such as a water tank) until it is needed for use in a building. These systems remain costly, however, and most U.S. sales have been in California where there have been income tax credits much larger than the federal residential renewable-energy tax credits that expired in 1985. Even so, most of the over two million collectors installed in the U.S. have been for water heaters and swimming pools rather than the much larger market for space conditioning. By contrast in a few small countries, such as Israel, Jordan and Cyprus, solar collectors heat between 25-65% of the water in homes.[26]

There are four major types of solar thermal systems: parabolic troughs, parabolic dishes, central receivers, and solar ponds. Parabolic troughs are the big success story, with a marginally economic and very effective system of several hundred megawatts built in the last decade in California's Mojave Desert. This system, operated by Luz International for Southern California Edison, has a capital cost of $2500/kWh and produces electricity at 10-12¢/kWh.[27] The parabolic troughs contain mirrors that focus sunlight onto oil-carrying receiver pipes that are heated up to 400°C, thereby creating steam to drive a turbine generator. These troughs are mounted on a north-south axis and track the sun. Parabolic dishes and central receivers are more expensive but still promising solar technologies. A parabolic dish uses a bowl-shaped reflector to focus sunlight onto a small area, and achieves temperatures of 1000-1500°C. Central receiver systems use sun-tracking mirrors (heliostats) to concentrate sunlight onto a central receiving tower, also reaching higher temperatures. To date, only small prototype plants using these two technologies have been built in the U.S. Finally, solar ponds, which trap heat between varying layers of varying density of saline water, are promising in just a few locations.

Photovoltaic cells, direct coverters of photons of light into electricity, have fallen in cost from $60/kWh in 1970 to about 25¢/kWh today. Consequently their market has steadily grown, which totalled 46 megawatts of capacity in worldwide sales in 1990.[28] Photovoltaic cells based on thin films of semiconductor material (such as amorphous silicon) may hold the most promise for low-cost, mass-production techniques. At present, however, this technology has not been proved superior to crystalline or concentrator-celled photovoltaics. The choice between the three options will be decided by a tradeoff between production and cost efficiency.[29]

Wind power is another renewable energy source that experienced rapid growth and great success in the 1980s. About 1660 MWe of wind-electric capacity was installed worldwide in the 1980s, 85% of this total being in California which now gets 3% of its electric power capacity from wind turbines.[30] California's wind machines are primarily located at Altamont, Solano, Techachapi, and San Gorgonio Passes, and mostly serve the customers of Pacific Gas & Electric Company. Most of their 7500 plus wind turbines (or "wind farms") are of the horizontal axis type, and are no more than 300 kilowatts of generating capacity each. Favorable sites for substantial additional wind power capacity exists in Central states such as Texas, Montana and the Dakotas.

A decade ago favorable tax policies and high prices paid by utilities for wind-based electricity spurred on the industry but encouraged faulty turbine sitings and designs. During the past five years the tax breaks have been eliminated and the wind industry consolidated, while its economics has markedly improved. In California, e.g., wind power is sold for 6¢/kWh (a nearly order of magnitude price decrease in a decade) and available at 50% of rated generating capacity during peak demand periods.[31] Yet most of the cost reduction and improved reliability of wind power systems can be attributed to organizational learning rather than to technological innovation. The small size of wind turbines and rapid construction time has resuited in standardized procedures and mass production, better turbine siting and maintenance.[32] These achievements should be replicated in Europe where several countries have huge potentials for using wind power, especially in Denmark, the Netherlands and Britain.[33]

Biomass energy sources were identified in the influential 1990 EPA report to Congress on global climate change as being the most important near-term renewable source of energy.[34] An important advantage of biomas over solar energy and wind power is that it is readily available over much of the earth's surface and offers a convenient form of energy storage, i.e., as chemical energy in plants. Indeed, biomass currently meets about a third of the energy needs in the developing world, although it is not used sustainably. If used on a sustainably basis, biomas could result in no net increase in CO_2 emissions, and much lower levels of other air pollutants.[35] Biofuel burning cookstoves, which are commonly used in the developing world, can be improved to cost-effectively cut their direct greenhouse gas emissions in half.[36] In the U.S., biomass energy is especially advantageous as a source of electric power and liquid fuels for motor vehicles.

Biomass is already widely used as a source of electricity and process heat in the pulp and paper industry. Wood waste in particular is used as feedstock for steam-turbine cogenerators, which is only economical on a small scale where low-cost biomass fuel can be readily obtained. For larger-scale applications up to about

100 MWe, gas turbines are the power technology of choice due to simplicity and economics. This is because of greater technological improvement in gas turbines over the past several decades. Biomass integrated gasifiers can be combined with STIG or ISTIG technology for cogeneration systems in several industries. Gasification would likely be done with air and steam at high pressure; the gas should also be cleaned of impurities that might damage the turbine blades before combustion.[37] The General Electric Company is in the process of testing this technology in the Northeast U.S. Another option for such cogenerators is the distillation of ethanol for use as an automotive fuel. This would be more cost-effective abroad where sugar cane production (and thus plentiful residues) is more prevalent.[38]

Ethanol and methanol, along with compressed natural gas, electric cars and hydrogen, are much discussed alternatives to gasoline in the U.S. Most ethanol in use is derived from corn and used in 90% gasoline, 10% ethanol blends; it is also costly. Methanol is used in its pure ("neat") form in race cars but more typically in an 85% methanol, 15% gasoline blend. While methanol can be derived from wood, near-term market conditions favor the use of gas or coal which could exacerbate global warming. The most promising alcohol fuel option may be ethanol derived from cellulosic biomass, such as herbaceous or woody energy crops, agricultural or forestry residues, or municipal solid waste. Aggressive research and development could make this technology cost-competitive with gasoline by 2000, and it could eventually displace a significant quantity of gasoline and foreign oil.[39]

The largest-scale development of hydroelectric power has taken place in North America, with Canada being the largest hydropower producer in the world. The Federal Energy Regulatory Commission has indicated that about half of the 147 GWe of hydro capacity in the U.S. has been developed.[40] Recently installed capacity generates very competitively at 3-6¢/kWh. Much of the remaining, small-scale dam sites, however, will probably not be developed due to environmental concerns and public opposition. Many of the remaining opportunities for hydroelectric power in the U.S. are at existing dam sites. One estimate of these upgrade or expansion opportunities is about 37 GWe.[41]

A much larger potential exists for the development of hydroelectricity in the developing countries, perhaps over 200 GWe in the next decade. In Latin America, Africa and Asia, only 5-10% of the potential hydro resource has been tapped.[42] As is the case with nuclear power in the West, however, dam projects in the developing world now commonly face the same capital constraints and public opposition. These conditions are especially prevalent in Brazil, India and the former USSR.[43] Large-scale hydroelectric development in Brazil has also contributed to deforestation in the Amazon, which emits more CO_2 overall in Brazil than does fossil fuel combustion.

Geothermal Energy Options

Geothermal energy, while not renewable, is another attractive source of clean energy. It is already being used for either direct heat or electricity generation in 20 countries. About 6 GWe of generating capacity has been installed in the world, with

418 Global Climate Change: Implications, Challenges and Mitigation Measures

just less than half being in the U.S., and significant capacity also existing in the Philippines, Mexico, and Italy.[44] Most of the U.S. capacity is located at The Geysers reservoirs in northern California, where over 2000 MWe of small units generate low cost power for Pacific Gas & Electric Company. Though there are many different kinds of geothermal energy, by far the most commercially advanced type is hydrothermal. These resources contain hot water or steam trapped in fractured or porous rocks at accessible depths (usually 100-450 meters). The exploitation of hydrothermal energy involves sinking wells, extracting the hot liquid-vapor mixture (and rarely dry steam), and using it for electricity generation or direct heat production.

Since the total resource base of geothermal energy makes it by far the largest nonrenewable energy source,[45] it cannot be ignored in a greenhouse world. While better known for their hydrogen sulfide emissions, geothermal plants have very low CO_2 emission rates, at about 5% of that from a coal-fired power plant and 8% of that from an oil-fired power plant.[46] In addition, the other environmental impacts of geothermal plants are generally very low,[47] with the exception of thermal pollution and potential deforestation (e.g., Hawaii). A large volume of waste brine is created, however, which can be reinjected (as in the U.S., Japan and New Zealand) or disposed of at the surface (elsewhere).

Two EPA Case Studies

The EPA conducted two case studies in the first half of 1990 to investigate the technological feasibility and cost of greenhouse gas reduction strategies for the U.S. Both studies focused on CO_2 or CO_2 equivalent emissions reductions from the energy sector, and examined potential results for the U.S. in 2000 and 2010. We assumed that these reductions would be required based on an international framework climate convention and accompanying protocols or legislation, but did not focus on the details of specific programs. It is possible, therefore, that the reductions could be achieved through a carbon tax, tradeable emissions permits, or a variety of other energy and environmental policies, which would have different implications for program or administrative costs. My focus here is on the market penetration of the various energy supply technologies examined in the case studies, not on their costs, although both studies focused on the cost-effectiveness of these technologies (in addition to energy efficiency) in the context of "what-if" scenarios.

The first case study, prepared by ICF, projected greenhouse gas emissions out to 2000 and 2010 in a base case, and then examined the feasibility and cost of an emissions reduction program. ICF used a spreadsheet approach to integrate and cross-check analytical results supplied by several other EPA contractors.[48] The overall findings were that natural gas was the most attractive energy option for lowering CO_2 emissions out to 2000, with renewables growing more by 2010. Gas use was found to have a potential to grow by about 35% from its 1988 level to a total of 24.62 quadrillion BTUs, after accounting for energy savings due to end use efficiency. Most of this growth would be due to increased gas use for electric power. The various renewable energy options were found to have a potential to grow by 61% from the 1988 actual level to a total of 5.13 quadrillion BTUs. This growth is again largely for electricity generation, from hydro and wind power. Solar and biomass (wood) fuel options were not seen as taking off until after 2000, when they were found to have very significant potential, especially solar and ethanol. Geothermal and nuclear power were not found to grow beyond their 2000 base cases, but expanded moderately by 2010.

The second case study, prepared by Brookhaven National Laboratory, used a dynamic linear programming model called MARKAL to optimize a network representation of the U.S. energy system to meet assumed demand for over 200 energy supply, conversion and efficiency technologies, at least cost.[49] MARKAL is subject to typical constraints found in an energy system linear program, in addition to an overall CO_2 constraint. Various levels of CO_2 constraints were examined; the discussion below focuses on results for a 20% reduction requirement below the 1990 level. In the MARKAL model competing energy technologies (both supply and demand-side) are chosen based on their ability to meet the constraints for the given level of energy demand in each of several sectors at least cost, assuming a 7% real discount rate. A 7% discount rate was also assumed in the ICF study.

The supply-side results of EPA's MARKAL analysis were similar to the findings of the ICF study for nuclear power, hydroelectricity, wind power, solar energy, and wood-based ethanol. None of the clean-coal technologies were selected very much in the model. Table 1 indicates the rough order of preference for the renewable energy technologies in the MARKAL analysis, taking into account shadow prices and the size of the potential contribution available from each energy option. MARKAL results, in contrast, were more pessimistic for natural gas and wood-based electricity generation, and more optimistic for geothermal energy. The gas results were especially pessimistic, with natural gas use declining in all but one of the policy and sensitivity analyses. This result was attributed to conservative assumptions about gas reserves.

TABLE 1

MARKAL Results: Principal Renewable Energy Technologies by Category

Electric Power Generation:	Demand Devices:	Processes:
Hydroelectricity	Firewood	Landfill gas production
Small-scale hydro	Solar water heat	Agricultural crops for pipeline quality gas*
Geothermal	MSW for industrial heat	
Wind (central)	Geothermal for industrial heat	Wood for liquid fuel*
Wind (local)	Solar for industrial heat	
Solar (central thermal)		
Solar (central photovoltaics)		
Ocean thermal electric		
Biomass*		
Municipal solid waste (MSW)*		
Wave power (central)*		

*Appears only in CO_2-constrained cases.
Source: Reprinted with permission, Table 5, from the book published by MIT Press, 1991 as cited in reference number 49.

CONCLUSIONS

In this chapter I have provided an overview of the major energy supply options that will be available over the next decade for reducing or offsetting CO_2 and CH_4 emissions. How much these options are relied on is a matter for public policy and the marketplace to decide. Most of the technologies that I have discussed would result in multiple benefits, e.g., they could simultaneously reduce greenhouse gas and acid rain emissions. It should be noted, however, that the number one short-run category of cost-effective options for reducing greenhouse gas emissions is energy efficiency, as confirmed by the recent report of the U.S. National Academy of Sciences.[50] Energy supply technologies should be viewed as part of a comprehensive greenhouse gas reduction strategy, their ultimate contribution toward being contingent on the level of reduction agreed to by policymakers. The MARKAL analysis, in particular, provides a useful framework for comparing the relative attractiveness of numerous energy options under different reduction scenarios.

The good news is that many energy supply choices are available, the most attractive ones in the short-run being natural gas, geothermal energy, hydroelectric power, wind power and wood-based biomass fuel. Over the longer term solar energy technologies such as photovoltaics will become more widespread, as may wood-based ethanol. Nuclear power has a chance to make a comeback, but will have to overcome more obstacles than are faced by the other alternatives. While further research and development will be important, the marketplace also needs a clear signal that these technologies will be needed as part of a comprehensive, long-term, U.S. and international strategy to combat the greenhouse effect.

REFERENCES

1. U.S. Department of Energy. 1991. *National Energy Strategy: 1991/1992.* U.S. Government Printing Office, Washington, D.C.
2. Fritsche, U., L.R. Rausch and K.H. Simon. 1989. *Environmental Analysis of Energy Systems: The Total-Emissions-Model for Integrated Systems.* Institute for Applied Ecology, University of Kassel, Kassel, Germany.
3. U.S. Agency for International Development. 1988. *Power Shortages in Developing Countries.* U.S. AID, Washington, D.C.
4. U.S. Department of Energy. 1989. *Clean Coal Technology Demonstration Program: Final Programmatic Environmental Impact Statement.* DOE/EIS-0146, Washington, D.C.
5. Energy Information Administration. 1991. *International Energy Annual 1991.* DOE/EIA-0219 (89), U.S. Department of Energy, Washington, D.C.
6. British Petroleum. 1991. *BP Statistical Review of World Energy.* BP, London.
7. Energy Information Administration. 1991. *Annual Energy Outlook.* DOE/EIA-0383(91), U.S. Department of Energy, Washington, D.C.
8. Williams, R.H. and E.D. Larson. 1988. Aeroderivative turbines for stationary power. *Annual Review of Energy* 13:429-489.

9. Intergovernmental Panel on Climate Change. 1990. *IPCC First Assessment Report.* IPCC Secretariat, World Meteorological Organization, Geneva.
10. Makarov, A.A. and I.A. Bashmakov. 1990. The Soviet Union, pp. 35-53. In: W.U. Chandler (Ed.) *Carbon Emissions Control Strategies: Case Studies in International Cooperation.* World Wildlife Fund, Washington, D.C. pp. 263.
11. ICF, Inc. 1990. *Methane Emissions to the Atmosphere from Coal Mining.* Report to the U.S. Environmental Protection Agency, Washington, D.C.
12. For an economic analysis of coal v. nuclear power using new plant designs, see Hudson, Randall. 1990. Economic comparison of electric generation alternatives. Oak Ridge National Laboratory, Oak Ridge, TN.
13. Keepin, W. and G. Kats. 1988. Greenhouse warming: comparative analysis of nuclear and energy efficiency abatement strategies. *Energy Policy* 16:538-561.
14. MHB Technical Associates. 1990. *Advanced Reactor Study.* Report to the Union of Concerned Scientists, Cambridge, MA.
15. 10 CFR Part 52. 1989. Early site permits; standard design certifications; and combined licenses for nuclear power reactors.
16. Reference 14, op. cit.
17. Blowers, A., D. Lowry and B.D. Solomon. 1991. *The International Politics of Nuclear Waste.* The Macmillan Press, London, pp. 348.
18. Carter, Luther J. 1987. *Nuclear Imperatives and Public Trust: Dealing with Radioactive Waste.* Resources for the Future. Washington, D.C., pp. 473.
19. Solomon, B.D. and D.M. Cameron. 1984. The impact of nuclear power plant dismantlement on radioactive waste disposal. *Man. Environment, Space and Time* 4:39-60.
20. Pasqualetti, Martin J. (Ed.) 1990. *Nuclear Decommissioning and Society: Public Links to a New Technology.* Routledge, London, pp. 237.
21. Williams, R.H. and H.A. Feiveson. 1990. Diversion-resistance criteria for future nuclear power. *Energy Policy* 18:543-549.
22. See, e.g., Idaho National Engineering Laboratory, *et al.* 1990. *The Potential of Renewable Energy: An Interlaboratory White Paper.* Solar Energy Research Institute, Golden, CO.
23. Brower, Michael C. 1990. *Cool Energy: The Renewable Solution to Global Warming.* Union of Concerned Scientists, Cambridge, MA, pp. 89.
24. Rader, Nancy. 1989. *Power Surge: The Status and Near-Term Potential of Renewable Energy.* Public Citizen, Washington, D.C.
25. U.S. Department of Energy. 1989. *Solar Buildings Program Summary: Fiscal Year 1988.* DOE/CH 10093-47. Government Printing Office, Washington, D.C.
26. Flavin, Christopher and Nicholas Lenssen. 1990. *Beyond the Petroleum Age: Designing a Solar Economy.* Worldwatch Institute Paper 100, Washington, D.C.
27. Reference 23, op. cit.
28. 1991. Photovoltaics-clean energy now & for the future. Rocky Mountain Institute *Newsletter* 7(Spring):5-7.
29. Andrejko, D.A. 1989. Assessment of solar energy technologies. American Solar Energy Society, Denver, CO.
30. McGowan, Jon C. 1991. Large-scale solar/wind electrical production systems -

predictions for the 21st century, pp. 727-736. In: J.W. Tester, D.O. Wood and N.A. Ferrari (Ed.) *Energy and the Environment in the 21st Century.* MIT Press, Cambridge, MA, pp. 1006.
31. Weinberg, C.J. and R.H. Williams. 1990. Energy from the sun. *Scientific American* 263:146-155.
32. Ibid.
33. Grubb, Michael J. 1990. The Cinderella options: a study of modernized renewable energy technologies, Part I-A, technical assessment. *Energy Policy* 18:525-542.
34. U.S. Environmental Protection Agency. 1990. *Policy Options for Stabilizing Global Climate.* Final Report to Congress, Washington, D.C. pp. 850.
35. Smith, Kirk R. 1987. *Biofuels, Air Pollution, and Health: A Global Review.* Plenum Press, New York.
36. Ahuja, Dilip R. 1990. Research needs for improving biofuel burning cookstove technologies. *Natural Resources Forum* 14(2):125-134.
37. Reference 31, op. cit.
38. Ogden, J.M., R.H. Williams and M.E. Fulmer. 1991. Cogeneration applications of biomass gasifier/gas turbine technologies in the cane sugar and alcohol industries, pp. 311-346. In: J.W. Tester, D.O. Wood and N.A. Ferrari (Ed.) *Energy and the Environment in the 21st Century.* MIT Press, Cambridge, MA, pp. 1006.
39. Lynd, L.R., J.H. Cushman, R.J. Nichols and C.E. Wyman. 1991. Fuel ethanol from cellulosic biomass. *Science* 251:1318-1323.
40. Federal Energy Regulatory Commission. 1988. *Hydroelectric Power Resources of the United States.* Government Printing Office, Washington, D.C.
41. Reference 23, op. cit.
42. Reference 34, op. cit.
43. See, e.g., Sagers, Matthew J. 1989. News notes (review of energy industries in 1988 and 1989 plan). *Soviet Geography* 30:306-352.
44. DiPippo, Ronald. 1991. Geothermal energy: electricity production and environmental impact, a worldwide perspective, pp. 741-754. In: J.W. Tester, D.O. Wood and N.A. Ferrari (Ed.) *Energy and the Environment in the 21st Century.* MIT Press, Cambridge, MA, pp. 1006.
45. Armstead, H.C.H. and J.W. Tester. 1987. *Heat Mining.* Spon, London.
46. DiPippo, Ronald. 1988. Geothermal energy and the greenhouse effect. *Geothermal Hot Line* 18(2):84-85.
47. Reference 44, op. cit.
48. ICF Inc. 1990. *Preliminary technology cost estimates of measures available to reduce U.S. greenhouse gas emissions by 2010.* Final report to the U.S. Environmental Protection Agency, Washington, D.C.
49. Morris, S.C., B.D. Solomon, D. Hill, J. Lee and G. Goldstein. 1991. A least cost energy analysis of U.S. CO_2 reduction options, pp. 865-876. In: J.W. Tester, D.O. Wood and N.A. Ferrari (Ed.) *Energy and the Environment in the 21st Century.* MIT Press, Cambridge, MA, pp. 1006.
50. National Academy of Sciences. 1991. *Policy Implications of Greenhouse Warming.* National Academy Press, Washington, D.C.

Global Climate Change: Implications, Challenges and Mitigation Measures. Edited by S.K. Majumdar, L.S. Kalkstein, B. Yarnal, E.W. Miller, and L.M. Rosenfeld. © 1992, The Pennsylvania Academy of Science.

Chapter Twenty-Nine

IMPLEMENTATION OF MITIGATION AT THE LOCAL LEVEL:
The Role of Municipalities

L.D. DANNY HARVEY
Department of Geography
University of Toronto
100 St. George Street
Toronto, Canada M5S 1A1

INTRODUCTION

The economically attractive potential for greenhouse gas emission reduction is enormous, as indicated elsewhere in this volume.[1] Stabilization of the atmospheric greenhouse gas concentration at less than the equivalent of a carbon dioxide doubling from pre-industrial levels will require, over the next few decades or less, a near phase-out of CFC production, at least a 50% reduction of carbon dioxide emissions, a 15-20% reduction in methane emissions, and a 70-80% reduction of nitrous oxide emissions.[2] On a longer time scale, as the rate of CO_2 absorption by the oceans and biosphere slowly decreases following the initial emission reduction, further CO_2 emission reductions will be required.[3] Inasmuch as improvement of living conditions in the developing world will initially necessitate some increase in their greenhouse emissions, emission reductions well in excess of 50% are likely required from the developed countries over the next few decades. Although deforestation is an important source of carbon dioxide, and agricultural activities are important sources of methane and nitrous oxide, energy use is responsible for two-thirds or more of projected future increases of greenhouse gases under typical business-as-usual scenarios.[3,4] The primary onus for greenhouse gas emission reduction therefore falls on energy use in industrialized countries.

424 Global Climate Change: Implications, Challenges and Mitigation Measures

The purpose of this chapter is to identify the important role that municipal governments can and must play in a coordinated strategy to reduce greenhouse gas emissions involving all levels of government, and to indicate ways in which municipalities can effect significant emission reductions within their jurisdictions even in the absence of initiatives from higher levels of government. Municipalities generally have at least partial control over land use through zoning regulations and official plans; are responsible for issuing building permits and approving major developments; exercise control over parking supply and rates, roads, and public transit; often own or regulate municipal power and natural gas utilities and district heating systems; having responsibilities concerning planting, maintaining, and protecting urban trees; play a central role in waste management; can influence the market through their own purchasing decisions; and can play an important educational role through the educational system and various community outreach strategies. This chapter deals with medium to large urban areas (population in excess of 100,000) with densities in excess of 1000 persons per square kilometre. Some of the measures discussed here are applicable to smaller municipalities, depending on local circumstances.

In addition to influencing greenhouse gas emissions through actions falling within their legal jurisdiction, municipalities can also play an effective advocacy role to higher levels of government, either by sending formal resolutions or by enacting bylaws in areas traditionally or legally the domain of higher levels of government. Examples include the actions of several municipalities to ban CFC's or require their recycling prior to regulations at the national level.

MUNICIPAL RESPONSES TO GLOBAL WARMING

Until recently, very few municipalities had made explicit commitments to reduce greenhouse gas emissions. In January 1990, Toronto City Council passed a resolution committing the City to a 20% carbon dioxide emission reduction from the 1988 level by 2005. Although the resolution is symbolic in nature, it has nevertheless been extremely useful in signalling to the entire city bureacracy that greenhouse gas emission reduction should be treated seriously. In March 1991, the city's *Special Advisory Committee on the Environment* presented a comprehensive set of strategies for achieving the 20% target while accomodating up to a 20% increase in the number of people living and working in the City,[5,6] as targeted in the revised official plan proposals (although over a longer time horizon). As discussed below, an important long term strategy to reduce greenhouse gas emissions is to create more compact, higher density urban areas as an alternative to continued urgan sprawl. The City's commitment therefore implies a 33% reduction in per capita CO_2 emissions.

The City of Vancouver is also developing programs specifically aimed at reducing carbon dioxide emissions,[7] and a greenhouse action plan has recently been presented to London-area municipalities.[8] Although few cities have yet to make commitments toward reduction of greenhouse gas emissions, several cities have embarked on programs of energy efficiency and conservation and automobile minimization which will have the net effect of reducing greenhouse gas emissions.[9] San Jose (California) has embarked on a "Sustainable Cities" project aimed at, among other things, reducing energy use by 10% compared to baseline projections,[10] which already incorporate some energy conservation.

Recently, a consortium of 12 cities have formed the *Urban CO_2 Project*. The twelve cities involved in the project are: Ankara, Copenhagen, Dade County (Florida), Denver, Hannover, Helsinki, Minneapolis/St. Paul, Portland, Saarbrucken (Germany), San Jose (California), and Toronto. The first meeting of delegates was held in Toronto (where the project is based) in June 1991, and will be followed by a series of meetings through to June 1993. During this time each city will develop a profile of its own CO_2 emissions, assemble data concerning cost, equity, employment, institutional, and other environmental implications, and formulate a stategic implementation plan for CO_2 emission reduction. It is hoped that each city will commit to a target in the range of a 15-25% emission reduction by 2005, and commit to assisting a partner city in establishing a CO_2 emission reduction program, effectively making it a 24 city project.

ISSUES

Before outlining municipal strategies for reducing CO_2 emissions, a number of important issues and questions are addressed below.

Sustainable Development Context

Greenhouse gas emission reduction is one component of the larger problem of sustainable development.[11] Solutions to global warming must therefore not undermine sustainable development in other ways, but rather, must be part of an integrated strategy which addresses other sustainable development issues. Among these are the continuing losses of agricultural land due to urban sprawl and the problem of accumulation and disposal of nuclear wastes. This implies that the primary emphasis in reducing greenhouse gas emissions must be on improving energy efficiency and creating more compact urban forms.

Which Gases Should Be Targeted?

Although carbon dioxide is the single most important greenhouse gas, emissions of other gases are also important and can be influenced by municipal actions. These include methane (CH_4) and nitrous oxide (N_2O), which are greenhouse gases, as well as NO_x, carbon monoxide (CO), and hydrocarbons, which are not greenhouse gases but contribute indirectly to increasing greenhouse gas concentrations.[12] Measures to reduce CO_2 emissions through more efficient use of energy or by use of renewable energy forms will tend to reduce emissions of non-CO_2 gases. However, there could be increases in emissions of one or more non-CO_2 gases if carbon dioxide emissions are decreased through fuel switching.

Of particular concern are emissions of methane, which occur in association with the extraction and/or distribution and use of oil, coal, and natural gas. Methane is 25-50 times more effective as a greenhouse gas than CO_2 on a molecule-per-molecule basis, and leakage of methane from natural gas extraction, transmission, and distribution could offset much or all of the CO_2 reduction resulting from using natural gas in place of coal or oil. However, in some instances consideration of CH_4 will enhance rather than diminish the greenhouse benefit of switching to natural gas. If high efficiency (>90%) natural gas furnaces are used in place of coal-derived electricity for space heating (with 33% generation efficiency), the primary energy requirements are reduced by a factor of about 2.7. Thus, for natural gas emission factors (mass CH_4 per unit of energy) no more than 2.7 times *greater* than for coal, absolute CH_4 emissions are reduced by switching from coal to natural gas. For bituminous coals and natural gas leakage rates estimated for the U.S., the emission factor for natural gas may in fact be *smaller* than for coal,[13] implying that methane emissions will be reduced upon switching to natural gas heating.

Given changes in absolute emissions of CO_2 and other gases, the calculation of the net greenhouse benefit is complicated by the fact that the atmospheric lifetimes of CO_2 and other gases are different, so that the relative greenhouse effect of different gases changes through time. Furthermore, there are different methods for calculating the relative effects, each with its strengths and weaknesses. There is no objective basis for choosing one method or one time horizon over another.

These difficulties provide further reasons for placing the primary emphasis in CO_2 emission reduction on measures which will simultaneously reduce emissions of several greenhouse gases or which, as a minimum, do not increase emissions of non-CO_2 gases. This can be achieved through measures which reduce overall energy demand, through fuel switching in selected applications, or when fuel switching is accompanied by end-use energy saving measures.

Some national governments have argued that credit toward greenhouse emission reduction targets should be allowed for reduction and eventual elimination of CFC emissions. Thus, instead of reducing CO_2 alone by a given amount, the net greenhouse forcing of all greenhouse gases, including CFC's, would be reduced by the targeted amount. This has been refered to as the "basket" approach. To the extent that municipalities are playing a role in the capture and recycling or destruction of CFC's, one could argue for applying the basket approach at the municipal level.

There are compelling arguments against the basket approach, whether at the national or municipal level. First, as indicated above, there is no single, objective method for calculating the relative greenhouse effectiveness of different gases, and even if agreement were to be reached concerning methodology, there is large uncertainty concerning the values of the input parameters required for the calculation (particularly the atmospheric lifetimes of different gases and the role of indirect chemical effects). Secondly, although the CFC's are strong greenhouse gases, the stratospheric ozone which they destroy is also an effective greenhouse gas. According to Ref. 14, the cooling effect in mid-latitudes of the northern hemisphere of ozone destruction during the 1970's masked about 50% of the incremental heating due to the CO_2 increase during the same period. Inasmuch as there is large uncertainty in these calculations, the net effect of CFC emissions could range from a modest heating to a modest cooling. Finally, the basket approach assumes that protection of global ecosystems require elimination of CFC's or reduction of CO_2 emissions or reductions of some other greenhouse gases, when what is required are reductions of CFC and CO_2 and other greenhouse gas emissions as rapidly as possible (see Ref. 15 and references therein).

Rather than taking credit for CFC emission reductions, municipalities should pay close attention to proposed HCFC and HFC substitutes, some of which are almost as effective as greenhouse gases as the CFC's which they replace.[16] As with CH_4, there are instances where measures to reduce CO_2 emissions can also reduce the need for CFC's and HCFC or HFC substitutes, as explained later.

Sectoral Allocation of Emission Reduction

Having adopted an overall CO_2 emission reduction target, the question arises as to how to allocate the emission reduction among various sectors (residential, commercial, industrial, transportation) or among various energy forms (electricity, natural gas, oil). Ideally, the reductions should be allocated such that the marginal cost of further reductions in each sector is the same, as such an allocation will minimize the total cost. Although one might attempt to do this as targets in the range of 50-75% emission reduction are reached, there are good reasons for applying less stringent, interim targets (such as a 15-25% reduction by 2005) uniformly across sectors and energy forms (with adjustments to account for fuel switching). First, the economically attractive emission reduction potential is likely to be at least 15-25% in all sectors, and each sector should strive for the full economically attractive potential irrespective of environmental benefits. Secondly, in any cost-effective strategy to achieve 50-75% emission reductions, each sector is likely to require emission reductions of at least 20%. Since atmospheric stabilization requires going beyond a 20% target as soon as possible, there is no longterm loss in requiring each sector to achieve a 20% emission reduction as a first step (indeed, longterm gain is more likely).

Creating Accountability

Having adopted sectoral subtargets, it is important that a single agency or individual be charged with the task of coordinating the measures needed within each sector to achieve the given target. In the case of electricity and natural gas use, the local electric and natural gas utilities, which generally have franchise monopolies, are the appropriate agencies to ensure that their targets are achieved. In the case of the transportation sector, the appropriate individual might be the Commissioner of Planning. Although specific agencies and individuals should be held accountable for achieving the sectoral subtargets, the local municipal council will have to work closely with the individual agencies to insure that its own actions facilitate the achievement of the sectoral subtargets, and to ensure coordination between sectors.

Determination of Emission Reductions from Electricity Savings

Inasmuch as electricity is supplied in most regions from a mix of power sources, with different sources providing baseload, intermediate, and peak power requirements, there is not a simple relationship between electricity savings and CO_2 emission reduction. In some instances, non-fossil fuel sources (such as nuclear and hydraulic) provide baseload demand while coal or oil-fired generation provides intermediate and peak demand and can constitute close to 100% of the electricity supplied on the margin. In this case, electricity conservation measures will lead to a disproportionately large CO_2 emission reduction in the short term. However, if the long term impact of electricity conservation is to reduce the utility's investment in non-fossil power plants, then there would be less long term CO_2 emission reduction. In the extreme case, if all of the electricity savings leads to less nuclear or hydraulic capacity than would otherwide be the case, there would be no longer term CO_2 emission reduction.

Fortunately, electricity conservation will lead to CO_2 savings even if utilities also cut back on non-fossil supply options such as nuclear. There are two reasons for this. First, even for utilities with a large nuclear/hydraulic component, some fossil generation is needed to meet the variable component of electricty demand. As overall demand shrinks, so will the variable component and hence the required fossil fuel supply. More importantly, most conservation measures diproportionately reduce peak demand relative to base demand, and hence will disproportionately reduce required fossil fuel generation. This is especially true for measures such as switching from electric resistance to natural gas space heating and measures which reduce air conditioning requirements. Cogeneration will also displace utility fossil fuel generation, to the extent that the variation in electricity supplied matches the variation in total demand.

A MUNICIPAL ACTION PLAN

In the following section a comprehensive set of strategies for greenhouse gas emission reduction by municipalities is outlined. Only qualitative features are presented here; further discussion and quantitative analysis for the City of Toronto can be found in Ref. 6. Coordination will be needed between the actions outlined in the following sections.

Strategies for Electricity Use

The regulatory framework for electric utilities in the United States and elsewhere increasingly requires electric utilities to pursue "least cost" planning in which demand and supply side measures are considered on an equal footing.[17,18] This approach is also known as Integrated Resource Planning (IRP), and in some cases requires that environmental credits be assigned to energy conservation measures (demand management). Thus, in some jurisdictions the regulatory framework is already favourable to strong action at the municipal level regarding efficient use of electricity, although further regulatory improvements are needed.[19] Very few utilities so far have embarked upon energy conservation programs, and in those instances where programs are in place, reductions in peak demand of only 5-10% *compared to baseline growth projections* have been targeted. Furthermore, emphasis has recently shifted from energy efficiency programs to programs aimed at shifting demand from peak to non-peak periods, which will result in little or no CO_2 emission reduction if fossil fuels supply electricity on the margin during both peak and non-peak periods. Dramatically stronger and more aggressive programs than those currently contemplated are required.

Among the measures which municipalities and municipal utilities can undertake with respect to electricity use are:

- rate reforms, including elimination of declining block structures, minimum monthly bills, and flat rates for water heaters where these exist;
- provision of financial incentives to electricity users to help in meeting the cost of replacing inefficient equipment with state-of-the-art equipment;
- strongly discouraging use of electric resistance heating in new buildings or renovations, possibly through large hook-up fees, and provision of incentives to convert from electric resistance to high efficiency natural gas heating or to heat pumps;
- prohibition of bulk metering of electricity in new residential buildings and gradual conversion of existing multi-residential buildings to individual metering (this results in electricity savings of 15-40% based on analyses performed for Ontario, Canada);
- incentive programs involving thermal envelope improvements in electrically heated buildings (where economically attractive savings should be at least 25%);
- encouragement of local cogeneration of heat and electricity, either as part of a district heating and cooling system or in individual apartment or commercial buildings;
- conversion of street lighting to state-of-the-art lighting; and
- development of programs to insure 100% penetration of state-of-the-art lighting in commercial and institutional buildings.

Replacement of coal-derived electric resistance heating with high efficiency natural gas space heating, or displacement of coal-derived electricity with electricity from natural gas cogeneration, will reduce CO_2 emissions by a factor of 3 to 4[6]. If coal is on the margin at all times, then similar savings will result if electric resistance water heaters are replaced with high efficiency natural gas heaters; if non-fossil fuel sources are on the margin part of the time, then much smaller CO_2 savings are likely to be achieved because of the ability of utilities to remotely turn water heaters on and off and their preference to run them during times of low electrical demand.

Preference should be given to thermal envelope improvements and reduction in hot water demands over fuel switching, even though the latter generally results in much larger emission reductions in the short term. This is because the use of natural gas is only a bridging fuel on the road to a sustainable energy system, in which hydrogen produced from renewable energy sources is likely to play a key role.[20] Furthermore, improving the thermal envelope first and fuel switching second, rather than the other way around, reduces total costs as a smaller heating system can be installed. Similarly, attention should be paid to reducing electrical, heating, and cooling loads before cogeneration systems are installed or HVAC (Heating-Ventilation-Air Conditioning) systems upgraded.

Natural Gas

Integrated Resource Planning concepts are just now beginning to be applied to the regulation of natural gas utilities. Adoption of an IRP framework for gas utilities would allow them to aggressively promote efficient use of energy in the same way as electric utilities, through appropriate price signals, financial incentives, and information and technical assistance programs. As IRP concepts are much less accepted in the gas sector than for electric utilities, municipalities can play a particularly useful role by intervening at regulatory hearings to argue that gas utilities should be required, and permitted, to aggressively promote energy efficiency and conservation.

Among the opportunities for significantly reducing current or projected natural gas use are:
- use of high efficiency (>90%) natural gas furnaces in place of standard gas furnaces (generally <70%);
- use of high efficiency (>90%) natural gas water heaters in place of standard gas heaters (50% efficiency);
- use of state-of-the-art windows, building insulation, ventilation heat recovery, and passive solar design features in place of standard practice.

Opportunities for combining natural gas space and water heating with passive solar features should also be actively promoted. For example, integrated natural gas space and water heaters for residential applications are now on the market, in which space heating is obtained by blowing air past a hot water coil, permitting inexpensive retrofitting in forced-air heating systems. Solar panels can be used to preheat water to 80°C even on cold, sunny winter days, with the natural gas serving as a backup. If installation of such a system is combined with thermal envelope upgrades and simple measures to reduce consumptive hot water use (such as low flow showerheads), CO_2 emission reductions of up to 90% can be readily achieved.

District Heating and Cooling

Modern district heating systems involve a central heating plant from which heat is distributed to surrounding buildings as hot water, with subsequent return flow forming a closed loop. In spite of typical distribution heat losses of 10%, such systems reduce energy use and hence CO_2 emissions by 10-12% because of the greater efficiency of central plants compared to on-site heating systems. District heating systems also result in lower NO_x emissions, both because of better regulation of central heaters and the fact that less natural gas is used.

Much larger energy savings are possible if district heating is combined with electricity cogeneration. In cogeneration, the marginal efficiency of electricity generation is typically 80-85% and is consumed on site or nearby, compared to 33% efficiency in central power plants and a further 5-10% distribution loss. In summer the waste heat from electricity cogeneration can be used for hot water requirements or for cooling purposes *without using CFC's* through absorption cooling methods. It is likely that existing CFC-based electric chillers will require retrofitting or replacement in order to be compatible with CFC replacements, and that electricity required to met a given cooling load will increase.[9] Given uncertainty as to when the current set of replacements will be phased out, building owners are likely to welcome the opportunity to connect to a CFC-free district cooling system if the site-specific economics is favourable.

District heating and cooling (DHC) systems are economically attractive in new developments or redevelopment projects of sufficent size that the infrastructure costs are largely offset by not requiring boilers and chillers in each building. DHC systems with electrical cogeneration can start small, wherever there is a cluster of buildings with sufficient electrical and heating or cooling loads to be economically viable, and grow from a number of nuclei on an opportunistic basis. Because an absorption-based district cooling system would displace electric chillers, which contribute strongly to peak summer electricity demand, there are significant avoided cost benefits to electric utilities. Expansion or installation of district cooling systems should therefore qualify for financial incentives from electric utilities under IRP principles.

As with other fuel switching options, connection of an individual building to a district heating and cooling system should be accompanied by thermal and cooling load reductions through lighting retrofits, building envelope improvements, and improved ventilation and heat recovery. There are at least two ways in which such improvements could be leveraged. First, expansion of the main district heating/cooling infrastructure could be financed through local improvement levies (which are applicable whether or not a building utilizes the new service), with the magnitude of the levy for a given building scaled by its peak heating and cooling demand. Secondly, electric utilities could offer to pay all or part of the cost of hooking up an individual building to the main pipes if the building meets minimum energy performance standards, with the incentive level based on the utility's avoided costs (adjusted to incorporate environmental benefits).

A district heating system could also be used as a broker of heat energy, collecting heat from whatever sources are available, upgrading it with heat pumps, and delivering it to wherever the heat is needed. Other sources besides electricity generation include wastewater, subways, dry cleaners, and bakeries. Heat recovery from wastewater (which is as warm as 10°C even in winter) is already practiced in Tokyo and some European cities.[21] In summer the heat pump can be operated in the opposite direction, removing heat from the district cooling system and transferring it to the wastewater. The wastewater thus serves as a heat source in winter and a heat sink in summer.

Buildings

In addition to promoting electric and natural gas utility reform, and adoption of district heating and cooling systems with cogeneration, there are further actions that municipalities can undertake to insure efficient use of energy in the building sector. Foremost in importance, municipalities should encourage or require new buildings to meet minimum energy performance standards. Initially, these can be based on the ASHRAE 90.1 standard of the American Society of Heating, Refrigerating and Air-Conditioning Engineers for commercial construction, and the Canadian R-2000 standard for residential construction. Buildings constructed according to these standards should use 30-40 percent less energy than the existing average building stock in most North American cities.

It is economically attractive, however, to go far beyond the ASHRAE 90.1 and R-2000 standards. Early strengthening of the standards, based on life-cycle cost analysis, is therefore required. In the meantime, municipalities can offer incentives to developers to go beyond the current standards, either through financial incentives in conjunction with utility demand management programs, or by accelerating the development approval process.

Transportation

Significant reduction of transportation-related automobile emissions requires major investments in rapid transit infrastructure, accompanied by an increase in the population density of urban centres in North America and better integration of housing, jobs, and urban amenities. The issue of urban intensification is discussed separately in the following section. Given that eventual carbon dioxide emission reductions of at least 75% will be required by developed countries for atmospheric stabilization, and that a doubling of the current average automobile fuel efficiency is perhaps the most that can be expected,[22] significant reductions in automobile vehicle kilometers travelled (VKT) will be required. This implies a fundamental lifestyle shift toward greatly reduced dependence on the automobile. Admittedly, such changes will not occur rapidly or easily.

Nevertheless, there are a number of actions that municipalities can undertake in the short term to reduce automobile use at relatively little cost. The most important of these is to make public transit more attractive by creating a comprehensive network of dedicated bus and/or streetcar lanes, from which private automobiles would be excluded, at least during rush hours. The competition between fully loaded transit vehicles and private automobiles, usually with only one occupant, represents inefficient use of limited road space. Setting aside dedicated lanes for transit vehicles, with frequent enough service that public transit will be heavily used, will result in more efficient use of existing road resources. Transit lanes could be accompanied by transit vehicle traffic light priority measures and proof-of-payment fare systems to permit more rapid loading and unloading of passengers. Such measures would significantly improve both the speed and reliability of public transit service in many cases, as well as increasing the peak pasenger carrying capacity of existing fleets.

Many cities are laced by a network of freight rail corridors which are generally underutilized. Opportunities exist through advanced computer switching and control technology to utilize these corridors for rapid public transit while still permitting freight uses; such practice is common in European cities. The freight demands on the rail corridors within cities could be reduced (to some extent) by building rail bypasses, the same way that expressway bypasses are built. Such bypasses could also reduce health and safety concerns associated with the transport of hazardous substances through urban centres. Opening up of rail corridors for public transit uses could be accompanied by redevelopment and housing intensification along the corridors.

High voltage power transmission corridors could also be utilized for rapid rail transit. Health effects of the associated magnetic fields on passengers would be negligible due to the short duration of exposure, although possible health effects on transit staff would necessitate a fully automated system (such as found today in Lille, France).

Other measures include increase in the price of parking and a gradual decrease in the public parking supply, while keeping provision of parking in new developments to a bare minimum. Several cities require major employers to develop auto-minimization programs while others are experimenting with high occupancy vehicle (HOV) lanes.[9] However, unless car-pooling schemes play a subordinate role within a comprehensive set of strategies in which the major emphasis is on public transit, the likely longterm effect will be to perpetuate the low density land use patterns which create automobile dependence in the first place.

Finally, a comprehensive network of dedicated bicycle lanes could encourage a significant fraction of commuters to travel by bicycle. In northern European cities such as Amsterdam and Copenhagen, 20-30% of all trips are made by bicycle. To be effective, this measure should be accompanied by other infrastructural support for cyclists, such as adequate and secure bicycle parking facilities (both at work and in multi-unit residences) and requirements for shower facilities in new commercial developments.

Fuel switching, from gasoline to natural gas, methanol, or electric vehicles, is receiving increasing attention as a measure to reduce urban air pollution.[23] This will increase greenhouse gas emissions in the case of methanol produced from coal or coal-fired electricity,[24] and gives only small reductions at best in other cases. Any automobile-dominant urban transportation system is inherently inefficient, and this inefficiency is not resolved by switching fuels. Fuel switching alone therefore does not contribute to longterm sustainable development. On the other hand, greatly reduced automobile use will simultaneously reduce greenhouse emissions, local and regional air pollution, and traffic congestion with its negative economic and social impacts.

In order to shift a significant number of commuters from automobile to public transit, regional transportation authorities will need to be created where they do not already exist. Such authorities would make strategic planning decisions, in consultation with local municipal planning departments, and insure the coordination and integration of local surface transit services with regional rapid rail services.

As dependence on the automobile decreases and the capacity of public transit to meet transportation demand increases, municipalities will be able to consider "decommissioning" some urban expressways. This seemingly radical step is already being considered in the case of the Gardiner Expressway in Toronto,[25] which runs parallel to a major rail transit corridor, although the main motivation is the urban redesign opportunity it provides rather than environmental benefits.

Land Use

The high per capita transportation energy use in North American cities is a land use planning problem, inasmuch as land use planning decisions have favoured low density suburbs in which distances between residence, work, and social amenities are large and public transit is not economically viable. This creates a dependence on the automobile for every day use which, coupled with large commuting distances, leads to high fossil fuel energy use.[26] The greatest long term impact on transportation energy use will likely result from a re-orientation of urban planning toward higher density development in which various urban land uses and services are located within short distance of one another. This can occur through redevelopment and intensification of the existing urbanized area.

Urban regions such as Stockholm, London, Paris, San Francisco, Denver, Melbourne, Sydney, Toronto, and many others have developed policies to intensify the existing urbanized area. The *Metro Toronto Strategic Plan* targets a population increase from 2.2 million to 2.5-2.8 million over the next 20 years within an already urbanized area.[27] Urban intensification does not imply high rise apartment buildings with their associated negative impacts on the surrounding communities, nor does it require disruption of existing neighbourhoods. The 1980's have seen several innovative and well-designed redevelopment projects which have involved a healthy mix of land uses, building heights, income levels, and tenure.[28] In the case of Toronto, the following options are being considered or actively promoted: conversion of existing single-family dwellings into multi-household units through renovation and/or additions; redevelopment of the city's main streets with 4 to 5 floors of residential accommodation above commercial space at grade; redevelopment of selected shopping malls and a racetrack, each surrounded by large parking lots, as medium density mixed commercial/residential land use with reduced parking provided underground, improved public transit linkages, and provision of amenities to the surrounding communities;[29] and redevelopment and rehabilitation of industrial land rendered surplus as a result of the shift to a more service-oriented economy. Urban intensification need not be restricted to residential developments; low density office parks may be as great or even greater a contributor to the North American low density urban form.

Apart from reducing transportation energy use, urban intensification can limit or stop the loss of farmland due to urban sprawl. It also addresses a number of economic and social problems associated with continuing urban sprawl and compounded by an aging population and declining average household size.[28]

Urban Forestry

According to recent studies,[30] urban shade trees in midlatitude cities can reduce residential air conditioning requirements by 15-30% and winter heating requirements by 15% or more, the latter as a result of wind sheltering. Increasing the urban shade tree population can therefore make a significant contribution to reducing air conditioning energy use as well as mitigating the expected warming over the coming decades. The cooling effect of trees arises both from direct shading of buildings and by cooling of ambient air; thus, buildings not shaded by trees will nevertheless benefit from the urban forest.

Recently there has been interest in urban tree planting as a response to global warming. It is important, however, to avoid a crash-program mentality. Each tree implies a substantial long term commitment for watering, pruning, and eventual removal. Careful attention must be given to species selection. As with humans, health care costs for trees increase substantially as they age. It is best, therefore, to develop an urban forest with a wide range of tree ages. This implies modest but longterm tree planting programs. Nevertheless, there is probably room for doubling or tripling the rate of tree planting in most municipalities. Equally important are strong measures to protect existing trees and modifications to urban design guidelines, particularly regarding infiltration of rainwater, so as to permit healthy tree populations.

Waste Management

Industrial emissions typically account for about 1/3 of total CO_2 emissions in developed countries. Hence, significant emissions are embodied in the manufactured goods which are consumed and eventually disposed of by urban populations. However, significant energy savings are possible when manufactured products are reused or recycled. For example, production of iron and steel from scrap uses 8 to 9 times less energy than production from raw ore, while recycling of newsprint uses about 1/4 less energy than production from virgin fibres.[31] Thus, municipalities can significantly reduce greenhouse emissions by aggressively promoting recycling and reuse of wastes.

Landfills are important sources of methane,[32] but much of the landfill methane can be captured and used for electricity generation, or sold, or burned off as carbon dioxide.

Education and Community Outreach

Finally, municipalities can play an important role in educating the public concerning individual responsibilities in reducing greenhouse gas emissions and in contributing to sustainable societies. Among the avenues available are the elementary and secondary school systems, community organizations and comprehensive community-based programs, municipal and utility billings, and advertising in various media. Municipalities, in conjunction with the electric and natural gas utilities, can hire and train members of local communities to go door to door within their community as energy auditors and line up work that can be done by trained energy efficiency contractors. A centralized approach is unlikely to be effective, particularly in diverse, multicultural cities.

In addition, municipalities can establish Round Tables on environment and economy involving local politicians, bureaucrats, business leaders, academics, and community and environmental organizations. Such round tables can contribute to building concensus concerning the need for action to assure sustainable development, and can greatly increase the likelihood of concrete community-based actions.[33]

COSTS OF GREENHOUSE GAS EMISSION REDUCTION

Space does not permit a quantitative analysis of the costs of municipal actions to reduce greenhouse gas emissions and, in any case, large uncertainties remain concerning costs of reductions on the order of 50-75%. Nevertheless, a few useful generalizations are possible.

Firstly, initial and large greenhouse gas emission reductions can be obtained at no net cost if discounted energy savings are compared against upfront captial costs.[34] Second, although later emission reductions will entail net cost, there will be significant benefits from reduced environmental damage and reduced risk. Many of the most important costs of unrestrained greenhouse gas emissions—such as species extinctions and ecosystem degradation—cannot be quantified in economic terms. Thirdly, the costs of greenhouse gas emission reduction will be spread among other benefits, such as reduced urban and regional air pollution and preservation of agricultural land.

Whatever the final cost of atmospheric stabilization, it is clear that the cost will be smaller the sooner we get started. The opportunities for reducing energy demand in new buildings are much greater, and much cheaper, than in existing buildings, while replacement of old equipment with the most efficient equipment available at the time will entail little extra cost if this occurs as part of the normal equipment turnover, rather than through accelerated turnover. There are also likely to be significant new manufacturing and export opportunities for those regions which act first to address the problem of greenhouse gas emissions.

Although most of the initial measures needed to reduce greenhouse gas emissions are economically attractive in their own right, there are a number of barriers to their implementation, including initial upfront costs. Energy service companies and electric and natural gas utilities can provide some or all of the capital requirements. Another potential source of capital, which requires careful consideration, are pension funds.

CONCLUSIONS

This chapter has presented a cursory overview of the strategies that municipal governments can use to reduce greenhouse gas emissions. As the examples presented here indicate, considerable power exists at the municipal level. Inasmuch as almost half of the world's population will live in urban centres by year 2000, the world's municipal governments can collectively bring about a substantial reduction of global greenhouse gas emissions and demonstrate to higher levels of government the courage and commitment needed to resolve this and other problems of sustainable development.

REFERENCES

1. Easterly (this volume), Fischer (this volume), Ross (this volume).
2. Intergovernmental Panel on Climate Change (IPCC). 1990. Policymakers' Summary, *In* Houghton, J.T., G.J. Jenkins and J.J. Ephraums (eds.), *Climate Change: The IPCC Scientific Assessment*, pages vii-xxxiii, Cambridge University Press, Cambridge.
3. Harvey, L.D.D. 1989. Managing Atmospheric CO_2. *Climatic Change* 15:343-381.
4. Harvey, L.D.D. 1990. Managing Atmospheric CO_2: Policy Implications. *Energy* 15:91-104.
5. Special Advisory Committee on the Environment (SACE). 1991. *The Changing Atmosphere: Strategies for Reducing CO_2 Emissions. Policy Overview.* 16 pages. Available at no charge from Resource Centre, City Hall, Toronto, M5H 2N2.
6. Special Advisory Committee on the Environment (SACE). 1991. *The Changing Atmosphere: Strategies for Reducing CO_2 Emissions. Technical Volume.* 106 pages. Available for $10 from Resource Centre, City Hall, Toronto M5H 2N2.
7. Task Force on Atmospheric Change (TFAC). 1990. *Clouds of Change: Final report of the City of Vancouver Task Force on Atmospheric Change. Volumes I and II.* Available from City of Vancouver, 453 W. 12th Street, Vancouver V5Y 1V4.
8. Hurdle, D. and M. Adams. 1991. *Global Warming: How London Boroughs and the Government can respond.* Available for £10 from London Boroughs Association, 23 Buckingham Gate, London SW1E 6LB.
9. Totten, M. 1990. *Energywise Options for State and Local Government. A Policy Compendium.* Center for Policy Alternatives, 2000 Florida Ave., NW, Suite 400, Washington D.C. 20009.
10. City of San Jose. 1991. *Sustainable City Strategy 1991-1992.* City of San Jose Office of Environmental Management, 777 N. First Street, Suite 450, San Jose 95112. 70 pages.
11. World Commission on Environment and Development (WCED). 1989. *Our Common Future,* Oxford University Press, Oxford, 383 pages.
12. Shine, K.P., R.G. Derwent, D.J. Wuebbles and J-J. Morcrette. 1990. Radiative forcing of climate, *In* Houghton, J.T., G.J. Jenkins and J.J. Ephraums (eds.), *Climate Change: The IPCC Scientific Assessment,* pages 45-68, Cambridge University Press, Cambridge.
13. Barns, D.W. and J.A. Edmonds. 1990. *An evaluation of the relationship between the production and use of energy and atmospheric methane emissions.* U.S. Department of Energy, Washington, DOE/NBB-0088P, 143 pages.
14. Lacis, A.A., D.J. Wuebbles and J.A. Logan. 1990. Radiative forcing of climate by changes in the vertical distribution of ozone. *Journal of Geophysical Research* 95:9971-9981.
15. Vellinga, P. and R. Swart. 1991. The Greenhouse Marathon: A proposal for global strategy. Guest Editorial. *Climate Change* 18:vii-xii.
16. Fisher, D.A., C.H. Hales, W-C.Wang, M.K.W. Ko and N.D. Sze. 1990. Model calculations of the relative effects of CFC's and their replacements on global warming. *Nature* 344:513-516.

17. Geller, H.S. 1989. Implementing electricity conservation programs: progress towards least-cost energy services among U.S. utilities. *In* Johansson, T.B., B. Bodlund and R.H. Williams (eds.) *Electricity: Efficient End-Use and New Generation Technologies, and their Planning Implications,* Lund University Press, Lund, pages 741-763.
18. Wiel, S. 1991. Efficiency in the USA: A state regulator's view. *Energy Policy,* April, 202-204.
19. Moskovitz, D.H. 1990. Profits and progress through least-cost planning. *Ann. Rev. Energy* 15:399-421.
20. Harvey, L.D.D. 1990. *A techno-economic assessment of hydrogen fuel.* Report to the Environmental Defense Fund, New York, 124 pages (available from the author).
21. Tokyo Metropolitan Government (TMG). 1990. *Urban Heat: Heat utilization system from sewage.* Sewerage Bureau, Tokyo Metropolitan Government.
22. Renner, M. 1989. Rethinking transportation. *In* L.R. Brown (ed.), *State of the World 1989,* W.W. Norton, New York, pages 97-112.
23. Russell, A.G., D. St. Pierre and J.B. Milford. 1990. Ozone control and methanol fuel use. *Science* 247:201-205.
24. DeLuchi, M.A., R.A. Johnston and D. Sperling. 1989. Transportation fuels and the greenhouse effect. *Transportation Research Record* 1175, 33-44.
25. Royal Commission on the Future of the Toronto Waterfront. 1990. *Watershed.* Available from Government of Ontario Bookstore, 880 Bay Street, Toronto M7A 1N8.
26. Newman, P.W.G. and J.R. Kenworthy. 1989. *Cities and Automobile Dependence: A Sourcebook.* Aldershot: Gower International. 388 pages.
27. Metro Toronto Strategic Plan. 1991. Available from Metropolitan Toronto City Hall, 390 Bay Street, Toronto, M5H 3Y7.
28. Canadian Urban Institute (CUI). 1991. *Housing Intensification: Policies, Constraints, and Options.* 56 pages. Available for $12 from Canadian Urban Institute, City Hall, Toronto, Canada M5H 2N2.
29. Canadian Urban Institute (CUI). 1991. *Charrette on Housing Two Million in the Greater Toronto Area by 2021. Proceedings.* 63 pages. Available for $17 from Canadian Urban Institute, City Hall, Toronto, Canada M5H 2N2.
30. Akbari, H. 1990. *Urban climate and heat island control strategies: Impacts on residential building energy use in Toronto, Canada.* Lawrence Berkeley Laboratory, Berkeley, California.
31. Harvey, L.D.D. 1990. *Carbon dioxide emission reduction potential in the industrial sector.* Report prepared for the *Ontario Select Committee on Energy.* Royal Society of Canada, Ottawa, 45 pages (available from the author).
32. Khalil, M.A.K. and R.A. Rasmussen. 1990. Constraints on the global sources of methane and an analysis of recent budgets. *Tellus* 42B:229-236.
33. National Round Table on the Environment and Economy (NTRTEE). 1990. *Sustainable Development and the Municipality,* 8 pages (1 Nicholas Street, Suite 520, Ottawa K1N 7B7).
34. Lovins, A. 1990. *The role of energy efficiency. In Global Warming: The Greenpeace Report.* J. Leggett (ed.), Oxford University Press, Oxford, pages 193-223.

Global Climate Change: Implications, Challenges and Mitigation Measures. Edited by S.K. Majumdar, L.S. Kalkstein, B. Yarnal, E.W. Miller, and L.M. Rosenfeld. © 1992, The Pennsylvania Academy of Science.

Chapter Thirty

FORESTRY RESPONSES TO MITIGATE THE EMISSION OF GREENHOUSE GASES

STEVEN M. WINNETT
Office of Policy Analysis
US EPA, PM 221
401 M Street, SW
Washington, D.C. 20400

INTRODUCTION

The concentration of greenhouse gases in the atmosphere is increasing on a yearly basis, strengthening concerns about the possibilities of changes in the Earth's climate. Manipulating forests offers one way to mitigate the potential effects of climate change on the planet. Specific strategies for manipulating forests include creating new forests (afforestation), reestablishing forests where they have previously been (reforestation), and managing existing forests to increase their ability to remove or sequester carbon from the atmosphere. Our interest in forests with regard to climate change is their role in the carbon cycle.

Trees and other plants absorb carbon and store it in their tissues. Wood and vegetation are roughly 50% carbon. This fixation process removes CO_2 from the atmosphere, and it is the only practical way we know at present of removing atmospheric carbon. By increasing the total mass of wood and vegetation, normally referred to as "biomass," we can to a degree remove some of the carbon that has been emitted to the atmosphere and store it in the form of carbon "sinks." Actions which store the carbon in forests and wood products keep the carbon out of the atmosphere as long as these sinks are intact. A tree-planting program designed to plant, improve and maintain one billion trees a year for 20 years in the United States would store 1.4-2.8 billion metric tons of carbon after 40 years! (A metric ton is approximately 1.1 English tons, and all tonnage figures in this paper are metric.) At 1989 levels, this is approximately equivalent to between 2-5% of the yearly emissions of carbon from the U.S. over that 40-year period. In addition to the carbon they actually store, some of these trees provide carbon benefits by shading buildings and reducing the energy needed (and fossil fuels burned) to cool them. They may also be used in place of fossil fuels to generate energy, further reducing a net source of atmospheric CO_2.

Forests store more carbon than any other terrestrial ecosystem. Old-growth Douglas fir forests (*Pseudotsuga menziesii*) in the Pacific Northwest may contain 612 tons of carbon per hectare,[2] while mature southern pine forests contain 200-275 tons per hectare,[3] and 50-year old northern hardwood forests 250 tons.[4] Primary tropical moist forests (the tropical equivalent of old-growth forests) may contain 320 tons per hectare, while the secondary forests which grow up following disturbances contain 240 tons per hectare or less[5] (Figure 1). A hectare is roughly 2.5 acres.

While the largest single source of greenhouse gas emissions is the burning of fossil fuels, deforestation and activities involving the conversion of forest land contribute a significant amount of CO_2 and other greenhouse gases to the atmosphere. Of the

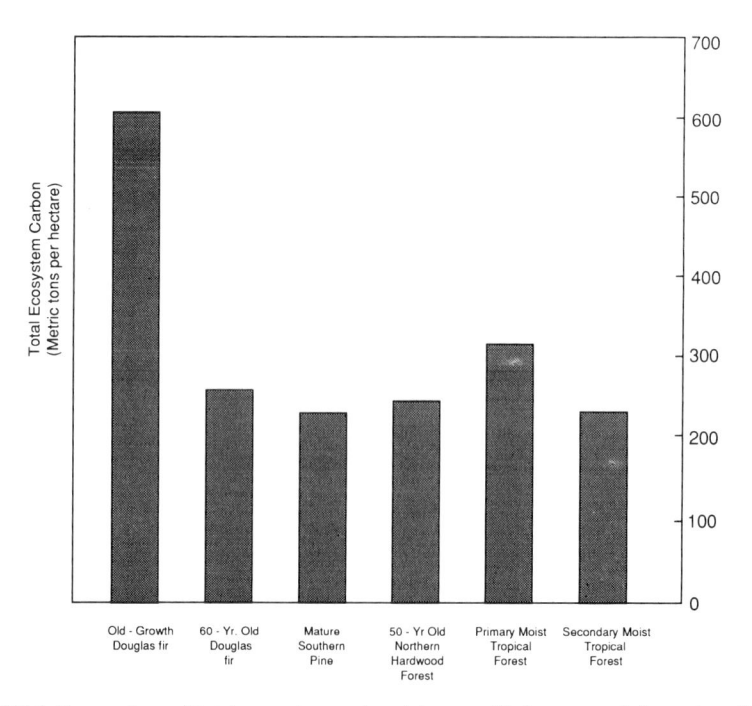

FIGURE 1. Comparison of total ecosystem carbon (above and below ground) for various forest ecosystems, in metric tons per hectare. Figures given are approximations or midpoints of ranges. Sources: Harmon et al., 1990[2]; Birdsey, 1990[3]; Larson et al., 1986[4]; Houghton et al., 1983.[5]

6.9 billion tons of carbon emitted globally every year, 18% comes from deforestation and changes in land-use.[6] In most cases, various population and development pressures are responsible for these destructive activities. Forestry and other programs involving vegetation may be able to reduce the effects of these pressures and so aid in slowing land conversion.

It is unrealistic to expect that planting trees can by itself stabilize or reduce the concentration of CO_2 in the atmosphere. Sedjo[7] estimated that a minimum of 470 million hectares of plantations would be needed to stabilize atmospheric CO_2 at its current level. This is an area 1.5 times the size of the forested area of the United States, and five times the size of plantation culture throughout the world.[8] To simply stabilize the loss of forest from deforestation at current levels would require a yearly reforestation of as much as 20.4 million hectares,[9] an area larger than the state of Washington. Trees and forests do represent part of the answer to mitigating the build-up of greenhouse gases.

This chapter examines the role of trees in mitigating the emission of greenhouse gases, and discusses the ways in which trees and forests can be or are being managed to reduce the effects of global change. The first section will discuss the role of trees in mitigating greenhouse gas emissions and will give an overview of the types of response strategies that are possible. The second section examines the implications of some specific forest management strategies and discusses the issues involved in their use when considering the goal of carbon storage and sequestration. The discussion will center on emission and fixation of carbon dioxide (CO_2), the most important of the greenhouse gases, and the gas whose concentration can be most affected by manipulating vegetative systems.

THE ROLE OF TREES

Trees play an important role in mitigating the emission of carbon dioxide, and we manipulate forested ecosystems in many ways to enhance the forest's performance in that role. No matter which type of strategy we apply to a given parcel of land, we are using forests to manage the carbon cycle in three basic ways. All of our management goals can be categorized by these three biophysical functions as follows.

Strategies Which Maintain Existing Carbon Sinks

We can maintain existing carbon sinks by protecting forests from catastrophic disturbance, from conversion to other land uses, and from degradation to a condition of lower biomass. By practicing sound forest management, we can insure the continuity and health of the forest ecosystem.

Regenerating a forest (or returning it to a fully stocked condition) as soon after harvesting is possible, helps insure the maintenance of the carbon sink over the long run, although the sink is depleted at the time of harvest. Conducting a harvest operation carefully so as to damage the remaining stand and ecosystem as little as possible helps to insure long-term health by reducing the opportunity for pests and pathogens to gain entry, and reduces the time until the stands return to a state of maturity and maximum carbon storage.

Establishing forest preserves or protected stands maintains the existing carbon sink and may actually increase it. Mature forests have many times the carbon in young stands, even though young stands may be increasing in biomass at a greater rate. If a plantation replaces harvested old-growth, it may take hundreds of years for the new stand to regain the levels of biomass contained in the old stand, and only if it is allowed to grow that long. In the interim, that carbon is either in the atmosphere or in wood products that may be shorter lived than the stand, and generally represents only a small percent of the stand's original carbon stock. By establishing economically based rotations which terminate stand growth after 60-80 years, managers further reduce the carbon sequestration. In the Pacific northwest, old-growth stands contain as much as 612 tons carbon per hectare, while a commercial stand will contain about 259-275 tons at 60 years.[2]

It is possible that the stand may never regain the level of biomass experienced before harvesting. This appears to be the case in much of the large-scale deforestation in the tropics. In this situation, the loss of seed sources, the degradation of the soil, and the changes in micro-climatic cycles induced by forest clearing and burning have caused permanent degradation of the forest. These ecosystems may not return to their condition before human disturbance without massive restoration efforts. Further, the plantations which often replace them will probably not contain as much carbon, nor be as stable a system as the original forest. Whereas intact primary forests may contain 277-317 tons carbon per hectare, the secondary forests which replace them after human disturbance may contain between 208 and 238 tons per hectare.[5] The plantation forests may contain only 27-68 tons per hectare.[10]

Forests suffer the stresses of disease, pests, acid and heavy metal deposition, ozone, and extreme variations in climate. These stresses may lead to loss of productivity and vigor of individual trees or whole stands, the death of individual trees, or the death of whole stands. The introduction of an oriental fungus (*Endothia parasitica*) in the 1890's lead to the total destruction of the American chestnut (*Castanea dentata*), then the dominant tree species in what is now the eastern oak-hardwood forest. Extreme variations in winter temperatures, along with a set of insect pests and pathogens, appear to be partially responsible for the current decline in sugar maple (*Acer saccharum*) in the northeastern hardwood forest.[11] Climatological and chemical stresses reduce the ability of a stand to resist the invasion of pests and pathogens. Any activities which increase resistance to these stresses increase the survival and maintenance of the forest ecosystem, and of the carbon sink.

The question of forest fire poses a different question, as fire is a natural process in forests and can be a beneficial aid in regenerating stands. Species such as lodgepole pine (*Pinus contorta*), found in the Rocky Mountains, require fire to open their cones and release their seeds. Many species intolerant of shade require, for their germination and growth, the bare, unshaded ground that fires naturally provide.

Destruction of a forest by a forest fire is an obvious loss of a carbon sink. However, preventing small fires from burning can lead to a build up of dead wood in the forest, which can eventually act as a fuel supply when wildfires do occur. This fuel load can turn a small, useful fire into a destructive inferno. Instead of a fire which may have burned low on the forest floor, reducing fuel loads and returning nutrients to the soil, the results may be a roaring blaze that rises high into the forest killing healthy, mature trees and releasing large amounts of carbon. It may be that it is more useful to allow small fires to burn than to risk the catastrophic loss that may follow preventing them. The choice is almost never clearly defined and no one has analyzed the tradeoff in detail.

Other non-forestry activities may have the effect of protecting intact forests against destruction. In the tropics, shifting agriculture is responsible for a great deal of the destruction of primary forest. Forest dwellers need to grow crops for food and they cut down forests for land to plant them.

Unlike most soils in the temperate region, most tropical soils are inherently infertile, and their productivity is tied up in the organic matter, roots, and vegetation of the forest itself. Removing the forest and planting crops causes most of the productive potential to be lost. When the soil fertility declines after a few years of cropping, farmers abandon the fields and cut down more forest, starting the process all over. Sustainable agricultural and agroforestry systems (combinations of trees and crops on the same land) have been proposed as substitutes for shifting agriculture. These systems would promote sustained productivity of the land, and a higher level of biomass through the combination of tree and herbaceous crops. Farmers could farm fields which would retain their productivity over time, and would not have to cut down additional forest for suitable cropland.

Another strategy which has been suggested for protecting tropical primary forests is developing products from the intact forest that could be consumed or sold for income. Peters *et al*.[12] estimated that the financial return from the sustained use of intact tropical forest could be greater than the income from the destructive farming and grazing practiced on the same land. Fruits, nuts, natural fibers, latex, medicinal herbs, and animals could be harvested in a sustainable manner, requiring that the forest remain intact for their long-term productivity. Markets would have to be developed for products that do not yet have significant economic value, and forest dwellers would have to be taught new methods of managing the intact forest. There are also methods of managing the forest in a benign manner developed by the native peoples in the past that are being rediscovered. These strategies have the potential to provide rural dwellers with food and income, and insure the maintenance and health of the standing forest.

Strategies Which Increase Carbon Sinks

Reforestation or afforestation programs increase carbon sinks by creating new forests. The United States' Conservation Reserve Program (CRP) has been successfully promoting planting forests on marginal crop and pasture land, increasing substantially the amount of carbon stored on these lands. Parks[13] estimated that there are 49 million hectares of these lands capable of growing trees in the United States, which could sequester 25 million tons carbon per year. Houghton *et al*[10] estimated that there are as much as 579 million hectares of land in the tropics (Latin America, Southeast Asia and Africa) that are capable of accepting forest plantations, and 356-499 million hectares suitable for agroforestry systems. The research of Trexler[14] indicates that the practical availability of land for reforestation and other forest activities may be substantially below the physical availability, due to social, economic and political constraints.

Existing forests can be managed to produce higher yields of wood, thus increasing carbon storage. Management practices include improving the stocking (number of trees per unit area) of understocked or overstocked stands, using greater amounts of fertilizer, and lengthening rotations to let trees grow larger. Hughes[15] estimates that highly intensive management of southern pine plantations, combined with lengthening the typical commercial rotation period could result in yields quadruple those in a typical natural pine stand, or about 560 cubic meters of wood per hectare by age 40. An extensively managed plantation on the same land in coastal North Carolina would be expected to produce only about 330 cubic meters per hectare.

Row[16] has estimated that of the 2.4 million forested hectares of U.S. military bases, there are about 161,000 hectares which could realize an increase in carbon storage of 1.3 to 1.4 tons carbon per hectare with instensified management. The treatments, which include thinning densely stocked, stagnate stands, removing trees with poor growth potential, and releasing trees with high potential for growth from competition, would result in a total carbon benefit of about 9 million tons of carbon over 40 years. Nationwide, Vasievich et al.[17] identified a potential increase of 273 million tons of carbon per year from improving the management of private non-industrial forest lands. Ten million hectares have the potential to respond vigorously to the types of stocking control described above. Seventeen million hectares would require regeneration activities such as replacement of the stand or replanting in order to realize significant carbon benefits. Most of this potential is in the southeast and south central part of the country.

Replacing decrepit, degraded, or unproductive stands with young healthy stands may substantially increase standing biomass. In some cases, replacement of the existing stand with species more suited to vigorous growth on the site may be needed. The Defense Department lands have some 448,000 hectares in stands of slash pine (*Pine* spp.), scrub oak (*Quercus* spp.), and other species poorly suited to the site or the climate. Replacing these woodlands with vigorous stands of species such as longleaf pine (*P. palustris*) on the southern deep sand sites could result in an increase of 11 million tons in the national carbon sink over 40 years.[16] There are a number of measures which create "informal" forests which result in an increased carbon sink. In the developing world, the establishment of farm and village forests can be a significant addition to the world's biomas supply. If even established as non-contiguous individual trees around a field or residence, or in orchards and near watercourses, these forests can perform as well as hectares of closed-canopy plantations.

In the developed world, informal forests can be established around farms as wind breaks and shelterbelts, as orchards, and along highways and transportation corridors. A complete program of windbreaks around agricultural fields, pastures, and farm homesteads, and on rural roads in the American midwest, could result in the sequestration of 1.0 - 1.5 million tons carbon per year directly.[18] The most visible and well-known application of informal forests is the planting of trees in communities and urban areas. Though we do not know how many tons of carbon could be sequestered in fully forested urban areas, with only one out of every four dead or dying trees being replaced in U.S. communities, there are many opportunities to increase this carbon sink.

Strategies Which Reduce the Sources of Emitted Carbon

The third class of responses reduces the sources of emitted carbon, usually from energy production. These strategies reduce the need for energy (and the carbon emitted from its production), substitute biomass fuels for fossil fuels, or offset the carbon emissions of energy facilities.

Trees can be planted to produce shade and wind shielding, which can reduce energy consumption for cooling and heating, respectively. The direct shading of trees has been shown to reduce the energy used to cool dwellings by as much as 50% under certain conditions,[19] while the effect of wind shielding has shown the capability for reductions of up to 40% in energy use for winter heating.[20] Trees also cool the air by evapo-transpiration (the equivalent of sweating and breathing in plants), and by lowering air temperatures, can reduce the need for cooling energy. The cooling provided by trees planted along streets and in parking lots can increase the efficiency of automobile engines and reduce the need for air conditioning, thus reducing the carbon emitted from stressed engines.[21] Including the potential of urban trees to reduce energy use as well as to absorb carbon, large-scale tree planting in urban areas could result in a carbon benefit of 3-15 million tons.[22]

Trees can be planted to produce fuels, either in the form of chipped or solid wood, or in wood-derived alcohols, which substitute for fossil fuels. By replanting the forest used for growing the fuel, the carbon emitted by burning the wood is then reabsorbed by the new forest. This fuel switching keeps the same carbon cycling through the biosphere and decreases the net addition of carbon from fossil fuels (which cannot be put back into the ground). The Oak Ridge National Laboratory has been experimenting with growing highly productive tree species on very short growing cycles to produce biofuel stock. Researchers have used plantations of silver maple (*A. saccharinum*), black locust (*Robinia pseudoacacia*), sycamore (*Platanus occidentalis*), sweetgum (*Liquidambar styraciflua*), and various poplars (*Populus* spp.) grown on 6-8 year rotations to produce stocks for new conversion processes which may someday equal fossil fuels for cost and efficiency in energy conversion.[23] Researchers estimate that if 14 million hectares of land were devoted to growing these short-rotation woody crops for energy under current technologies, 67 million tons of carbon from fossil fuels would be displaced in the direct production of electricity.[23] Use of a larger land base, advanced technologies and high growth rates could result in a displacement equal to more than 80 percent of the carbon from fossil fuel released by electricity production in the United States.

Trees can also be planted to absorb and offset the carbon liberated by the production of energy from utilities which burn fossil fuels. An electric utility company building a power plant in Connecticut initiated a program to plant trees in Guatemala to offset the carbon that would be emitted over the projected 40-year life of the plant.[24] The program eventually designed has the goal of sequestering 15.8 million tons of carbon through a program of integrated land management. The program mandates 68,000 hectares in agroforestry, 13,000 hectares in new plantations, 38,000 hectares in logged and managed forests, and 20,000 hectares in protected forest reserves.

Finally, increasing the use of wood products in place of concrete and steel can reduce the burning of fossil fuels used in the process of producing these two energy-intensive materials. This strategy has an additional benefit in that the carbon in the wood is sequestered for the entire life of the product or structure, and for as long as it takes the wood to decay once it is discarded. Opportunities include replacing steel studs with wooden ones, sheetrock with plywood, cement telephone poles with wooden poles in places like Europe, and wooden beams for steel in structures like small bridges. These changes could help increase the global carbon sink and reduce the use of production processes which liberate large amounts of carbon.

MANAGING FORESTS FOR CARBON

We have explored why manipulating forests is a good idea from the perspective of climate change, and what some of the benefits are that can be accomplished. This section discusses some of the specifics of how forests can be managed. In traditional forest management, there are many ways in which practices can be changed in small or large ways to maintain, enhance or deplete carbon in the system. The system being managed and the forester's management goals frequently dictate the use of a specific set of management practices, and the use of each practice has implications for the carbon dynamics of the system. Examining several of these management systems in greater detail will illustrate some of the issues surrounding the choice of a strategy and how those choices affect the carbon in the system.

Large plantations are an attractive way to increase carbon sinks. They offer the possibility of growth rates increased through genetic improvement of already fast-growing species like the southern loblolly pine (*P. taeda*), the use of advanced technologies in management practices, and optimization of growth conditions. These species are typically planted in monocultures, that is, without any other major species in the ecosystem. These monocultures are relatively easy to manage due to their single-species composition, and have very high growth rates. They are also much more susceptible to outbreaks of disease and insect pests than are stands with a mix of species. A pest or pathogen that predates on a single species can devastate a monoculture plantation since every tree is a potential host. These effects are mitigated in a mixed stand as the pest cannot so easily move from tree to tree, and cannot destroy the entire stand. While the outbreak of the gypsy moth (*Porthetria dispar*) in the summer of 1981 is a case of one pest preying on all species of a mixed-species forest, the gypsy moth usually targets only the oak species in a hardwood forest, leaving most other hardwoods and all the conifer species alone. The more biologically diverse ecosystem of the mixed stand may also offer more predators or biological agents against the pathogen.

Individual species also have optimum ranges of temperature and moisture within which they grow and survive best. These ranges differ from species to species. In the case of catastrophic drought, flood, and cold spells or heat waves, certain species may suffer extraordinarily as they endure climate which is suboptimal for their survival. If the entire stand is composed of that species, the entire stand can suffer damage or be lost. In a mixed stand, the species better adapted to the change in climate will continue to function normally and perhaps flourish, insuring the survival of part of the stand. These species may, through their presence, mitigate the effects of the change on the local microclimate, helping to reduce the deleterious effect on those species most susceptible.

Plantations consisting of a single species have replaced naturally occurring mixed-species forests in some areas. In some cases, the hardwood component of a hardwood-pine has been eliminated, removing species that offer biological diversity and different chemical and biological inputs to the system. There is a question whether naturally occurring systems are inherently more stable and benign than introduced systems managed to draw nutrients from the soil for rapid growth. While it is too soon yet to judge whether these monocultures are having a degrading effect on the system's long-term productivity, it may be wise to consider the potential if we are concerned about reducing the stress on a system. Clearly there is a balance that must be achieved between growth (amount and speed of carbon uptake) and system stability (the maintenance of the carbon sink). More research needs to be done to determine what the tradeoffs are, if they are significant, and how to resolve them.

The tradeoff between plantations and managed mixed stands is important for another reason. Plantations are normally managed on an "even-aged" basis, where all individuals are planted and harvested at the same time. This type of system is thinned at various key points in the rotation, harvested at maturity by clearcutting, and regenerated by planting seedlings. Douglas fir and loblolly pine are prime examples of systems managed in this manner. Mixed-species stands like the hardwood forests of the east and northeast are normally managed on an "uneven-aged" basis. That is, there are various age classes represented in the stand, and the stands tend to have multiple stories, an attribute that increases the biomass production due to the increased leaf area available for CO_2 uptake.[25] Even-aged stands normally have a canopy with a single layer.

Uneven-aged stands are usually managed with a selection harvest system. Trees are selectively harvested from the stand at periodic intervals, and at least half the stand is kept intact after each harvest operation. Trees are normally cut when they have reached some target size. Because the stand is never cut completely, more carbon is kept sequestered all through the life of the stand than in a clearcutting operation (Figure 3). These stands normally do not grow as fast as plantations, but more carbon is kept stored in the residual stand and the system is kept intact to a greater degree. Uneven-aged forests can be managed for levels of biomass at least as great as comparable plantations by increasing the size that trees need to be before they are cut, and lengthening the time between cuttings.[25]

An important consideration in forest management for carbon storage is the soil. Soils contain more carbon than is present in the entire above-ground terrestrial ecosystem.[22] Carbon is incorporated into soil through natural processes involving soil microorganisms, roots, and the deposition of leaves and other vegetation. Though some ecosystems may have greater carbon in biomass above ground, many forest soils contain more carbon than the tree they support. The carbon in northern spruce-fir forests is twice that achieved in the mature forest above it (Figure 2a).[3] This is partially due to the slow decomposition of organic matter caused by cold temperatures. Warmer temperatures speed the rate of organic decomposition, and

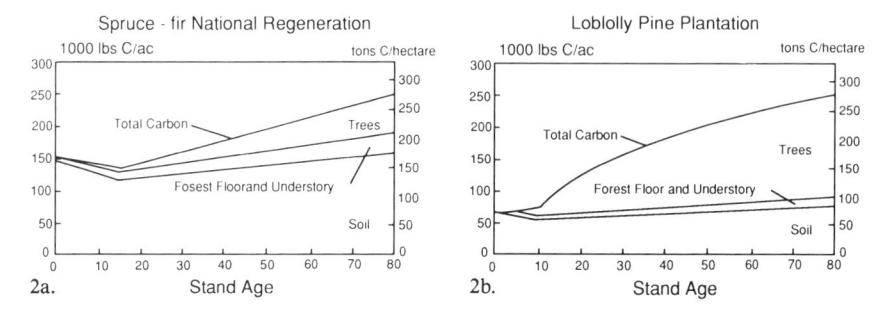

FIGURE 2 (a) - (b). Carbon storage by forest ecosystem component for a spruce-fir forest in the Northeast and a loblolly pine plantation in the Southeast. Both are forests that have been clear cut and regenerated. Source: Birdsey, 1990.[3]

SILVICULTURAL SYSTEMS

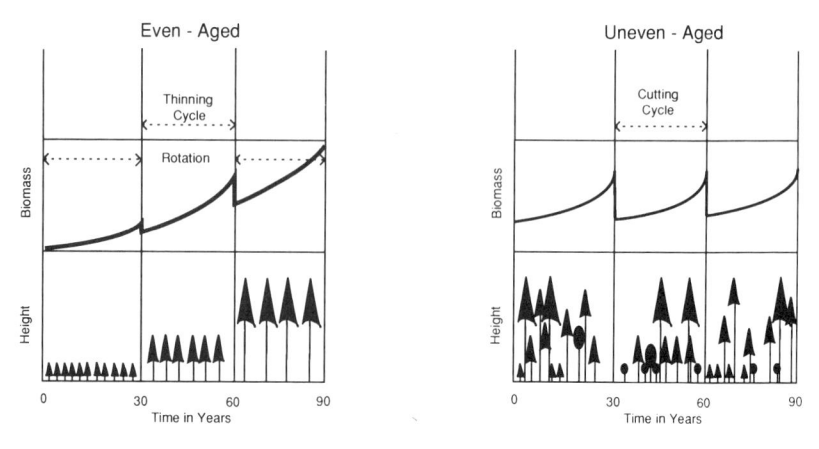

FIGURE 3. Comparison of even- and uneven-age approaches to silviculture, showing stand structure and biomass dynamics. The frequency of cutting and thinning have been assigned arbitrarily. Source: Helms, 1991.[25]

ecosystems in warmer climates tend to have a lower percent of the carbon that is available for use in the soil (Figure 2b).[3]

Soils begin to lose carbon when they are warmed, or exposed to sunlight. Harvesting operations that disturb the soil and open the forest floor to sunlight expose both the organic litter on the forest floor and the soil carbon below to increased rates of decomposition and loss. Harvest operations also expose soils to erosion from wind and water flow. This is particularly a problem with steep slopes, such are found in areas like the Pacific Northwest and Himalayas of Asia. Loss of soil carbon not only removes part of a carbon sink but it can reduce the future productivity of the soil. Some of the steep slopes harvested in the Pacific Northwest have never successfully regenerated despite replanting efforts.

Clearcutting operations remove most, if not all, of the above-ground biomass, and are frequently followed by burning and mechanical preparation of the land for planting. Soils are exposed to the sun, heated by fire, and are physically moved to expose the mineral soil below the organic litter layer. The net result is that much organic matter is lost, and decomposition of the remaining organic matter is expedited. It should be noted that not all the carbon in the soil profile is susceptible to loss from disturbances. Selective harvest operations, and other types of harvests that leave trees standing, retain the root systems that keep soils from eroding, and they retain overstory and understory vegetation that shade the soil. These cooler soils loose less carbon through decomposition. Likewise, any activity which increases cooling or shading of the soil will promote slower decomposition rates and help retention and accumulation of soil carbon. Practices which increase levels of herbaceous and understory vegetation accomplish this goal, and add additional carbon in the form of leaves and roots to the organic layer.

CONCLUSION

This chapter has discussed many types of activities related to forests and forestry. These strategies have benefits for directly sequestering carbon from the atmosphere, reducing the amount of carbon released from energy consumption, and reducing the net addition of atmospheric carbon from the use of fossil fuels. The strategies differ as to the speed with which they sequester carbon and the amount they can sequester over time, and there are costs and benefits to the use of each. This paper has largely avoided questions of the practical and economic feasibility of these strategies, but our social and economic systems impose limitations which severely complicate the use some of the response strategies discussed here. Such a discussion issues is beyond the scope of this chapter.

Although it is clear that forests are not the whole answer to the CO_2 question, there are many ways in which forestry responses can contribute to the solution. This chapter has discussed the range of options available for using and enhancing the role of vegetation in the carbon cycle, and has examined a few issues of forest management in detail. Much is still unknown about the potential benefits, costs, and effects of the strategies discussed here, and more research is needed to evaluate them completely. Fortunately, understanding of the problems and issues is constantly expanding, and the flow of new research and ideas for response strategies indicate a continued interest in finding ways to apply forests to the climate change problem.

For a more thorough discussion of the range of responses possible, the reader is directed to the World Resource Institute's recent publication, "Minding the Carbon Store."[22] Helm's excellent paper, "Management Strategies for Sequestering and Storing Carbon in Natural, Managed, and Highly Disturbed Coniferous Ecosystems" is a progressive approach to the question of managing forests for carbon.[25]

REFERENCES

1. Andrasko, K.J, K.E. Heaton and S.M. Winnett. 1991. Evaluating the costs and efficiency of options to manage global forests: a cost curve approach. In: *Proc. Wkshp. to Explore Options for Global Forest Management*, Bangkok, Thailand, April 1991. pp. 349.
2. Harmon, Mark E., William K. Ferrell and Jerry F. Franklin. 1990. Effects on carbon storage of conversion of old-growth forests to young forests. *Science* 27(4943):699-701.
3. Birdsey, Richard A. 1990. Potential changes in carbon storage through conversion of lands to plantation forests. In: *Proc. N. Am. Conf. on Forestry Responses to Climate Change*. Climate Institute, Washington, DC, 1992. pp. 446.
4. Larson, Bruce C., Clark S. Binkley and Steven M. Winnett. 1986. Biomass harvesting: Ecological and econonic consequences. In: C.T. Smith, C.W. Martin and L.M. Tritton (Ed.) *Proc. 1986 Symp. on the Productivity of Northern Forests Following Biomas Harvesting,* USDA Forest Service Northeast Ex. Sta. NE-GTR116:75-81.
5. Houghton, R.A., J.E. Hobbie, J.M. Melillo, B. Moore, B.J. Peterson, G.R. Shaver, and G.M. Woodwell. 1983. Changes in carbon content of terrestrial biota and soils between 1860 and 1980: A net release of CO_2 to the atmosphere. *Ecol. Monogr.* 53:235-262.
6. Intergovernmental Panel on Climate Change. 1991. Climate change: the IPCC response strategies. World Meteorological Organization and United Nations Environment Program.
7. Sedjo, R.A. 1989. Forests: a tool to moderate global warming? *Environment* 31(1):14-20.
8. Walstad, John D. 1991. Feasibility of large-scale reforestation projects for mitigating atmospheric CO_2—Ecological considerations. In: J.K. Wingum and P.E. Schroeder (Ed.) *Proc. Intern. Wkshp. on Large-Scale Reforestation*. U.S. Environmental Protection Agency. EPA/600/9-91/014. pp. 148.
9. World Resources Institute. 1990. World Resource 1990-91. Oxford Univ. Press. New York. pp. 383.
10. Houghton, R.A., J. Unruh and P.A. Lefebvre. 1991. Current land use in the tropics and its potential for sequestering carbon. In: *Proc. Wkshp. to Explore Options for Global Forest Management,* Bangkok, Thailand, April 1991. pp. 349.
11. Auclair, N.D., R.C. Worrest, D. Lachance and H.C. Martin. 1990. Global climate change as a general mechanism of forest dieback. In: *Proc. Ann. Mtg. of the Am.*

Phytopathological Soc. and the Can. Phytopathological Soc. August 1990, Michigan. In Prep.
12. Peters, C.M., A.H. Gentry, and R.O. Mendelsohn. 1989. Valuation of a Amazonian rainforest. *Nature.* 339:655-656.
13. Parks, Peter J. 1990. Costs and benefits of converting marginal cropland and pastureland to forests. Duke University, Durham, N.C. pp. 30.
14. Trexler, Mark C. 1991. Estimating tropical biomass futures: a tentative scenario. In: *Proc. Wkshp. to Explore Options for Global Forest Management,* Bangkok, Thailand, April 1991. pp. 349.
15. Hughes, Joseph H. 1991. A brief history of forest management in the American south: Implications for large-scale reforestation to slow global warming. In: J.K. Wingum and P.E. Schroeder (Ed.) *Proc. Intern. Wkshp. on Large-Scale Reforestation.* U.S. Environmental Protection Agency. EPA/600/9-91/014. pp. 148.
16. Row, Clark. 1992. Enhancing the Management of Forests and Vegetation on Department of Defense Lands. U.S. Environmental Protection Agency. Washington, DC. pp. 74.
17. Vasievich, J.M., R.J. Alig, E.A. Hansen and K.J. Lee. 1992. Opportunties to increase timber growth on timberlands. In: N.L. Sampson and D. Hair (Ed.) *Forests and Global Change,* Volume 2. American Forests. Washington, DC. In prep.
18. Brandle, J.R., T.D. Wardle and G.F. Bratton. 1992. Opportunities to increase tree planting in shelterbelts and the potential impacts on carbon storage and conservation. In: N.L. Sampson and D. Hair (Ed.) *Forests and Global Change,* Volume 1. American Forests. Washington, DC. pp. 285.
19. Parker, John H. 1989. The impact of vegetation on air conditioning consumption. In: K. Garbesi, H. Akbari and P. Martien (Ed.) *Controlling Summer Heat Islands.* Lawrence Berkeley Laboratory, Berkeley, CA. LBL-27872:45-51.
20. Heisler, G.M. 1989. Tree plantings that save energy. In: Phillip D. Rodbell (Ed.) Make our Cities Safe for Tree: Proc. 4th Urban Forestry Conf. American Forestry Assoc., Wash. DC. pp. 258.
21. Environmental Protection Agency. 1992. Cooling our communities: The guidebook on tree planting and light-colored surfacing. U.S. Environmental Protection Agency. Washington, DC. pp. 217.
22. Trexler, Mark C. 1991. Minding the carbon store. World Resources Institute. Washington, DC. pp. 81.
23. Wright, L.L., R.L. Graham, A.F. Turhollow and B.C. English. 1992. Opportunities for short-rotation woody crops and the potential impacts on carbon conservation. In: N.L. Sampson and D. Hair (Ed.) *Forests and Global Change,* Volume 1. American Forests. Washington, DC. pp. 285.
24. Lashof, Daniel A. and Dennis A. Tirpak. 1990. Policy options for stabilizing global climate: Report to Congress. U.S. Environmental Protection Agency. Washington, D.C. pp. 449.
25. Helms, John A. 1992. Management strategies for sequestering and storing carbon in natural, managed, and highly disturbed coniferous ecosystems. In: N.L. Sampson and D. Hair (Ed.) *Forests and Global Change,* Volume 2. American Forests. Washington, DC. In prep.

Global Climate Change: Implications, Challenges and Mitigation Measures. Edited by S.K. Majumdar, L.S. Kalkstein, B. Yarnal, E.W. Miller, and L.M. Rosenfeld. © 1992, The Pennsylvania Academy of Science.

Chapter Thirty-One

ENERGY POLICY RESPONSES:
Concerns About Global Climate Change*

THOMAS J. WILBANKS
Oak Ridge National Laboratory
P.O. Box 2008
Oak Ridge, TN 37831-6184

INTRODUCTION

Responses by energy decisionmakers to concerns about global climate change are complicated by the fact that economic and social development depends considerably on the availability of abundant and affordable energy services. If reducing carbon emissions and other contributors to the likelihood of climate change means reductions in these energy services, the energy sector is placed in a difficult position—in danger of failing to deliver *some* requirements for improving the quality of life in order to avoid *other* threats to the quality of life.

Avoiding such a dilemma is the central challenge to energy policy in an era of concern about global environmental change. Fortunately, several "no regrets" alternatives are available as a basis for energy-sector contributions to the global response in the next several decades, although the long term will need more far-reaching alternatives than those currently on hand. It is not yet clear, however, whether these nearer-term alternatives will in fact be pursued vigorously enough for the energy sector to transform itself from the main problem for global climate change to a leading part of the solution.

---

*Research supported in part by the Office of Energy, Science and Technology Bureau, U.S. Agency for International Development (A.I.D.). The judgements expressed are not necessarily those of A.I.D.

THE CENTRAL CHALLENGE TO ENERGY POLICY

The Energy-Development Relationship

It is quite clear that energy services are closely related to the quality of life, both as cause and effect. According to available historical data, for example, economic growth of 1% per capita in a developing country has been associated with an increase in the primary consumption of "modern" fuel sources (i.e., electricity and liquid/gas fossil fuels) of 1.3 to 2% per capita as a conservative approximation (Oak Ridge National Laboratory, 1985). Taking a multiplier of 1.3 as a conservative approximation, Figure 1 indicates how human appetites for energy services combine with economic and population growth to create growing demands on the energy systems of the world's countries.

We have learned in recent years that these curves represent demands for energy services such as comfort, convenience, and mobility, not necessarily demands for energy products themselves such as kilowatts of electricity or liters of diesel fuel. In some situations, efficiency improvements can make it possible to decouple energy service growth from primary energy consumption, although consumers (especially at lower levels of well-being) may choose to invest some of the resulting benefits in additional energy services rather than reduced energy consumption (Wilbanks, 1991a).

But the continuing demand for energy services worldwide seems inexorable, barring social and cultural changes that are difficult to foresee. Because the available measurements and forecasts are of energy consumed rather than energy services

ENERGY REQUIREMENTS FOR DEVELOPMENT

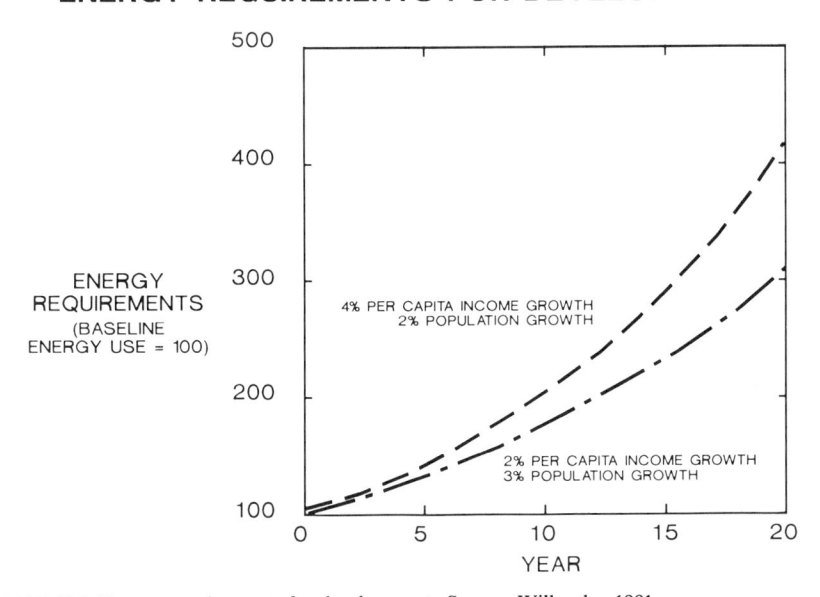

FIGURE 1. Energy requirements for development. *Source:* Wilbanks, 1991a.

provided, and because the relationship between energy service levels and the quality of life can vary with population distribution, cultural patterns, and other factors, depicting this expectation for services quantitatively is not a simple matter. Even so, virtually all current forecasts of global energy requirements—most of which include substantial reductions in the relationship between energy consumption and economic production—show a continuing association between energy consumption and development. For example, Figure 2 shows the trends projected for the 1989 World Energy Conference for "moderate" and "limited" scenarios, based on the expert knowledge of thirty regional specialists (Frisch, 1989). Forecasts prepared for the May 1991 Senior Expert Symposium on Electricity and the Environment (Helsinki) look farther out to the year 2050, showing a wide range of uncertainty between a moderately optimistic growth-oriented future (TG) and an all-out efficiency-oriented future (TE) (Helsinki Symposium, 1991); the growth in energy requirements, however, is substantial in either case and monumental in the TG case (Figure 3). As one further example, a recent analysis for the U.S. Working Group on Energy Efficiency, using a methodology originally developed for the Environmental Protection Agency (Lashof and Tirpak, 1990), shows global primary energy consumption doubling between 1985 and 2025, barring unprecedented efforts to improve energy efficiency (Levine et al., 1991).

Particularly striking is the need in developing countries for additional energy services in order to close the prevailing North-South Gap. Their energy consumption is typically five to ten times lower than that of the industrialized countries, and rising incomes will mean growing demands for energy-using devices that are much more common in more affluent areas: personal highway vehicles, appliances in the home and office, labor-saving equipment in industry, and the like.

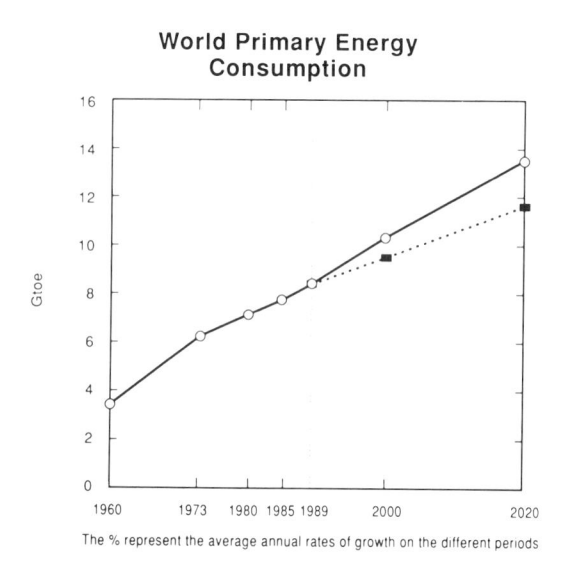

FIGURE 2. World primary energy consumption. *Source:* Frisch, 1989.

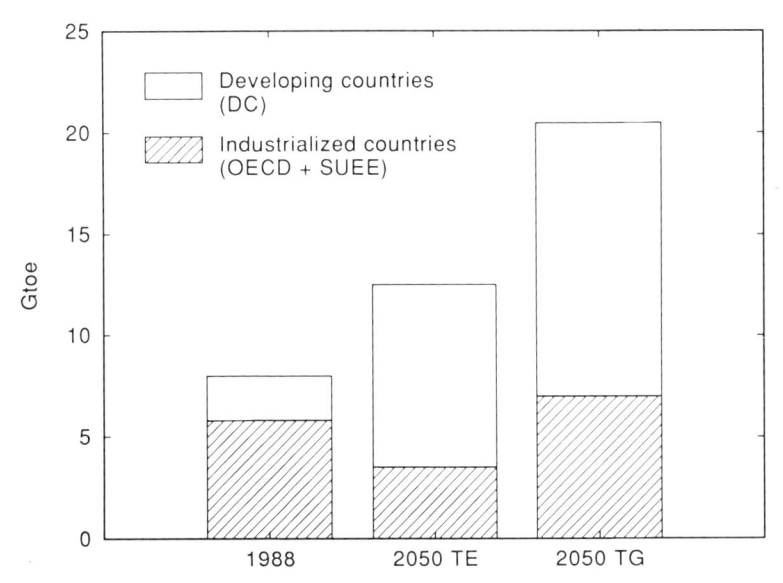

Energy Policy Responses: Concerns about Global Climate Change 455

FIGURE 3. World energy demand. *Source:* Helsinki Symposium, 1991.

As a consequence, energy growth rates in developing countries are far higher than in the rest of the world. The World Energy Conference projections, for instance, indicate that about 60 percent of the increased consumption in the moderate case will come from developing countries, with another 20 percent from Eastern Europe. Figure 4, taken from Levine *et al.*, 1991, shows the same kind of expectation. Other things equal, of course, this is *good* news, not bad, because it indicates that the development process is working in the lower-income parts of the world.

The Energy Service-Fossil Fuel Relationship

As these trends unfold, current forecasts indicate that fossil fuels will remain the dominant source of energy for the global economy, especially the developing countries, certainly over the next three decades and almost as certainly for the next half-century and more. The World Energy Conference forecasts, for example, show oil and gas consumption rising gradually to the year 2020, while coal consumption rises relatively steeply. By 2020, these three fossil sources will still represent two-thirds of global energy supply (Figure 5).

Developing countries are a central reason for this expectation. Besides accounting for most of current global deforestation, developing countries and Eastern Europe are already important sources of carbon emissions; in terms of national totals, China ranks third (behind the United States and the USSR), with India seventh and Poland eighth.[*] But the more serious concerns are about the future.

---

[*]Developing countries argue that accounting by national totals, unweighted by population and other factors, reflects an affluent country bias: e.g. Center for Science and Environment, 1991.

456 Global Climate Change: Implications, Challenges and Mitigation Measures

A recent survey of electric power sector expansion plans for the 1990's in 70 developing countries indicated plans to add 384 MW of generation capacity during the

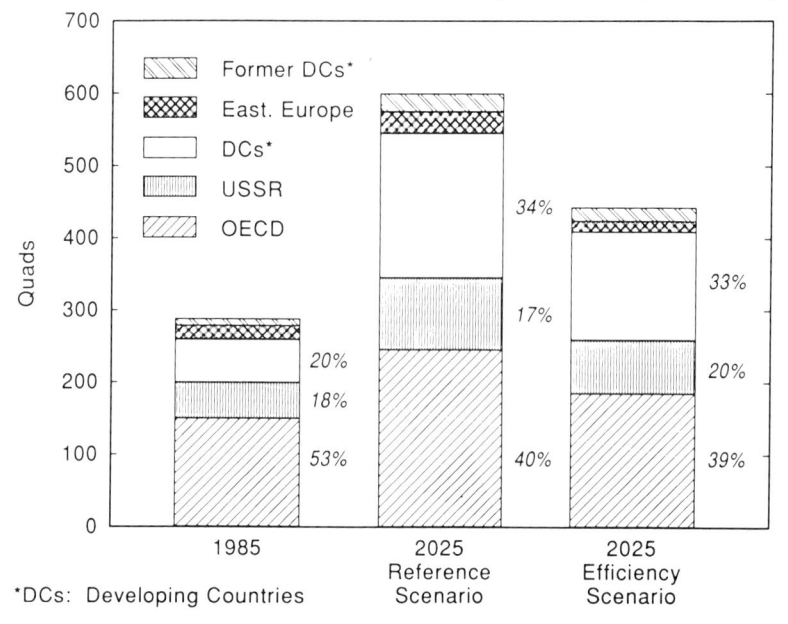

FIGURE 4. World primary energy consumption by regions. Source: Levine et al., 1991.

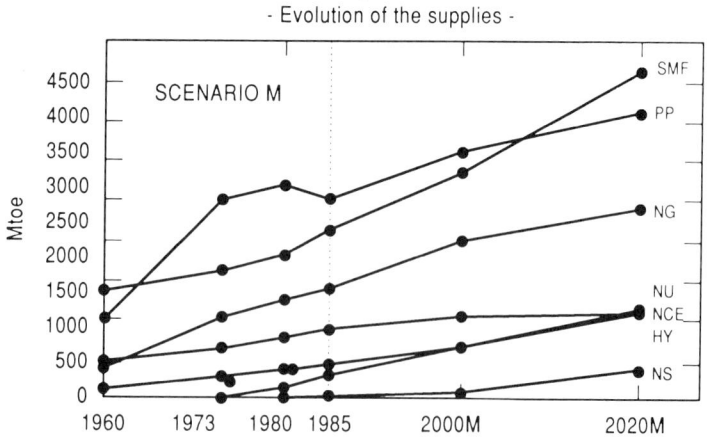

FIGURE 5. Structure of the world energy consumption by fuels. *Source:* Frisch, 1989. Curves are solid mineral fuels (SMF), petroleum (PP), natural gas (NG), nuclear (NU), noncommercial sources (NCE), hydropower (HY), and new sources (NS).

decade, half based on coal (much of the rest on large-scale hydroelectric facilities) (World Bank, 1990). In the World Energy Conference figures, all of the doubling in coal used by 2020 (in the moderate growth case) occurs in the Third World and Eastern Europe. As a result, developing countries will soon become the largest contributors to total global carbon emissions, in terms of regional aggregates, and their rising curve of energy needs will become the main reason that the Earth's total emissions keep climbing (Figure 6).

If this is true, reducing global greenhouse gas emissions from current levels will prove almost impossible. Fossil fuel use accounts for about three-quarters of global CO_2 emissions and more than half of total global greenhouse gas emissions. It is difficult to see how *total* emissions can be stabilized while *fossil fuel* use grows, but the available energy forecasts suggest that it is difficult to envision global economic development—especially in developing countries—without a growth in fossil fuel use. For instance, the issue paper on energy and electricity supply and demand for the Helsinki Symposium concluded that "total CO_2 emissions within the electric power sector worldwide cannot be reduced dramatically without socially and economically unacceptable curtailments in electricity services in many countries" (Helsinki Symposium, 1991). Reductions in emissions from industry may also be difficult, and transportation may be the least tractable of all the sectors.

For more than a decade and a half, we have known that over the long term the resource limitations of the liquid and gas fossil fuels face us with one of the greatest scientific and institutional challenges of our time: shifting away from these energy sources while at the same time we meet the rapidly rising demands of developing

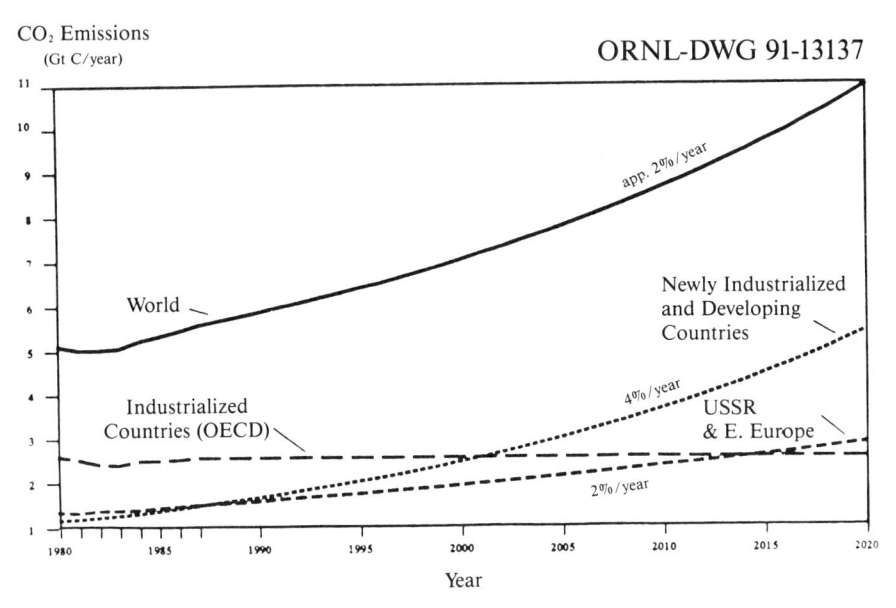

FIGURE 6. Fossil-fuel emissions from newly industrialized and developing countries could soon overtake emissions from developed countries. From Fulkerson et al. (1989).

countries for energy services (Wilbanks, 1988). Meeting this challenge without massive increases in the use of coal in the near and mid-terms, the period when we are essentially limited to the resource/technology combinations that are known today, is the central problem for an energy policy response to the global climate change issue.

PERSPECTIVES ON ENERGY POLICY RESPONSES

Current views about how to solve this problem range widely from country to country and constituency to constituency (e.g., Morrissette and Plantinga, 1991; World Meteorological Organization, 1991), depending on attitudes toward (a) other economic and social priorities and (b) the significance of uncertainties about climate change and its impacts.

The Current Variety of Perspectives

This diversity greatly complicates any effort to design and implement an effective energy policy response, because it so often seems to amplify differences in views rather than similarities. To illustrate, one can point to differences in perspectives between industrialized and developing countries—and within each of the two groups. And, within the United States, there are clear differences between nongovernmental environmental groups and most spokespersons for private business. Obviously, these differences have a strong bearing on any discussion of policy responses.

For example, as a broad generalization, the industrialized countries tend to approach the issue from a set of concerns about the ability of humanity to keep "Spaceship Earth" under control, which may pose threats to human survival. Industrialized countries differ from one another in how urgent they consider this threat to be, considering all the uncertainties, with the Northern European countries expressing a relatively strong sense of urgency and other countries, most notably the United States, less prepared at this point to consider any threats to be real and imminent (see below).* But the mainstream view—as expressed in such international actions as the Montreal Protocol on Substances that Deplete the Ozone Layer, establishment of the IPCC process, reorientation of the IGBP effort toward global climate change, and the creation of the Global Environment Facility (GEF) within the World Bank—is that some kind of significant policy response is a high priority for the global community.

Developing countries, on the other hand, generally say to the industrialized countries: (1) *you* caused the problem; don't expect us to bear the brunt of solving it; (2) show us that you are prepared to lead by example, adopting and demonstrating yourselves the options you must want us to use; and (3) in this way and others, either give us alternatives that meet our development needs without contributing to global

---

*Also see Schmidt, 1991.

climate change, or else pay us to use alternatives that are not as attractive to us economically (IFIAS *et al.*, 1990). Even within the developing countries, however, there are differing points of view. For example, the small island nations—concerned about such impacts as a rising sea level and intensified tropical storms, while they contribute very little to total fossil fuel emissions—tend to share the sense of urgency of the Northern Europeans. The major oil-exporting developing countries are suspicious of any industrialized country effort to discourage fossil fuel use. The populous upwardly-mobile countries with domestic coal resources, such as China and India, appreciate the need to find alternative paths for development but feel that development is a higher priority than preventing climate change, if a choice must be made.

Cutting across national similarities and differences, at least within the industrialized countries, are differences among major constituencies in their views. This variety can be illustrated by contrasting views of environmental interest groups and private industry. Environmental groups, who try to speak for the environment, argue that if the world waits for unambiguous evidence that the planet faces a catastrophe, it may well be too late. The prudent course of action is to act as if a threat to global survival is at hand. If it is, we will help to avert the threat; if it is not, we will have to become more sensitive to the fragility of the global environment, and other good will result. Industry groups, who try to speak for the economy, argue that expensive responses to uncertain dangers are likely to undermine economic growth and development. The prudent course of action is to wait until the dangers, and the costs and benefits of different response strategies as well, are much better understood. If a major response is not required, unnecessary economic costs will have been avoided. If a major response turns out to be required, the world will be able to focus the effort more accurately, increasing its effectiveness.

These kinds of differences have understandably led to a variety of views about energy response strategies. Although each side recognizes the validity of some of the concerns of the other, each sees the other imbeded in a complex mixture of selective blindness and vested interest; and the resulting suspicions are a part of the problem. The challenge can be illustrated by sketching two points of view—both with significant and heart-felt advocates—that are close to opposite poles along a continuum of supportable positions.

One Extreme: Painful Problem-solving

Toward one pole is the view that greenhouse gas emission reduction targets must be set—and met—in the interest of global well-being, regardless of the economic cost, because global survival is at stake. The core of this approach is reducing the world's fossil fuel emissions, although it addresses deforestation as well.

Advocates of this point of view believe that "... anything approaching 'business as usual' from the ... energy [sector] ... now threatens our very existence on this planet. Some fundamental changes in approach and thinking are needed among energy industries and governments" (Boyle, 1991). The general course of action should be to set emission reduction targets that will first stop the increase in atmospheric concentrations of greenhouse gases and then reduce these concentrations, using policy tools as needed to meet the targets—such as "greenhouse taxes" (e.g., taxes on carbon emissions of up to $50 to $100 per ton), strong incentives for the use of alternative transportation fuels, and investment assistance for renewable energy sources.

Policy targets might include reducing carbon emissions from fossil fuel use to 30 percent of current levels by 2020, along with eliminating CFC's, halting deforestation, and reducing emissions of other greenhouse gases (Kelly, 1990). Even though the economic impacts on some areas may be substantial*, calling for unprecedented international financial initiatives to assure equity, advocates argue that the costs are modest compared with the impacts of failing to act. Meanwhile, they argue that energy resources and technologies are readily available to displace fossil fuels; the problem is that our political and institutional infrastructures are caught in a dangerous pattern of thinking that our energy futures must evolve incrementally from the past (e.g., Flavin and Lenssen, 1990).

The strong point of this perspective is that it is the best the energy sector can do to avert a global catastrophe. It requires nothing that is not technologically feasible and, if a catastrophe lies ahead, it can transform the world's future in a matter of one or two generations.

The weak point of this perspective is that it requires a willingness to accept higher energy costs that has very little political support across the globe, in either industrialized or developing countries. Making it work calls for a kind of leadership that would expose most political leaders in most countries to the loss of their leadership positions. The reason is that, based on current information, the majority of the world's population is more concerned about the health of its economy than about risks associated with global climate change.

Another Extreme: No Action While Uncertainties Remain

Toward the other pole is the view that, even if increasing concentrations of greenhouse gases are a scientific fact, the *impacts* of such increases are so uncertain (as are the costs and benefits of possible response options) that no significant policy response is justified at this time. The need, instead, is to support research activities to reduce the uncertainties, along with steps to create processes for dealing with better information as it emerges.

Advocates of this approach believe that emotion related to the greenhouse warming issue has outpaced the knowledge about it. They point to a combination of (a) continuing scientific disagreement about the likelihood of adverse impacts from greenhouse gas concentrations (e.g., Balling *et al*, 1991) and (b) estimates that limiting greenhouse gas emissions would involve very high costs** and changes in

---

*Flavin (1989:68) suggests, for example, that China might be required to cut its projected economic growth rate from 3.5 percent per year to 1.5 percent.

**For example, Manne and Richels (forthcoming) have estimated that limiting global CO_2 emissions to about 18 percent above current levels well into the 21st Century would require carbon taxes in the range of $250 per ton of carbon (equivalent to about $125 per ton of coal and $30 per barrel of oil), reducing the world's Gross Domestic Product by 3 to 10 percent. Other estimates support this view, although such studies as Goldemberg *et al*. give a sharply different picture (e.g. Darmstadter, 1991).

way of life. Moreover, they see the attainment of many of the targets being suggested as requiring heavy government interference in economic markets, reducing economic vigor and efficiency throughout the economy, and eventually enforcement mechanisms that are likely to be distasteful. Given all this, they believe that (other than the CFC protocol) energy responses should emphasize data gathering, research and development (R&D), and participation in frameworks for continuing international deliberation (e.g., Bush, 1991a and 1991b). Otherwise, normal market processes should continue to operate.

The strong point of this perspective is that it reduces the likelihood that scarce capital resources will be spent on problems that turn out not to be real. A further strength is that is supports the reduction of scientific and policy uncertainties related to the energy/climate change issue.

The weak point of this perspective is that it increases the likelihood that global climate change will occur, with economically and socially damaging effects in many areas. R&D and information gathering suggest a period of some years, if not decades, before uncertainties are reduced dramatically and actions are taken; and, meanwhile, greenhouse gas concentrations and energy systems based on fossil fuels will have continued to grow.

ELEMENTS OF A RATIONAL MIDDLE GROUND

Surely, somewhere between these extremes must be a middle ground, offering the prospect of a significant energy-sector response to the global climate change issue without requiring politically untenable proposals for accepting discernable costs in the face of real uncertainties. The starting point is to look for actions that would be good for economic and social development even if there was no concern about global climate change (IPCC, 1990). In other words, is it possible that some strategies are available that, at the same time, offer major environmental benefits and make economic sense under current policy and market conditions?*

The answer is yes, beyond question. Within the past year or two, a wide range of sources have reported that a combination of so-called "no regrets" options can indeed make a difference in the next decade or two, buying the world relatively cheap insurance against risks associated with climate change (e.g., Lashof and Tirpak, 1990; NAS/NRC, 1991; Office of Technology Assessment, 1991a; National Governors' Association, 1990; World Resources Institute, 1990; Helsinki Symposium, 1991).

The central elements of the proposed energy strategy are (a) making a major push to realize potentials for cost-effective energy efficiency improvements and (b), where supply-side alternatives to fossil fuel are not competitive, promoting the use of natural gas as the lowest impact of the fossil fuels (i.e., the smallest emitter per unit of energy services delivered):

---

*Gilbert White has suggested that the main near-term significance of the climate change issue may in fact be to redirect attention to development strategies that are viable in their own right but that have been underemphasized so far.

Potentials for Energy Efficiency Improvement. Where energy efficiency is concerned, in nearly every country an enormous number of opportunities can be identified that are relatively low-cost compared with energy supply options, attractive for both economic and environmental reasons (Wilbanks, 1991a), and available for implementation relatively quickly.

The most thorough exploration of efficiency improvement as the focus for energy strategies is a widely-noted study by Goldemberg *et al.* (1988; also see 1985, 1987a, and 1987b). These authors paint a picture of astonishing potentials: for example, advances in the average standard of living of developing countries to that of Western Europe with only a little more total energy consumption than today, in kilowatts per capita.*

Other studies are also encouraging. In the United States, for example, the Electric Power Research Institute (EPRI) estimates that the use of the most efficient currently available end-use technologies could save up to 44 percent of the electricity the U.S. would otherwise be using by the year 2000 (EPRI, 1990), and other estimates of maximum savings range as high as 75 percent (e.g., Fickett *et al.*, 1990). Potentials seem to be especially bright for improvements in the efficiency of lighting, electric motor systems, and cooling/refrigeration.

In Europe, stimulated by the findings of the Brundtland Commission (World Commission, 1987), several countries have looked carefully at the potential for energy efficiency improvement (along with renewable energy system options) to reduce carbon dioxide emissions. For instance, the Danish Brundtland Energy Plan forecasts that CO_2 emissions from that country can be reduced by 50 percent over a 40 year period (Morthorst, 1991). Interestingly, an "economic" scenario for Denmark shows less than half the CO_2 emissions associated with the "reference" scenario while total energy system costs are *reduced* by about 10 percent. This illustrates the powerful appeal of efficiency improvement in many cases: looking good for both environmental and economic reasons.

In developing countries and Eastern Europe, the prospects are also bright, considering possible efficiency improvements across the board—in supply and conversion systems, distribution systems, and end uses (Wilbanks, 1991b). Generally, except possibly in some parts of Eastern Europe (Levine *et al.*, 1991), efficiency improvements will not obviate the need for further supply facilities, but it can reduce supply requirements to a level more likely to be financially viable and environmentally acceptable. It should be recognized, however, that the rate of efficiency improvement depends considerably on the rate of economic growth, which affects the ability to invest in good ideas as well as the rate of capital stock turnover.

Emphasizing Natural Gas. Besides efficiency improvement, the most accessible low-cost option for reducing greenhouse gas emissions in many countries is substituting natural gas for coal and oil products. Natural gas is a relatively low-impact fossil fuel; compared with coal, it contains only a little more than half the carbon per unit of energy produced. In countries such as the United States and the Soviet Union, gas reserves are readily available, and new discoveries are being

---

*In GJRW usage, this terminology is shorthand for kilowatt-years per year per capita.

reported in developing countries as well. Clearly, natural gas is relatively difficult to store, and it is relatively difficult to transport where pipelines are not feasible; but the state of the art of gas combustion technologies has advanced rapidly in the past decade, and efficiencies in both combustion and end use have risen impressively. In many areas, increased use of gas in electric power generation, industrial production, and perhaps (in the longer run) in the transportation sector is technologically feasible and economically within reach without major cost increases. The principal uncertainties are about the size of the recoverable resource base, the prospects for international trade, and the costs of developing a national distribution system in countries not now using natural gas very extensively.

Other "no regrets" options can also be added as part of a national response strategy, including an acceleration in the use of renewable energy resources and technologies where they are economically attractive (or very nearly so)* and attention to institution-building to encourage environment policy dialogues, both between and within nations.

Accelerating the Use of Renewable Energy Resources. In the past half-decade, the international experience with renewable energy sources and technologies has turned around dramatically. Whereas an objective review of the experience through the early 1980's, at least in developing countries, would have featured disappointments associated with excessive optimism, a lack of attention to local needs and institutional capabilities, and a tendency to focus on technology performance rather than sustainable infrastructure, the more recent experience has been quite different. Markets for solar electric technologies—photovoltaics and windpower—are expanding rapidly, and biomass appears to be the best of all the supply-side options for pleasant surprises in the next decade or two (e.g., Helsinki Symposium, 1991). Renewable energy options (other than large-scale hydroelectric power, which faces environmental impact concerns of its own) start from a very small base in terms of percentage contributions to global energy supply, which means that major expansions over the next several decades may still have a limited impact on the global picture, especially with demands for energy services growing. But many applications of renewable energy systems at relatively small scales are economically, environmentally, and socially preferable to conventional strategies; and the more closely energy needs are scrutinized at a detailed scale, the better renewable energy systems usually look. In many cases, in fact, the only real economic competition for renewable energy options is efficiency improvement, not large-scale nonrenewable supply options.

Promoting Environmental Institutional-Building. Besides the characteristics of energy resource and technology options themselves, energy choices are affected by the political and social context of decisionmaking, especially in a world where democratization is gaining strength. In most cases, where that context

---

*The Global Environmental Facility, recently established by the World Bank, is intended to finance projects in developing countries that are beneficial for the global environment but not quite economically competitive without a modest subsidy. Many of these projects will be focused on the use of renewable energy resources.

includes strong advocates of environmental management, the choices will show more environmental sensitivity (Wilbanks, 1990b). Even if the main concerns in most countries are with local and regional environmental and health issues rather than global change, effective *endogenous* environmental institutions will improve the prospects for constructive dialogue about global issues, and effective *international* environmental institutions will offer a very important alternative to governmental frameworks for pursuing the dialogue (e.g., Gerlach and Rayner, 1988).

Although the central appeal of this general strategy is that it avoids difficult choices between environmental and economic objectives, it does depend on one kind of cross-cutting policy reform: putting alternatives to fossil fuel use, whether on the demand or supply side, "on a level playing field" with fossil fuels in both industrialized and developing countries. From an economic point of view, it is certainly undesirable to turn "no regrets" strategies into "regrets" strategies by subsidizing them so strongly that they impose significant costs on a developing economy. But from *both* economic and environmental points of view, it is equally undesirable to subsidize conventional energy alternatives, explicitly or unobtrusively, so that "no regrets" strategies cannot compete in terms of economic and financial incentives. Removing subsidies that have evolved over several decades and are supported by powerful parties at interest is not a simple matter, of course, but in most countries it is a key to finding a new path that will be self-sustaining.

Finally, it should be recognized that the energy sector is not unique in offering more or less "no regrets" responses to concerns about global climate change. Energy choices are not the only source of pressure on the global environment; in fact, they are often secondary effects of other choices related to population growth and economic expansion. Consequently, energy choices are not the only response options. For example, "no regrets" forest and other resource management policies are also important, especially relative to deforestation (e.g., NAS/NRC, 1991; IPCC, 1990; Lashof and Tirpak, 1990). Population policy has so far been underemphasized in the policy dialogue because of its social sensitivity, but its importance is potentially more far-reaching than energy strategies as such (Wilbanks, Petrich *et al.*, 1990). And economic restructuring is a significant part of a "no regrets" response from many countries as well. A focus on the energy sector—as important as it is in the global climate change picture—should not overshadow the fact that a coordinated multi-sector approach, related to an overall national and international consensus about priorities, is far more likely to make a difference than a few isolated policy initiatives (Lashof and Tirpak, 1990).

U.S. POLICY DIRECTIONS RELATED TO A MIDDLE GROUND

If this is the most sensible general direction for the *world* to take, given the whole set of risks and uncertainties, what should the *United States* be doing in its energy policy as a part of the global response? Many things are worth considering seriously, but U.S. policy should emphasize the following commitments (Wilbanks, 1989 and 1990a; also see Rayner, 1991):

(1) *Leading by example.* Where any kind of alternative path or innovative option is concerned, the industrialized countries need to be prepared to take the lead in demonstrating what they can do, both because they have the most technical and financial resources and because they have been responsible for causing most of the problems to date. In this vein, the United States should take it upon itself to show how greenhouse gas emissions, including CO_2, can be stabilized and reduced through the kinds of market processes that the country advocates. This includes policy initiatives to add incentives for energy efficiency improvement and the use of low-impact supply sources, along with enhanced public-private sector cooperation in demonstrating good ideas.* For example, within the public sector itself the U.S. government can increase its emphasis on efficient energy use in public buildings (Office of Technology Assessment, 1991b; Hopkins, 1991). Another part of the U.S. responsibility is to urge attention to environmental impacts by national and multinational public-sector lending institutions (see #6 below).

(2) *Encouraging technical cooperation with developing countries and Eastern Europe to improve energy efficiencies.* Many people have suggested that the payoff for a U.S. public-sector dollar invested in energy efficiency improvement may be greater in Eastern Europe or in developing countries than in the United States. Certainly, global trends cannot be changed significantly without changes in the projected energy supply and use trends of such key countries and regions as China, India, Southeast Asia, and Eastern Europe. With this in mind, the United States should be promoting technical cooperation and assistance in these areas (Levine *et al.*, 1991). Combined with leadership in the industrialized world in improving their own energy efficiencies, this is the most promising option for the *near-term* for reducing energy-sector pressures on the global environment.

(3) *Developing and demonstrating innovative energy alternatives.* For the longer term, however, further energy technology alternatives will be needed on both the supply and demand sides. In the developing country context, for instance, cost-effective electric power generation technologies are needed at moderate or larger scales that are not based on fossil fuels or large-scale hydroelectric facilities; and a new generation of appropriate electricity and transportation fuel end-use technologies is needed, meeting local requirements for robustness, simplicity, reliability, and low costs. This is essentially an R&D challenge, and it is the best way to make a difference in the *longer term.*

(4) *Supporting the development of institutional frameworks.* Along the lines of its declared support for developing processes and frameworks, the United States

---

*The central policy issue is how much and in what ways to intervene in existing markets. Where the operation of a particular market is relatively close to "perfect," major interventions are hard to reconcile with a "no regrets" posture. On the other hand, where markets include considerable imperfections—e.g., information gaps, existing subsidies, incomplete recognition of social costs, etc.—policy interventions are a part of making a market system operate more efficiently, not a departure from depending on the market mechanism; see especially the recent calls for full social cost pricing of energy sources (e.g., NAS/NRC, 1991). In fact, a market approach can be incorporated within the framework of policy interventions themselves: e.g., an international market in emissions permits (Epstein and Gupta, 1990).

should help the rest of the international community to develop effective institutions for weighing the importance of environmental objectives in determining energy-sector strategies. One example is the pressing need in many developing countries for competent, effective nongovernmental environmental interest groups if local democratic processes are to keep the environment in perspective (Wilbanks, 1990b; Wilbanks, Hunsaker, Petrich, and Wright, 1991). Our commitment should include initiatives to assist such institution-building at the enterprise, national, and international scales, wherever it is needed.

(5) *Contributing in positive ways to policy dialogues.* In line with this commitment, the United States should be trying very hard to become a more positive contributor to dialogues about the global response. It is one thing to moderate premature moves in the direction of responses that are not supported by scientific evidence. But it is quite another to become so enmeshed in procedural debates that the U.S. abdicates its position as an innovative, progressive force in identifying *good* ideas for substantive action in the near term.

(6) *Promoting innovative financial mechanisms, rooted in the private sector.* Finally, the United States should be seeking ways to relate (a) the availability of financing for energy sector investments to (b) the consistency of investment proposals with the middle ground, "no regrets" strategy outlined above. Obviously, our policymakers fear strong international pressures to transfer very large shares of financial resources from industrialized to developing countries in connection with a global response to the climate change issue, when our own domestic budgets are already painfully tight and our overall budget deficits must still be reduced dramatically. Capital resources are scarce in lending regions as well as borrowing regions. But real progress in reducing pressures on the global environment will simply not happen without capital investment. One key to unraveling this dilemma is to reduce investment requirements (see efficiency improvement above). Another is to emphasize technical and information assistance so that decisions based on *available* financial resources will be better-informed. But, in the end, energy-sector decisions will gravitate toward the options easiest to finance. If these are the "no regrets" options outlined above, the result will be a significant response. If they are *not* these options, then other kinds of policy initiatives related to concerns about global climate change will face heavy sledding. For examples of related policy options, see Environmental and Energy Study Institute, 1991.

These directions are not dramatically opposed to current U.S. policy. In fact, they have developed largely out of analyses by the U.S. energy research community and consensus-oriented institutions such as the U.S. National Research Council and the Office of Technology Assessment of the U.S. Congress; and several of them are reflected in the current policies of the executive branch. The issues are less of basic philosophy than of emphasis and leadership. But the United States should not let itself go so preoccupied by what we do *not* know (and cannot know from the scientific information available at this point) that it fails to see that a great deal *is* known and that it fails to recognize that several kinds of policy initiatives focused on substance make eminent good sense right now, in addition to policy initiatives focused on frameworks and processes.

CONCLUSION

The fact is that, in the early 1990's, energy policy responses that sacrifice economic development for the sake of avoiding climate change are probably unrealistic in most countries, if not all, regardless of their pros and cons. The challenge is to do our best to understand what is possible without such a sacrifice, if we are innovative and determined, and to push the limits of such a *realistic* response.

For the near term, at least, we understand that a great deal can be done, even before we begin to push the limits of social and economic acceptability. The array of energy options that are good for reducing greenhouse gas emissions but that also make sense for other reasons is impressive indeed. In many cases, this reflects a kind of perverse serendipity; our very inefficiency offers us an attractive—if depletable—resource. For the longer term, a great deal more innovativeness will be required, but this is nothing new. Concerns about global warming simply add urgency to a transition that was going to be necessary in any case.

In the meantime, the main question for energy sector responses is not in fact of a capacity to contribute to reducing global greenhouse gas emissions but of political and institutional will to realize this capacity. And nowhere is this question more important in the United States, as one of the very most important shapers of international currents. In this sense, like it or not, it is possible that our own actions will go a long way toward determining the global response—and determining whether in the long run we will be considered a wise and innovative partner in worldwide problem-solving or one of the principal scapegoats for global pain and suffering. The choice is up to us.

REFERENCES

Balling, R., *et al.* 1991. *Global Climate Change: A New Vision for the 1990's,* Laboratory of Climatology, Arizona State University.

Boyle, S., 1991. Remarks to the Senior Expert Symposium on Electricity and the Environment (Helsinki), May 17, 1991 (representing Greenpeace International).

Bush, President G. 1991a. "Remarks at the Opening Session of the White House Conference on Science and Economics Research Related to Global Change," 26 Weekly Comp. Pres. Doc. 585, April 17, 1990.

Bush, President, G. 1991b. "Remarks at the Closing Session of the White House Conference on Science and Economics Research Related to Global Change," 26 Weekly Comp. Pres. Doc. 592, April 18, 1990.

Center for Science and Environment. 1991. "Global Warming in an Unequal World," New Delhi.

Darmstadter, J. 1991. "Estimating the Cost of Carbon Dioxide Abatement," *Resources,* Resources for the Future, No. 103 (Spring 1991):6-9.

Electric Power Research Institute (EPRI). 1990. "New Push for Energy Efficiency," *EPRI Journal,* 15-3:4-17.

Environmental and Energy Study Institute. 1991. "Partnership for Sustainable Development: A New U.S. Agenda for International Development and Environmental Security," Washington.

Epstein, J. and R. Gupta. 1990. "Controlling the Greenhouse Effect: Five Global Regimes Compared," Occasional Paper, The Brookings Institution.

Fickett, A., C. Gellings and A. Lovins. 1990. "Efficient Use of Electricity," *Scientific American*, 263(3):64-74.

Flavin, C. 1989. "Slowing Global Warming: A Worldwide Strategy," Worldwatch Paper 91.

Flavin, C. and N. Lenssen. 1990. "Beyond the Petroleum Age: Designing a Solar Economy," Worldwatch Paper 100.

Frisch, J.R. 1989. *World Energy Horizons: 2000-2020,* Paris: Editions Technip.

Fulkerson, W., et al. 1989. *International Impacts of Global Climate Change, Testimony to the U.S. House Appropriations Subcommittee on Foreign Operations, Export Financing, and Related Programs,* ORNL/TM-11184, Oak Ridge National Laboratory.

Gerlach, L. and S. Rayner. 1988. "Culture and the Common Management of Social Risks," *Practicing Anthropology,* 10(3):15-18.

Goldemberg, J., T.B. Johansson, A.K.N. Reddy and R.H. Williams. 1988. *Energy for a Sustainable World,* Wiley Eastern Ltd, New Delhi.

Goldemberg, J., T.B. Johannsson, A.K.N. Reddy, R. and H. Williams. 1985. "An End-Use Oriented Global Energy Strategy," *Annual Review of Energy.*

Goldemberg, J., T.B. Johansson, A.K.N. Reddy and R.H. Williams. 1987a. *Energy for a Sustainable World,* World Resources Institute, Washington, DC.

Goldenberg, J., T.B. Johansson, A.K.N. Reddy, R.H. Williams. 1987b. *Energy for Development,* World Resources Institute, Washington, DC.

Helsinki Symposium. 1991. "Energy and Electricity Supply and Demand: Implications for the Global Environment," Senior Expert Symposium on Electricity and the Environment, Key Issues Paper No. 1 (SM-323/1.

Hopkins, M. 1991. *Energy Use in Federal Facilities,* Alliance to Save Energy, Washington: January 1991.

IPCC. 1990. *First Assessment Report,* Vol.I: Overview and Policymakers Summary, Intergovernmental Panel on Climate Change, August 1990.

IFIAS et. al. 1990. "Climate Change and Energy Policy in Developing Countries," Report of an International Workshop, Montebello, Quebec, Canada, July 29-August 1, 1990. Human Dimension of Global Change Programme, International Federation of Institutes for Advanced Study, Ottawa.

Kelly, M. 1990. "Halting Global Warming," in Leggett, 1990, *op. cit.*

Lashof, D. and D. Tirpak. 1990. *Policy Options for Stabilizing Global Climate,* Report to U.S. Congress, U.S. Environmental Protection Agency, Office of Policy, Planning and Evaluation.

Leggett, J. 1990. *Global Warming: The Greenpeace Report,* New York: Oxford.

Levine, M., S. Meyers and T. Wilbanks. 1991. "Energy Efficiency and Developing Countries," *Environmental Science and Technology,* 25:584-89.

Levine, M., *et al.* 1991. "Energy Efficiency, Developing Nations, and Eastern Europe;

Report to the U.S. Working Group on Global Energy Efficiency, Washington, D.C. June 1991.

Manne, A. and R. Richels. Forthcoming. "Global CO$_2$ Emissions on Reductions: The Impact of Rising Energy Costs," *The Energy Journal.*

Morrisette, P.M. and A.I. Plantinga. 1991. "The Global Warming Issue: Viewpoints of Different Countries," *Resources,* Resources for the Future, No. 103 (Spring 1991):2-6.

Morthorst, P.E. 1991. "Potentials for Electricity Savings in Western Europe," in Proceedings of the Senior Expert Symposium on Electricity and the Environment (Helsinki), Background Paper 1-4.

NAS/NRC. 1991. *Policy Implications of Greenhouse Warming,* Washington: National Academy of Sciences/National Research Council.

National Governors' Association. 1990. "A World of Difference: Report of the Task Force on Global Climate Change," Washington.

Oak Ridge National Laboratory. 1985. "Energy for Development," working paper for the U.S. Agency for International Development.

Office of Technology Assessment (OTA), U.S. Congress. 1991a. *Changing by Degrees: Steps to Reduce Greenhouse Gases,* Washington: U.S. Government Printing Office.

Rayner, S. 1991. "Prospects for CO$_2$ Emissions Reduction Policy in the U.S. Energy Sector," in M. Grubb, ed., *Energy Policy and the Greenhouse Effect,* v.2, Royal Institute of International Affairs, Aldershot (UK): Dartmouth, forthcoming.

Schmidt, K. 1991. "Industrial Countries' Responses to Global Climate Change," Special Report, Environmental and Energy Study Institute, Washington, D.C., July 1991.

Wilbanks, T. 1988. "Impacts of Energy Development and Use: 1888-2088; in *Earth '88: Changing Geographic Perspectives,* Washington: National Geographic Society.

Wilbanks, T. 1989. Presentation to Workshop on the U.S., Developing Countries, and Global Warming, Office of Technology Assessment, U.S. Congress, September 1989.

Wilbanks, T. 1990a. "Energy Strategies for Developing Countries: Ethics and the U.S. National Interest," GTE Lecture Series on Energy, Economics, and the Environment, University of Oklahoma, April 1990.

Wilbanks, T. 1990b. "Implementing Environmentally Sound Power Sector Strategies in Developing Countries," *Annual Review of Energy,* 15:255-76.

Wilbanks, T. 1991a. "The Case for Energy Efficiency Improvement as a Global Strategy," in A. Zucker and M. Kuliasha, eds., *Technologies for a Greenhouse-Constrained Society,* forthcoming 1991.

Wilbanks, T. 1991b. "The Outlook for Electricity Efficiency Improvements in Developing Countries," in Proceedings of the Senior Expert Symposium on Electricity and the Environment (Helsinki), Background Paper 1-9.

Wilbanks, T., C. Petrich, *et al.* 1990. "Global Climate Change and the Programs of the U.S. Agency for International Development," report prepared by the Oak Ridge National Laboratory for the Directorate for Energy and Natural Resources, Bureau of Science and Technology, U.S. Agency for International Development, February 1990.

Wilbanks, T., D. Hunsaker, C. Petrich and S. Wright. 1991. "Potentials to Transfer the U.S. NEPA Experience to Developing Countries," in J. Cannon and S. Hildebrand, eds., *The Scientific Challenges of NEPA,* forthcoming 1991.

World Bank. 1990. "Capital Expenditures for Electric Power in the Developing Countries in the 1990's, Energy Series Paper No. 2, Industry and Energy Department.

World Commission on Environment and Development. 1987. *Our Common Future*, New York: Oxford.

World Meteorological Organization. 1991. "Climate Change: World Leaders' Viewpoints," WMO-748, Geneva.

World Resources Institute. 1990. *World Resources, 1990-91.* Prepared in collaboration with the United Nations Environment Programme and the United Nations Development Programme. New York: Oxford University Press.

Global Climate Change: Implications, Challenges and Mitigation Measures. Edited by S.K. Majumdar, L.S. Kalkstein, B. Yarnal, E.W. Miller, and L.M. Rosenfeld. © 1992, The Pennsylvania Academy of Science.

Chapter Thirty-Two

ENERGY PRODUCTION, ECONOMY AND GREENHOUSE GAS EMISSIONS IN HUNGARY:
Possible Regional Consequences of Global Climate Changes

E. MÉSZÁROS and Á. MOLNÁR
Institute for Atmospheric Physics
H-1675 Budapest, P.O. Box 39
Hungary

INTRODUCTION

The composition of the atmosphere plays an important part in the control of climate on the Earth. This is caused by the fact that the transfer of short-wave solar and long-wave terrestrial radiations in the air is affected by atmospheric constituents. On the other hand, the composition is controlled by material flows in nature, known as biogeochemical cycles. During geological times, a natural equilibrium was developed between atmospheric sources and 'sinks' of different components leading to a constant composition.

This natural equilibrium is now jeopardized by human activities which release a large amount of material (pollutants) into the atmosphere. Fortunately, man is not able to modify the atmospheric concentration of main constituents (oxygen and nitrogen) owing to their huge quantities. However, the levels of so-called trace substances, contributing less than 0.04% of the volume of dry air, can considerably be increased. The modification of the amount of gases (e.g. carbon dioxide, methane, nitrous oxides), absorbing thermal radiation emitted by the Earth's surface, is particularly dangerous. The residence time of these greenhouse gases is relatively long (10-100 years) compared to the characteristic mixing time of the atmosphere. Consequently, they are well-mixed in the entire atmosphere and contribute to world-wide problems like global warming.

This does not mean, however, that anthropogenic warming is solely a global issue. For solving this international problem all nations have to determine the extent of their greenhouse gas release and should study the possibilities of mitigation measures. Moreover, the interest of each country is to estimate future impacts of global warming in its area and to take all feasible steps to avoid or reduce harmful effects.

The aim of this paper is to discuss these two questions for a small country, Hungary. Since greenhouse gas emissions are in close relation to energy production as well as agricultural and industrial activities of a certain area, the economy of the country will also be discussed in some detail. It is also to be noted that only the emissions of carbon dioxide and methane are presented in this work.

CARBON DIOXIDE EMISSION

Biogenic sources

Carbon dioxide is the most important nutrient for plants, 90% of their dry matter come from CO_2 fixed by photosynthesis.[1] A part of carbon dioxide fixed is emitted by the plants by respiration. The release by respiration can be estimated on the basis of net primary production by different kinds of vegetation.[2] If we assume that the total quantity of carbon assimilated is shared equally by net assimilation and respiration and then consider the area of Hungary covered by different plants and forests, the Hungarian CO_2 emission (expressed in C) by respiration of vegetation is 27.6 Tgyr^{-1}. 70% of this emission is due to agriculture related to human activities.

The human and animal respiration also results in carbon dioxide emission. This emission can be determined by measurement of the quantity of CO_2 exhaled by humans and different animals. The results of such measurements show that for mammals, the CO_2 produced is proportional to the weight of the body. Taking into account the results of Freeman[3] as well as the number of people and animals in Hungary, we calculated a value of 2.4 TgCyr^{-1} for this emission. The major portion of this CO_2 source is caused by the respiration of man (28%) and pigs (33%). The magnitude of this latter calculation is due to the importance of pigs in Hungarian animal husbandry (their number is 8.7 million).

Soil is also an important source of biological carbon dioxide. This source consists of three parts: microbiologial activity in soil, root respiration and decomposition of organic matter on the surface. In the literature empirical relationships can be found for estimating the magnitude of this complex source. These relationships are given for different vegetation types[1], including cultivated areas.[4] Considering this information and the area of different types of vegetation cover, the calculation results in a CO_2 emission between 38.6 and 47.7 TgCyr^{-1}. The uncertainty comes from the fact that a rather wide range of emission rates for agricultural soils (10-150 gCm^{-2} yr^{-1}) is given in the literature (e.g. 4). On the basis of our calculations, we can conclude that in Hungary, deciduous forests control the emission of CO_2 (73-90%).

Energy production and industry

As is well-known, the combustion of fossil fuels is the most important worldwide anthropogenic carbon dioxide source. This emission depends on the quantity and carbon content of the fossil fuels used as well as on the fraction of carbon which can be oxidized.[5] For this reason, the emission factors published in the literature vary slightly.

The first column of Table 1 gives the energy produced in Hungary in 1987 by burning fossil fuels of different types. It should be noted that the sum of the values tabulated does not give the total energy use of the country. This is due to the fact that 110 PJ was produced in 1987 by a nuclear power plant, while an important quantity of electricity (106 PJ) was imported from the Soviet Union. The second column of the table shows the CO_2 emissions published in ref. [6]. It can be concluded that an important part of the total emission is caused by the use of solid fossil fuels (44%). Taking into account the population, a value of 2.24 tCyr^{-1} per capita can be calculated for Hungary. This is about two times more than the world average, but not too large compared to the values for developed countries.[5]

The Hungarian CO_2 emission per energy unit (17.5 MtC/EJ) is rather similar to the global mean value. The Hungarian energy demand per capita (127 GJyr^{-1}) is acceptable considering the values for other countries. The problem is caused by the inefficient use of energy produced. This is illustrated by the very high CO_2 emission per capita during the production of goods with a value of one U.S. dollar. This figure is around 1000 gCyr^{-1}, which is considerably higher than the corresponding values for developed countries.

Finally, it is to be noted that the cement industry and the treatment of waste materials also emit carbon dioxide into the atmosphere. Calculations show, however, that for Hungary, the contributions of these sources can be neglected compared to the release of carbon dioxide during energy production (0.6 and 0.1 TgCyr^{-1}).

Atmospheric Budget

As mentioned, vegetation is a net carbon dioxide 'sink'. As with CO_2 released

TABLE 1
Hungarian energy structure and carbon dioxide emissions[6]

Fossil fuel	Energy (PJ)	Emission (TgCyr^{-1})
Solid	359	10.5
Liquid	391	8.0
Gaseous	389	5.4
Total	1139	23.9

during respiration, the net 'sink' can be determined on the basis of the mass of dry matter formed in plants.[2] Considering the appropriate data, a value of 55.0 TgCyr^{-1} is calculated for the country. A negligible 'sink' is provided by the removal of CO_2 from the air by precipitation. However, this factor (0.03 TgCyr^{-1}), is not considered in the following discussion.

Table 2 summarizes the results of our study. One should say, however, that the figures given should be considered with some caution. While the error of the CO_2 emission by fossil fuel burning is not greater than $+/-15\%$, the uncertainty of other terms reaches probably $+/-50\%$. In spite of this, it can be accepted on the basis of values tabulated, that atmospheric carbon dioxide generation vs removal is positive over Hungary: about two times more carbon is released than removed. This means that even if the energy production were entirely stopped, biogenic sources would produce more carbon dioxide than the quantity removed by the vegetation. Unfortunately, it is not possible to determine the role of man in the unbalance of biological sources. Agricultural activities produce an important amount of carbon dioxide at present, but we do not know the biogenic emission of this area before the beginning of intensive agriculture. Considering the fact, however, that an important part of the territory of the country was covered by forests about a thousand years ago (the present forested area is only 17%), one can speculate that deforestation has played some part in the development of the present situation.

METHANE EMISSION IN HUNGARY

In the case of methane, it is again very difficult to differentiate natural and anthropogenic sources. This may mean that a large part of biological methane release is due to human activities.

Animal husbandry is one of these biological sources, which modifies considerably the emission of this gas. If we know the quantity of feedstuff consumed by ruminants, their methane production can be calculated. In Table 3, the input data necessary for this calculation, as well as the results obtained, are tabulated. It follows from this information, that ruminants in Hungary emit into the air a CH_4-C quantity of 0.125 Tg per year. In spite of the fact that the Hungarian live-stock farming in this category of animals is dominated by sheep, an important fraction of methane emission is caused by cows.

TABLE 2

Atmospheric Carbon Dioxide Budget Over Hungary

	Sources (TgCyr$^-$)	Sinks (TgCyr^{-1})
Vegetation	27.6	55.0
Man and animals	2.4	
Soils	38.6 - 47.7	
Energy	23.9	
Total	92.5 - 101.6	55.0

TABLE 3
Estimation of methane production from fermentation of ruminants

Ruminants	Feed uptake[16] (dry matter kg/d)	CH$_4$ prod.[17] (gCH$_4$/day per capita)	Number of animals 10^3	CH$_4$ yield (MgCyr^{-1})
Calves	2.8	64	214	3.8
Heifers	8.3	173	395	18.7
Cows	10.6	219	219	45.4
Beef cattle	13.0	266	266	26.3
Sheep	1.7	43	2337	27.5
Deer	3.0	68	55	1.1
Fallow deer	2.0	49	14	0.2
Roe deer	1.0	29	227	1.8
Moufflon	1.0	29	9	0.1
Total				124.9

Under anaerobic conditions, methane is also formed in soils—mostly in marshy areas. Concerning human activities, the most important global methane source of this kind is provided by paddy fields. The strength of this release depends on many factors, as discussed by several authors.[7,8] These factors include the content of organic matter and temperature of soils as well as the physiological activity of the vegetation. Methane emission also can be observed in the case of fresh waters, meadows, forests and different cultivated lands. Calculations made on the basis of the information on emission factors available[7,8] and of the area of appropriate surface in Hungary, show that the total methane emission from soils is between 0.05 and 0.1 TgCyr^{-1}.

The uncertainty of these figures makes it evident that the error of such calculations is at least a factor of 2. Due to the structure of agriculture, the emission from paddy fields is not too important in this country: it is less than about 0.02 Tgyr^{-1} expressed in carbon. On the other hand, a rather large amount of CH$_4$ is released into the air by reeds, lakes, fish ponds and water catchments. The upper limit of this emission is around 0.05 TgCyr^{-1}.

Methane molecules are also released into the air from solid and liquid waste materials. In 1986, the total quantity of solid wastes in Hungary was equal to 18.5 x 10^6 m^3. According to Bingemer and Crutzen,[9] under anaerobic circumstances, about 80% of the organic matter is transformed and results in biogas, 50% of which is CH$_4$. Such conditions are created if the solid wastes are deposited in landfills or in open dumps. Taking into account this information and the composition of the waste quantity given above, a yearly methane formation of 0.189 Tg can be calculated. It can be assumed, however, that only a fraction, a value of 0.7 of this mass,[10] is released to the air, a CH$_4$-C emission of 0.132 Tgyr^{-1} is obtained for this source strength.

For liquid waste materials, similar calculations were carried out. In these calculations, the organic dry matter content (0.3 kgm^{-3}) and biodegradability (of 3% of the wastes), were taken into account. The results indicate that this methane source's strength has an order of magnitude of 10^{-3} TgCyr^{-1}, which can be neglected when compared to the magnitude of other terms.

The main component of natural gas used for energy production is methane. Due to this fact, during the mining and distribution of natural gases, some methane is released into the air. According to the literature[11] and information given by the Hungarian Gas Industry Trust, methane release during mining is equal to 2%. Since 86% of natural gas exploited in Hungary (7.1×10^9 m^3) is CH_4, the loss in this way can easily be calculated. In Hungary, an important part of natural gas (4.8×10^9 m^3: containing 90% CH_4) is imported via pipe-line systems from the Soviet Union. At the joining points of pipe-lines, about 3% of the gas is lost.[11] This figure can also be applied for the transport of natural gas mined in Hungary. Taking into account all these values, we calculate a total CH_4-C release of 0.234 Tgyr^{-1}.

Coal and lignite mining also provides a methane source, since coal contains a certain amount of this gas. This amount averages 5 m^3 per ton of coal,[7] 50-100% of this methane comes to the air during coal mining. For methane content of lignites, we estimated the same value of 5 m^3/ton and assumed that 25-75% of this methane escapes into the air. The total quantities of coal and lignite exploited yearly in Hungary are 15.3 Mt and 7.0 Mt, respectively. This results in a CH_4 emission between 0.025 and 0.055 TgCyr^{-1}.

Considering the results discussed, one can conclude that the total methane emission in Hungary is around 1 TgCyr^{-1}. About a quarter of this emission is due to natural gas production and distribution (see Table 4). It can also be stated that about 100 times more carbon dioxide-carbon is released into the air than carbon in methane form. This means that mitigation measures must be centered on CO_2 emission. This conclusion remains valid even if we take into account the fact that, for the same concentration increase, methane is a more efficient greenhouse gas than carbon dioxide.

For the determination of atmospheric methane burden, the 'sinks' of this gas should be estimated. It is well known that methane does not utilize significant direct sinks, although CH_4 uptake of some types of soil might be important. Methane is primarily removed by oxidation (via OH radicals) in the troposphere. In the stratosphere, CH_4 is also oxidized by OH radicals and excited by oxygen atoms and they play non-negligible role in the removal of chlorine containing hydrocarbons. Determination of the yield of these processes is so uncertain, mainly for a small country like Hungary, that we did not undertake doing it.

TABLE 4

Details of Hungarian Methane Emission

Source type	Emission ($tgCyr^{-1}$)
Enteric fermentation of animals	0.125
Soils, water surfaces, marshes, etc.	0.054 - 0.093
Solid wastes	0.132
Natural gas production and distribution	0.234
Coal and lignite mining	0.025 - 0.055
Total	0.570 - 1.209

POSSIBILITIES OF THE REDUCTION OF CARBON DIOXIDE EMISSION

On the basis of the above discussion it is obvious that the main possibility of the reduction of the CO_2 emission in Hungary is via *energy saving* and *conservation*. Considering the value produced, the energy demand and consequently the quantity of carbon dioxide emitted are too high. There is no intention here to discuss, in detail, the possible ways of rational energy use. It is clear, however, that the transformation of the total economic system is needed. After essential political changes, our main aim is exactly to create an effective industry with much less energy demand. For this, however, important capital investment and good economic policy are necessary. We note in this respect, that in our fund-deficient country, during the time period 1981 and 1985, 40% of industrial investment was used in the branch of energetics, which restrained the transformation of industrial structures.

The CO_2 release into the atmosphere can also be reduced by the *diversification of energy sources*. This can be obtained by the application of sources with lower or zero specific emissions e.g. nuclear and hydroelectric power plants, electric energy import, geothermal and solar energy use. According to our experts it would be possible by the year 2010 to provide 80% of the electrical energy consumption in this country via nuclear energy.[6] However, it is very probable that all significant part of the population would be against this solution, not only because of a desire to protect the atmosphere but also because of problems due to the storage of radioactive waste materials.

In Hungary, 29% of electric energy is presently imported. Considering that this is a very high proportion, nearly unparalleled in the world, further increases of imported electric energy are not desirable from technical or economic points of view.

The construction of high capacity hydroelectric power plants is not possible in Hungary, considering geographic and hydrologic conditions. On the other hand, conditions in Hungary are relatively good for use of geothermal energy. However, it is estimated that this energy type is able to cover only 1-2% of total energy demand. The same is true concerning solar energy. According to the results of calculations, the details of which are not given here, solar energy would make possible energy savings of 2.0-2.5% in the next 20 years.

According to our calculations, in the future, among traditional fuels, natural gas is preferred to coal: the same energy production, about 60% of CO_2 is emitted into the atmosphere using natural gas, with the case of coal.

An interesting possibility of reduction of CO_2 emissions is to use the *biomass* for energy production. As unpublished material of the Hungarian Academy of Science shows, the carbon content of the biomass in Hungary is 25 Tg. About half of this quantity is a by-product which could be burned to produce energy. Since the carbon content of the biomass is due to the absorption of atmospheric CO_2, biomass burning does not increase further the carbon dioxide concentration in the air.

Another theoretical possibility is provided by *afforestation*. This possibility is offered by the fact that 10% of the area of Hungary is not suitable for agricultural cultivation. If this area e.g. were planted by acacia trees, this would make it possible, by burning the wood, to produce primary energy of 190 PJ, which is 11% of the total energy demand planned for the year 2010. Further, afforestation would increase the CO_2 absorption capacity of the vegetation in Hungary. However, such an action would make an important change in the present environment of the country.

The possibility of a reduction in methane is unfortunately limited. It is well known that the largest changes of global and national CH_4 production are of anthropogenic origin. Increasing agricultural (raising cattle, growing rice, etc.) and industrial activities represent the major sources of atmospheric methane. Thus, it is difficult to control or even reduce their magnitude. As can be seen in Table 4, natural gas production and distribution are the most important sources of methane. During distribution a large amount of methane is released, which can probably be reduced by modernizing the pipe-line network. Another significant amount of methane is released by wastes. It's strength could be decreased by more reasonable utilization of wastes (municipal and agricultural) and biogas production in closed systems.

CO_2 CONCENTRATIONS AND ENVIRONMENTAL CONSEQUENCES OF GLOBAL WARMING IN HUNGARY

Of the most important greenhouse gases, only carbon dioxide is measured regularly in Hungary.

The CO_2 concentration in the air has been monitored since 1981. The observations are carried out at a background air pollution monitoring station *K-puszta* by means of a Siemens Ultramat 3-gas analyser. The station is located in the middle of the country ($\varphi = 46°58'N; \lambda = 19°35'E; H = 125$ m above sea level) at a cleaning of a forested area on the Hungarian Great Plain.

According to the results obtained, the annual average atmospheric CO_2 level in 1981 was 351.6 ppm, while the corresponding value for 1990 was equal to 367.7 ppm. This is equivalent to an annual increase of 1.79 ppm or 0.51%. These figures are comparable to those gained at the Mauna Loa Observatory in Hawaii (U.S.A.) where the increase between 1958 and 1984 varied between 0.2 and 0.5% per year.[2] However, the absolute concentration values observed in Hungary are somewhat higher than the concentrations reported in Hawaii. This variance is obviously due to the nearby anthropogenic/biogenic sources.

An important role of present Hungarian research in atmospheric science is to estimate the possible climatic consequences of the CO_2-induced global changes in the country. This is a very difficult task since even the most sophisticated climate models are unable to give reliable results for such a small territory. For this reason, global changes during the last century were compared statistically with the variations of the meteorological parameters observed in Hungary. The results of this study indicate that if the average temperature in the Northern Hemisphere were higher by 1 K, the summer and winter temperatures in Hungary would be higher by 1.5 K and 2.0 K, respectively.[3] Under the same conditions, the precipitation amount during the vegetation period in this country would be less by 60-100 mm (10% more than the average). At the same time, the number of hours with sunshine would increase and the probability of drought would be higher by about 60%. It goes without saying that such changes would modify disadvantageously water management and agricultural production of the country.

The effects of climate changes on water management can be estimated on the basis of a hydrological study made by using past data.[14] The results of this study indicate that an increase of 0.5 °C in average temperature and a decrease of 5% in yearly precipitation affect all watersheds in the country. Thus, in the watershed of the Zagyva River, the run-off would decrease from 58 mm to 47 mm (relative decrease: 19%). In the watersheds of the Danube and Lake Balaton, the relative decrease of the run-off is around 10%. Moreover, the increase of the temperature makes evaporation of the lake more intensive.

The effects of increasing carbon dioxide concentration on plant production is twofold. First, more available CO_2 results in a more efficient photosynthesis. Second, higher temperature and less precipitation amount could be very bad for agriculture.

This is particularly true in Hungary where the most important climate factor for agriculture is the amount of precipitation. Without going into details (which are quantitatively not known), we note that it will probably be necessary to increase the ratio of plants (e.g. unbearded wheat) with higher drought-resistance to other cultures (e.g. spring wheat). This will be indispensable since the organization of a more effective irrigation system will be hindered by the shortage of water. Anyway, much more research is needed to determine quantitatively the impact of changes in temperature and precipitation on Hungarian agriculture and to work out a strategy for reducing harmful effects.

A little more information is available for the impact of climate change on our forests.[15] This information indicates that in the case of a temperature increase of 1.5 K and of a precipitation decrease of 30% (both for the vegetation period), forest ecosystems of different types will move to northern and higher territories. The area of forest steppe will be increased, while that of beaches will be decreased. It is probable that climate change will be favourable for turkey oak and unfavourable for pinewoods on the Hungarian Great Plain. It follows from this short discussion that a special forest plantation program is desireable to moderate these effects.

CONCLUSIONS

Hungarian carbon dioxide emissions due to energy production can be reduced by energy conservation. On the other hand, the release of methane can be mitigated by the production of biogas from waste materials and by modernization of the pipeline network for natural gas.

Since water management, agriculture and forestry of this country are climate dependent, Hungary is very much interested in the climatic changes due to the increasing concentration of greenhouse gases, and it supports all international efforts aiming to reduce their unfavourable effects.

REFERENCES

1. Fung, I.Y., C.J. Tucker and K.C. Prentice. 1987. Application of advanced very high resolution radiometer vegetation index to study atmosphere-biosphere exchange of CO_2. *J. Geophys. Res.* 92:2999-3015.
2. Bolin, B., E.T. Degens, S. Kempe and P. Ketner. 1977. (eds.) *The global carbon cycle. SCOPE 13.* John Wiley and Sons Ltd. Chichester - New York - Brisbane - Toronto.
3. Freeman, B.M. 1973. Metabolic energy and gaseous metabolism. *Poultry Physiology.* Acad. Press. 282.
4. Hampicke, U. 1980. The effect of the atmosphere-biosphere exchange on the global carbon cycle. *Experiencia* 36:776-780.
5. Rotty, R.M. 1987. A look at 1983 CO_2 emissions from fossil fuels (with preliminary data for 1984). *Tellus.* 39B:203-208.
6. Lévai, A. and E. Mészáros. 1989. Energy production and carbon dioxide emissions in Eastern Europe with special reference to Hungary. *Időjárás.* 93:196-204.
7. Ehhalt, D.H. 1985. Methane in the global atmosphere. *Environment.* 27:10.
8. Holzapfel-Pschorn, A. and W. Seiler. 1986. Methane emission during a cultivation period from an Italian rice paddy. *J. Geophys. Res.* 91:11803-11814.
9. Bingemer, H.G. and P.J. Crutzen. 1987. The production of methane from solid wastes. *J. Geophys. Res.* 92:2181-2187.
10. Jager, J. and J. Peters. 1985. Messung der Oberflachenemission von Deponiegas. *Stuttg. Ber. Abfallwirtsch.* 19:337-345.
11. *CONCAWE: Volatile organic compound emissions: an inventory for Western Europe.* report No 2/86. 1986. Den Haag.
12. Pueschel, P. 1986. *Man and the Composition of the Atmosphere.* UNEP-WMO, Geneva.
13. Mika, J. 1988. Regional features of a global warming in the Carpathian Basin (in Hungarian). *Időjárás.* 92:178-189.
14. Nováky, B. Climate change and its hydrological effects. (manuscript).
15. Pálvölgyi, T. and T. Szedlák. 1990. Possible impacts of global warming on the forrestal ecosystems in the Carpathian Basin. *Ambio.* (submitted).
16. Kakuk, T. and J. Schmidt. 1987. *Animal Feeding* (in Hungarian) Mezõgazdasági Kiadó. Budapest.
17. Baintner, K. 1957. *Animal Feeding* (in Hungarian). Mezõgazdasági Kiadó. Budapest.

Global Climate Change: Implications, Challenges and Mitigation Measures. Edited by S.K. Majumdar, L.S. Kalkstein, B. Yarnal, E.W. Miller, and L.M. Rosenfeld. © 1992, The Pennsylvania Academy of Science.

Chapter Thirty-Three

GLOBAL CLIMATE CHANGE AND HUMAN POPULATIONS IN THE CENTRAL AND WEST AFRICAN REGION:
An Examination of the Vulnerabilities, Impacts and Policy Implications

STELLA C. OGBUAGU
Department of Sociology
University of Calabar
Calabar, Nigeria

INTRODUCTION

The scientific nature of global climate change tends to becloud the priority that should be accorded with population as a significant contributor'to the build up of the greenhouse effect and a major recipient of the potential impacts of climate change. Thus, in many international agenda (for example, the 1990 Second World Climate Conference) population attracts peripheral mention. Yet, without the anticipated disastrous consequences of global warming on human beings (their survival and continuity) discussions on climate change would probably not have assumed the proportions they have attained in international conferences, workshops and seminars. It is argued therefore that intensified global activities on climate change in recent decades are necessitated largely by the threat of climate change to human life. Based on this premise, it is pertinent to examine the impacts of global climate change on human populations in order to determine what policies to adopt to "save" mankind from the gravest challenge to its survival it has yet faced.

Although, climate change remains a controversial issue, scientists have generally agreed that the earth is warming at a tempo that is unprecedented in human history. Such warming, they suggest, will have mixed impacts on different parts of the globe. The developed world with better capabilities for handling the anticipated impacts would probably fare much better than the developing world with heavy technological and financial disabilities.

The West and Central African Region which stretches from Mauritania in the North to Angola in the South, belongs to one of the poorest Regions of the World. It is also a region that is experiencing huge increases in population that demand space, finances and other resources for sustenance. This region is in dire need of studies for proper assessment, monitoring and understanding of the potential impacts of climate change on its peoples and their socio-economic facilities if they will not become victims of a global challenge.

Environmental Consequences of Population Growth in the Region

The Population of the region, which is estimated at about 269 Million (United Nations, 1988), is growing at the rate of 2.9 percent per annum. Its high rate of growth makes it a young population with a dependency ratio of about 47-50 percent. The demands of such a rapidly growing population are matters of great concern to administrators and leaders of the respective countries. Writing on the incorporation of population considerations in conservation matters, Munro (1990)[3] highlights the importance of stabilizing populations in order to prevent further deterioration of the biosphere.

In the region, efforts at stabilizing the populations have not been very successful. For instance, Nigeria with the largest population of nearly 110 million people is only just beginning to implement its population policy launched in 1989. That policy streamlines the benefits of planning the family essentially through voluntary application of family planning facilities. Other countries in the region are more or less in similar situations thereby making it less likely that their populations would decline significantly before the close of this century.

What this situation means for the region is the continued depreciation of existing resource base without complementary replacement. The growing population leads to the over-use of lands, the crowding of cities and towns, increased violence and political unrest as well as other socially undesirable trends that tend to threaten human society. These issues remain fore-boding with the possibility of further worsening if environmental problems that would be associated with climate change are not handled adequately.

Other Major Areas of Concern

Primary factors that will be accentuated by climate change in this region are floods, droughts, seawater intrusion and run off water. Flood has remained a yearly monster to farmers and citizens particularly in the low lying areas of the region.

In the coastal areas of say, Nigeria and Ghana, flood menace particularly during the rainy season is a nightmare to most residents.

In Ghana, along the Takoradi/Sekondi zone, many citizens consider moving to drier parts of the country a reasonable option. In Nigeria, tales of woe of flooded residential areas, homes lost to flood water, highways rendered unusable not only by water but also by flood action — deep gullies across national expressways and even human beings drowned in turrents of floods become common place during the rainy season months (April - October). For example, in September 1991, the national television showed pictures of the overflow of the Benue river in the middle belt of Nigeria. Thousands of citizens living along the riverine areas were dispossessed of their houses, life-long accumulated property, and of even the lives of their relatives by floods.

This news item was underscored by newspaper report of the incident. Writing under the caption "Lightening kills 4 brothers," Umar and Zhegu (1991) describe the experience of the residents following this event. In their words, "the flood, described as unprecedented in recent years... rendered over 2000 families homeless... about 90 percent of food crops like rice, maize, guinea corn and millet were destroyed." In another incident, the Federal Government of Nigeria is reported to have approved N15 million (fifteen million Naira, the equivalent of US $1.5 million) for another community — Awoye in Ondo State of Nigeria to fight erosion. The report by Oyelegbin (1991)[5] asserts that the money was to help resettle inhabitants of the riverine community hit by erosion, which has caused 20,000 people to flee the town. Some school children were also killed at the height of the sea surge last year. The problem has also rendered "100,000 people in the over 70 communities in the predominantly riverine areas homeless..." Stories of even wider dimension are told about parts of Borno State (Mongonu and others) where the local river had crashed down existing weak barriers along its banks, flown into people's homes and farms causing much devastation in its trail. As a result of these mishaps, speculations are rife that harvest of crops would be very poor. In point of fact, it is feared that crop yield has been so adversely affected that there is the threat of famine for the rest of the year.

Similar tales in the past years reinforce the argument that this region remains one of the most vulnerable areas of the world to climate variations, fluctuations and change. In 1987, the late arrival of rains for farming caused much havoc on flocks and farm crops. That year harvest yield was extremely low such that local food crops doubled and tripled in price in the coming year. In 1988, it was floods that caused losses in human life and property. In 1990, when the rains were late in coming, farmers were distraught with fear and concern. In Nigeria, if it is not dryness that is the concern, it is flood. Each year, millions of the Naira (Nigeria's local currency) are lost to flood and flood action and thousands of residents are dislocated while livestock and human beings lose their lives.

These go to confirm Ibe and Quelennec's (1989:6)[6] observation "erosion is a prevalent phenomenon along the Atlantic coast of West and Central Africa. In many places, the rate of coastline retreat and the resulting environmental degradation and economic loss is on such a scale as to be alarming." In the same vein Tebicke (1989)[7] expresses the view that the concentration of industry along the coast of Sub-Saharan Africa is fraught with as much danger for the socio-political stability of Africa as if sea-level were to rise even modestly.

Incidents such as the ones mentioned above illustrate the unfavourable dependence of Third World countries on natural forces for food and maintenance of life-sustaining resources. Failure or change in the "normal" natural events such as weather sends a tremor through them upsetting the balance between man and the ecosystems that sustain him.

Frighteningly, the lowlying coastal areas in most of the region's countries contain well over 60 percent of the population, socio-economic installations as well as most of the important capital cities and towns. In other words, these are areas of high human activity much of which adds to the build up of human waste, pollution and unsanitary conditions. The incessant move in of new migrants added to the quite high rate of natural increase in the cities aggravate the environmental problems with no real solutions in sight.

The other side of the environmental problem experienced by countries in the region and which is also exacerbated by growing numbers is drought. The Northern parts of many of these countries lie in the dry zones of the continent narrowly espacing being in the sahel region. They experience rainfall only during three months of the year usually between June and October. Outside of this period, heat waves, sand storms and human misery prevail. It is so hot and dry that livestock for lack of or inadequate grazing grounds shrivel up and die. Human beings suffer from excessive loss of water and illnesses such as cerebral meningitis. Human activities in agriculture which expose the soil to the scorching heat, cutting of trees for wood and for construction further depreciate the quality of the soil rendering it more susceptible to weather actions. This heightens the problem of desert encroachment. Thus, in the Northern parts of Nigeria (which border the Lake Chad) reports constantly show depreciation in soil quality. Even Lake Chad itself is said to have gone down in volume.

The situation is threatening enough to require attention. If the current high rate of population growth is left unchecked, environmental condition in this part of the region will continue to deteriorate. In the event of temperature increase in response to global warming, this area would most probably be in jeopardy. It is not surprising, therefore that van den Oever (1990)[8] discussing the link between population and the environment suggests "new-style conservation strategies acting as an umbrella policy for integrated population and environment action". It, thus, becomes important to understand the population/climate interface in order to plan well for the well being of the human society.

Mention must also be made of other sources of concern which include salt water intrusion and run off water. Salt water intrusion is already a problem in some sections of the coastal areas. In Lagos, Calabar and Port-Harcourt in Southern Nigeria, fresh water supply constitutes a major handicap for many citizens precisely because of this problem. In these cities, one does not have to dig deep to meet the water level. Surrounded by sea water, wells dug in home compounds taste salty. The irony of the situation is that water surrounds the people but they face serious fresh water problem.

In Calabar, for example, few sources of good drinkable water exist. For most citizens, the dry months of the year (November to March) present harrowing experiences. Hundreds move around the town in search of water. Often times, they are forced to resort to drinking water from any source. In other parts of the country where people do not have access to good water supply, guinea worm infestation plagues the populace. This sort of health problem is likely to be accentuated if there is further salt water infilteration as a result of golbal warming and sea level rise. In that event, large numbers of people would be adversely affected. In the same way, increase in run off water definitely exacerbates soil erosion, reduces soil quality and depletes soil nutrients with detrimental impacts on species dependent on existing conditions.

From the foregoing, it is obvious that the impact of climate change on human populations if current scenarios prevail, would be great and largely grave. Much of the consequences would be precipitated by growth and pressure of the populations in their efforts to accommodate, feed and sustain the teeming young people under conditions of diminishing resources.

Human Settlements

Settlements in this Region assume three broad patterns viz: the densely populated "seaward' zone, the sparsely populated "middle belt" and the densely populated northern parts known as the "Sudan Savanna." In the different climate/vegetation zones populations are likely to have adjusted to the prevailing conditions with regard to social, economic and political values and habits. They have developed building models and structures that are appropriate for their type of environment. For example, riverine populations take more to fishing than terrestrial residents and grasslands attract herders and grain growing people. Such "comfortable" adjustments stand the threat of being disturbed if significant changes occur in the climate/population equilibrium.

Forest Depletion

Human beings derive much of their food and energy from forests. This dependence is heightened where much of the populace is engaged in agriculture. Furthermore, most of the people use wood for much of their domestic cooking needs, construction and building requirements. It is thus, obvious that the rapid increase in population accelerates the pace of exploition, consumption and depletion of forest resources. Through the heavy and often unplanned exploitation of forests, soil erosion, environmental pollution as well as deforestation are accentuated to the detriment of man. The effects of careless and constant cutting down of trees without adequate replanting has been adequately discussed by various scholars[9] (see for example, Meadows, 1974; Hayes, 1977; Eckholm, 1979; and Hays, 1979; IPCC WG II, 1990). From the observations of these scholars, there is the need to control the unwholesome practice of forest degradation, otherwise, the consequences for future generations will be critical.

Food and Nutrition

Concerns have also been expressed about food production in the event of climate change (see for example, Parry and Chen, 1990, and Report of IPCC Working Group II, 1990)[10]. The dependence of the West and Central African Region on very simple technologies for food production will be severely challenged when climate variations and sea level rise alter existing conditions. The shift in production zones will necessitate the development of new technologies appropriate for the new conditions. This region has neither the technology nor the financial and technical resource to immediately respond to emergency situations. Experience so far had been one of constant losses in farm products and subsequent famine threats. Without adequate warming and preparation to mitigate climate change impacts, the teeming population will face food problems at scales not yet experienced in the region. Food depletion and change will affect the nutrition of the populace and ultimately their health which is one of the poorest in the World.

Human Health

Evidence abounds that human health can be greatly affected by climate. For instance, patients with some form of tuberculosis are known to get better in warmer climates than in the cold temperate climates. Furthermore, extremes of climate render affected areas virtually uninhabitable.

Although mortality is generally falling in Central and West Africa, it still has one of the highest rates in the world. Despite increased funding for health programmes by the respective governments and by International Agencies such as the World Health Organization, the World Bank and the United Nations Children's Fund, mortality rates are still above 15 deaths per 1000 population. These are high rates compared with the developed world where death rates have fallen mostly below 10 deaths per 1000 population. The region's life expectancy at birth of about 50 years is also one of the lowest in the world. Just as its infant mortality rates of over 60 per 1000 live births rank very high in relation to much of the rest of the world.

These indices of the state of the region's population show that the health of the people is in dire need of improvement. Yet, it is difficult to imagine how this improvement can be achieved. The situation as it currently exists is that the rapidly growing population is posing a challenge to governments to provide medical doctors and other medical personnel, hospital facilities and medicines for an ever increasing number of users. The result in many of the countries is that medical staff are grossly inadequate to effectively man even the limited existing medical facilities. Thus, vast areas particularly in rural communities have neither hospitals nor good medical clinics. Simply put, the populations suffer from inadequate medical attention and hence the heavy toll on human lives. That is why it is often frightening as it is frustrating to hear of numerous minor medical cases that end in could have been prevented deaths. In effect, life in this region still remains largely hazardous from infancy to the grave. Discussions on this issue show that this awesome situation would be greatly worsened if the anticipated climate change is to occur by 2025 as is generally predicted.

Part of the explanation for this view is the increase in pests and other vectors of diseases that climate change would most probably occasion. For example, Kellog and Schware (1982)[11] see bacillary dysentery, schistosomiasis, hookworm-induced health problems, yaws and cerebral meningitis increasing as a result of climate change. In the same vein, the United Nations (1973)[12] observes that warmer temperatures may lead to increase in pests that are injurious to human health and activities.

Besides pests and diseases that cause serious health hazards for the people of this region, poor environmental sanitation and low nutritional levels also take their tolls. Sanitary conditions in most of these countries, with Nigeria as a notable example, are well below acceptable standards. The environments, particularly in the urban areas, are literally inundated or deluged with refuse. People sometimes ease themselves in the open with minimum concern for the environment and the health of other people and gutters are usually clogged and unsightly. Such practices added to the over-crowding in usually sub-standard housing units, pose real serious health problems for the people. These overwhelming problems are likely to be further intensified by climate stresses.

The same observations can be made about morbidity in the region. Although it is not easy to get exact morbidity statistics for the countries, one can guess that morbidity is high by knowing that malaria, cholera, yellow fever and other such debilitating illnesses are rampant. With high morbidity, loss in man work hours is also high. Thus, Kellog and Schware (1982) estimate that for every 1°C increase in temperature, 2-4 percent work is lost. In the West and Central African region, many of the illnesses such as malaria may not kill outright but they sap the victim's energy thereby rendering him listless and unable to perform his activities at full capacity.

This situation, particularly with regard to malarial attacks, is getting worse. Malaria parasites seem to have become immuned to many of the common medications that have traditionally been used to cure malarial attacks. Medical opinion is tending to suggest that in many cases several doses of different malarial medications have to be applied to give relief to a patient.

It needs to be emphasized that the control of malaria needs research and researches generally require enormous funds. Given that most of the governments of the countries in this region are engrossed in the difficult tasks of coping with their teeming populations with scarce or limited resources, it will be yet a while before they will have the leverage to fully undertake such major health programmes. Fears are that the situation might become aggravated and get out of hand if there are complications as a result of the anticipated climate change.

Migration

Much of the migratory trends in this region are national, that is rural-urban migration. International migration is limited in scope except when relations among some of the nations become strained or reach a low ebb. In that event, they sometimes resort to the repatriation of the citizens of their respective countries. Such was the case when Nigeria and Ghana and more recently Mauritania and Senegal had strains in their bilateral relations.

Internal migration in the West and Central African region is usually in response to what Todaro and Stilkind (1981)[14] refer to as "city bias and rural neglect." Many young, educated and able people tend to leave the unattractive rural areas of the country to move to the few urban areas which have a disproportionate share of social and economic infrastructures. This influx of rural people to the urban areas, in addition to the natural increase in the populations of the cities, creates conditions of excessive growth which strains existing social amenities and infrastructures. It leads also to over-crowding in stolid and sub-standard housing units found in festering shanty locations which often accentuate crime and worsen unemployment problems. A coastal city like Lagos in Nigeria manifests many of these features. A recent report on Lagos observed that it is "a city under stress" largely from over population and unsanitary conditions. This observation is underscored by another assertion that the influx of migrants into Lagos as well as its natural growth rate pose "overwhelming" problems.

These observations can easily be made by any new comer to Lagos. The stench from the Lagoon as well as the litter on the streets draw immediate attention to the serious debasement of the environment. In recent times, however Nigerian governments have tried to make the country's environment a cleaner place by decreeing a national clean up day (usually the last Saturday of each month). Although, this effort is commendable and indicates a new awareness of and desire for clean surroundings, the fact remains that much still needs to be done in this and other related issues.

As serious as these conditions are contemperaneously, there is every fear that they would be worsened if nothing is done to ease them off prior to any significant change in climate or sea level as had been predicated for the next century. Experience with the drought which occurred in the Sahelian region between 1967-73 and in the early 1980s, show that increased migration usually takes place when there is a shift in climate conditions. If the anticipated changes in climate become a reality, there is every likelihood that the countries in the region especially the coastal areas would be greatly affected. This is partly because people from the severely affected areas would move to other unaffected or least affected parts of their countries or even across national boundaries. This would further aggravate the already poor living conditions of the people, upset their socio-economic values and possibly their political stability.

Impacts on Socio-Economic Installations

The vulnerability of the West and Central African region to climate alterations shows up significantly in its socio-economic installations and facilities. Energy disruptions by floods and wind actions; seaport, roads and airports dislocations in response to weather conditions and the disorganization of the flow of goods and services by weather variations draw attention to the need to prepare effectively for countering anticipated disruptive impacts of climate change and sea level rise. The economic loss to these nations as a result of say, flooding, drought and wind storms point to the fact that caught unprepared, they would fare terribly badly if climate change at predicted levels occurs.

Preparedness of the Region for Climate Change

It would be presumptious to argue that the countries of this region have done nothing at all in anticipation of projected climate change and sea level rise. What may be right to say is that much of the action taken so far have been primarly in reaction to events that have occured. Thus, when for example, floods occur as is often the case in Nigeria as well as in some of the other countries, government aid is given to victims depending on the scope and intensity of the disaster. In some cases, relief aids are sent to ameliorate the sufferings of the afflicted. In other cases, victims have to cope on their own.

As preventive measures, national governments often build embarkments to forstall sea levels rising in high risk areas beyond containable limits. For example, in Lagos and Calabar (coastal towns) as well as Bonny and Opobo islands, the Nigerian government had erected concrete walls to stop further encroachment of the sea into the surrounding lands. Thus far, however, what had been accomplished is limited in scope because the magnitude of the problems that may arise from anticipated climate change has not been properly assessed and so have not been adequately addressed. Instead, policies and programmes of governments tend to focus attention on what they consider to be politically relevant and expedient issues such as economic development and rapid population growth. In effect, there appears to be insufficient realization of the fact that for adequate and meaningful economic development to take place, the factor of climate change must be well understood and integrated into national planning and action. This is why it is considered necessary to suggest a few policy options.

Recommended Options

The high vulnerability of populations in the region is not going to be solely the effect of future climate change by itself, but more likely the inability or lack of capacity to anticipate and plan against the problems associated with it. In this regard, the desired measures must start with an increasing awareness of the dynamics of present and future man-climate relationships. This will be followed by the formulation of national and regional policies that will give broad directions to specific measures against the impact of climate change.

Awareness

The starting point is the realization that difficult-to-manage problems are likely to be associated with climate change. Such awareness needs to be created through available mass communication media not only among political leaders and bureaucrats, but also among the generality of the populace. An increasing awareness on the part of the people of the close relationship between human actions and the environment through climate change, fluctuations, and variations is a prerequisite for other response measures that may be adopted to mitigate the adverse consequences of climate alterations.

Policy Impact

The formulation and adoption of a broad course of action is also a necessary basic condition. This is required not only at the national level but also at the regional level. The required policy will spell out the broad objectives intended to be achieved, the strategies, (technical and institutional) that will be adopted as well as the legal provisions that will back these up.

Information System

Whatever strategies that are adopted must be supported by sound statistical and qualitative information. It is therefore essential that information systems be set up at the national and regional levels. Such a system will be backed up with adequate research inputs by experts from differing disciplinary backgrounds.

Monitoring

Although, globally much effort has been directed at monitoring, predicting and warning of the potential impacts of the expected climate change, it is regretable that in this region very little has been done in this direction. In point of fact, it is likely the region still depends largely on weather information from sources external to it for its daily meteorological activities. It is even more disquieting that navy units in the region rely mainly on foreign meteorological services for their sea operations.

There is, therefore, the great and urgent need to update the monitoring capacity of each member state as well as create viable regional bodies with the necessary resources to meaningfully undertake this onerous task.

Prediction and Warning

Alongside monitoring should be the development of predictive capacity and warning systems. Given the continuing rapid growth of population and the consequent pressure on land as well as on other resources, it is absolutely important that citizens in the most vulnerable areas should be warned very much ahead of time of the expected occurrences to enable them to make necessary adjustments and plans to mitigate the impact of such climate changes. While different nations beef up their activities in this regard, regional sharing or exchange of information is very much needed. This will, without doubt, aid in the frugal management of resources.

Strategies

Strategies, both preventive (for example, land use planning) and anticipatory, should be adopted as appropriate in order to mitigate the anticipated impacts of climate change.

Trade and Financial Issues

As severally observed, the West and Central African region is poorly located in the world economic system. Many of the countries are financially strapped. For example, two of its members (Nigeria and Cote D'lvoire) are among the fifteen most heavily indebted countries of the World. They need financial relief and technical assistance to handle their socio-economic problems. Without such assistance, it is hard to see how they can meaningfully and realistically deal with the added problems of global warming and climate change.

CONCLUSION

In this paper, an attempt was made to show that population processes and climate factors share some important relationships. Each set of the factors apparently affects the others. It is also argued that the current rapid population growth rate of the countries of the West and Central African region among other things, accentuates urban problems, makes heavy demands on governments finances and thereby makes desired resource investment and reinvestment difficult goals to attain. Furthermore, the view is presented in such a way that the pressure of population on land largely leaves it impoverished with such consequences as lowered food production and nutrition.

It is further observed that given the rapidly expanding populations, any significant change in climate or rise in sea level would cause disaster on human and material resources of the area. It is thus, suggested that governments should, among other measures, step up their monitoring, predictive and warning capabilities. They should, in addition, form relevant and necessary regional bodies for the management of any crises that might result from anticipated climate change.

Finally, it is suggested that while the countries of this region strengthen their "houses" by providing administrative, legal and political frameworks for preparing for climate change, the International Community should come to their aid with financial and technical assistance.

REFERENCES

1. The Second World Climate Conference. 1990. Held in Geneva. October 29- 7 November.
2. United Nations, 1988. *World Population Prospects.* (New York).
3. Munro, D. 1990. "A Strategy for Sustainability." *Earthwatch* No. 40. P.I.
4. Umar, Y., and D. Zhegu. 1991. "Lightening kills 4 brothers." *National Concord.* Thursday, September 19.
5. Oyelegbin, R. 1991. "Ondo gets N15m to fight Erosion." *The Guardian* Wednesday, September 18.
6. Ibe, A.C., and R.E. Quelennec. 1989. *Methodology for assessment and control of coastal erosion in West and Central Africa.* UNEP. Regional seas Reports and studies No. 107.
7. Tebicke, H.L. 1989. "Climate change, its likely impact on the energy sector in Africa." Remarks presented at IPCC WG II section 5. Lead Authors meeting, Tsukuba, Japan, 18-21 September.

8. Van den Oever, P. 1990. "Leading the Way." *Earthwatch* No. 40 p. 6.
9. Meadows, D. et. al. (eds.). 1974. *The Limits to Growth.* (New York: Universe Books Publishers); Hayes, D. 1977. "Energy for Development: Third World Options." *Worldwatch* Paper No. 15 December; Eckholm, E. 1978. "Disappearing Species: The Social Change." *Worldwatch* Paper No. 22 July and IPCC Report. 1990. *Potential Impacts of Climate Change: Agriculture and Forestry.* Working Group 11.
10. Parry, M. and Z.J. Chen. 1990. "The Impact of Climate Changes on Agriculture." In Second World Climate Conference. *Abstracts of Scientific/ Technical Papers and Reports of IPCC Working Group 11.*
11. Kellogg, W., and R. Schware. 1982. *Climate Change and Society.* (Boulder, Colorado: West View Press, Inc.).
12. United Nations, 1973. *The Determinants and Consequences of Population Trends.* (New York).
13. Kellogg, W., and R. Schware. *Op. cit.*
14. Todaro, M., and J. Stilkind. 1981. *City Bias and Rural Neglect: The Dilemma of Urban Development:* (New York: The Population Council).

Global Climate Change: Implications, Challenges and Mitigation Measures. Edited by S.K. Majumdar, L.S. Kalkstein, B. Yarnal, E.W. Miller, and L.M. Rosenfeld. © 1992, The Pennsylvania Academy of Science.

Chapter Thirty-Four

REGIONAL EFFECTS OF GLOBAL WARMING:
Israel and Eastern Mediterranean Basin

ARIEL COHEN*
Chairman of the Israel National Committee on Global Warming
The Hebrew University of Jerusalem
and
Department of Physics and Atmospheric Science
Drexel University
Philadelphia, PA 19104

INTRODUCTION

When the State of Israel was born, it became known that the prime minister Mr. David Ben-Gurion was looking for a chief scientist with one hand. When asked for the reason of his request, he said: Up until now, whenever I needed the advice of a scientist, the answer I would get was: 'On one hand - On the other hand.' Israel atmospheric scientists, running mesoscale numerical models aimed at predicting expected rise in temperature in the coming 50 years, found themselves forced to agree (in order to avoid accusation of having two hands) on an average increase of 4 degrees. The only disagreement seems to be whether it is going to be measured in degrees Fahrenheit or Centigrade.

One way or another, an increase of 4, or 2 degrees Centigrade, represents a drastic

---

*Parts of this report are based on a summary of the workshop, prepared in collaboration with Prof. M. Magaritz and Dr. M. Graber, a summary of which was approved by the members of the Israel National Committee on Global Warming.

change in all aspects of life. Therefore, in recent years, the effect of future climate change on the environment has been placed at the forefront of scientific as well as political activity worldwide. Awareness of the impact of expected changes on all ecosystems has led both heads of state and presidents of national academies of science to place the issue at the top of their list of priorities.

Israel has formed a National Committee on Regional Effects of Global Warming, including members from universities; research institutes such as the meteorological service, the hydrological service, institute of oceanography & limnology, and the geological service; ministries of environment, science, health, and agriculture; Soreq and the Negev nuclear research centers; industry; public and private companies; and a mayor representing the interests and needs of towns near the sea shore.

In order for the State of Israel to plan its immediate and long term action on this issue, and to enable it to participate in global efforts by acceding to the relevant international conventions and protocols, the Israel National Academy of Sciences via its President and the State of Israel, via the Minister of Environment, decided to organize the first workshop in Israel on the regional implications of these expected changes.

The following is a summary of this workshop, which took place at the Weizmann Institute of Science, Rehovot, Israel from 29 April - 2nd May, 1991, with the participation of several visitors from the United States and Europe and about 250 scientists from Israel. This summary outlines the objectives of the Workshop, and includes an abstract of the findings presented at the Workshop. The recommendations suggested by the national committee on how to enhance activities in Israel related to climate change as well as operational recommendations, are also presented.

OBJECTIVES

The objectives of the workshop were to determine the requirements, the necessity and the importance of climate change research in Israel, by encouraging:
- Evaluation of the reliability of forecasts of climate change in our region as indicated by present global and regional models.
- Joint examination of the appropriate research disciplines by members of the scientific community in Israel, in order to understand all aspects of the expected environmental impacts (i.e., the disciplines of energy, ecology, water resources, air resources, meterology, geology, oceanography, agriculture, economics).
- An intensive search of the existing literature and databases for research studies on past climate changes in our region, in order to draw conclusions regarding future climate change.

ABSTRACT OF THE FINDINGS PRESENTED AT THE WORKSHOP

A. As mentioned above, global models indicate a rise in temperature of 2°C to 4°C in the Eastern Mediterranean Basin, similar to the expected average global rise.

This temperature increase refers to future conditions expected about 50 years from now, when the concentration of greenhouse gases in the atmosphere will be equivalent to double the present concentrations of CO_2.

B. Although global models show an increase (mainly in summer) in average global precipitation, this increase has a very high local variability. Thus, different global models predict a different distribution between summer and winter precipitation. Some models predict an increase, whereas others predict a decrease in the precipitation expected in the future in the Mediterranean Basin.

C. Local meso-meteorological models indicate a possibility of an increase in future precipitation in Israel.

D. It was emphasized that the rise in temperature is followed by an increase in evaporation from the ground, independent of the type of ground cover. Thus, a decrease in the total quantity of water charging the groundwater aquifers is to be expected despite the possible increase in future precipitation.

E. On the national level, there are two possible strategies that can be adopted in order to reduce the risks and damages that might occur:

1) The European approach, led by Holland according to which work on a national scale should begin immediately (in effect it has already begun) in order to minimize anticipated adverse effects. This approach incorporates plans that are expected to contribute to the national economy as well as taking future threats into account at the present time. For example, Holland has already launched projects to protect her sea front by expanding into the sea, which are based on the expectation of a higher sea level in the future compared to the existing sea level of today.

2) The American approach emphasizes the need for additional research before taking any actions which have economic implications. This approach has led to a significant increase in resources allocated in the U.S.A. to climate research in comparison to other scientific research fields.

F. Compared to other countries, there is an abundance of data and findings in Israel which can sustain detailed research on past climate changes, in addition to changes in the environment and water resources. The past covers various time scales, ranging from a few decades to thousands of years ago. This type of data should enable researchers to calibrate and verify regional climate models, and thus global climate models as well.

G. Human activities during past decades, which have led to alterations of the landscape and ground cover (i.e., natural vegetation, agriculture, forests) have caused, on the one hand, changes in the reflectivity of the ground to solar radiation and in local precipitation; and on the other, in evaporation from the ground. These changes have affected the national water balance. It appears that efforts to cover arid or semi-arid areas with vegetation led to an increase in the water balance, whereas planting trees in areas that enrich aquifers led to a decrease in the water balance.

H. Studies on the cloud transformation from marine to continental types over the Mediterranean were presented, and implications of the findings on Israel's

national effort to increase precipitation by cloud seeding were reviewed.
I. Reported recent temperature changes in our region (including the regions of the Black Sea, the Adriatic Sea and the Eastern Mediterranean Basin) indicate an opposing trend to that reported globally. Thus, in the 1970's, a significant decrease in the annual average temperature was observed in our region, in contrast to an increase in the annual temperature averaged over the globe.
J. It was emphasized that conclusions drawn from climate data obtained from areas that have undergone urbanization should be assessed with caution.
K. The fauna and flora in Israel are sensitive to climate change but appear to be able to adapt to it. It was suggested that global changes in flora and fauna should be monitored as an indicator for changes in the water balance and climatic conditions.

RECOMMENDATIONS

A. Short- and long-term research projects involving scientists from different disciplines should be enhanced and encouraged in order to establish a data base in Israel that will serve as input to the various climatic models. This recommendation refers to research which provides data on climatic change that have occurred in the past, are occurring now and will occur in the future.
B. Establishing a detailed research program on the local and regional water cycle is of vital importance to Israel. Under this heading, the following research fields should be included: cloud development; the impact of ground cover changes on the albedo on the one hand, and on evaporation on the other, and their impact on the recharge of groundwater aquifers in Israel. These issues should be studied in view of future changes expected in the water level of inland water bodies, and in view of the future rise in the Mediterranean Sea—estimated today to be between 8 and 30 cm in the year 2030.
C. The following projects should be initiated:
   a. An in-depth study of the impact of land use and land use intensity on the national water balance and energy consumption.
   b. An in-depth study of the impact of afforestation on the national water balance.
   c. Studies of the expected impact of sea level rise on the shoreline and beaches of Israel.
   d. A study of the impact of climate change on the quality of life in Israel, taking into account the required changes in the near future as result of accession to international environmental conventions (i.e., the impact on the economy, energy demand, type of energy, water demand).
   e. A study of the expected impacts on agriculture as a result of the possible decrease in available water quantity and the increase of CO_2 concentration in the air.

f. A study of the changes in the structure and function of natural ecological systems, nature reserves and open spaces, and the limits of expansion of plants and animals to be caused by the expected global changes.
g. Incorporation and promotion of the subject within the educational curriculum.

OPERATIONAL RECOMMENDATIONS

A. To re-evaluate existing priorities, giving priority to interdisciplinary environmental research and to recognize environmental research as an independent field similar to physics and chemistry.
A1. To allocate suitable resources to such projects proportional to the funds allocated to interdisciplinary environmental research projects by, for example, the USNSF and the EC.
B. To increase the awareness of the various government ministries in Israel to the above mentioned requirements, and place at the disposal of the relevant government ministries funds, the size of which, compared to other high priority research projects (such as space research), will be adjusted to the proportions common in the world and in the U.S. government.
C. To establish in the relevant bodies (especially in the Meteorological Service, Hydrological Service, and the Israel Oceanographic and Limnological Research Institute) a base of historical data, and of current measurements as well, of the meteorological, hydrological and oceanographic variables that directly affect the regional aspects of climate change.

The Hebrew University of Jerusalem has been conducting research in different aspects of Atmospheric Sciences for the past several decades, playing a major role in the development of the research effort in atmospheric dynamics, and atmospheric physics, all over the country.

The study of future changes in the atmosphere and their implications to Israel, requires the enhancement of the potential capabilities of remotely measuring atmospheric parameters from satellites and from the ground. The atmospheric varying concentrations of desert dust particles are among the most important unknown parameters in the future calculations of the Earth's radiation balance. The remote determination of such number densities can improve the forecasts for the future warming estimates discussed above. The remote sensing research group at the Hebrew University Department of Atmospheric Sciences has been engaged in such efforts for several years, i.e. the development of the laser radar field for ground base, airborne, and satellite borne measurements of the atmospheric particulate concentrations in various altitudes. The need for accurate scattering theories for the analysis of such remote measurements, has created an on going collaboration between scientists from Jerusalem and scientists from Pennsylvania (Drexel University). It will be the result of such international collaborations that the world will be able to decide on the appropriate measures in order to protect itself from any future global change.

Global Climate Change: Implications, Challenges and Mitigation Measures. Edited by S.K. Majumdar, L.S. Kalkstein, B. Yarnal, E.W. Miller, and L.M. Rosenfeld. © 1992, The Pennsylvania Academy of Science.

Chapter Thirty-Five

GLOBAL CLIMATE CHANGE:
The California Perspective

JAMES M. STROCK
Secretary for Environmental Protection
California Environmental Protection Agency
Air Resources Board
2020 L Street, P.O. Box 2815
Sacramento, CA 95812

INTRODUCTION
CALIFORNIA'S ROLE IN GLOBAL CLIMATE CHANGE

It may seem odd that a single state, representing only a small fraction of only one nation, sees a major role for itself in the resolution of an environmental problem so complex and so truly international as the threat of global climate change. Global climate change is an issue that must be dealt with by international efforts; what possible effect can a small entity like California have?

The answer is "quite a lot!"

California may be physically small, but we are a diverse and dynamic society, a major player on the world economic scene, and, because of our state's physical and economic diversity, a microcosm of the industrialized world. By planning for our own future, we can make a significant contribution to the development of new technologies and new lifestyles that could be adapted to many different settings and have impacts far beyond California.

CALIFORNIA IN NATIONAL AND GLOBAL PERSPECTIVES

In order to understand how California can influence international trends, it helps to keep in mind how the state fits into the national and international scene. Our importance and visibility are not based on physical size, but on our economic size and our global visibility.

Size

In area California is small, when viewed in a global perspective, about the same size as Japan. Our borders encompass roughly 160,000 square miles (410,000 sq. km.), less than 5 percent of the land area of the United States, less than 2 percent of North America, and about 1/4 of one percent of the Earth's land area.

In population, California is a little more significant. At about 30 million, we're the most populous state in the U.S. (about 12 percent of the total), and a little larger than Canada. On a global scale we are less than 3 percent of the developed world, and only about 1/2 of one percent of the world's population.

Economy

But, it is our economic strength and role in world trade that make California's global mark; if California were a separate country, here's how it would be characterized (Fay & Fay, 1991):

— California's gross annual product, about $3/4 trillion, would rank us seventh largest among the nations of the world.
— Our exports rank us eleventh among the nations of the world, and imports eighth. The Los Angeles-Long Beach port complex accounts for 19 percent of U.S. port activity; San Francisco-Oakland another 10 percent.
— Per capita productivity, at about $25,000, would be fourth highest in the world.

And California's economy is very diverse:

California is highly urbanized (over 90 percent), but we are also a major producer of basic resources, such as lumber, petroleum, rice, and cotton, much of this for world export.

California agriculture produces over 200 commercial crops; although we have only 3-1/2 percent of U.S. farms, we produce over $17 billion annually, more than 11 percent of U.S. farm production (CDFA, 1988).

At $14 billion per year, California manufacturing is about 13 percent of the U.S. total (Fay & Fay, 1991).

Energy

All this economic power is literally fueled by a voracious appetite for energy, about 6-1/2 quadrillion BTUs per year (CEC, 1987). California has a diverse energy base, including petroleum, nuclear, natural gas, coal, geothermal, hydroelectric, solar and wind power, but we also own the most cars and drive the most miles per capita of any state in the U.S. Obviously, anything that influences energy costs or supplies has the potential to seriously affect our economy and our future.

We Californians are already acutely aware of our energy dependence, and we have been working for the last two decades both to diversify our energy base and to increase our energy efficiency. As a result, we can claim higher energy efficiency than most of the U.S.; with our annual usage of roughly 225 million BTU per capita we are about 30 percent below the U.S. average. However, we are still far behind most of the developed world (*e.g.*, we use about 50 percent more energy per capita than western Europeans) (CEC, 1990a).

International Stature

California is a very special place. In the early 19th century, California was a sparsely populated outpost of Spanish colonial Mexico, but the discovery of gold and the rush it spawned transformed a remote frontier into a world-famous destination for immigrants, adventurers, and visionaries. The history of California is peppered with grand projects from the laying of the first transcontinental railroad to the building of the greatest irrigation systems in the world; settlers came with dreams of wealth, empire, and utopian communities.

While the frontier has gone, the vitality, the special visibility, and the global attraction to immigration have not. In some ways, California is still thought of as a land apart; for many, we're the object of jokes about our "new age" mentality; we're the land of Hollywood, of California cuisine, and the "California life style." But we're also the home of Silicon Valley and its burgeoning computer industry, we're a major center of aerospace technology, scholars from every state and nearly every foreign land compete to enter our universities, we're still the preferred destination of hundreds of thousands, tourists and immigrants alike, from all over the world, and our innovative environmental protection programs are being imitated in the majority of the world's industrialized nations.

This special international recognition gives California a leadership role; innovation here is often the seed of national and global trends. Coupled with our economic clout as a major producer and a major market for global commerce, we have influence on global economic and social trends that far exceeds our modest size.

CALIFORNIA'S CONTRIBUTION TO GLOBAL CLIMATE CHANGE

Like the rest of the industrialized world, California is contributing to the atmospheric buildup of greenhouse gases by burning fossil fuels. California currently emits about 310 million metric tons of carbon dioxide (CO_2) annually, about the same as France. This is about 6-1/2 percent of U.S. emissions, or 1-1/2 percent of global CO_2 production (U.S. emissions are 23 percent of global emissions) (CEC, 1989). California's CO_2 emissions are much more globally significant than those for methane (CH_4) production, which are about 2.3 million metric tons per year or about 0.03 percent of global CH_4 emissions (6.7 billion tons per year) (NAS, 1991).

At roughly 11 tons per person per year, Californians produce a little more than half the CO_2 of the typical resident of the other 49 states, owing primarily to our minimal coal use and our large reliance on natural gas; but even at this level, we are no better than Germany (11 tons per person) or England (10.3), and well behind Japan (7.7) (CEC, 1991).

What are the prospects for significantly improving California's energy efficiency still further?

Energy Production and Use

California's economy is strongly influenced by current and historical patterns of energy production. Early development of hydro-electric power and exploitation of local oil deposits set the tone (the state still produces about half the oil it consumes), followed by aggressive development of western natural gas supplies.

California's major end use energy budget breaks down as follows (CEC, 1987):

Use Type	Percent Supply by Sector				Percent of Total
	Petroleum	Gas	Electric	Coal	
Transportation	99+				50
Residential	3	68	29*		12
Commercial & Residential	41	36	22*	1	38

*includes all generation modes

California's electric generating capacity is distributed over a mixed resource base (CEC, 1987):

SOURCE	PERCENT
Natural Gas	24
Coal*	22
Nuclear	20
Geothermal	15
Hydropower	14
Biomass	4
Wind	0.3
Solar	0.3
Distillate Oil	0.1

*Generated almost entirely outside California.

Our fossil fuel dependence is obvious, although unlike much of the industrial world, we are relatively less dependent on coal. Our current emphasis on petroleum for transportation and industry makes us relatively vulnerable to its pronounced variability of price. This is somewhat balanced by our use of domestic natural gas by utilities, residences, and industry, but this fuel, too, is likely to increase in price over the next few decades.

The message is becoming clear that the old economics which shaped our energy system are not the best mix for today, and pose considerable economic risks in the future. Over the next several decades, we will have to revolutionize our energy economy. We see this as an opportunity to re-think the basis of our energy supply system, and an ideal time to include consideration of not only our own local interests, but global environmental and economic concerns as well.

POTENTIAL IMPACTS OF GLOBAL CLIMATE CHANGE ON CALIFORNIA

The global or continental scale of most published research on climate change makes it very difficult for us to get a clear picture of the effects it could have on any particular region. In order to understand the potential risks to California from global warming, the California Energy Commission (CEC, 1989, 1991) has recently prepared an analysis of some of the potential impacts of global climate change on California.

Global climate changes could affect water supplies, energy, agriculture, forestry, natural habitat, outdoor recreation, air quality, public health, the stability of our bays and beaches, and the entire economy of California. The more significant potential impacts of global climate change on California, based on an assumed 3°C temperature rise, are discussed below. There is simply not yet sufficient understanding of the earth's climate system to know exactly how or when the impacts will occur, or, indeed, if they are ever to occur.

WATER

California is perhaps uniquely dependent on storing vast amounts of water and moving it long distances to meet our demands. Two-thirds of the state's rain and snow falls on the northern one-third of the state. Statewide approximately 80 percent of the annual precipitation occurs in the five months November through March. Consumer demand is at a minimum during the rainy winter season, and at a maximum during the dry summer season. Thus, collecting and storing the winter precipitation and later transporting it to where it is needed are critical to the life and livelihood of Californians (Kahrl, 1978).

Even if the amount of precipitation does not change, global warming may effectively diminish surface water supplies in California. Higher winter temperatures could cause more precipitation to fall as rain and less as snow (Gleick, 1988). Unlike the historical pattern, less water would be retained in the snowpack, and the winter rainfall would run off immediately, filling downstream reservoirs earlier in the season. Because of flood-control requirements, a substantial portion of the early runoff could not be stored. In spring, when the reservoirs ordinarily fill, reduced snowmelt runoff would result in less water to store for summer and fall use. In addition, projected sea level rise may require a greater volume of releases from upstream reservoirs to repel salt water intruding into the Sacramento/San Joaquin River Delta from which much of California's urban and irrigation water is drawn (USEPA, 1988; Gleick, 1988).

Higher average temperatures would increase agricultural and urban demands for water through increased evaporation and irrigation demand (Gleick, 1988). Winter flooding might be more frequent and cover a greater area because of increased winter rainfall (as opposed to snow) and the attendant runoff, coastal lowlands may be more vulnerable due to rising sea levels.

In addition, global warming may magnify water quality problems by reducing spring and summer flow in streams and rivers and limiting their ability to dilute pollutant loading.

AGRICULTURE

Farming and associated industries add some $93 billion to the California economy (over 15 percent of the total). California's agricultural operations cover 30.6 million acres, about 30 percent of our total land area. The State's 9.3 million irrigated acres contribute the overwhelming majority of California's farm production.

Farming, of course, is vulnerable to many environmental factors. Global warming would offer farmers a mixed bag; there could be both negative and positive effects. Possibly the most serious problem warming may present to farmers is an increase in weather variability. An increase in summer temperatures may make some crops untenable in their current growing areas. At the same time, higher temperatures may increase the need for irrigation.

On the other hand, farmers might benefit from a greenhouse-induced warming. Increased levels of CO_2 may increase plant growth and the efficiency with which many crops use water; and, while warming will bring higher summer temperatures, it will also bring a longer growing season.

Historically, farmers have shown an ability to adapt to a changing climate. Because of our vast distribution system, agriculture in California is perhaps more flexible than in many other parts of the world. The crucial question is whether California agriculture can accommodate more total climate change in the next 50 years than farmers have faced since the dawn of agricultural societies (CEC, 1991).

OCEAN

Sea level rise due to global warming is expected to occur as a result of thermal expansion of the ocean waters and melting of the earth's glaciers and polar ice fields. The current rate-of-rise would cause ocean levels in the San Francisco Bay Area to rise by 10 to 13 cm (4 to 5 inches) in the next 50 years. A recent study by the Environmental Protection Agency estimates that if temperatures rise 3°C (5.4°F) by the year 2050, a one-meter sea level rise could result by the year 2100 (USEPA, 1988).

Rising sea levels could affect California's freshwater supply, exacerbate flooding danger and cause more severe storm damage and coastal erosion.

Over half of the freshwater supply serving central and southern California is tranferred through the Sacramento-San Joaquin River Delta (DWR, 1983). Progressively lower summer flows due to global warming, as well as rising sea levels, may exacerbate the problem of salt water intrusion. It is estimated that, without other protective measures, a one-meter (3.28 foot) rise in sea level would require doubling reservoir releases to repel the intruding salt water from the river deltas (USEPA, 1988; Gleick 1988).

Coastal erosion may also increase due to sea level rise. Higher seas provide a higher base for storm surges and increase the potential for more destructive storms. In addition, coastal erosion and shoreline retreat could leave hundreds of hazardous waste sites and landfills under water or exposed to waves (Flynn, 1986). Rising sea level would also cause contamination of coastal fresh water-aquifers as increased hydrostatic pressures caused sea water to migrate inland.

NATURAL HABITATS

California supports over 250 plant and animal species which are listed as threatened and endangered, and approximately 700 species that are candidates for listing. Warmer temperatures, sea-level rise and changes in water availability could result in substantial impacts to endangered species; major changes in coastal and interior wetlands could endanger the species that occupy these habitats.

AIR QUALITY

The present costs of air pollution to California are substantial, affecting human health, natural ecosystems, agriculture, materials and visibility (Rowe, et al., 1986). Changes in temperature, atmospheric ventilation, solar radiation and precipitation may affect air quality in California in both adverse and positive ways. However, the magnitude of these possible effects is unclear.

Higher temperatures and increased ultraviolet radiation (due to stratospheric ozone depletion) will likely increase summer ozone in urban areas. Change to local wind patterns could aggravate pollution problems, or they could help flush pollutants from urban areas. On the plus side, shorter, warmer winters may moderate problems with carbon monoxide in our cities. An increase in low cloud cover in the summer could decrease ultraviolet radiation and thus reduce ozone levels in urban areas.

HUMAN HEALTH

Global warming of 3°C would likely result in both increased morbidity (prevalence of illness) and mortality among California citizens. The elderly and the very young would be the most severely affected. Most of the adult deaths would result from exacerbation of coronary heart disease and cerebrovascular disease primarily in persons over 65 years of age, as a result of increased heat stress.

Global warming may also indirectly lead to an increase in the number of premature births and perinatal deaths (deaths occurring before, during, or just after birth). Increases in the number of preterm births and perinatal deaths are generally associated with the warmer summer months.

In addition, global warming could result in conditions more favorable to the spread of vectors such as mosquitoes and ticks. This could lead to reintroduction or increased incidence of disease such as malaria, dengue fever and encephalitis.

CALIFORNIA'S RESPONSE TO GLOBAL CLIMATE CHANGE
ENERGY CONSERVATION AND CLIMATE CHANGE —
A 'NO REGRETS' POLICY

As with most of the industrialized world today, we have historically optimized the economic efficiency of our energy supply, rather than the thermodynamic efficiency of our energy use. Over the last century that was a reasonable policy, since our population was modest and we had abundant energy resources in California and the West. As we enter the next century, the tables are turning; we find our burgeoning population and economy beginning to feel the pinch of limited natural resources and limited waste disposal capacity. Unlimited growth in supplies at consistently low prices cannot be sustained; our future economic expansion must be increasingly fueled by creating innovative ways of supplying and using energy that will reduce the resource inputs per unit of economic output.

Our interests in increasing energy efficiency are not rooted in any particular moral or ethical ideal. They emerge naturally from a pragmatic recognition that all aspects of California's economy are dependent on having abundant supplies of basic inputs and services such as labor, energy, materials, transportation, and waste disposal. Guaranteeing these supplies takes planning and careful allocation of resources (for example, labor supplies depend on housing, transportation, education and public health services). Planning for the future is already a routine function of government; the challenge now is to redirect these existing processes to respond to a new concern, the fact that fossil fuel combustion has long-term global consequences.

To the extent that energy conservation is in our near-term economic interest, the side benefit of reducing greenhouse gases makes energy conservation a "no regrets" policy. The actual CO_2 emission reductions achieved by such a policy will not be large at first, however any reduction is helpful, and the earlier reductions are begun, the greater the overall benefit for that level of effort. We expect that we can at least shift from increasing CO_2 emissions to stabilizing or perhaps decreasing emissions with this policy. The experience gained in energy management, combined with the capital base and market momentum given to energy-saving products and technologies, will provide both a new growth industry for California and a technology base on which to draw for further CO_2 reductions, should they become necessary.

The "no regrets" approach allows us to confront the global climate change threat even though our scientific understanding of climate change is not sufficiently refined to accurately predict the magnitude of the problem; if we ultimately discover that CO_2 emissions aren't a big problem, we retain the many economic advantages of more efficient energy use. If we discover that drastic CO_2 reductions are needed, we will have already established a considerable technology base to apply to the problem.

In a later section, I discuss how we have already begun to implement global change consideration into environmental planning processes in California.

CALIFORNIA'S POTENTIAL FOR CHANGE

If we are to pursue energy conservation, we must understand how much change is possible within the existing economic and social structure, and set appropriate goals. I am convinced that considerable savings can be achieved without unacceptable social or economic dislocations. To explain why I believe this, the following sections briefly review examples of how energy efficiency can be improved in various sectors of California's economy.

Industry

In order to give a picture of how practical it is to expect fundamental changes in energy efficiency, let's take as examples four of California's largest industries: petroleum and chemicals, aerospace, electronics and computers, and food processing.

The chemical and petroleum industries, while generally capital intensive and having some of the oldest industrial facilities in the state, have historically chosen periodically to upgrade various portions of their facilities to meet changing market demands, new environmental regulations, or to remain cost-competitive as new processes enter the market. In many cases, major retrofits are economically supportable due to relatively high product value. In other cases, reconstruction to improve energy efficiency or reduce pollution can have significant economic benefits.

Producing and marketing transportation fuel is a very large sector of our economy, so there is concern that rapid advances in energy conservation in the transportation sector could reduce cash availability for modernization; on the other hand, siting for significant expansion of refining facilities is difficult to accomplish due to many factors, so phased programs to moderate demand can extend the life of existing facilities, making refitting more economically attractive.

Aerospace manufacturing is a unique combination of high value design, tooling, and component fabrication with space- and labor-intensive final assembly. Overall, it is not particularly energy intensive, but certain of its manufacturing processes may be. For example, energy economies associated with environmental management of very large buildings can be very cost effective. Equipment modernization is not generally a problem in aerospace operations since products and processes are continually evolving.

The aerospace industry may contribute its greatest energy savings through redesigning their products for greater in-use energy efficiency, since the energy invested in building aircraft is usually quite small compared to their lifetime fuel consumption. Strong trends in this direction already exist due to current market forces in air transport, and to a lesser degree in military aircraft.

Aerospace technologies also offers considerable opportunity for technology transfer to other sectors of the economy. For example, thermal insulation, weight-saving design and materials, and other advances pioneered in aerospace applications can improve energy efficiency for a wide range of products and processes.

Electronics and computer manufacturing is generally thought of as a "clean" industry, and for the assembly and software development aspects of the business, this is quite true. However, component manufacture is somewhat different, employing considerable inputs of electricity, water, lithographic chemicals, cleaning solvents, plastic resins, etc. In this industry, direct savings in energy consumption can be made, but other significant environmental benefits, including energy savings, will probably come from new processes.

Process revision is cyclical in this industry, as new products are introduced very frequently. Manufacturers will continue to seek new processes for economic reasons, among which are the need to increase yield (the ratio of working microchips to total output coming from a production line), the need to make smaller, more complex devices, and the extreme competitive pressure in this international business. In addition, global cessation of use of chlorofluorocarbon solvents, which cause stratospheric ozone depletion, as well as being greenhouse gases themselves, is driving a current round of process revision.

Food processing looms large in California's economy, serving both the state's agricultural production and its urban consumer market. At first look, it does not appear to be an area ripe for significant energy savings, since the energy demands of cooking, freezing, etc. are relatively fixed. Upon closer examination, there are substantial opportunities for both direct energy savings and diversion away from fossil fuel energy sources. Direct energy demand reduction has come from several areas, such as improved plant design, better refrigeration systems, and use of cogeneration to produce both heat and electricity from a single boiler.

Biomass-fired boilers and cogeneration systems using agricultural wastes for fuel have already come into use in California (for example, in an almond cannery powered by steam generated from burning the almond hulls). This type of enhanced utilization of agricultural products is especially desirable, since it not only displaces fossil fuels, but reduces air pollution caused by burning these wastes in the field.

These four industries represent only a small fraction of the total range of energy-consuming technologies in use in California, but the general pattern of potential for significant improvement in energy consumption and decreased greenhouse gas emissions applies broadly. While dramatic energy savings cannot be expected in all industries, we are confident that significant energy conservation can be achieved throughout the industrial sector.

Housing and Consumer Products

Energy conservation or other environmental management programs that impact industry, utilities, or other elements of the state's infrastructure are largely invisible to the general public, and therefore relatively easy to implement. Unfortunately, the "invisible" changes discussed in the preceding section are not, of themselves, going to generate the degree of energy conservation necessary to achieve large energy savings or significant reduction of CO_2 emissions. Residential and transportation energy consumption combined represent more than 60 percent of California's energy demand and more than 80 percent of fossil fuel use. This forces any comprehensive energy conservation programs to focus on personal energy consumption.

California has already taken a number of steps to control designs of consumer products, add energy efficiency elements to building codes, and to promote adoption of energy-saving technologies such as solar water heating and waste heat recovery systems; we anticipate further progress in these areas (CEC, 1991).

Personal Energy Consumption

The one area we haven't yet addressed is discretionary personal energy consumption, such as how warm we keep our home, or what kind of car we drive, and how much we drive it.

Personal energy consumption is potentially a very contentious subject, striking at the heart of individual lifestyles, self-image, and even identity. In business and government, the issues surrounding adoption of environmental regulations usually focus on engineering and economics; in dealing with personal behavior, we must, in addition, confront potential conflicts of cultural values, perceptions of personal freedom, and even civil rights. We cannot ignore these issues, but we must not be intimidated by them, either.

To the greatest possible degree, these programs should be based on encouraging personal choices that serve general societal interests, rather than on strict regulation; in this way people can balance their individual desires for certain kinds of activities against economic or other incentives or disincentives.

Transportation

The largest element of personal energy use in California, by far, is transportation, and that is dominated by use of gasoline in private cars. Transportation accounts for about 86 percent of the state's petroleum consumption. In terms of statewide CO_2 emissions, transportation produces 58 percent of the total (43 percent directly, and 15 percent from fuel production and distribution) (CEC, 1990b).

California Transportation Fuel Consumption

MODE	PERCENT
Autos - Light Trucks	52
Heavy Trucks	16
Aircraft	15
Ships	14
Rail	2
Public Transit	1

With about 22 million motor vehicles and 160,000 miles of public roads, we have not only an enormous investment in auto transport, but also face great inertia if we attempt to change the system too rapidly. The challenge then is to make the existing system more efficient, not only in terms of fuel consumption, but in the equally valuable measures of time spent in transit, and passenger and freight capacity.

There is much wasted capacity in our transportation system — cars with only a single occupant, buses less than half full, "deadhead" runs to return passenger and freight vehicles to loading points, etc. Of course, some of this capacity is unavailable for other uses, such as tank trucks returning to lading points, or empty seats in cars used for family vacations, but there is much "free" additional capacity available that we should exploit.

Most near-term progress in improving transportation and energy efficiency will come from efforts to manage existing capacity. Many of these are relatively low cost and "low tech". Private passenger vehicle capacity utilization can be greatly increased by carpooling and vanpooling, designating special lanes for high occupancy vehicles, and manipulation of parking supply or cost (*e.g.,* limited free parking at destinations associated with severe congestion). Roadway capacity improvement can come from vehicle restrictions like rush-hour bans on heavy trucks, flow management by restricting or metering access, and demand spreading (*e.g.,* staggered work hours).

In the intermediate term (5 to 10 years) the vehicle fleet itself can be significantly changed by regulatory requirements and/or economic incentives. These might include tighter fuel-efficiency standards, fuel taxes, weight- or fuel economy-based registration fees, road use taxes, or other incentive/disincentive measures now being discussed.

With longer term programs (10 to 50 years), we can change some of the fundamental causes of high vehicle usage; by extension and expansion of public transportation lines; encouraging land use practices that promote access to public transit; and guiding new development to reverse the current trend of increasing commute distances and increasing numbers of vehicle trips per person.

POLICY FRAMEWORK FOR UNILATERAL AND COOPERATIVE ACTION IN CALIFORNIA

In confronting global climate change, we may have to go well beyond the scope of the "no regrets" energy conservation policies discussed above. Should that occur, local economic shifts and reorganization, played out against a background of changing inter-regional economic relationships, will pose major challenges for policymakers.

GLOBAL VS. LOCAL SCALES OF ACTION

Undoubtedly, some local entities (*i.e.,* individual states or countries) will elect to act unilaterally, either to assert rights to local control within a framework of global cooperation, or to move more quickly than regional or global programs. Having taken that course in previous environmental management programs, particularly air pollution control, California's experience leads us to offer some general observations about the relationship between local and universal actions.

Problem Solving vs. Problem Displacement

Local actions to address greenhouse emissions must be considered not only in the context of local goals, but in the larger framework of global controls.

Consider a hypothetical state regulation requiring industries to reduce CO_2 emissions by half within five years. Even if such a large reduction over such a short time were technically feasible, it would impose an intolerable economic burden because it only applies to a few producers, leaving them at a competitive disadvantage compared to industries located elsewhere. Rather than reduce emissions by costly plant modifications, most owners would prefer to relocate some or all of their production, thus acheiving the local jurisdiction's CO_2 reduction, but by transferring it out of state, rather than actually eliminating those emissions.

The Linkage Between Local and Global Actions

The point of the foregoing example is that local and global programs must be linked, so that regulatory actions both reduce global emissions *and* promote orderly economic restructuring. Unilateral actions must be undertaken in the context of exerting leadership or demonstrating new technologies, and always with a view to influencing a global effort.

If a global greenhouse gas emission control program is established, it will necessarily be quite different from state and national emission controls as we know them. The diversity in economic and social structures among major CO_2 and CH_4 producing nations, coupled with the disparities in levels of development and resource bases, makes it impossible to device a "one-size-fits-all" approach to controls. It seems more likely that the nations of the world would establish broad general goals that would be implemented by regulations appropriate to each nation that would be implemented by nationally autonomous regulations (basically an elaboration of the current international agreements on CFCs and, in Europe, CO_2).

And let's not forget that a goal-oriented international program would create interesting opportunities for development and marketing of efficient greenhouse industrial technologies, transportation systems, housing designs, and agricultural systems (both new cultivars and growing methods). These are fields in which California has already started to develop next-generation technologies. Government can provide a nurturing environment for these efforts, and use their power to open new markets necessary for moving new ideas from the laboratory to commercial development.

THE LIMITS OF REGULATORY POWER

Regulatory power is often viewed as a negative—the power of the state to restrict behavior. Admittedly, it can be a clumsy tool to promote change. Wise application of regulatory authority requires sensitivity to the sometimes unpredictable nature of social and economic behavior. Simple-minded prohibitions often backfire, leading to substitution of one undesirable activity for another, or merely displacing it to another jurisdiction. The challenge of environmental regulation is to devise a system that minimizes environmental degradation without stifling technological and social change.

As we move to control CO_2 and CH_4 emissions, which have no direct, local impacts, and which are, of themselves, not a threat to public health, it will be preferable to acknowledge human and economic diversity, and with it the idea that managing the environment means guiding *most* behavior; only rarely does it necessitate 100 percent bans.

This principle is exemplified in our current regulatory systems by such concepts as the "bubble" for industrial air pollutant emissions, corporate average fuel economy (CAFE) rules, and setting lower limits on the size of facilities subject to air pollution permit requirements. Such rules do allow for some high-emission industrial processes, some "gas guzzlers," and some uncontrolled pollutant sources, but they succeed in controlling *most* of the problem without imposing unmanageable burdens on the regulated parties. One must recognize that there are increasing costs and returns in the effort to implement controls on 100 percent of sources.

This is not to say we will forego vigorous enforcement of environmental regulations, rather that we should concentrate on creating a system that is largely self-enforcing; one in which we have translated the needs of the public good into regulations that make individual goals parallel to our societal goals. When viewed in this context, the combination of economic incentives (or disincentives, such as fees or taxes) and generic performance goals appear to provide a means to steer economic development in environmentally beneficial directions.

LEADERSHIP, INNOVATION, AND TECHNOLOGIC AND SOCIAL CHANGE

Environmental regulation does not occur in a vacuum, but in the larger context of government activity including provision of basic services, promotion of economic welfare, law enforcement, etc., and all this is only a part of the larger economic and social life of the community or the nation.

An important element of government's ability to steer economic and social change is found not in regulation, but in the power of the government to persuade, and to lead by example. In confronting the social, economic, and technologic changes that will occur in the next several decades, as we enter a period of reduced per capita energy consumption, government has a responsibility to use its authority not only to limit undesirable behavior (specifically "excess" energy use), but to conform its own activities to be the model for the future.

Government can support significant research and development programs that may be too speculative or too long-term for private funding; it can participate in technology demonstrations to overcome skepticism facing new ideas; or it can use its considerable consumption to create demand for new technologies, thus making private investment in innovative technologies more attractive.

Government alone has the ability to openly manipulate the marketplace, through tax policy, technology specification, and other means. In this way, too, change can be accelerated.

Government, alone or in partnership with private individuals or organizations, can marshall the resources of the mass communication media to direct public attention to a problem, or even change public opinion (the current efforts in regard to drugs and tobacco show how far public opinion can turn with the benefit of sustained public awareness programs).

Given the great economic changes we can expect in the next half century, these may not be discretionary actions, but essential government participation in preserving and enhancing the economy.

MODELS FROM AIR POLLUTION CONTROL IN CALIFORNIA

The general concepts discussed above may sound radical, but they are largely routine in government today, not just in California, but throughout the developed world. If California is in any way unique, it is in the manner we have employed these tools of economic and technologic management to meet our environmental problems. In the next few paragraphs, I offer some examples from our recent past as proof that we can play the role we have defined for ourselves.

Catalytic Converters — Forcing Technological Innovation

California has long been an international leader in automotive emission controls, beginning with the first new-car emission standards for 1965 model cars. As these regulations have evolved over the last quarter century, they have on several occasions required auto manufacturers, within a few years, to produce vehicles that are significantly cleaner than was possible with existing technology. This type of regulation, requiring new designs that have not yet been demonstrated in commercial applications, we call "technology forcing".

One of our most significant technology forcing regulations took effect in 1975, with the advent of oxidation catalysts in auto exhaust systems. At the time that regulation was adopted, many people, both in the auto industry and outside it, protested that no economically viable catalyst system could be built, and there was talk by some of withdrawing from the California market if these "unrealistic" regulations weren't withdrawn. As we all know, the exhaust catalyst has since become ubiquitous in automotive emission control systems.

There are two fundamental reasons for the success of these regulations: first, they were based on sound engineering evaluations which indicated that it was highly likely that existing laboratory technologies could be brought to market in the appropriate timeframe; and second, the competitive nature of the automobile business made it unwise for any company to forego investing in redesign in the hopes that the state would relent; California sales were too important to risk losing market share to a manufacturer who elected to meet the standards.

In considering the potential for developing "climate-friendly" technologies, we hope that we can primarily rely on market forces and economic incentives, but if sufficient progress is not being made, we have the ability to change the nature of the market by regulatory actions, thus forcing technological change that might not otherwise appear for many years.

Low-Emission and Zero-Emission Vehicles

In September of 1990, the California Air Resources Board adopted significant new vehicle emission standards (to take effect over the period 1994-1998) that allow for manufacturer fleet averaging of emissions, with specific provisions for using various types of low-emission vehicles, and, eventually zero-emission vehicles, to compensate for emissions from continued use of more conventional engine-fuel combinations. Beginning in 1998, manufacturers will be required to produce zero-emission vehicles for California (2 percent of sales in 1998 and 10 percent in 2003).

This is a somewhat different kind of technology forcing regulation, in that it gives the manufacturer some flexibility to choose the technologies and the optimal mix of them. From this we expect to see not only a limited number of zero-emission vehicles (probably electric) brought to market, but also cleaner-burning fuels, such as methanol and natural gas, and significantly improved internal-combustion engine vehicles even in the dirty end of the permissible range.

Moreover, the new low-emission and zero-emission vehicles will start to bridge the gap to future generation vehicles that are significantly more energy efficient, and that can wean us from our singular dependence on petroleum as the source of our transportation energy.

These new regulations demonstrate how existing programs can yield significant benefits in planning for a less petroleum-dependent future well *before* petroleum supply restrictions or a global warming crisis would force such changes.

The South Coast Air Quality Management Plan —
Integrating Energy and Environmental Planning

Los Angeles is almost as famous for its smog as for its sunshine, beaches, and Hollywood. The reputation is not unjustified. The air pollution problem in southern California is one of the greatest environmental challenges California has faced. The challenge is to find a way to manage air quality in a region with nearly 14 million people and meteorological conditions that provide a perfect pollutant trap. Through state-of-the-art automotive and stationary source controls, we have managed since the 1960s to improve air quality substantially in the region, while population and industrial output have more than doubled.

There is still more to be done. And, since we soon will have implemented virtually all the possible "conventional" emission controls on industrial and commercial sources and putting in place the most stringent motor vehicle emission controls anywhere, we have reached a point of diminishing returns on source-only control programs. The next step is to address not only the performance of our technologies, but how we use them.

The South Coast Air Quality Plan (SCAQMD, 1990) is a landmark effort that looks not only at traditional sources, but also addresses consumer products, alternative fuels, and source use patterns (most significantly, automobile use and traffic management).

In creating this plan, we have come to the point where we can no longer deal separately with energy planning, transportation planning, land-use planning, and environmental management. That this synthesis has come about in response to air pollution is not surprising—combustion exhaust gases are one waste we cannot confine to dumps or pipe to treatment plants; we are immersed in our own effluent, and the magnitude of these gaseous wastes is enormous.

The South Coast's plan was created to address air pollution, and is not intended as an energy management plan *per se*, but the linkage between energy-demanding activities and air pollution is so direct that the plan's energy impacts became a significant factor in its formulation and in assessing its costs and benefits. The extension to include discussion of its global pollution impacts was but a short step.

A full discussion of the origin and content of the South Coast plan is beyond the scope of this paper; I cite it here because I view it as a harbinger of the coordinated energy-environmental management programs of the future. The South Coast plan brought together municipal, county, and state elected officials, regulatory agencies such as California Energy Commission, California Public Utilities Commission, and local and regional planning staffs, as well as the "traditional" air quality agencies (South Coast Air Quality Management District and California Air Resources Board).

We have learned much from the process, but two points particularly impress me: this kind of effort demanded new perspectives and new approaches to cooperation among traditionally separate government agencies, and they rose to the occasion; and, most importantly, early and continued involvement of elected officials and the public, difficult as it seemed at the time, has led to a more daring plan, rather than to political game-playing and finger pointing.

CURRENT ACTIVITIES AND POLICIES

California Energy Commission (CEC) Global Warming Study

The California Energy Resources Conservation and Development Commission (better known as simply the California Energy Commission, or CEC), has a statutory responsibility to oversee California's energy supply. Their duties include demand forecasting, monitoring utility resource plans, energy facility siting and licensing, setting energy efficiency standards for buildings and consumer products, promoting energy conservation, and performing research into the economics, technologies, and environmental effects of energy development and consumption.

In early 1988, then-Governor George Deukmejian signed a bill that required the Energy Commission, with assistance from other state agencies, to carry out a study of the probable impacts of global climate change on California. The Commission's final report, (CEC 1991) includes recommendations for appropriate responses to avoid, reduce, and mitigate these impacts.

The Commission's three reports (CEC 1989, 1990b, 1991) attempted to quantify, within the limits of reported scientific understanding, California's contribution to global climate change, the probable impacts on the state, and the potential of a wide variety of strategies to control greenhouse emissions or to mitigate probable greenhouse effects. Much of the material I have presented in this chapter substantially relies on those three reports.

As we have seen, despite the inherent imprecision of projections of greenhouse impacts, it is clear from the CEC report (CEC 1991) that there is a significant risk of severe and widespread negative effects on California. Moreover, this report presented estimates of the relative costs and effectiveness of a wide variety of both emission control options and measures to mitigate greenhouse effects.

The Commission's report established an invaluable information base from which to make policy judgments and to judge the relative merits of various responses to the greenhouse threat.

Assigning values to the social costs of global climate change and the benefits of mitigating it has proved to be a very controversial and complex subject. First, as we have seen, social costs of climate change are difficult to quantify because of the uncertainties regarding the future economy and the timing of climate change, and the extent and actual distribution of environmental impacts. Second, because the environmental benefits of reduced emissions of greenhouse gases are distributed globally, the "true" benefits to California from reduced emissions are difficult, if not impossible, to define. These analytical uncertainties should not be construed as meaning that the costs should be left unaddressed, rather, they argue for conservative approach to policymaking while we work to reduce some of the remaining scientific uncertainty about warming.

By focusing on measures which reduce greenhouse gas emissions at little or no additional cost, or which provide solutions to other high priority environmental problems, the near-term expenditures required to finance initial greenhouse gas emission reductions can be evaluated or justified based on near-term goals; the long-term climate management benefits that accompany any particular action are then secondary considerations which influence choice among options to reach near-term goals.

CALIFORNIA MEASURES TO REDUCE CARBON EMISSIONS

In discussing energy conservation, I mentioned many of the possible actions to reduce CO_2 emissions that the CEC has studied. It would be outside the scope of this paper to list all the options covered in the CEC reports, but I will note a few findings that are of particular interest.

516 Global Climate Change: Implications, Challenges and Mitigation Measures

Measures To Reduce Carbon Emissions From Transportation

Transportation CO_2 emissions can be compared across transportation modes (auto, bus, rail, etc.) by calculating emissions per passenger mile. Such analysis yields some unexpected results (CEC, 1991):

TRANSPORTATION MODE	AVERAGE RIDERSHIP	CO_2 EMISSIONS (pounds/passenger mile)*
Transit Bus (Diesel)	13.1	0.59
Trolly Bus (Electric)	14.9	0.29
Light Rail (Electric)	25.9	0.29
Heavy Rail (Electric)	21.3	0.26
Commuter Rail (Electric)	36.7	0.29
Ferry Boat (Diesel)	498.0	0.73
Vanpool (Gasoline; 10mpg)	8.2	0.26
Auto (Gasoline):		
Driver Only, 21 mpg		1.06
Driver + 1 Passenger, 21mpg		0.51
Driver Only, 35mpg		0.62
Driver + 1 Passenger, 35mpg		0.29

*Adapted from CEC, 1991

If we compare these numbers, it becomes clear that the intuitive impression that mass transit is more carbon-efficient is not necessarily correct. A typical 43 passenger transit bus is very efficient if it is full, but at current modest ridership levels, a 35-mpg car with multiple passengers is more efficient. While, in the long run, mass transit is more carbon-efficient, capturing these benefits requires either high ridership (diesel buses) or a fixed-route facility (wires for electric buses, tracks for rail, etc.). Construction and expansion of modern public transit is going forward throughout California, but we are still a long way from a network that can supplant auto travel as the primary transportation mode. Since we cannot expand our mass transit system overnight, it seems appropriate to look to increasing auto occupancy, which, with high-mileage cars, can approach the lower end of the mass transit efficiency range.

Electricity

The amount of carbon emissions resulting from electric generation is a function of the fuel mix used to generate the power. In the late 1970's, California was highly dependent on oil and gas for electric generation, with these fuels supplying 60 to 80 percent of the state's electricity needs. The fuel diversity policy established by our Energy Commission was successful in reducing the share of oil and gas in electric generation to 33 percent in 1990. The current generation mix in the California electric system is one of the most diverse in the world. Over 50 percent of California's electricity supply comes from non-fossil fuels that emit no greenhouse gases. Current trends point to an increasing use of natural gas generation, which only produces about 3/4 of the CO_2 per kW-h of petroleum, or about 2/3 that of coal.

Conservation can also make considerable contributions. Existing CEC programs to reduce residential and commercial electricity demand are expected to produce statewide savings of about 12 percent (CEC, 1991), and these programs will probably be expanded in the future.

CALIFORNIA MEASURES TO REGULATE METHANE EMISSIONS

The relatively high global warming potential (about 20 times that for CO_2) and short atmospheric life span of methane suggest that reductions in methane could have a significant and immediate impact on mitigating global climate change. Since methane emissions are linked to human activity, future emissions will likely increase as the California population increases unless we move quickly to limit them.

Landfills

Decaying organic wastes in landfills are the single greatest source of methane emissions in California, accounting for about 42 percent of the total. There are several ways to significantly reduce landfill methane.

Landfill gas emissions can be collected and burned to produce energy. Depending on system design and landfill characteristics, recovery efficiency can range from 20 to 70 percent (Crawford, 1985). The resulting byproducts of combustion have a lower greenhouse impact than raw methane.

Landfill gas emissions can also be reduced by composting or burning municipal solid waste. These technologies divert waste from being landfilled and instead process it to derive useful products such as energy. Although processing the waste still generates CO_2, the net emissions from composting and waste-to-energy facilities are less than the waste would have produced in a landfill. Recycling and reduction of waste at the soure are important measures that will also reduce methane emissions.

Livestock

Methane emissions are generated from animals during digestion and from large scale manure disposal. Reducing emissions directly from the animal (12 percent of total emissions) appears to have limited prospects in the short term, but reducing emissions from manure (24 percent of emissions) has already been successfully implemented at some sites.

To encourage methane recovery from animal waste, the EPA has a program devoted to the worldwide development of projects for the purpose of recovering and using methane. This program is generating information that will be valuable to future efforts in California, where the dairy and livestock industries are large and growing larger.

CHLOROFLUOROCARBONS (CFCs)

Control of CFCs is the leading "success story" in the development of international cooperation to address a global air pollution problem. International agreements to phase out CFCs (discussed elsewhere in this volume) set out a broad scheme to eliminate these chemicals early in the next century; we in California are doing our part to speed that process, and to reduce both CFC demand and CFC releases in the interim.

CFCs are especially effective greenhouse gases. Although CFCs constitute only minute atmospheric concentrations relative to CO_2, they contribute up to 20,000 times more to warming than CO_2, per molecule (USEPA, 1989b). Also, CFCs have an extended atmospheric lifetime of about 100 years on average. California's use of CFCs represents about 5 percent of the world total (Citizens for a Better Environment, 1988).

Automobile air conditioning represents about 19 percent of the total ozone depleting potential in the United States. The technology for recovery, reconditioning and reuse of CFCs in these systems is well established.

Auto manufacturers are rapidly phasing out CFCs in new vehicle designs. By the middle of this decade, most new cars will contain refrigerants that are relatively benign, both as ozone depleters and as greenhouse gases. That this shift away from CFCs is occurring well in advance of the deadlines called for in the Montreal and London protocols shows that American industry can meet the challenge.

In addition to this, substitutes for the CFCs used as solvents and in the production of polyurethane foam are rapidly coming to market. This will reduce CFC emissions still further over the next decade.

INTEGRATING GLOBAL RESPONSIBILITY INTO PLANNING AND REGULATORY ACTIVITIES: PREVENTIVE GOVERNMENT

California today finds itself at a transitional moment: we recognize that we face new, complex problems in environmental management, and we are beginning to deal with some of these issues, but we are doing it within an older, more compartmentalized government structure. The future challenges of integrated environmental management will not fall neatly under the jurisdiction of any single agency.

For the near-term, we will look to improve communication and coordination among existing units of government, and we will streamline our chain of command so that policy decisions and regulatory actions are free of unnecessary institutional boundaries. In the more distant future, we will have to devise organizations that are both technically robust and highly flexible—able to deal with a wide range of social, economic, and environmental issues with sophistication and in-depth expertise. This will be a major challenge for California.

California's growing concern and involvement in confronting global pollution is an evolutionary process, growing out of our existing institutional structure. As we become more adept at managing our environmental impacts, the organizational and analytical skills we develop here will be exportable; in this way we can contribute to a growing global capability to manage our environment.

INTEGRATED RESOURCE MANAGEMENT

The long-term future of all aspects of environmental management lies in integrated approaches. The present regulatory structure, with a growing number of independent, sometimes overlapping programs, each directed at a particular environmental problem, such as "airborne toxics," or "municipal solid waste" cannot continue to expand indefinitely. The integrated management approach looks at our human environment as a set of interrelated systems, and seeks to manage these systems (e.g., transportation, food production, energy supply) to reduce resource demand, minimize waste production, and prevent, rather than remedy, pollution.

As an example, consider California's water supply system, which is based on massive transport of water from the north to the south, using both existing rivers and man-made systems. It is a prime candidate for integrated planning, and, in fact, we are moving in that direction already.

Water resource planning includes consideration not only of urban and agricultural needs, but the energy used to transport water, effects on water quality, wildlife, urban growth, and the potential for conservation and waste-water reclamation to generate additional supplies. We are even looking into revising current allocations to better serve multiple purposes, rather than a single end use (for water management, by law and tradition based on rights established by historical use, this is a radical concept, indeed).

COMPREHENSIVE GROWTH MANAGEMENT

In the long run, California must confront its natural limitations. California's spectacular growth in this century has been based in part on the fact that we have managed to build what nature did not provide—the great harbors of Los Angeles and Long Beach, the manmade rivers of the California Water Project. As we enter the next century, we will find that we cannot sustain this rate of growth indefinitely. While we may go to heroic lengths to transport water, we cannot make it rain more; we cannot create vast new acreage of fertile agricultural soils; we cannot change the winds to better disperse air pollution. Even as we make our lifestyle more energy efficient or more frugal with water, our growing population makes our demand ever greater.

The greatest future challenge to environmental management will be growth management. Future growth will have to be better planned, both locally and statewide. There will need to be a continuing effort by local and regional governments to direct new growth, rather than continuing the present system in which localities compete with one another. Those who set these policies will have to give equal consideration to impacts on existing communities and to meeting the needs of new ones; this will lead, in turn, to structural changes in our urban landscape that will make it both more efficient in use of resources, and more hospitable to its inhabitants.

THE NEED FOR "PREVENTIVE GOVERNMENT"

In developing his policies, Governor Wilson uses an approach he characterizes as "preventive government". In social policy, this means preventive programs such as ones that keep children in school now to avoid creating a larger future problem of unskilled drop-outs unable to find a place in our workforce; in environmental policy, it means preventing pollution, rather than cleaning it up later; and in economic policy, it means anticipating change and preparing for it.

This is the key to California's intense interest in the problem of global climate change—if we can foresee a problem, we can plan for it; to do otherwise is to ignore government's duty to the people. We have looked at the potential for global climate change to impact us directly, and it is sobering to consider how widespread the effects may be. We have considered how international responses to global climate change may impact us economically. It seems only right that we confront this challenge head-on, preparing ourselves for unavoidable changes, and laying the technological and organizational groundwork for continued economic success in a future that will be quite different from the world of today. I urge others to watch us, learn from us and follow our example.

REFERENCES

California Department of Fish and Game. *At the Crossroads - A Report on the Status of California's Endangered and Rare Fish and Wildlife,* 1980.

California Department of Food and Agriculture. *California Agriculture,* 1988.

California Energy Commission. *Biennial Fuels Report,* 1987.

_____ , *1990 Energy Development Report,* August, 1990a.

_____ , Global Climate Change: *Potential Impacts and Policy Recommendations,* December 1991, P500-91-007.

_____ , *The Impacts of Global Warming on California, Interim Report,* August 1989, P500-89-004.

_____ , *1988 Inventory of Greenhouse Gas Emissions,* 1990b.

Citizens for a Better Environment. *A Fragile Shield Above the Golden State,* 1988.

Crawford, John and Paul G. Smith. *Landfill Technology.* Butterworths. 1985.

Department of Water Resources. Bul. 160-83. *The California Water Plan, Projected Use and Available Water Supplies to 2010,* December, 1983.

Fay, J.S. and S.W. Fay. *California Almanac,* Pacific Data Resources, Santa Barbara, CA. 1991.

Flynn, Timothy. "Implications of Sea Level Rise for Hazardous Waste Sites in Coastal Floodplains." *Greenhouse Effect and Sea Level Rise.* Eds. Barth, *et al.* 1986.

Gleick, Peter. "The Implications of Climate Change for California." Pacific Institute for Studies in Development, Environment, and Security, November 21, 1988.

Kahrl, W.L. eds. *The California Water Atlas, State of California,* 1978.

National Academy of Sciences. *Policy Implications of Greenhouse Warming,* 1991.

Rowe, *et al. The Benefits of Air Pollution Control in California,* California Air Resources Board, Contract No. A2-118-32, 1986.

South Coast Air Quality Management District. *Air Quality Management Plan, South Coast Air Basin,* 1990 Draft.

United States Environmental Protection Agency. *How Industry is Reducing Dependence on Ozone-Depleting Chemicals.* A Status Report Prepared by the Stratospheric Ozone Protection Program, Office of Air and Radiation, 1989a.

_____ , *Policy Options for Stabilizing Global Climate.* Draft Report to Congress. Office of Policy, Planning, and Evaluation. February, 1989b.

_____ , *Potential Effects of Global Climate Change on the United States.* Draft Report to Congress. Joel B. Smith and Dennis A. Tirpak, Eds. Office of Policy, Planning, and Evaluation. Office of Research and Development. October, 1988.

Global Climate Change: Implications, Challenges and Mitigation Measures. Edited by S.K. Majumdar, L.S. Kalkstein, B. Yarnal, E.W. Miller, and L.M. Rosenfeld. © 1992, The Pennsylvania Academy of Science.

Chapter Thirty-Six

COMBATTING THE CLIMATE CHANGE - A CHALLENGE FOR INTERNATIONAL COOPERATION IN THE '90s*

AIRA KALELA

Special Advisor
Ministry of the Environment
Ratakatu 3, 00120
Helsinki, Finland

INTRODUCTION

The Second Climate Conference in November 1990 can be considered as a major starting point for development of an international regime to combat climate change. Ministers and high level representatives from almost 140 countries agreed that climate change is a common concern for all mankind and committed themselves to actively control this change.

Leading scientists now are in agreement about some basic findings. Still, a considerable number of issues related to the warming process, its regional variations, impacts, feedbacks and social, economic and political consequences need to be further clarified. Scientists agree that emissions resulting from human activities are substantially increasing the atmospheric concentrations of greenhouse gases, resulting, on the average, in additional warming of the globe's surface. They also agree on climate trends with certain variations. Four different trends are illustrated by various counter measure scenarios. These include proposals for preventive measures and adaptation.

---

*This article was written in September 1991, when the negotiations on the Climate Convention were in the midway. It reflects the expectations of the author for the Convention and international cooperation based on the Convention. The negotiations were concluded in May 1992. The article should be read in the above mentioned context.

We have to thank the World Climate Programme and the Intergovernmental Panel on Climate Change for considerably increasing our understanding of the complex issues at stake. Governments now consider that threats for serious irreversible damages are so great that scientific uncertainty cannot be used to postpone cost-effective measures to prevent the damages. Governments futher agree on the urgency and have set the target of signing the Climate Convention in connection with the United Nations Conference on Environment and Development in June 1992. A year has already passed since the major breakthrough was made in the form of an agreement to start negotiations and giving the guidelines for this work. The progress made so far has not matched the mandate to work effectively and without delay.

We need firm commitments to agree in a relatively short period, before the United Nations Conference on Environment and Development, on the arrangements for intensive cooperation to combat this climate change. The Climate Convention can provide such arrangements. In addition, it should include effective measures to counteract and adapt to climate change. The main elements of this cooperation and the commitments are presented in the following.

Shared responsibility and partnership

The common concern about climate change should lead to shared responsibility. This means that all Governments should participate in the negotiations for a treaty and shoulder their responsibilities in reaching the agreed targets.

The shared responsibility should, however, be differentiated according to the level of development and other specific circumstances of each country. The main goal of Finland in these negotiations will be a widely acceptable Convention which effectively counteracts the climate change. Effectiveness would mean, that all countries whose contribution to the warming process is considerable, should consequently shoulder their responsibilities in a corresponding manner. As it is mainly the developed countries that are causing the climate change, they should make the first commitments both to counteract the change and to assist developing countries to fulfill their obligations.

Long-term objectives

The international agreement should set as the ultimate global objective the stabilization of greenhouse gas concentrations at a level which prevents dangerous anthropogenic interference with the climate. Such a level should be reached within a period allowing ecosystems to adapt without disturbances. It is most essential to maintain such climate conditions that, in particular, food production and the quality and structure of forests can be guaranteed and enhanced.

The Intergovernmental Panel on Climate Change has presented four scenarios for response strategies: high and low emission scenarios, and two control policies. The two latter scenarios, which would progressively increase control measures, would probably lead to a slowing down of the average increase in global mean temperature to about $0.1°C$ per decade. Some scientists indicate that this might allow the ecosystems to adapt naturally to the change.

An enormous challenge is the combination of climate policies with such economic policies as foster sustainable growth. Here again, the main responsibility rests with the industrialized countries, as their economic activities are the main cause of the climate change. At the same time the development priorities of the developing countries should be considered together with global climate policies. A strengthened and sustainable economic growth and an alleviation of poverty in developing countries should be coupled with the necessary resources for local and regional counteraction and adaptation measures in these countries. This would be essential in the long-term objectives with which all governments should agree to cooperate.

Financial Resources and Mechanisms

New and additional financial resources should be made available especially to developing countries to enable them to meet their obligations under the Convention. Technological cooperation should also be strengthened for the transfer of the best applicable environmentally sound options as well as for the development of local technologies. A mechanism functioning smoothly and effectively and including a clearing house, should be set up. This clearing house should help governments address climate change in the most cost-effective way.

The system should include a multilateral fund and other multilateral and bilateral arrangements. All arrangements should be based on mutual partnership and cooperation. Policy decisions on the mechanism, its funding targets and appropriation of funds should be made by the Parties in such a way that each Party has an equal say in decision-making. The expertise and resources of existing organizations and programmes should be used to the maximum in the planning and implementation of special projects under the mechanism. The arrangements of the Global Environment Facility and the Montreal Protocol Interim Fund provide useful examples which should be developed further. The Global Environment Facility could also provide arrangements for administering the funds.

Targets for Joint Comprehensive Action

Long-term objectives should be turned into targets. They should be set for long and short term action. The targets should be defined over a period of time, starting with those elements where our knowledge is best and supplemented later by others, as scientific and technological developments allow.

The international community should further turn targets into commitments. General commitments should include measures to limit, reduce, modify and control human activities that result, or are likely to result, in adverse effects on the climate. These measures should include both reduction of all greenhouse gas emissions and maintaining and enhancing sinks, in particular forests, oceans and coastal areas. Sources and sinks should be dealt with in an integrated way both in the Draft Convention and in its protocols.

The commitments should cover the net emissions of greenhouse gases wherever this is possible. It might turn out that for practical reasons, the commitments may at the beginning need to be made separately for sources and sinks. The targets should be so specific as to meet the requirements of the negotiation mandate to work out an efficient framework convention.

The concept of net emissions has been widely discussed by experts in the negotiating process and at other meetings. It is likely that calculation or even estimation of the whole carbon budget of a country will be difficult. However, the capacity of the biomass as carbon sinks can be calculated. Therefore such countries which take measures to increase their sinks should have the right to take credit for these measures, if they can prove, in quantitative terms, the increase in their sinks.

Stabilization and Reduction of Greenhouse Gases

One of the first steps should be the stabilization of carbon dioxide emissions. A considerable number of the industrialized countries have already indicated their willingness to take action. By the year 2000, they aim to stabilize, in general at the 1990 level, their emissions of carbon dioxide and other greenhouse gases not controlled by the Montreal Protocol. We hope that this commitment can be made by as many industrialized countries as possible. Differentiation of the commitments, in particular the time frame, can be considered if special circumstances so require. This may, in particular, be the case in some Central and Eastern European countries.

The stabilization of carbon dioxide emissions should be followed by further reductions. Proposals have been made e.g. for a 20 percent reduction by 2005. Some governments have already made such commitments. Several delegations have advocated, during the ongoing negotiations, that the message concerning joint targets should be clear and loud so that various sectors of society have sufficient time to modify their developments.

The Convention should include a review mechanism. New targets should be negotiated well before the agreed targets have been reached.

Developing countries as well should make their first commitments concerning carbon dioxide emissions. Their commitments should include minimization of the growth of emissions.

The Draft Convention should be flexible in such a manner that each party can choose its own measures to reach the targets. Energy saving and increasing energy efficiency should be applied in all countries. E.g. the Finnish estimates indicate that with current technology there is a potential for 15-20 percent savings in energy consumption in Finland. A 15 percent saving can be achieved at a reasonable cost. Further savings would require considerable investments. At the same time, prognoses indicate a growth which is higher than the saving capacity. This will lead to a search for alternative, environmentally sound energy sources.

The substitution of fossil fuels with cleaner energy sources would, in most countries, be at least a partial solution. However, the main long-term solution can only be the efficient development of renewable, environmentally sound, alternative energy sources. The Climate Convention should lead to such international cooperation that industrialized countries actively develop even such energy sources that are not necessarily applicable in these countries themselves. The cooperation should bring about a global energy programme, or arrangements where new, renewable energy sources—hydro, solar, wind, geothermal, wave, biomass—are developed and used where this can be done in a cost-effective manner. E.g. Finland participates in the development of solar energy in China.

The possibilities to reduce other emissions than carbon dioxide should be studied and measures taken step by step. Improved management and maintenance of landfills, coal mining, gas drilling, venting and transmission could reduce methane emissions cost-effectively with available technology within a relatively short time.

The strengthening of the Montreal Protocol of Substances that Deplete the Ozone Layer also helps in slowing down global warming. All parties have agreed to phase out their use of the major ozone-depleting substances by 1998. Several governments have already set earlier dates—in fact, the main part of the ozone-depleting substances could be phased out by 1995. In Finland this will be done by 1994.

Maintenance and Enhancement of Sinks

The first measures to limit greenhouse gas emissions should be accompanied by action to maintain and enhance sinks, in particular, forests. Even though our exact knowledge on sinks is not on the same level as that on carbon dioxide emissions, there are several reasons which speak strongly in favour of taking the first steps here as well. Sustainable forest management is very important for economic development, as a source of energy, and for the prevention of erosion and desertification in many countries. Forests are also a valuable source of biodiversity. Furthermore, both means and technology are immediately available.

We should make differentiated commitments to protect and enhance greenhouse gas sinks and reservoirs through the adoption of appropriate and environmentally sound policies. These should cover the management of forests in a sustainable manner and the application of appropriate agricultural techniques with the aim of reducing the degradation of forests, wetlands and other ecosystems including the marine ecosystems. Local socio-economic needs and land use patterns should be taken into account and the ecological balance maintained in the areas concerned.

It will be important to maintain and, where possible, enhance the amount of biomass in forest ecosystems and promote their health and ability to absorb carbon dioxide. We should also counteract the causes of threats against forest ecosystems, such as the emissions of pollutants, soil degradation, environmentally damaging use, or damaging forest fires.

An important aim is to stabilize the forest area and where possible to increase it. We should also cooperate to further scientific research and monitoring and the exchange of information on the role of ecosystems as sinks and reservoirs and on sustainable management of ecosystems.

Country Programmes

Many governments have presented or are in the process of preparing country strategies or programmes responding or adapting to the climate change. Finland would like to welcome very much the preparation of such programmes. They should be among the first commitments in the Convention. The strategies should cover inventories of the emissions of various greenhouse gases and of relevant sinks. Likewise they should indicate the targets for emissions and sinks as well as options and actual measures for achieving these targets. These strategies should be reviewed by a special panel or executive committee set up by the Parties. The reports on the reviews should be presented to the meetings of the Parties.

As the country programmes would be among the first important steps, we propose that an interim arrangement be made for their review during the period when the Convention has not yet entered into force. The first strategies could be prepared within a year after agreement is reached on the Draft Convention, and they could be followed by other reviews. The programmes should be updated later within a given period of time.

Systematic Research and Monitoring

Addressing all scientific uncertainties will be essential in international cooperation. We have to commend the Intergovernmental Panel on Climate Change for improving our understanding on the physical, chemical and biological processes related to climate change and on its impacts and response strategies. Research needs to be intensified in all these three main sectors. The most essential area where we lack information is the cost of the various options in the response strategies and the cost of inaction.

Finland's capacity to participate in international scientific cooperation has increased considerably with the establishment of a comprehensive atmospheric research programme in 1990. This research programme under the Ministry of the Environment is administered by the Finnish Academy of Science. Over 100 scientists participate in the programme. So far, it is the biggest Finnish research programme related to environmental issues.

We have made considerable progress in preparing inventories of greenhouse gas emission and sinks. OECD has made a considerable contribution in strengthening international cooperation in this field. As our understanding of the total carbon budget still needs to be improved, Finland has taken an initiative to organize an expert meeting to develop a methodology for estimating the carbon budget of forest ecosystems in May 1992.

A well-functioning international observation network is a prerequisite for monitoring the change in the chemical and physical composition of the atmosphere. Over the past few years, we have been concerned about some indications of delays and even stagnation in the development of a global monitoring network both in developed and in developing countries. Finland has therefore launched a 10 million US dollars programme as a contribution to the improvement of global atmospheric monitoring networks.

Resolve and Leadership

People around the world look to their governments to show resolve and necessary leadership to make it possible to avoid the catastrophes of climate change and a global warming. This resolve was indicated more than a year ago. It should not be lost. The cause of hesitation is not scientific uncertainty. The cost of the countermeasures lie at the root of the hesitation.

There are no specific calculations of the response costs, but they may run up to billions of dollars. However, this should be seen as an input in the development of national economies and as an integral part of the overall development of societies.

The cost of inaction may rise even higher. As yet there are no calculations of what will be the costs of lost revenues, food shortages, loss of forests and agricultural land as well as destruction of human, animal and plant habitats. Human life and misery cannot be measured in monetary terms. New environmental refugees fleeing from lost arable land, destroyed habitation, and sea level rise may cause unforeseen social instability and lead to conflicts between and among nations. These costs may be much greater than the cost for action.

Representatives of some 140 governments agreed at the Second World Climate Conference that lack of full scientific evidence should not postpone action. Our policies should be based on prevention and precaution.

We have to understand that we are not negotiating arrangements dealing with environmental issues or the sustainable use of natural resources. We are working out a new regime which will not leave untouched our basic production and consumption patterns. We are considering the bases of global economic development. Changes will be required in our energy policies, forest management, agricultural policies and transportation. Prices, in particular energy prices, need to be reconsidered and the environmental costs included. Economic instruments have to be developed in order to avoid trade distortions.

No government wants catastrophe or uncontrolled change. Recent developments in the Middle East and Europe indicate that our international economic, political and ecological systems are vulnerable. We should work with determination for a controlled change which strengthens our economic, ecological, social and political security.

Global Climate Change: Implications, Challenges and Mitigation Measures. Edited by S.K. Majumdar, L.S. Kalkstein, B. Yarnal, E.W. Miller, and L.M. Rosenfeld. © 1992, The Pennsylvania Academy of Science.

Chapter Thirty-Seven

A CANADIAN PERSPECTIVE ON CLIMATE CHANGE*

IAN BURTON and DEBORAH HERBERT

Climate Adaptation Branch
Canadian Climate Centre
Atmospheric Environment Service
Environment Canada
373 Sussex Drive
3rd Floor Block E
Ottawa, Ontario, Canada K1AOH3

INTRODUCTION

In most parts of Canada, winters are long and cold. In Toronto, the average dates of the first and last "killing" frosts are October 5 and May 8, respectively. The average growing season is therefore only 149 days, and in many other regions of the country, it is much less. Whatever the adverse effects of climate change elsewhere, it is clear that for Canada, at least, there will be some mitigating benefits and opportunities as well as costs and risks. As might be expected, therefore, there are some members of the Canadian public who welcome the prospect of warmer weather. In 1990, 3 out of 10 Canadians polled believed that climate change would involve a net benefit for Canada (Standing Committee, 1991). At a scientific meeting in Windsor, Ontario, it was reported that the climate of southern Ontario may come to resemble that of Kentucky (Sanderson, 1988). The prospect is hardly displeasing. After all, short winters, blue grass, racehorses, and bourbon are by no means symbols of hardship, privation, or environmental destruction.

---

*The authors are solely responsible for the views expressed in this paper, which do not necessarily represent the position or views of the Government of Canada.

Nevertheless, the overwhelming weight of scientific opinion in Canada and elsewhere is that the net effects of climate change will be negative for Canada, and perhaps extremely detrimental (Standing Committee, 1991). Just how severe the effects may be, it is not possible to say with confidence but a series of impact studies by the Canadian Climate Centre has shown that some major costs are likely. The costs, benefits, and their causes vary regionally: sea level rise will be a main agent of change in the Atlantic region while changes in temperature and precipitation will cause most of the impacts expected elsewhere in Canada.

THE COSTS OF CLIMATE CHANGE IN CANADA

Well over 100 communities in the four Atlantic provinces are at some risk of flooding from sea level rise; Charlottetown, Halifax, and four other municipalities will require extensive protection to avoid serious damage (Stokoe, 1987; P. Lane, 1986). Although some communities on the Pacific coast, including Vancouver (Smith, 1989) but not including the important Fraser Delta, may be less vulnerable to flooding than those on the Atlantic coast, there will probably be an increase in winter precipitation, resulting in more frequent landslides and floods (Canadian Climate Program Board (CCPB), 1991).

Climate change will have a direct impact on Canada's natural resource base, and this impact is likely to be negative. Natural resource industries account for a large share of the Canadian economy: agriculture is an important activity in most provinces, especially on the Prairies, and forestry accounts for almost as much export revenue as fisheries, mines, agriculture and energy put together. One of Canada's most valuable assets is its abundance of rivers, lakes, and marine systems, providing water supply for consumption, agriculture, and hydro-electricity production. The impacts on agriculture, forestry, water resources, and hydro-electricity production in Canada, among others, have been estimated and are discussed below.

Climate change is likely to have a profound effect on Canada's agricultural productivity and potential. While warmer temperatures and a longer growing season will enhance production potential in northern arable areas, production in the south, the current agricultural zone in Canada, may well be constrained by increased water stress and probability of drought (Arthur, 1986; Williams, 1988; Land Evaluation Group, 1985, 1986). Canadian forest resources may also be hard hit by climate change. Warmer temperatures and water stress will cause existing forest zones to shift northward; it is likely that forest area will be lost because die-out in the southern fringes of existing forest zones will occur faster than migration of new forests to replace them (Intergovernmental Panel on Climate Change, 1990). In fact, it has been suggested that the boreal forest zone in Canada may completely disappear west of James Bay (Rizzo, 1990). This will be exacerbated by an increased probability of forest fires in dry areas as well as by increased activity of pests and diseases (CCPB, 1991; Wheaton *et al.*, 1987). Climate change will affect Canada's water resources from both the demand and supply sides: increased temperatures will, other things being equal, induce greater consumption of water and will also increase evaporation (which may be nullified in some areas by increased precipitation) (Howe, 1986). Warmer temperatures and reduced stream flow may create anaerobic conditions in some rivers and lakes and may increase pollution concentrations. A positive potential effect of climate change will be the reduction of sea and lake ice: it will facilitate expansion of shipping and transportation in the Great Lakes and Atlantic and Arctic Oceans as well as reduce the costs of offshore oil operations (Stokoe, 1987; CCPB, 1991; Howe, 1986). Impacts on hydro-electric energy production will vary from region to region, but it has been estimated that production at James Bay and Labrador plants will increase (Stokoe, 1987; Singh, 1988); this energy could be exported to areas, like the Great Lakes region, which experience a decrease in production (Howe, 1986).

Unfortunately, many of the impacts of climate change in Canada are unknown, or are highly uncertain. The information cited above is the result of several excellent impact studies; however, these studies are only useful first approximations of what Canadians should expect because they have typically focused on one or a few sectors in one particular region. Moreover, the studies have large uncertainties and are largely qualitative, primarily due to the lack of reliable climatic and economic forecasts. These studies have also concentrated on changes in climate means rather than climate variability or extremes.

CANADA'S COMMITMENT TO LESS DEVELOPED NATIONS

While the net costs for Canada may be considerable, they are likely to be much lower than for many other countries, especially some of the developing nations with large low-lying deltas (Egypt, Bangladesh, Vietnam, etc.) and the small island states. In addition to costs incurred in Canada, therefore, it is expected that Canada will also contribute to the international effort at mitigation of this global threat, perhaps by expanding assistance to developing countries on a direct bilateral basis and/or participating in international financial agreements. As an example of Canada's commitment, the government of Canada has already pledged $1 million to enable developing countries to participate in understanding climate and climate change.

It is feared that a larger contribution to global greenhouse gas emissions will come in the future from countries such as India and China as they expand and accelerate their development, using domestic coal reserves. Canada is also a coal-rich nation, and one area where Canada may be of particular help in solving the problem of global warming is in the development and transfer to China and India of "clean coal" technology to reduce emissions of CO_2 from coal and to improve its efficiency in use.

CANADA'S CONTRIBUTION TO GREENHOUSE GAS EMISSIONS AND LIMITATIONS

In developing its policy on global warming and greenhouse gas emissions, Canada, like other nations, must take its own circumstances into account. Canada is a high energy user. In fact, it has the highest per capita and per unit output levels of energy use of the major developed countries. There are three reasons for this. First, as previously mentioned, the climate of Canada is cold, which means that proportionally more energy must be used for space heating in Canada than in warmer regions. Second, the low density and scattered distribution of Canada's population mean that Canadians have high transportation requirements; until there is stronger penetration of alternative fuels in the energy market, fossil fuels will continue to be virtually the only fuels used for transportation in Canada. Third, and most importantly, Canada has large fossil fuel reserves as well as excellent hydroelectric sites. Because Canada is also rich in natural resources, it has been economically attractive to carry out primary processing of those resources, especially minerals, in Canada; these processing industries are more energy-intensive than other industries and have helped to create an energy-intensive economy. The abundant supply of fuel reserves at relatively low cost in Canada has also, until recently, served to encourage high consumer use of energy. Improvements in energy conservation and efficiency in Canada have been identified by many people as the least-cost method to reduce greenhouse gas emission levels in the short term.

Fortunately, the carbon content of Canada's energy use is lower than that for many other nations, because of the importance of hydroelectric and nuclear sources in Canada's energy use mix (the provinces of Ontario, New Brunswick, and Prince Edward Island depend on nuclear fuel for a significant portion of their energy requirements and hydroelectricity supplies more than 90 percent of the energy needs of Newfoundland, Quebec, Manitoba and British Columbia). Canada contributes about 2 percent of the total global emissions of carbon dioxide, and, of the OECD countries, contributes the fifth highest total CO_2 emissions, and ranks second and eighth in terms of CO_2 emissions per capita and per unit of GDP, respectively (OECD, 1991).

Canada's shares of world emissions of the other important greenhouse gases are similar. Canada's share of nitrous oxide and methane emissions is estimated to be between 1 and 2 percent; its share of CFC emissions is about 2 percent. Canada ranks ninth and fourth out of the OECD countries in terms of total and per capita greenhouse gas emissions, respectively (OECD, 1991). Tables 1 and 2 provide a ranking of the top ten greenhouse gas emitters among the OECD member countries.

Both government and industry in Canada are committed in principle to controlling greenhouse gas emissions. Canada is a signatory to the Montreal Protocol on Substances That Deplete the Ozone Layer, and under that Protocol, has committed itself to fully phasing out CFC production and new consumption by 1997 (which is actually 3 years earlier than the date prescribed in the Protocol). At the Second World Climate Conference, Canada committed itself to stabilizing its greenhouse gas emissions at 1990 levels by the year 2000 (Government of Canada, 1990). Unfortunately, the large proportion of energy-intensive industry in the Canadian economy (as well as the geographical and climate factors mentioned above) implies that further emissions restraints might have a damaging effect on Canada's competitive position if they were taken on a unilateral basis. Eric Haites, who has extensively analyzed Canada's opportunities to reduce CO_2 emissions, stated that the World Conference on The Changing Atmosphere (Toronto Conference) goal of reducing emissions to 80 percent of 1988 levels by 2005 "can be achieved in Canada, but only by doing everything technically possible to reduce carbon dioxide emissions. It is a very ambitious target." (Haites, 1990).

There is scope within Canada to limit emissions through the development and commercial application of low-carbon fuels, such as natural gas and "clean" coal, and alternative fuels, such as hydroelectric, nuclear and renewables as well as through improvements in energy use. Ontario Hydro, the government-owned utility which supplies electricity to the province of Ontario, estimates that it will need to build three new nuclear power plants, at a total cost of $15 billion, in order to supply forecasted electricity demand (in the year 2005) and observe the 20% reduction in

TABLE 1
Carbon Dioxide Emissions From Energy Use

Emissions per Unit GDP (kg/$1000 US, 1985)		Emissions Per Capita (tonnes)	
Portugal	428	United States	5.8
Australia	404	Canada	4.8
Ireland	392	Australia	4.3
Netherlands	380	Finland	3.7
Belgium	370	Netherlands	3.4
United States	324	Denmark	3.4
United Kingdom	317	Belgium	3.2
Canada	316	West Germany	3.2
Spain	302	United Kingdom	2.9
Finland	302	Sweden	2.5

Source: OECD. *Environmental Indicators,* Paris, 1991: p. 17.

TABLE 2
Total Greenhouse Gas* Emissions, Late 1980s

Emissions per Unit GDP (kg/$1000 US, 1985)		Emissions Per Capita (tonnes)	
New Zealand	1375	Australia	11.0
Portugal	1231	United States	10.0
Greece	1200	New Zealand	9.3
Turkey	1132	Canada	9.2
Ireland	1037	Norway	8.5
Australia	1035	Netherlands	6.4
Spain	709	Ireland	5.9
Netherlands	705	Denmark	5.8
Canada	608	Finland	5.5
Belgium	604	United Kingdom	5.4

*Carbon dioxide, methane and CFCs; mesured in Equivalent Carbon Dioxide Heating Effect.
Source: OECD. *Environmental Indicators,* Paris, 1991: p. 19.

CO_2 emissions recommended at the Toronto Conference (Ontario Hydro, 1989). According to another analysis, economically attractive measures to limit CO_2 emissions in Canada as a whole would reduce emissions by just under 12 percent of 1988 levels by 2005; these measures would also provide a net benefit (after recovery of the necessary investment) of $108 billion from energy savings (Haites, 1990).

In fact, the Canadian Petroleum Association, the Canadian Gas Association, and the Coal Association of Canada, which represent most enterprises involved in their respective industries, have each acknowledged the problem of global warming and have expressed that they are committed to research and policy to help ameliorate it (Government of Canada, 1989). Global warming may indeed represent an excellent trade opportunity for the Canadian energy sector: it is in the interest of the companies and associations involved to develop "cleaner" technologies, or commercialize existing technologies, and commercialize alternative fuels, because that technology can then be sold abroad, and, as mentioned earlier, transferred to countries like China and India. A good deal of valuable work in this area has already been undertaken by the industry itself.

DIRECTIONS OF RESEARCH AND POLICY IN CANADA

The National Action Strategy on Global Warming has been proposed by the Canadian Council of Ministers of the Environment as a framework for policy response to the issue. The National Action Strategy has three aspects: emissions limitation, adaptation to climate change, and reduction of the uncertainty of climate change through scientific research. Measures suggested by the proposed National Action Strategy to reduce greenhouse gas emissions include energy efficiency improvements, development of alternative forms of fuel, reforestation, dissemination of information to Canadians on how to lower their energy consumption, and reduction of emissions from agriculture (for instance, through measures to stabilize and increase the organic content of soil to reduce CO_2 levels in the atmosphere) (Canadian Council of Ministers of the Environment, 1990). The federal government has proposed new legislation in the form of the *National Energy Efficiency and Alternative Energy Bill.* The Bill specifies minimum efficiency standards and requires efficiency labelling for equipment which is traded across provincial and national borders, and provides for expansion of the National Energy Use Database (Government of Canada, 1991). A program, called *Energuide,* is already in place to label home appliances according to their energy efficiency. The federal government has also instituted a program to plant 325 million trees in Canadian communities over the period from 1991 to 1996 (Government of Canada, 1990).

The National Action Strategy pledges to assist Canadians "anticipate and prepare for the potential effects of any warming that might occur" (CCME, 1990). To fill in the gaps not yet covered by the Candadian climate change impact studies thus far undertaken, the National Action Strategy provides for three five-year multidisciplinary projects to assess the socio-economic impacts of climate change in the Great Lakes and St. Lawrence River Basin, the Prairies, and the MacKenzie

River Basin (CCME, 1990). In addition, guidelines to be developed will ensure that the implications of climate change are considered in major projects (Government of Canada, 1990). Finally, the government will determine, by 1996, policy changes that may be needed to deal with a rising sea level at both the Atlantic and Pacific coasts of Canada (Government of Canada, 1990).

As outlined in the Green Plan, the federal government's central environmental policy package, and the proposed National Action Strategy, the Canadian government is increasing its commitment to research on global warming. In addition to the three five-year multidisciplinary projects, the government will publish annual reports on the state of the Canadian climate, operate a climate change detection network consisting of volunteer-run stations across all of Canada's climatic zones (to be in place by 1996), and initiate, by 1992, a national program of ocean research relating to climate change (Government of Canada, 1990).

The Canadian Climate Centre has, in addition to undertaking the climate change impact studies mentioned above, developed a General Circulation Model (GCM). It is one of three second-generation models in existence; it has higher spatial resolution as well as more accurate simulation of oceanic and sub-ice transport and optical feedback effects than earlier models (Canadian Climate Board, 1991). The Canadian Climate Board has also recommended that the Canadian Climate Centre GCM be augmented by a supercomputer and dedicated staff, but allocation of funds to this end is pending (CCPB, 1991). Canada continues to contribute substantially to international research on climate change.

PARTICIPATION OF ALL CANADIANS IN LIMITATIONS AND ADAPTATIONS

The National Action Strategy is a draft agreement for discussion and adoption. It proposes the full participation of all the provincial, territorial, and municipal governments as well as the private sector, environmental organizations, and citizens' and aboriginal groups (CCME, 1990). As jurisdiction over environmental issues is shared between the federal government and the provincial and territorial governments, the Government of Canada cannot unilaterally achieve a stabilization of greenhouse gas emissions. Furthermore, Canadian citizens, environmentalists, and aboriginals are important stakeholders in the climate change issue, and are insisting on full participation in determination of Canadian policy on the issue. Participation of such groups and organizations will help to ensure that the policy is effective, once it has been agreed upon. It is also understood that the majority of emissions reductions and adaptation strategies will be undertaken by the private sector. The National Action Strategy seeks to encourage private sector support by providing information to and asking for input from firms and individuals. Moreover, national and provincial Round Tables on Economy and Environment have allowed for interaction between government and the private sector on environmental issues (Government of Canada, 1990).

Since a changing climate means that past climatic norms will not be an accurate guide to the future, the Climate Change Adaptation Branch has been established within the Canadian Climate Centre. The Branch is endeavouring to determine how Canadians can best adapt to changes in climate, climatic variability and extreme events. This information, along with climatic data for any area of Canada, will be transmitted to Canadians for use in investment and other decisions in which weather events and norms play a part.

CONCLUSION

Despite the cold climate of Canada, global warming is a cause for concern for most Canadians. Canada contributes to the international effort to ameliorate and understand the effects of climate change and has taken a large step towards domestic policy on the issue. These efforts will continue in some or all of the following directions. As high energy users, it is important that Canadians consider how to constrain their consumption of energy. Research and development activities toward this end have been undertaken in the private sector and are encouraged by governments. As a member of the world community, Canada will continue to play a role in negotiations for a Framework Convention on Climate Change. Existing information on the impacts of climate change in Canada is more qualitiative than quantitiative and certainly falls well short of what is needed. There is a role for adaptation to climate change and variability and to do this Canadians need to have the information necessary to incorporate climate variables into investment and operational decisions.

It is expected that, through the National Action Strategy, effective decisions will be taken to deal with climate change. If they are to be truly effective, such decisions will necessarily reflect the concerns of all groups and governments within and, to some extent, outside of Canada. In her book, *Survival,* Canadian author Margaret Atwood said that the essence of Canadianism is endurance and that survival is a Canadian's idea of epic achievement. Climate change, as distinct from the harshness of today's climate, is a threat to Canadian well-being; Canadians are already closely investigating what can be done now, in terms of adaptation and limitation responses, to mitigate the effects of this threat, and thus continue to be able to rely on Canada's environment as an integral part of their lives and livelihoods.

REFERENCES

Arthur, Louise. *et al.* 1986. *Towards a Socio-Economic Assessment of the Implications of Climate Change for Agriculture in Manitoba and the Prairie Provinces: Phases I and II.* Manitoba: Department of Agricultural Economics, University of Manitoba, for the Atmospheric Environment Service, Environment Canada.
Atwood, Margaret. 1972. *Survival, A Thematic Guide to Canadian Literature.* Toronto: Anansi.

Canadian Almanac & Directory. 1991. Toronto: Canadian Almanac & Directory Publishing Company Limited.

Canadian Climate Program Board. 1991. "Climate Change and Canadian Impacts: The Scientific Perspective". *Climate Change Digest,* CCD 91-01, Downsview: Atmospheric Environment Service, Environment Canada.

Canadian Council of Ministers of the Environment. 1990. *National Action Strategy on Global Warming.* Ottawa.

Government of Canada (Hansard). 1989. *Minutes of the Proceedings and Evidence of the Standing Committee on Environment,* Issues 22, 28, 35, 38, and 41. Ottawa.

Government of Canada. 1990. *Canada's Green Plan.* Ottawa.

Government of Canada. 1991. "New Federal Act Promotes Efficiency and Alternative Energy" (News Release). Ottawa: Energy, Mines, and Resources Canada.

The Greenprint for Canada Committee. 1990. *Greenprint for Canada: First Annual Report Card.* Ottawa: Greenprint for Canada Committee.

Haites, Eric. 1990. *Canada: Opportunities for Carbon Emissions Control;* Richland: Pacific Northwest Laboratory.

Howe, D.A., et al. 1986. *Socio-Economic Assessment of the Implications of Climatic Change for Commercial Navigation and Hydro-electric Power Generation in the Great Lakes-St.Lawrence River System, Phase II.* Windsor: Great Lakes Institute, University of Windsor, for AES, Environment Canada.

Intergovernmental Panel on Climate Change. 1990. *Scientific Assessment of Climate Change.* World Meteorological Association and United Nations Environment Programme.

Land Evaluation Group. 1985. *Socio-economic Assessment of the Implications of Climate Change for Food Production in Ontario,* Publication No. LEG-22. Guelph: Department of Geography, University of Guelph, for AES, Environment Canada.

Land Evaluation Group. 1986. *Implications of Climate Change and Variability for Ontario's Agri-Food Sector,* Publication No. LEG-26. Guelph: Department of Geography, University of Guelph, for AES, Environment Canada.

Lamothe & Périard Consultants. 1988. "Implications of Climate Change for Downhill Skiing in Quebec". *Climate Change Digest,* CCD 88-03, Downsview: AES, Environment Canada.

OECD. 1991. *Environmental Indicators.* Paris.

Ontario Hydro. 1989. *Task Force on Greenhouse Effect,* Report 678SP. Ontario: Ontario Hydro.

P. Lane & Associates. 1986. *Preliminary Study of the Possible Impacts of a One Meter Rise in Sea Level at Charlottetown, Prince Edward Island.* For AES, Environment Canada.

Rizzo, Brian. 1990. Ref. in Standing Committee on Environment, *Out of Balance: The Risks of Irreversible Climate Change,* Ottawa, Canada.

Sanderson, Marie. 1988. "Effects of Climate Change on the Great Lakes". *Transactions of the Royal Society of Canada,* Series 5, Vol. 3: pp. 33-44.

Singh, Bhawan. 1988. "The Implications of Climate Change for Natural Resources in Quebec." *Canadian Climate Digest,* 88-08.

Smith, Jamie. 1989. *Possible Impacts of Mean Sea Level Rise on Downtown Vancouver.* Waterloo: University of Waterloo, for AES, Environment Canada.

Standing Committee on Environment. 1991. *Out of Balance: the Risks of Irreversible Climate Change.* Ottawa.

Stokoe, Peter K. *et al.* 1987. *Socio-economic Assessment of the Physical and Ecological Impacts of Climate Change on the Marine Environment of the Atlantic Region of Canada: Phase I.* Halifax: School for Resource and Environmental Studies, Dalhousie University.

Wall, G. *et al.* 1985. *Climatic Change and Its Impact on Ontario Tourism and Recreation, Final Report.* For AES, Environment Canada.

Wheaton, E.E. *et al.* 1987. *An Exploration and Assessment of the Implications of Climatic Change for the Boreal Forest and Forestry Economics of the Prairie Provinces and Northwest Territories: Phase One,* SRC Technical Report No. 211. For AES, Environment Canada.

Williams, G.D.V. *et al.* 1988. "Estimating Effects of Climatic Change on Agriculture in Saskatchewan, Canada". In *The Impact of Climate Variations on Agriculture: Volume 1* (M.L. Parry *et al.*, eds.), Netherlands: UNEP.

Global Climate Change: Implications, Challenges and Mitigation Measures. Edited by S.K. Majumdar, L.S. Kalkstein, B. Yarnal, E.W. Miller, and L.M. Rosenfeld. © 1992, The Pennsylvania Academy of Science.

Chapter Thirty-Eight

UK POLICY ON CLIMATE CHANGE*

SIMON OLIVER
Department of the Environment
Global Atmospheric Division
Room 103, Romney House
London, SW1P 3PY, England

THE GREENHOUSE EFFECT

The energy which drives our weather and climate comes from the sun. About a third of the energy that reaches the Earth is absorbed, heating the atmosphere, the oceans and the land. The warm Earth radiates infra-red energy back into space, but some of it is absorbed by gases in the atmosphere. This is similar to the effect of glass in a greenhouse, hence the gases involved are called the greenhouse gases, and the process known as the greenhouse effect. Greenhouse gases occur naturally in the atmosphere and keep the temperature of the Earth some 30°C warmer than it would otherwise be. Without them the Earth would be too cold to support life.

Water vapour is the most important natural greenhouse gas and its concentration depends on the Earth's temperature. The other principal natural greenhouse gases are carbon dioxide (CO_2), methane, nitrous oxide and ozone. Their concentrations depend on the balance between processes which produce them and those which absorb them. Emissions resulting from human activities are adding to the production of these gases and at the same time reducing the sinks such as forests which absorb CO_2 so that atmospheric concentrations are increasing. In addition the chlorofluorocarbons (CFCs), powerful greenhouse gases, are entirely man-made.

The Inter-governmental Panel on Climate Change (IPCC) Working Group I on the science of climate change, chaired by the UK, brought together over 300 of the world's leading scientists in this field, and reported in May 1990. Its report

---
*Contributed in August, 1991.

concluded that, allowing for the many uncertainties involved, increases in greenhouse gas concentrations caused by man's activities will result in additional global warming. If we go on as we are, global mean temperature will rise by around 0.3°C per decade, faster than anything we have seen for the last 10,000 years. This would result in rises of about 1°C by 2025 and 3°C by the end of the next century. Global warming will lead to unpredictable changes in the world's climate, and to a rise in sea levels due to thermal expansion of the oceans and some melting of land ice. This may threaten some low island and coastal zones by the middle of the next century, and affect fresh water supplies in many countries. We are already committed to climate change from the historical build-up of greenhouse gases in the atmosphere. We still need to do a good deal more work on the possible impacts of climate change on natural ecosystems, forestry, agriculture and other human activities. The UK will be playing a full part in further research on impacts and on the science of global warming, particularly through the work of the Hadley Centre for Climate Prediction and Research.

UK POLICY

The UK announced in May 1990, that provided other countries played their part, it was ready to set itself the demanding target of returning CO_2 emissions to the 1990 levels by the year 2005. This would be a significant first step in responding to the threat of climate change and would be achieved against the background of a substantial forecast increase in UK energy demand and CO_2 emissions. It is a realistic and achievable target which will contribute to the international response to climate change without imposing significant damage to the UK economy.

The UK target is the result of detailed work to assess the feasibility and costs of possible measures to limit CO_2 emissions in the UK, and the level of emissions which would otherwise have been expected in the future. The date of 2005 was chosen for two reasons. First, the IPCC has indicated it will take until then to achieve significant improvements in our understanding of the climate. Second, a longer time scale means that new technologies can be developed and investments made in the costly replacement of domestic and commercial energy consuming plants and products at the end of planned life times rather than prematurely.

Compared with the targets suggested by some other countries, the UK's target may not seem as ambitious. But it is not practical to suppose that every country, making comparable efforts, can stabilize its emissions at exactly the same timescale. All countries come to this issue from different starting points, e.g. some countries will start from a point of lower energy use, and may need room for development.

MEASURES TO MEET OUR TARGET

The Government set out possible measures to meet the UK target in the White Paper on the Environment "This Common Inheritance" published on 25 September

1990. The Government is already taking many practical steps which make sense in their own right as well as contributing to controlling UK emissions of greenhouse gases. Measures which reduce carbon dioxide emissions include improving energy efficiency, promoting renewable energy sources and reducing transport emissions. Promoting forestry increases the carbon dioxide absorbed and stored by trees. Britain, together with the European Community, is committed to phasing out CFCs by 1997.

The Energy Efficiency Office's very successful programmes to help promote energy efficiency will be stepped up and its budget has been substantially increased. Local authorities, in their housing repair and maintenance programmes, and through our Estate Action Programme, improve insulation in local authority housing. We are running the Green House Demonstration Programme to provide examples of effective energy efficiency in local authority housing. For private sector housing the new home renovation grants regime enables local authorities to help people less well off, in housing which is not energy efficient. In May we held a series of seminars to inform local authorities about the importance of energy efficiency in council buildings and how improvements could be made. Government Departments aim to make annual energy savings in their buildings worth up to £45 million within 5 years - a 15% increase in energy efficiency. Combined heat and power schemes will be promoted; energy labelling of houses and appliances encouraged; ways to further strengthen new building energy efficiency standards will be sought and the UK will press for new minimum efficiency standards for domestic and industrial appliances across Europe.

The Government is concerned to promote methods of electricity generation which produce less greenhouse gases and is spending a lot on research and development into renewable energy sources: over £24 million is being made available in 1991/1992, and around 80 new projects are being initiated annually. A review of renewable energy strategy in the 1990s and beyond was announced in August. The electricity producers believe that renewables and a switch away from coal to new high-efficiency gas-fired plant will keep emissions from power stations at the current level despite growth in demand.

The Government wants to see far more passenger and freight traffic travelling by rail. The pace of investment of all forms of railway has already been stepped up, and significant improvements to the Freight Facilities Grant scheme to encourage freight to move from road to rail in appropriate cases were announced in May. But investment in roads remains essential for economic, safety and environmental reasons. Emissions are at their worst in stop/start traffic so our national road programme aims to relieve congestion and to provide bypasses for hard-pressed towns and villages. Improved traffic management, stricter parking controls and better public transport will all play a part. The Government also announced in May a wide-ranging study of urban traffic congestion including the possible role of road pricing. Greater fuel efficiency is being encouraged in a variety of ways: the MOT test will be extended to improve tuning and enforcement of speed limits will be improved. Duties on fuels were substantially increased in the 1991 Budget, providing a strong incentive to use less fuel. We will consider whether further changes to the balance between fuel and vehicle taxes are needed. The Budget also raised company car taxation by 20%; this has now risen nearly four-fold over the last four Budgets.

We will continue to encourage tree planting both in Britain and support projects to help conserve the tropical forests. The UK has funded about 180 forestry projects on-going or in preparation, covering over 30 countries and with a total cost of about £108 million. This includes over 50 projects managed by British non-governmental organizations and 40 research schemes.

INTERNATIONAL ACTION

On 29 October 1990 the European Community (EC) Ministers agreed a concerted position on climate change. This important agreement endorsed the work of the IPCC and agreed that the Community, with its member states, were willing to take actions aimed at stabilizing CO_2 emissions at 1990 levels by the year 2000 in the Community as a whole. The agreement acknowledged the UK's conditional target for CO_2 emissions. It also identified areas of policies which should be pursued, such as energy efficiency and the use of renewable energy sources.

In December 1990 the United Nations General Assembly established an Intergovernmental Negotiating Committee (INC) to prepare a Framework Convention on Climate Change. The INC has met three times and will continue its work to complete the Convention in time for it to be signed during the United Nations Conference on Environment and Development in June 1992. The UK is playing a prominent part in these negotiations, and much of the discussion has been based on text produced by the UK. To provide the means for all countries to participate in the Convention the EC has proposed a process of "pledge and review". This would involve all Parties in devising and implementing national strategies containing measures to limit emissions and protect sinks. Strategies would be monitored and reviewed by an expert body established under the Convention. The EC has made clear that this would be complementary to specific commitments on emissions by developed countries.

Global Climate Change: Implications, Challenges and Mitigation Measures. Edited by S.K. Majumdar, L.S. Kalkstein, B. Yarnal, E.W. Miller, and L.M. Rosenfeld. © 1992, The Pennsylvania Academy of Science.

Chapter Thirty-Nine

JAPAN'S ACTION TO COMBAT CLIMATE CHANGE

SHINICHI ISASHIKI[1] and KAZUTO SUDA

[1]Director, Global Environmental Affairs Division
United Nations Bureau
Ministry of Foreign Affairs
2-2-1 Kasumagaseki, Chiyoda-KU
Tokyo, Japan

INTRODUCTION

In the 1960s and 1970s Japan was often referred to as a typical country troubled with serious air pollution. Later, as is now well known, both public and private sectors made enormous efforts and most serious problems have been solved or, at least, mitigated dramatically. Today, Japan is embarking on a new endeavor: environment action to combat global climate change. This effort is composed of three initiatives. The first is formulation and implementation of an action program, from 1991 to 2010, to limit CO_2 emission of Japan. The second is active participation in the negotiations on the Framework Convention on Climate Change. In the last two sessions held in June and September 1991, Japan made concrete proposals which turned out to stimulate the discussion. The third initiative is an attempt to reduce the concentration of CO_2 in the atmosphere by the end of the next century. Innovative technologies, such as fixation of CO_2, would play an important role in this very long-term operation. Japan has already established an institute to start with development of such technologies.

ACTION PROGRAM TO ARREST GLOBAL WARMING

This action program is a ministerial-level decision, made by the Council of Ministers for Global Environment Conservation, on October 23, 1990. Preparation started immediately after the Conference on Atmospheric Pollution and Climate Change held in Noordwijk, the Netherlands in November in the previous year, where many ministers emphasized urgent need to take action to combat climate change. After heated debate on the scientific evidence of global warming, the feasibility of preventive and mitigating measures and other issues, the Conference made a declaration with a paragraph in which countries recognized the need to stabilize CO_2 emission, in the view of many industrialized countries, by the year 2000. Japan maintained a cautious attitude because it had no concensus or program at that time to implement such a target of stabilization. The United States pointed out scientific uncertainties. Many Europeans, on the other hand, advocated immediate action and succeeded in making the above-mentioned declaration. Japan was criticized because of its attitude. Although it had been known, at least to experts, that the Japanese economy had achieved highly energy efficient ways of production and that the level of CO_2 emission was relatively low. Measured in terms of CO_2 emission per capita, or per GNP, the figures are one of the lowest among the OECD countries. In August, 1990, the Intergovernmental Panel on Climate Change (IPCC) made public its first report, in which renowned scientists had reached a concensus on the prediction that, the mean temperature of the earth would increase by $3°c$ by the end of the next century, unless appropriate countermeasures be taken and that, such global warming would cause a series of adverse impacts on many parts of the world. In November, 1990, on the occasion of the Second World Climate Conference, the environment ministers were expected to discuss how to tackle climate change, which scientists had already given some evidence on, although uncertainties remained on the details of the phenomenon.

In the meantime, in the Japanese government, the ministries and agencies concerned were engaged in intensive discussions on the scientific uncertainties, the feasibility of setting specific targets, measures to implement any decisions and so forth. After long and heated debate, a concensus emerged along the following line:
- In spite of remaining uncertainties, global warming is very likely to take place unless countries take action;
- Japan, although having achieved an energy efficient economy with relatively low CO_2 emission, should take further action to contribute to the collective action of countries in the world to counter global warming; and
- Targets of stabilization could be clear expression of firm will but can be meaningful only if they should be supported by feasible implementation measures.

Finally, the Council of Ministers for Global Environment Conservation adopted the "Action Program to Arrest Global Warming" on October 23, 1990. (See Annex I for the skeleton of the Program) It sets two targets. One is as follows: "The emission of CO_2 should be stabilized on a per capita basis in the year 2000 and beyond at about the same level as in 1990, by steadily implementing a wide range of measures under this Action Program, as they become feasible, through the utmost efforts by both the government and private sectors." The other reads as follows: "Efforts should also be made, along with the measures above, to stabilize the total amount of CO_2 emission in the year 2000 and beyond at about the same level as in 1990, through progress in the development of innovative technologies, etc., including those related to solar, hydrogen and other new energies as well as fixation of CO_2, at the pace and in the scale greater than currently predicted." The Action Program contains a number of concrete measures: urban and regional structure with low CO_2 emissions, transport system with low CO_2 emissions, production structure with low CO_2 emissions, energy supply structure with low CO_2 emission, and realization of life style with low CO_2.

In the Second World Climate Conference held November 6 and 7, 1990, the Japanese delegation was proud of introducing the Action Program. In contrast with the scene in Noordwijk in the previous year, the positive and concrete action of Japan was highly appreciated by the delegations as well as NGOs.

THE PLEDGE AND REVIEW

In the second session of the negotiations on the Framework Convention on Climate Change, held in June 1991, the delegations began, at last, discussion on the substance of the convention, after a long and strenuous debate on the procedural and organizational matters. Japan, together with the United Kingdom and France, proposed the pledge and review process as an implementation mechanism of commitments to be provided for in the convention. The thrust of the proposal is that each contracting party makes public a pledge, consisting of its past performance, strategies to limit greenhouse gas emission and targets or estimates for such emission as the result of the strategies. Then a review will be conducted periodically for each country/regional group, by a team of experts from different countries/regional groups and the report will be submitted to the permanent review committee. (See Annex II for the detail.)

The formula is expected to accommodate the very different situations of the countries and hence their various policies and measures, while ensuring that their activities be placed under international scrutiny. NGOs, however, criticized the "pledge and review" as the "hedge and retreat", misinterpreting that the real purpose of the proposal was to justify no or little action.

In the third session, held in September 1991, the Japanese delegation clarified the idea and also made a proposal on the commitment as follows:
• All Parties shall limit emissions of greenhouse gases; and Industrialized countries, in particular, shall make the best efforts aimed at stabilizing emissions of CO_2 or CO_2 and other greenhouse gases not controlled by the Montreal Protocol, as soon as possible, for example, by the year 2000 in general at 1990 levels recognizing the differences in approach and starting point in the formulation of objectives.

As a result, the criticism by NGOs receded while developing countries became concerned with the review. They thought that it could be interference in their internal affairs. Many developing counties, however, supported the review of implementation of the convention as an established practice of many international agreements. It seems that whether a review is acceptable for developing countries or not will depend on the specifics of the concept, which will be an important topic of the next session.

The texts drafted by the bureaus of the two working groups of the Intergovernmental Negotiating Committee on Climate Change include provisions as follows:
- common commitments for both developed and developing countries;
- other commitments which are differentiated between developed and developing counties;
- submission of reports on the measures parties have undertaken to implement their obligations; and
- review of the progress by the Parties in implementing all their obligations and actions under the convention.

The text reflects well the debate stimulated by the proposal of the pledge and review and the wording used is now more traditional legal terminology. The Japanese proposal contributed, after all, to development of the basic structure and content of the convention.

THE NEW EARTH 21

The third initiative is an action program called "The New Earth 21", which will pursue the recovery of the earth from the past accumulation of CO_2 and other greenhouse gases toward the end of the twenty-first century. The first fifty years will be a transition period when environmentally friendly technologies will be developed and introduced. (See Annex III for the detail.) The second fifty years are the period when future generations will draw upon the results of the first fifty years to achieve the recovery of the planet. The government of Japan advocates to promote this ambitions long-term action program in cooperation with other countries interested in the formidable, yet promising, enterprise toward the next century.

Annex I: Summary of Action Program to Arrest Global Warming

The Action Program to Arrest Global Warming is a decision which was made by the Council of Ministers for Global Environment Conservation, on October 23, 1990.

I. Background for the Action Program and its Significance
   The Action Program has been formulated to define Japan's basic position to take part in the formation of an international framework dealing with global warming.
II. Basic Elements to be Taken into Account for Promotion of Measures to Cope with Global Warming.
   1. Formation of environmentally-sound Society
   2. Compatibility with economy's stable development
   3. International coordination
III. Targets under the Action Program
   The targets for the limitation of greenhouse gas emissions shall be set as follows:

(1) The Government of Japan, based on the common efforts of the major industrialized countries to limit CO_2 emissions.
   a. The emissions of CO_2 should be stabilized on a per capita basis in the year 2000 and beyond at about the same level as in 1990, by steadily implementing a wide range of measures under this Action Program, as they become feasible, through the utmost efforts by both the government and private sectors.
   b. Efforts should also be made, along with the measures above, to stabilize the total amount of CO_2 emission in the year 2000 and beyond at about the same level as in 1990, through progress in the development of innovative technologies, etc., including those related to solar, hydrogen and other new energies as well as fixation of CO_2, at the pace and in the scale greater than currently predicted.
(2) The emission of methane gas should not exceed the present level. To the extent possible, nitrous oxide and other greenhouse gases should not be increased. With respect to sinks of CO_2, efforts should be made to work for the conservation and development of forests, greenery in urban areas and so forth in Japan and also to take steps to conserve and expand forests on a global scale, among others.

IV. Duration of the Action Program

The Action Program covers the period from 1991 to 2010 with 2000 set as the intermediate target year. During this period, the Action Program should be reviewed, as necessary, for its flexible response to international trends, accumulated scientific findings and so on.

V. Necessary Measures
  1. Measures to limit CO_2 emissions
     a. Formation of urban and regional structure with low CO_2 emissions
     b. Formation of transport system with low CO_2 emissions
     c. Formation of production structure with low CO_2 emissions
     d. Formation of energy supply structure with low CO_2 emissions
     e. Realization of life style with low CO_2
  2. Measures to reduce emissions of methane and other greenhouse gases
     a. Measures against methane
        i) Measures in waste management
        ii) Measures in agriculture
        iii) Measures in energy production and use
  3. Measures to enhance CO_2 sinks (forests and other greens)
     a. Adequate management of domestic forests and greens in urban areas.
     b. Regional use of timber resources
  4. Promotion of research and observation/monitoring
     a. Research and survey
     b. Observation/monitoring
  5. Development and dissemination of technology
     a. Technology for limiting the emission of greenhouse gases
     b. Technology for absorption, fixation, etc., of greenhouse gases

c. Technology for adaption to global warming
6. Promotion of public awareness
Efforts should be made to disseminate the outline of the Action Program and precise information, to promote environmental education, to support and subsidize voluntary actions for such purposes.
7. Promotion of international cooperation
    a. Comprehensive support for arresting global warming
    b. Promotion of technology transfer
    c. Support to conservation and development of tropical forests and other sinks of carbon dioxide
    d. Promotion of cooperation in research and the development of appropriate technology
    e. Promotion of international cooperation in the private sector
    f. Considerations of prevention of global warming in international cooperation projects
VI. Promotion of the Action Program
    (1) Each ministry and agency should implement measures for realization of the items in V.
    (2) The Council of Ministers for Global Environment Conservation should report each year on the progress of implementation of the measures. On the basis of such reports, the promotion of the Action Program should be reexamined as necessary.
    (3) The central government should provide support to local governments.
    (4) Each ministry and agency should strive to work for broad dissemination of the Action Program and take measures necessary to support the efforts of the business and other parties in these lines.

Annex II: The Pledge and Review Process

The following is intended to describe the Pledge and Review process as a possible mechanism to implement commitment defined on the basis of the convention.
1. Pledge
Each country (or regional group) makes public a pledge, consisting of its past performance strategies to limit greenhouse gas emissions and targets or estimates for such emissions as the result of the strategies.
    (1) Participaing countries (or regional groups) must make Pledges as soon as possible (within three months) after the ratification (entry into force) of the Convention. The pledges will be made public as soon as possible.
    (2) The pledges should include concrete response measures to be taken in each sector, e.g. energy, industry, agriculture, forestry, etc.
    (3) Regional pledges by reginal group must, in principle, be accompanied by subpledges of each country.
    (4) Necessary technical and procedural matters, etc., for the pledging should

be spelled out in the Convention or an annex of the Convention.
2. Review
   A review will be conducted peridically for each country/regional group by a team of experts from different countries/regional groups and the report will be submitted to the permanent review committee.
   Note: Each country/regional group will submit an interim progress report during the interval.
   (1) The review team will send questionnaires and carry out on-site surveys.
   (2) Based on answers to questionnaries, on-site surveys and consultations with the government, the review team will draft a report, including evaluations of the current state of the implementation and appropriateness of the pledge.
   (3) The results of the review, i.e. the report, will be made public through appropriate procedures by the review committee and the Conference of the Contracting Parties. (By making public the report, it is hoped that international opinion will encourage posititive action by participating countries.)
   (4) The report may include recommendations on the measures to be taken by the country/regional group. In the case of a developing country/regional group, it may include recommendations to the international community for assistance.
   (5) Necessary institutional and procedural matters, etc., for the implementation of the review should be spelled out in the Convention or an annex of the Convention.
   (6) Guidelines for the review will be established by the Conference of the Contracting Parties of the review committee.
3. Considerations for Developing Countries
   (1) Developing countries will be entitled to delay its pledge by 1 year.
   (2) Developing countries may request assistance for their country studies.
   (3) The need for external financing and technology transfer should be adequately considered in reviewing pledges made by developing countries.

Annex III: New Earth 21

Of the first half-century, the first decade, starting now, will be dedicated to intensified scientific research to reduce uncertainties and better energy efficiency through the increased use of available technology, both in developing countries and developed countries. Conventional measures to enhance sinks, i.e. reforestation and better forestry managements will receive full attention, as will technology development for the coming decades.

In the second decade, there will be a reduction in the use of fossil fuels through the introduction of non-fossil fuels, i.e. safe nuclear power plants, and new or renewable energy sources.

The third decade will see the spread of non-greenhouse gas substitutes for chlorofluorocarbons, carbon dioxide fixation and reutilization technology, and revolutionary, low-energy production processes.

The fourth decade is the decade of big advances in absorption. There will be substantial net gains from the reversal of desertification through the use of biotechnology, as well as more conventional means to enhance sinks such as reforestation. Oceanic sinks will also be enhanced.

The fifth decade will be the era of future generation technologies, such as fusion, orbiting solar power plants, magma electricity generation, energy applications of superconductive technology, and other new forms of energy technology, which may make fossil fuels unnecessary.

Global Climate Change: Implications, Challenges and Mitigation Measures. Edited by S.K. Majumdar, L.S. Kalkstein, B. Yarnal, E.W. Miller, and L.M. Rosenfeld. © 1992, The Pennsylvania Academy of Science.

Chapter Forty

UNITED NATIONS SPONSORED 1992 EARTH SUMMIT IN RIO DE JANEIRO:
Where Do We Go From Here?

E. WILLARD MILLER[1], MICHELLE A. BAKER* and SHYAMAL K. MAJUMDAR[2]

[1]Department of Geography
The Pennsylvania State University
University Park, PA 16802
and
[2]Department of Biology
Lafayette College
Easton, PA 18042

INTRODUCTION

The United Nations Conference on Environment and Development (UNCED) was held in Rio de Janeiro, Brazil in June in order to define goals for environmental protection and economic development for the next 10 to 20 years. Over 130 heads of state and delegates from 160 nations attended the conference (Speth 1992). The conference was organized by the International Council of Scientific Unions, which in the past has worked to integrate economic and environmental concerns (Marton-Lefevre 1992).

---

*Present Address: Department of Biology, University of New Mexico, Albuquerque, NM 87131.

The main goals of the Earth Summit were to devise plans for controlling global climate change, protection of biodiversity, examination of population growth and financial support of developing nations (Speth 1992). There were three main objectives: the problems of environment and development, the scientific understanding of the global ecosystem, and how science can contribute to environmental and developmental problems. These objectives were subdivided into sixteen themes for discussion. Section one themes emphasized population and use of natural resources, agriculture, land use and soil degradation, industry and waste, energy and health. Section two was concerned with the atmosphere and climate, marine and coastal systems, terrestrial systems, freshwater resources and biodiversity. Themes in section three included quality of life, public awareness, science and the environment policies for technology and institutional awareness. These became the cornerstone for An Agenda of Science for Environment and Development into the 21st Century or ASCEND 21 (Marton-Lefevre 1992).

The meeting marked the 20th anniversary of the UN Stockholm Conference on Human Environment, in which the United States played a major role. Before the conference in Rio began, the United States' environmental policy was perceived to be the major obstacle to a successful meeting (Speth 1992). Many hoped that the conference would bring a breakthrough in international politics and economy, sparking cooperation and understanding between nations (Kildow 1992).

Understanding the global environmental problems that we face: staggering human population growth, impacts of overconsumption, deforestation, climate change, and loss of biological diversity requires cooperation between scientists, government and the general public. Public awareness of scientific principles must be expanded to ensure protection of the environment. Science must take an active role in education and development in order to enlist the help of the public in solving environmental problems (Marton-Lefevre 1992). At the summit, ASCEND-21 urged cooperation between scientists and development agencies to address the concerns of the environment and economic development. It also called for extensive international research into the Earth's capacity to support life and ways to reduce population growth and overconsumption of resources (Marton-Lefevre 1992).

As mentioned in R.A. Reinstein's introduction to this volume, prior to the meeting, UN negotiators outlined plans that would address issues concerning the global environment, including treaties to decrease carbon dioxide emissions and preserve biological diversity. These major goals of the meeting in Rio were not fully supported by the United States (Associated Press). In Rio, the U.S. refused to sign the plan on global warming until the cap on carbon dioxide emission (a major greenhouse gas that contributes to global warming) was removed from the treaty (Associated Press). The United States also refused to sign the UN plan to protect biological diversity. Other agenda of the conference such as the development of the "Earth Charter" were not effective. Instead, the delegates produced a non-legally binding "Rio Declaration" and the Agenda 21, which outlined appropriate (non-legally binding) actions for governments and development agencies. The delegates also made a "Statement on Forest Principles" which was also not legally binding (Associated Press and Chemecology 1992).

The United States government policies on the environment have been criticized as being "misguided" since the US refused to make commitments to specific targets and timetables to reduce carbon dioxide emissions and specific plans for financial aid to developing countries (Speth 1992). Some critics of the development of sound environmental policies contend that supporting environmental and development issues will not advance United States' political and economic growth, when in reality, helping developing countries would expand US exports and expand the trade and investment foundation. Recognizing environmental concerns would make the US more competitive in international markets, where energy efficiency and reduced consumption of fossil fuel are important (Speth 1992). If the United States refuses to curb its overconsumption of fossil fuels to reduce carbon dioxide emissions, can it be expected that developing nations will reduce their consumption of these resources?

In general, the Earth Summit was perceived to be a success. Despite the lack of US support, the Biodiversity treaty was passed at the conference. The Global Warming Treaty was signed, with US support after the cap on carbon dioxide emissions was removed. A statement protecting forests and a plan of action for protecting the environment in developing countries were also made (Chemecology 1992). Despite these successes, a basic question must be raised: will the Rio conference be anything more than a platform for discussion? The expectation for a major breakthrough in international politics driven by unsolved problems of the environment may have been too much to expect from one meeting. But will this Earth Summit begin a process for the awakening of peoples and governments to the fact that action is required to protect a life-sustaining world? Will a plan evolve to effectively address the problem of doubling of the population of the Earth in the next several decades? There is strong evidence that the Earth Summit laid a solid foundation for the future. Each nation must now begin to implement plans to develop a livable world for the future. The impetus of the Earth Summit must not be lost, and there must be a continuation of the talks in future Earth Summit Conferences to achieve the desired goals.

REFERENCES CITED AND FOR FURTHER READING

Abelson, Philip H. Editorial: agriculture and climate change. *Science.* 257:9.

Associated Press. June 1992. Historic 'Earth Summit' opens this week. *The Express-Times,* Easton, PA.

Associated Press. June 1992. U.S. staves off summit critics with $1.4B fund. *The Express-Times,* Easton, PA.

Celso do Amaral e Silva, Carlos. 1992. Environmental protection, a view from Brazil. *Environmental Science and Technology. 26(6):1079-1080.*

Chemecology. 1992. Business, industry have voice at Earth Summit. *Chemecology.* July/August 1992.

Kaufman, Ron. 1992. Rio document spurs debate: is science an ecological foe? *The Scientist* 6(15):1.

Kildow, Judith T. 1992. The Earth Summit - we need more than a message. *Environmental Science and Technology.* 26(6):1077-1078.

Marton-Lefevre, Julia. 1992. Mobilizing international science. *Environmental Science and Technology.* 26(6):1085-1088.

Rubin, Edward S., Richard N. Cooper, Robert A. Frosch, Thomas H. Lee, Gregg Marland, Arthur H. Rosenfeld and Deborah D. Stine. 1992. Realistic mitigation options for global warming. *Science.* 257:148-149, 261-265.

Speth, James Gustave. 1992. On the road to Rio and to sustainability. *Environmental Science and Technology.* 26(6):1075-1076.

Valencia, Isabel M. 1992. Lessons for UNCED '92: a consortium of universities looks at the big picture. *Environmental Science and Technology.* 26(6):1081-1082.

Subject Index

A

Aboriginals, 535
Acid deposition, 295
Acid rain, 96
"Action Plan to Arrest Global Warming", 544, 546, 547
Adaptation, 292, 293, 295, 299, 331, 336, 338, 341
Adiabatic lapse rate
 -moist, 74
 -dry, 74
Adriatic Sea, 496
Afforestation, 439, 477, 496
Agenda of Science for Environment, 552
AIDS, 363
Air conditioning
 -automobiles,
 -role in mitigating heat-related mortality, 371
Air masses, 347
 -tropical, 347
 -polar, 347
Air pollution, 243, 389, 400, 403
 -effect on health, 243
Albedo, 3, 4, 82, 94, 112, 149-151, 180, 193-196, 200
 -clouds, 94, 194
 -planetary (reflectivity), 64, 67, 78, 79
 -surface, 193-196, 200
Animal husbandry, 474
Anthropogenic alteration, 107
Anthropogenerated sulfate, 107, 111-113
Anticyclones, 211
Arctic warming, 106
Atmosphere autovariation, 211
Atmospheric carbon dioxide levels, 227, 228
Atmospheric emissivity, 67, 70
Atmospheric radiative transfer, 70
Atmospheric window, 72
Autonomous increases in energy efficiency, 138
Average absolute temperature, 231

B

Bangladesh, 531
Beach creation, 265
Beach erosion, 273
Beach restoration projects, 261, 262
Biodiversity, 526, 552
Biodiversity Treaty, 553
Biogeochemical cycles, 471
Biological pump, 122
Biomass, 439, 442, 444, 447-449, 463, 501, 525
Biomass energy sources, 416, 420
Biotic response, 119, 120
Black-body, 65, 66, 68, 71
"Black rain", 243, 252
Black Sea, 496
Boreal forests, 275, 277-279, 282, 283, 294, 295
Boreal lakes, 296
Boundary forcing function, 83
 -energy output, 83
 -temperature distribution, 83
"Brightness temperatures", 166
British Columbia, 532
Brown tides, 298

C

C3 plants, 352, 353, 358
C4 plants, 352, 353
California, 498-500
 -California Energy Commission, 502
 -contribution to global climate change, 500
 -economic size, 499
 -economy, 499
 -electric generating capacity, 501
 -energy bases, 499, 500
 -Environmental Protection Programs, 500
 -influence on global economic & social trends, 500
 -national and global perspectives, 498, 499
Canada, 529
Canadian Climate Board, 535
Canadian Climate Center, 530, 535
Canadian Climate Center (CCC) model, 154, 155, 159, 163
Canadian Council of Ministers of the Environment, 534
Canadian Gas Association, 534

Canadian Petroleum Association, 534
Carbohydrate accumulation, 350
  -rate of, 350
Carbon
  -annual budget, 121
  -biosphere sink, 121
  -budget, 525
  -cycle, 117, 118, 119-122, 129
  -dissolved inorganic, 120-122
  -dissolved organic, 120
  -net loss, 120
  -particulate organic, 120
  -terrestrial pools of, 118, 119
Carbon cycle, 449
  -role of vegetation, 449
Carbon dioxide, 189, 205, 303-305, 309-312, 331-333, 350-356, 358, 381, 385, 386, 389, 402, 405, 410-412, 416-420, 423, 427, 435, 349-441, 447-49, 471-474, 476-479, 495, 500, 503, 526, 531, 532, 539-542, 543
  -absorption of, 423, 478
  -atmospheric concentration, 117, 118, 120-123, 129, 148, 190, 192-196, 198, 204, 205
  -carbon-14, 118, 126-128
  -climate, 147, 149
  -climate change experiments, 190
  -concentration of, 495, 496
  -doubled CO_2 experiments, 153, 154, 159, 190, 194-197, 200, 201, 205
  -economic activity, 130, 131, 133, 139
  -emissions, 135-140, 177, 317, 319, 324, 424, 425, 427-430, 432, 452, 455, 457, 465, 472-474, 476, 477, 479, 543, 552
  -emitted by California, 500
  -energy return on investment, 131, 138
  -fertilization, 282-284
  -fertilization effect, 385, 386, 403, 405
  -Gaussian forcing, 126-128
  -global climate threat, 131
  -impact of high levels on plants, 353
  -industrial emissions, 121, 132, 134
  -partial pressure, 121, 122
  -reduction of, 334, 424, 425-427, 462
  -relation to electricity savings, 428
  -sinks of, 547
  -strategies to reduce emissions, 445
Carbon monoxide, 425
Carbon/nitrogen ratios, 353
Carbon sinks, 439, 442,
Carbon tax, 418, 460
Caspian Sea, 242
Cement industry, 473

Central and West African region
  -agriculture, 485
  -deserts, 484
  -diseases, 484, 487
  -drought, 482-484, 486, 488
  -erosion, 483, 485
  -floods, 482, 483, 488
  -food and nutrition, 486, 491
  -forest depletion, 485
  -fresh water problems, 484, 485
  -heat waves, 484
  -human health, 486
  -human settlements, 485
  -migration, 487, 489
  -runoff water, 482, 484, 485
  -sand storms, 484
  -sea level rise, 484-486, 488, 491
  -seawater/salt water intrusion, 482, 484
  -socio-economic installations, 488
  -temperature rise, 484
  -wind storms, 484
Central receiver systems, 415
CFC production, 423, 424, 426, 427, 431
  -HCFC and HFC substitutes, 427
China, 526, 531
Chlorofluorocarbons, 101, 102, 332, 460, 461, 532, 539, 541, 549
Cholera, 363, 364
Circulation of the atmosphere, 7
Clean Air Act, 338
Climate, 256
  -effect on sea level position, 256
Climate change, 209, 210, 212, 213, 222, 324, 346-349, 351-358, 377, 522, 539, 540
  -adaption to, 357
  -CO_2-induced, 213
  -convention, 523
  -effect on agriculture, 346, 348-350, 354, 355
  -effect on temperate regions, 358
  -effect on tropical regions, 358
  -empirical studies, 211
  -impacts on human mortality, 381
  -rate of, 292
  -research, 494
Climate Change Adaptation Branch, 536
Climate data, 229
Climate modelling experiments, 145-152, 159, 189, 190, 191, 213
  -perturbation, 213
Climate system, 1, 7, 9, 11, 174, 177, 190, 205, 210, 213, 227, 228
  -components of, 1, 2, 7

Index 557

Climate variability, 43, 209, 210, 212-215, 222
  -empirical studies, 211
  -nature and causes, 210
Climate variables, 214, 217
  -sea ice boundaries, 214
  -sea surface temperatures (SSTs), 21, 217
Climatic
  -anomalies, 82, 88, 91
  -change, 14-16, 20, 25
  -discontinuity, 1
  -fluctuations, 15, 16
  -models, 79, 95
  -optimum, 50
  -oscillations, 15
  -system, 82
  -trend, 1
  -vacillation, 15
  -variability, 94
Climatic factors, 258
  -negative feedbacks, 258
Climatic variation, 354, 357
Climatology, 54
Clouds, 77-80, 193, 194
  -cirrus clouds, 78
  -cloud cover, 193, 194, 196, 205
  -convective, 194, 196
  -development, 496
  -effects on climate, 150, 151, 153
  -ice cloud, 154
  -optical properties, 78
  -stratiform, 194
  -stratosphere, 192
  -stratus clouds, 78
  -troposphere, 154, 192
  -water cloud, 154, 194
Cloud scheme, 154
  -cloud water variable (CW), 154
  -relative humidity (RH), 154
Coagulation, 245
  -of smoke, 245
Coal, 130-132, 135-137, 139, 140, 429, 460, 461, 499, 501, 531
  -consumption, 455, 457, 458, 462, 476, 477
Coal Association of Canada, 534
Coalbed methane, 412
Coastal communities, 271
  -geomorphology, 271
  -of Long Island, 271
Coastal development, 261
Coastal dynamics, 265, 266
Coastal erosion, 259, 260, 262
Coastal flooding, 259

Coastal morphology, 266
Coastal recession, 266
Coastal shore, 272
  -stabilization, 272
  -protection, 272
Coastal wetlands, 296
Cold snaps, 215
Combustion, 245
Community outreach, 435
Conference on Atmospheric Pollution, 544
Conservation, 138, 139, 424
Continental border geography, 228
Coral bleaching, 297, 298
Coral reefs, 297
Council of Ministers for Global Environment Conservation, 544, 546, 548
Cyclones, 211
Cyclonic storms, 347
Cyclonic flooding, 261
  -in Bangladesh, 261
  -as a result of sea level rise, 261
  -as a result of third world population growth, 261

D

Daily mortality, 372
Daily precipitation, 217, 221
Daily variability, 215, 219
Decade-to-decade climatic variations, 32, 33
Deforestation, 118-121, 129, 423, 319, 441, 442, 455, 459, 460, 464, 476
  -as a source of carbon dioxide, 423
Dendrochronology, 19
Dengue, 364-366, 369
Deserts, 295
Destruction of oil wells, 242-245
  -environmental impacts, 244
Development into the 21st Century
  -Agenda 21, 552
Dew formations, 3
Diesel fuel, 453
Diseases, 484, 487
  -bacillary dysentery, 487
  -cerebral meningitis, 484, 487
  -cholera, 487
  -hookworm-induced health problems, 487
  -malaria, 487
  -schistosomiasis, 487
  -yellow fever, 487

Dispersal, 294, 295, 299
Distillate oil, 501
District heating and cooling systems, 430, 431
Diurnal temperatures, 4, 106, 107
Droughts, 209, 215

E

Earth Charter, 552
Earth Radiation Budget Experiment, 79
Earth's energy budget, 2-7, 9
Earth-sun geometry, 9
Eastern Europe, 525
Eastern Mediterranean Basin, 494, 496
Economic development, 489
Economy and Environment, 535
Ecosystems, 526
 -aquatic, 296
 -terrestrial, 294
Egypt, 531
Eigenvector, 40
Electric power generation technologies, 465
Electric power sector expansion plans, 456
Electricity generation, 541
Electromagnetic radiation, 287
El Niño, 16, 30, 31
El Niño-Southern Oscillation (ENSO), 9, 29-33, 210
Emission, 73
Encephalitis, 367
 -Japanese, 364
 -St. Louis, 364
Energy
 -balance, 2
 -conservation programs, 429
 -consumption, 386, 454, 455, 462
 -demand, 137-140, 453, 540
 -efficiency, 96, 454, 541, 542
 -flows, 1, 3, 12
 -improvements, 462
 -intensity, 131, 135, 138-140
 -perspectives, 458
 -policy, 453, 458
 -prices, 135, 137-140
 -production, 473
 -radiative, 175
 -requirement, 454
 -responses, 458
 -services, 452-454
 -sources, 457, 463
 -supply, 136, 462, 465

Energy efficiency, 424, 425, 500
 -of California, 500, 501
Energy production, 501
 -hydroelectric power, 501
 -natural gas supplies, 501
Environmental disasters, 245
 -oil spills, 245
 -volcanic eruption, 245
Environmental interest groups, 459
Environmental observations, 245
Environmental Protection Agency (EPA), 145, 146, 155, 374, 377, 381, 416, 418, 419, 454, 503
 -Climate Change Division, 381
 -report on global climate change, 416
Equatorial regions, 4, 7
ERDAS GIS software, 266
Erosion, 265-267, 347, 449
 -as a result of harvest operation, 449
European Community (EC), 497
European Environment Agency, 321
Evaporation, 5, 6, 495
Evapotranspiration, 6, 51, 53-61, 275, 284, 335, 336, 445
Evapotranspiration regimes, 351

F

F-test, 219
Fauna and flora, 496
Feedback mechanisms, 84
"Fertilizer effect" of carbon dioxide, 107
Finland, 523, 526, 527
Finnish Academy of Science, 527
Floods, 209
Fluorescent light bulbs, 139
Foraminifera, 17
Forams, 17
Forests, 395, 399-401, 524, 526, 530, 547
 -effect on climate, 275
 -redistribution of tree species, 276
 -timber, 274, 278
Forest fire, 442, 526
Forest growth model, 284
Forestry, 439, 441, 449
Forty-fifty day oscillation, 29, 33
Fossil fuels, 117-119, 122, 126, 127, 128, 129, 131, 455, 457, 459, 461, 462, 465, 473, 500, 501, 549
 -combustion, 113, 120, 121, 130, 132
 -emissions, 120, 132, 134, 148

Framework convention of climate change, 533, 542, 543, 545
Fresh water supplies, 272
Freshwater wetlands, 296
Frozen-grid experiments, 39, 40
Fuel cycle, 411, 412
Fusion, 550

G

Gas consumption, 455
Gas emissions, 352
General circulation of the atmosphere and oceans, 4
General circulation patterns, 212
Geographic latitude/longitude matrices, 229
Geographic output, 230
Geophysical Fluid Dynamics Laboratory (GFDL), 191, 196, 198, 200, 201, 204
Geopotential height, 214
Geothermal energy, 417-420, 477
Geothermal power, 499, 501
GFDL model, 304-306, 309, 312
GIS, 264, 266, 267, 272, 273
  -in analysis of the effects of sea level rise, 264, 272, 273
GISS model, 304, 305, 308, 309, 312, 381
Glacial erratics, 16
Global agriculture, 346
Global change, 227
Global (General) Circulation Models (GCMs), 38, 41, 43, 44, 47, 51-54, 60, 68, 82, 86, 88, 94, 100, 101, 105, 106, 110, 113, 148-150, 177, 189-198, 200, 201, 204, 205, 209, 213-215, 222, 227-9, 233, 234, 277, 279, 284, 535
  -atmospheric GCM, 191-195, 200, 204, 205
  -atmosphere-ocean model, 121, 123, 128-130, 200, 201, 205
Global climate, 83, 244, 288
Global climate change, 95, 166, 168, 177, 274, 275, 278, 452, 461, 464, 431-491, 498, 500, 502
  -adaptation, 95, 97
  -awareness of, 489
  -the California perspective, 498, 500
  -in the Central and West African region, 481, 483-491
  -effect on agriculture, 502
  -effect on air quality, 502
  -effect on economy, 502
  -effect on energy, 502
  -effect on forestry, 502
  -effect on natural habitat, 502
  -effect on outdoor recreation, 502
  -effect on public health, 502
  -effect on stability of bays and beaches, 502
  -effect on water supplies, 502
  -impacts on human population, 481, 482, 486, 488, 489
  -precipitation, 276, 277, 279
  -prevention, 95-97
  -recommended options, 489
  -strategies, 490
  -temperature, 276, 277, 279
Global Climate Models (GCM), 495
Global economy
  -growth and development, 453, 457, 459
Global ecosystems, 425
Global energy supply, 455
Global environment
  -catastrophe, 460
  -change, 452
  -European Community, 320
  -facility, 524
  -Federalist approach, 321-323, 327
  -protection of, 327
  -survival of, 459
Global environmental problems, 552
Global-mean:
  -precipitation, 92
  -sea level, 92
  -surface warming, 92
  -temperature, 37-41, 43, 44, 47, 64
Global mean temperature (GMT), 335
Global precipitation, 351
Global scale climatic variations, 27-29, 31, 32, 33
Global temperature change, 90
Global temperature record, 179, 182-185
Global warming, 28, 31, 33, 43, 45, 81, 89-91, 96, 100, 101, 103, 110, 113, 154, 165, 166, 171, 177, 179, 184, 185, 189, 201, 204, 205, 207, 210, 256, 261, 262, 264, 265, 275, 277-279, 281, 283, 284, 287, 292, 295, 315-317, 319-321, 315-317, 319-321, 325, 326, 330-334, 336-341, 347, 350, 354, 357, 371, 372, 376, 381, 384-388, 390, 395, 404, 406, 425, 435, 467, 472, 478, 481, 484, 485, 491, 494, 502, 503, 540, 543
  -action plan, 543
  -anthropogenic, 256
    -physics of, 63
    -signal of, 179, 183
  -biophysical impact, 350
  -consequences on human beings, 481
  -federation, 321

-heat stress-related mortality, 371, 372, 374, 376, 377, 380
-impacts on human health, 371, 372, 376, 388
-response, 467
-in tropical regions, 354
Global Warming Treaty, 553
Goals of the Earth Summit, 552
Goddard Institute for Space Studies (GISS), 191, 196, 198, 200, 204
Grasslands, 295
Great Lakes, 531
Greenhouse
  -atmospheric, 63, 65, 66, 68
  -climate, 24
  -effect, 3, 10, 37, 38, 43, 47, 86, 87, 101, 107, 110, 112, 113, 145, 148, 165, 245, 264, 335, 338, 344
  -gas forcing, 394, 472, 481, 484, 485, 491, 539
  -gases, 37, 38, 41, 74, 79, 87-90, 94, 96, 102, 175, 177, 182, 185, 209, 211, 213, 227, 257, 258, 320, 330-334, 355, 439-441, 472, 495, 500, 539
    -concentration of, 423, 540
    -costs for reduction of, 436
    -emissions of, 410-412, 414, 416, 418, 420, 423-426, 433, 435, 436, 440, 441, 457, 459, 460, 465, 467, 524, 526, 539, 541, 545
    -carbon dioxide, 85-88, 91, 92
    -methane, 96
    -nitrous oxide, 96
  -reduction, 418, 420, 457, 462
  -taxes, 460
Greenhouse-induced warming, 503
Greenhouse warming, 245, 460
Green Plan, 535
Ground heat flux, 3
Groundwater aquifers, 495, 496

H

Hadley Cells, 7, 8
Hard-rock coasts, 260
Heat exhaustion, 376
Heat stroke, 376
Heat waves, 209, 215, 374, 376
Herbaceous vegetation, 449
High-latitude precipitation, 92
Historical climate data, 212
Historical Climate Network (HCN), 104, 105, 107, 183)
Historical temperature record, 38

Holdridge classification of "life zones", 335, 336
Horizontal temperature changes, 3, 4, 8, 9
Human
  -activity and climate, 1
  -health program, 381
  -mortality, 371, 372, 374
  -pollution, 81
Human-induced changes, 40
Human mortality, 371, 372, 374
  -global warming, 371, 372, 376, 381
  -rates of, 374
Humidity, 76, 77, 182, 194, 196
Hydrocarbons, 425
Hydroclimatic conditions, 51
Hydroelectric facilities, 457, 465
Hydroelectric power, 414, 415, 417, 420, 501, 532
Hydroelectric power plants, 477
Hydrologic cycle, 1, 5, 7, 310, 311, 347
Hydrologic variables, 54
Hydrological service, 497
Hydrology, 215, 228
Hydropower, 390, 394, 395
Hydrothermal energy, 418

I

Ice, 531
Ice Age, 256, 259
Immobile coasts, 260
  -Rocky/armored coasts, 260
Income effect, 131, 135
India, 531
Indian monsoon, 16
Infrared
  -absorption, 73, 107
  -radiation, 333
  -radiative flux, 68, 71
"Insulation effect", 66
Interannual variability, 210, 212-215, 222
  -internal and external forces, 210
  -of precipitation, 219
  -of temperature, 212, 214-217, 219, 221, 222
Inter-governmental Negotiating committee (INC), 542, 545
Inter-governmental Panel on Climate Change (IPCC), 37, 42, 43, 89, 100, 123, 170, 171, 523, 539, 540, 542, 544
Interhemispheric differential, 107
International Council of Scientific Unions, 89, 551

Intertidal salt marshes, 259
Intertropical Convergence Zone (ITCZ), 7
Intra-zonal variations, 236
 -zonal mean data, 236
Irrigation, 348, 351, 357
Isotopic daiting, 18, 19
Israel, 494-496
 -National Academy of Sciences, 494
 -Oceanographic and Limnological Research, 497

J

JABOWA, 275, 278, 280
James Bay, 530
Japan, 543, 548
Jaundice, 367
Jet streams, 9

K

Kentucky, 529
Kuwait, 242, 243
 -"Black Rain", 243, 252
 -firefighters, 243
 -oil fires, 245, 251, 252
 -oil wells, 246, 253
 -shoreline, 243

L

La Nina, 30, 31
"Laboratory earth", 81
Land surface conditions, 228, 230
Land temperature uncertainties, 38, 39, 42
Latent heat flux, 3, 5, 7
Layer emissivities, 69, 70
Leishmaniasis, 364
Lichens, 19
Lignite, 476
Little Climatic Optimum, 22
Little Ice Age, 23, 25, 50
Local Heat Flux (Q-Flux), 152, 153
Long-wave terrestrial radiation, 471
Low-lying deltas, 531

M

MacKenzie River Basin, 535
Magma electricity generation, 550
Manitoba, 532
Marine pollution, 319
Maritime continent, 30
Materials flow, 5
Methane, 332, 333
Mathematical models, 81
Mean global warming, 357
Mean precipitation, 212
Mean temperature, 212
Mean zonal comparisons, 233
Mesozoic, 20, 24
Meteorological service, 497
Meteorological variables, 92
METEOSAT, 246, 248, 250
Methane, 332, 333, 410, 412, 420, 472, 474-476, 478, 479, 500, 526, 532, 539, 547
Methane emissions, 423, 425, 426, 474-476
Microclimate, 447
Micro-wave Sounding Unit (MSU), 39, 40, 41, 165, 166, 168, 174, 176, 177
Middle East, 528
Midlatitude cells, 8, 9
Migration, 293, 294, 296, 298, 299, 381, 487, 488
 -internal, 487
Migration of tree species, 276
 -rate of, 276
Military conflict, 243
Mitigation, 330, 331, 340, 341
Mitigation measures, 472, 476, 479
Mixed-species forests, 447
Mix effect, 131, 132, 135
Model temperature range, 235
Model verification, 84-87
Moisture availability, 347, 350
Moisture stress, 351
 -on flowering, pollination, 351
Montreal Protocol Interim Fund, 524
Montreal Protocol on Substances that Deplete the Ozone Layer, 458, 532, 545
Monsoon, 239, 251
Monsoon circulation, 4, 9
"Monthly mean" atmosphere, 168
Mortality/gender relationships, 376
Mortality/race relationships, 376
Multilateral fund, 524
Multilayer model, 69, 7
Municipal action plan, 428
 -buildings, 431

-district heating and cooling, 430
-education and community outreach, 435
-land use, 434
-natural gas, 430
-transportation, 432
-urban forestry, 434
-waste management, 435
Municipalities, 423, 424, 429
-role in the reduction of greenhouse gas emissions, 422, 424-426, 428

N

National Action Strategy on Global Warming, 534
National Center Atmospheric Research (NCAR), 109, 110, 146, 147, 163, 191, 196, 198, 200, 201, 204, 205
National Committee on Regional Effects of Global Warming, 494
National Energy Efficiency and Alternative Energy Bill, 534
National Energy Use Database, 534
National Oceanographic and Atmospheric Administration (NOAA), 191
Natural ecological systems, 497
Natural gas, 131, 132, 136, 242, 243, 410, 411, 420, 429, 430, 461-463, 500, 533
Natural gas production, 476-479
Natural gas supplies, 501
Net greenhouse benefit, 425
Netherlands, 544
New Brunswick, 532
"New Earth 21", 545, 549
New Generation Climate Models, 109
Newfoundland, 532
Nitrous dioxide emissions, 423, 425
Nitrous oxide, 332, 472, 532, 539, 547
Nonfossil energies, 137
Non-fossil fuels, 549
Nuclear energy, 499, 501
Nuclear fission power, 412, 419, 420
Nuclear Non-proliferation Treaty, 414
Nuclear power, 410-414
Nuclear power plant, 473, 477, 549
Nuclear reactors, 414
Nuclear sources, 532
Nuclear wastes, 425
-accumulation and disposal of, 425
Nuclear weapons, 414

O

Ocean carbon models, 124
-Advective-Diffusive Model, 124, 128, 130
-Box-Diffusion Model, 123, 124, 128, 129, 130
-Out-Crop Model, 124, 127, 128, 129, 130
Ocean shorelines, 260
Oceanic circulation, 310
Oceanic sinks, 550
Oceans, 151, 524
-atmospheric coupling, 109, 110
-carbon cycle, 118, 120-123
-currents, 151-153
-models, 151, 191, 193, 195, 201, 204, 205
-sea surface temperature, 155
-slab ocean, 151-154
-temperature uncertainties, 40
-thermal lag, 102
-water temperature, 152, 153
Oil, 131, 132, 135
-prices, 135-137
Oil consumption, 455
Oil fields, 243-245, 251, 253
Oil terminals, 243
Oil wells, 242-244, 246, 250
-destruction of, 242-244
-environmental impacts, 244
One-layer model of earth-atmosphere system, 65-70, 88
Ontario, 529, 532
OPEC, 132, 135, 138
Orbiting solar power plants, 550
Organic decomposition, 448
-effect of temperature, 448
Organic sediment, 259
Orography, 228
Oxidation, 476
Ozone, 539
Ozone layer, 176, 193

P

Pacific/North American pattern (PNA), 30, 31
Paleoclimatic data, 189
Paleoclimatic evidence, 43
Paleomagnetic methods, 19
Paleozoic, 20
Parameterization, 83, 84, 192-194, 196
-parameters, 84
Periodic climatic variations, 210
Periodicity, 15
Permafrost, 294

Persian Gulf conflict, 242-244, 246, 252
  -environmental impacts, 242-244
Persistent climate regimes, 28, 29, 33
Pests, 353
  -effect of climatic change, 353
Petroleum, 499, 501
Photorespiration, 352
Photosynthesis, 332, 350, 352, 353, 479
Photovoltaic cells, 415
Planetary
  -atmosphere, 63
  -temperature, 27, 28
Planck's Law, 71, 72
Pleistocene, 21
Pliocene, 24, 25
Polar
  -Cell, 8
  -Esterlies, 8
  -Front Jet, 9
  -regions, 4
  -Winter Surface Warming, 92
Polar/Tundra, 294
Pollen analysis, 276
Pollution control, 331
Pollution rights, 318, 325
Pollution system, 101, 110
Popular vision, 100-103, 107, 109, 113
Population growth, 489, 491
Precipitation, 6, 7, 9, 56, 57, 60, 145, 147, 150, 151, 153, 157, 195, 229, 231, 233, 502
  -patterns of, 234
  -seasonal, 233
  -in tropical regions, 347
  -in temperate regions, 347
  -zonal, 233
Pressurized fluidized bed coal combustion (PFBC), 411
Prince Edward Island, 532

Q

Quasi-biennial oscillation (QBO), 29, 32, 33
Quebec, 532

R

Radiation, 193
Radiative equilibrium, 64-67, 74, 75, 80
Radiative gases, 258
Radioactive waste, 414
Rain forests, 295
Rainfall, 276
  -relationship between temperature, 276
RAND, 229-235
Random fluctuations, 210
Rayleigh scattering, 64, 70, 71
Red tides, 297
Reduction of sea ice, 92
Reforestation, 439, 534, 549
Regional climate models, 495
Regional impacts of climatic variations, 27, 28, 31, 33
Regional warming, 292
Renewable energy sources, 526, 541, 542, 549
Respiration, 472, 474
Retinitis, 367
Ridge-trough system, 31, 32
Rift Valley fever, 367-369
Rio Declaration, 552
Runoff, 6, 7

S

Salt-water intrusion, 259, 272, 502
Sandy shoreline, 260, 262
Sand storm, 248
  -in Saudi Arabia, 248
Satellite-based sensor systems, 182
Satellites, 242
  -geosynchronous, 250
  -METEOSAT, 246, 248, 250
  -Polar, 250
Saturated parcel convection, 75
Schistosomiasis, 363, 366, 367, 369
Sea ice, 146, 149, 151-154, 190, 191, 196, 200, 201, 204
Sea level pressure field, 203, 204
Sea level rise, 256-268, 387, 459, 496, 502, 528, 530
  -as a result of glacial melting, 258
  -as a result of thermal expansion, 258
  -coastal erosion, 262
  -due to coastal subsidence, 265
  -economic impact, 273
  -effects on land use, 268
  -fluctuation, 256
  -flooding, 259, 261
  -impact on coastal environments, 262
  -impact on recreation facilities, 273

-impact on shorelines of Israel, 496
-implication and responses, 259
-wave-induced erosion, 259
Sea surface temperature, 40-42, 46, 177, 194
Seasonal cycle, 191-193, 215
   -annual, 190
   -diurnal, 146-148, 152-154, 190, 191, 193
Second Climate Conference, 522, 528
Second World Climate Conference, 532, 540
Second-order interactions, 330
Sensible heat flux, 3
Sheltered coasts, 260
   -small bays, 260, 262
   -estuaries, 260, 262
Ships of opportunity, 40
Shore erosion problems, 260
Shoreline elevation, 267
Shoreline flooding, 268
Shoreline position, 259
Short-wave solar radiation, 471
Siberian forest, 279
Sinks, 524, 526
Sky light, 3
Small Island States, 531
Smallpox, 363
Smoke, 242-245, 249
   -coagulation, 245
   -dispersal, 246
   -oil fire smoke vs. forest fire smoke, 245
   -properties of, 245
   -reduction of sunlight, 252
   -satellite observations, 246
Socio-economic installations, 488
   -vulnerability to climate alterations, 488
Soil, 353, 439-449, 472, 473, 475, 476
   -moisture storage, 51
   -organic litter layer, 449
   -water surplus, 51
Soil microorganisms, 448
Solar
   -absorption, 64-66, 68-71
   -continental dryness/warming, 92
   -cycle and climatic variations, 32, 33
   -energy, 477
   -model constant, 147
   -power, 499, 501
   -radiation, 2-4, 7, 9, 11, 73, 79, 83, 147, 148, 150, 210, 348, 350, 495
   -radiative transfer problem, 71, 73
Solar electric technologies, 463
   -photovoltaics, 463
   -windpower, 463

Solar energy, 414, 415, 419, 420
Solar ponds, 415
Soot, 252
Southeast Trades, 7
Southern Oscillation (SO), 30
Spectral truncation schemes, 192, 195, 196
SPOT, 264, 265, 267, 269
St. Lawrence River Basin, 534
Statement on forest principles, 552
Station location, 38, 39
Steam injected gas turbine (STIG), 412, 417
Stochastic fluctuations, 82, 210, 211
Stratosphere, 169, 176, 177, 242-245, 251
   -anomalies, 170
   -lower, 169, 175
   -temperatures, 104
Stratosphering cooling, 92
Streams, 296
Sub-grid scale phenomena, 83
Submergence, 259
   -coastal wetlands, 259
"Summer dryness", 311
Sunspots, 9
Surface
   -energy, 2, 3
   -hydrology, 195
   -temperature, 43, 65, 66, 68, 69, 84, 88, 91, 173, 198, 200, 202
   -topography, 195
   -types, 4, 7
   -warming, 149
Sustainable growth, 524
Synoptic evaluation, 374, 376
Synoptic procedure, 374
   -automated air mass-based, 374
Synoptic scale weather processes, 211
Syphilis, 363

T

Tectonic activity, 260
   -earthquake, 260
   -glacial rebound, 260
Teleconnections, 30, 31, 33
Temperate agriculture, 347
   -effect of climate change, 347
Temperate forests, 294, 295
Temperate zones, 348, 354
   -specific studies on, 354
Temperature, 229, 231, 232, 235, 350, 372, 354
   -absolute, 231

-differences from RAND dataset, 234
-long-term fluctuations, 372
-range changes with latitude, 234
-rise in, 495
-zonal, 233
Temperature drop, 251
 -in Bahrain, 251
"Temperature threshold", 351, 353, 372, 374
Temperature zones, 348
"The Green Revolution", 348
Terrestrial climate, 145
Terrestrial radiation, 2-4, 6
Thermal
 -emission, 64-68, 70, 71
 -energy changes, 3
 -radiative transfer problem, 71, 73
 -regimes, 350
Thermal expansion, 503
Thermodynamics
 -first law, 63
 -second law, 130, 131
Thermometric record, 37-39, 43
Three cell model, 7
Three-dimensional atmospheric models, 88, 89, 190
Three-dimensional oceanic models, 88
Timber, 272, 278, 400
 -conservation of, 277
Topographical effects, 264
 -sea level rise, 264
Topography, 211, 228-230, 236, 239, 250, 262, 265, 269
Toronto, 529
Trace gas concentrations, 101, 102
Transient carbon dioxide, 88
Transient phase of warming, 88, 89
Transpiration, 353
Transportation, 432
Transportation fuels, 460
 -alternatives, 460
Tree planting, 542
Tropical forests, 295
Tropical zones, 351
Tropopause, 69, 170
Troposphere, 168, 171, 173, 177
 -anomalies, 170
 -upper, 169
Tropospheric circulation, 311, 312
Tropospheric Ozone, 389

U

Ultraviolet-B, 296
Understory vegetation, 449
United Kingdom Meteorological Office Model (UKMO), 109, 154, 191, 196, 200, 204
United Nations Conference, 523
United Nations Conference on Environment and Development, 551
United Nations Stockholm Conference on Human Environment, 552
United States, 544
 -action program to arrest global warming, 544
United States Environmental Policy, 552, 553
Urban heat island, 39, 41, 101, 104, 165, 179, 181-185
Urban warming, 181, 183-185
U.S. Global Change Program, 190
U.S. Working Group on Energy and Efficiency, 454
USGS quadrangle map, 266, 267
USNSF, 497
UTM, 266
UVB flux, 107
UV radiation, 73

V

Vampire bat rabies, 364
Vancouver, 530
Variability, 16
Vietnam, 531
Volatile organic compounds (VOC), 390
Vostok core, 102
Vulcanism, 169

W

Warming regimes, 27, 28
Waste management, 435
Waste materials, 473
Water, 3-5
 -budget, 5-7, 9
 -deficit, 51
 -runoff, 51
 -vapor, 72, 73, 76, 77, 79, 193
Water pollution, 394, 403

-GFDL model, 52-60, 155, 159, 385, 389, 395, 399, 402, 404
-GISS model, 52-59, 146-149, 152, 155, 159, 163, 228, 230, 231, 234-236, 276, 278, 282, 385, 386, 389, 395, 399, 402, 404
-grid cells, 230
-OSU model, 52, 56, 58, 59, 155, 163, 204
-UK-MET model, 52-56, 58, 59, 155, 159
Water quality, 496
Water stress, 530
Waterlog, 259
-wetland plants, 259
Wave refraction, 259
Weather maps, 168
Weather/mortality relationships, 372, 374-377
Weather prediction model, 83, 146, 191
Weizmann Institute of Science, 494
Wetlands, 262, 268, 271
-destruction of, 273
-effect of accelerated sea-level rise, 261, 262
-effect of flooding, 273
Wind electric capacity, 416
Wind electric systems, 414, 415
Wind power, 416, 419, 499, 501
Wind streams, 211
Wind systems, 7
Wind turbines, 28, 29, 33
World Climate Program, 523
World Conference on the Changing Atmosphere, 533
World Energy Conference, 455, 457

Y

Year-to-year climatic variations, 29, 33
Yellow fever, 364
Younger Dryas, 21, 22, 24

Z

Zero dimensional, 83
Zoonoses, 363